Fresh Water and W

The Handbook of Natural Resources, Second Edition

Series Editor:
Yeqiao Wang

Volume 1
Terrestrial Ecosystems and Biodiversity

Volume 2
Landscape and Land Capacity

Volume 3
Wetlands and Habitats

Volume 4
Fresh Water and Watersheds

Volume 5
Coastal and Marine Environments

Volume 6
Atmosphere and Climate

Fresh Water and Watersheds

Edited by
Yeqiao Wang

CRC Press
Taylor & Francis Group
Boca Raton London New York

CRC Press is an imprint of the
Taylor & Francis Group, an **informa** business

CRC Press
Taylor & Francis Group
6000 Broken Sound Parkway NW, Suite 300
Boca Raton, FL 33487-2742

First issued in paperback 2022

ISBN 13: 978-1-03-247439-7 (pbk)
ISBN 13: 978-1-138-33756-5 (hbk)

DOI: 10.1201/9780429441042

Library of Congress Cataloging-in-Publication Data

Names: Wang, Yeqiao, editor.
Title: Handbook of natural resources / edited by Yeqiao Wang.
Other titles: Encyclopedia of natural resources.
Description: Second edition. | Boca Raton: CRC Press, [2020] | Revised edition of: Encyclopedia of natural resources. [2014]. | Includes bibliographical references and index. | Contents: volume 1. Ecosystems and biodiversity — volume 2. Landscape and land capacity — volume 3. Wetland and habitats — volume 4. Fresh water and watersheds — volume 5. Coastal and marine environments — volume 6. Atmosphere and climate. |
Summary: "This volume covers topical areas of terrestrial ecosystems, their biodiversity, services, and ecosystem management. Organized for ease of reference, the handbook provides fundamental information on terrestrial systems and a complete overview on the impacts of climate change on natural vegetation and forests. New to this edition are discussions on decision support systems, biodiversity conservation, gross and net primary production, soil microbiology, and land surface phenology. The book demonstrates the key processes, methods, and models used through several practical case studies from around the world" — Provided by publisher.
Identifiers: LCCN 2019051202 | ISBN 9781138333918 (volume 1 ; hardback) | ISBN 9780429445651 (volume 1 ; ebook)
Subjects: LCSH: Natural resources. | Land use. | Climatic changes.
Classification: LCC HC85 .E493 2020 | DDC 333.95—dc23
LC record available at https://lccn.loc.gov/2019051202

Visit the Taylor & Francis Web site at
http://www.taylorandfrancis.com

and the CRC Press Web site at
http://www.crcpress.com

Contents

Preface ix
About The Handbook of Natural Resources xi
Acknowledgments xiii
Aims and Scope xv
Editor xvii
Contributors xix

SECTION I Fresh Water and Hydrology

1 Aquifers: Groundwater Storage 3
Thomas J. Burbey

2 Aquifers: Recharge 11
John Nimmo, David A. Stonestrom, and Richard W. Healy

3 Artificial Reservoirs: Land Cover Change on Local Climate 17
Faisal Hossain and Ahmed M. Degu

4 Drainage and Water Quality 27
James L. Baker

5 Endorheic Lake Dynamics: Remote Sensing 33
Yongwei Sheng

6 Eutrophication 45
Kyle D. Hoagland and Thomas G. Franti

7 Evaporation: Lakes and Large Bodies of Water 49
John Borrelli

8 Evaporation and Energy Balance 53
Matthias Langensiepen

9 Field Water Supply and Balance 59
Jean L. Steiner

10 Hydrologic Cycle 65
John Van Brahana

11 Hydrologic Modeling of Extreme Events 69
Jazuri Abdullah, Jaehoon Kim, and Pierre Y. Julien

12 Hydrology: Environmental 81
Thomas Boving

13 Hydrology: Urban 87
C. Rhett Jackson

14 Impervious Surface Area: Effects 93
Carl L. Zimmerman and Daniel L. Civco

15 Infiltration Systems and Nitrate Removal 101
Jose A. Amador

16 Low-Impact Development 107
Curtis H. Hinman and Derek B. Booth

SECTION II Water Management

17 Groundwater Contamination 115
Charles W. Fetter

18 Groundwater: World Resources 119
Lucila Candela

19 Irrigation: River Flow Impact 127
Robert W. Hill and Ivan A. Walter

20 Land Change and Water Resource Vulnerability 133
Gargi Chaudhuri

21 Pesticide Contamination: Groundwater 141
Roy F. Spalding

22 Pesticide Contamination: Surface Water 147
Sharon A. Clay

23 Stormwaters: Management 153
John C. Clausen and Michael E. Dietz

24 Stormwater: Urban Modeling 165
Miguel A. Medina, Jr. and Cevza Melek Kazezyılmaz-Alhan

25 Surface Water: Nitrogen Enrichment 191
Tim P. Burt and Penny E. Widdison

26 Surface Water: Nitrogen Fertilizer Pollution 197
Gregory McIsaac

27 Transpiration 205
Thomas R. Sinclair

28 Transpiration: Water Use Efficiency 211
Miranda Y. Mortlock

29 Water Deficits: Development 215
Kadambot H. M. Siddique and Helen Bramley

30 Watershed Hydrology and Land-Use and Land-Cover Change (LULCC)........221
Yuyu Zhou

31 Rain-Use Efficiency: Remote Sensing.. 225
Fang Huang

32 Water Management: South Asia.. 233
Soni M. Pradhanang and Marzia Tamanna

SECTION III Water and Watershed Monitoring

33 Water Quality and Monitoring.. 257
John C. Clausen

34 Algal Bloom Monitoring: Remote Sensing.. 265
Darryl J. Keith

35 Mapping Impervious Cover in Catchments Using High Spatial Resolution Aerial Imagery.. 277
Jessica Morgan, Yeqiao Wang, and Naomi Detenbeck

36 Analysis of Impervious Cover in Riparian Zones..291
Jessica Morgan, Yeqiao Wang, and Naomi Detenbeck

37 Cyanobacteria in Inland Waters: Remote Sensing.. 303
Lin Li

38 Bibliometric and Visualized Analysis of Water Footprint Research........... 323
Peili Duan, Lijie Qin, and Peng Yin

39 Calculation Method of Crop Water Footprint under Rain-Fed Condition 335
Lijie Qin and Jianqin Wang

40 Optical Properties of Colored Dissolved Organic Matter (CDOM): Poyang Lake..341
Jian Xu and Dan Gao

Index 347

Preface

Fresh Water and Watersheds is the fourth volume of *The Handbook of Natural Resources, Second Edition (HNR)*. This volume consists of 40 chapters authored by 55 contributors from 7 countries. The contents are organized in three sections: *Fresh Water and Hydrology* (16 chapters); *Water Management* (16 chapters); and *Water and Watershed Monitoring* (8 chapters).

Fresh water is a vital resource to the survival of all living organisms. Hydrologic cycle is one of the most basic concepts of water science. The dominant processes of the hydrologic cycle include evaporations from open bodies of water into the atmosphere; evapotranspiration by plants of soil water into atmospheric vapor; condensation of water vapor into liquid or solid particles; precipitation as the condensed water or ice falls from the atmosphere back to the surface; infiltration of the water into subsurface reservoirs; baseflow contribution to streams from groundwater; streamflow recharging of groundwater; streamflow runoff if the precipitation rate exceeds the infiltration rate; surface and subsurface flow back to oceans or intermediate reservoirs; and storage. Hydrologic cycle is complex because it involves all forms of water of the hydrosphere and it is affected by many natural and anthropogenic factors. Whether water added to the soil surface as precipitation or irrigation infiltrates or becomes surface runoff is critical to water quality. Groundwater constitutes an important source of water supply for domestic, industrial, and irrigation uses, which is not subject to multiannual and seasonal fluctuations. Evaporation is a major component of the hydrologic cycle. Precise documentation of evaporation from lakes and other water bodies is required for wise management of water resources. Climate change, human alternation of hydrological regimes by engineering structures, land use change and farming practices can all contribute to pollution and contamination of water resources.

With the challenges and concerns, the 40 chapters in this volume cover topics in *Fresh Water and Hydrology*, including ground water storage, aquifer recharge, drainage an water quality, field water supply and balance, hydrological cycle, environmental hydrology, urban hydrology, hydrological modeling of extreme events, evaporation and energy balance, endorheic lake dynamics, eutrophication, infiltration system and nitrate removal, effects of impervious surface areas, and low-impact development; in *Water Management*, including world resources of groundwater, groundwater contamination, impacts of irrigation in river flow, land change and water resource vulnerability, pesticide contamination of groundwater and surface water, stormwater management, urban stormwater modeling, nitrogen enrichment of surface water, nitrogen fertilizer pollution of surface water, transpiration and water use efficiency, water deficits, watershed hydrology and land use and land cover change, rain-use efficiency and water management; and in *Water and Watershed Monitoring*, including water quality monitoring, remote sensing of algal bloom, impervious surface area mapping in catchments, impervious surface areas in riparian zones, remote sensing of cyanobacteria in inland waters, calculation of crop water footprint under rain-fed condition, optical properties of colored dissolved organic matter, and bibliographic and visualized analysis of water footprint research.

The chapters provide updated knowledge and information in general environmental and natural science education and serve as a value-added collection of references for scientific research and management practices.

Yeqiao Wang
University of Rhode Island

About The Handbook of Natural Resources

With unprecedented attentions to the changing environment on the planet Earth, one of the central focuses is about the availability and sustainability of natural resources and the native biodiversity. It is critical to gain a full understanding about the consequences of the changing natural resources to the degradation of ecological integrity and the sustainability of life. Natural resources represent such a broad scope of complex and challenging topics.

The *Handbook of Natural Resources, Second Edition (HNR)*, is a restructured and retitled book series based on the 2014 publication of the *Encyclopedia of Natural Resources (ENR)*. The *ENR* was reviewed favorably in February 2015 by CHOICE and commented as *highly recommended for lower-division undergraduates through professionals and practitioners*. This *HNR* is a continuation of the theme reference with restructured sectional design and extended topical coverage. The chapters included in the *HNR* provide authoritative references under the systematic relevance to the subject of the volumes. The case studies presented in the chapters cover diversified examples from local to global scales, and from addressing fundamental science questions to the needs in management practices.

The *Handbook of Natural Resources* consists of six volumes with 241 chapters organized by topical sections as summarized below.

Volume 1. Terrestrial Ecosystems and Biodiversity
Section I. Biodiversity and Conservation (15 Chapters)
Section II. Ecosystem Type, Function and Service (13 Chapters)
Section III. Ecological Processes (12 Chapters)
Section IV. Ecosystem Monitoring (6 Chapters)

Volume 2. Landscape and Land Capacity
Section I. Landscape Composition, Configuration and Change (10 Chapters)
Section II. Genetic Resource and Land Capability (13 Chapters)
Section III. Soil (15 Chapters)
Section IV. Landscape Change and Ecological Security (11 Chapters)

Volume 3. Wetlands and Habitats
Section I. Riparian Zone and Management (13 Chapters)
Section II. Wetland Ecosystem (8 Chapters)
Section III. Wetland Assessment and Monitoring (9 Chapters)

Volume 4. Fresh Water and Watersheds
Section I. Fresh Water and Hydrology (16 Chapters)
Section II. Water Management (16 Chapters)
Section III. Water and Watershed Monitoring (8 Chapters)

Volume 5. Coastal and Marine Environments
Section I. Terrestrial Coastal Environment (14 Chapters)
Section II. Marine Environment (13 Chapters)
Section III. Coastal Change and Monitoring (9 Chapters)

Volume 6. Atmosphere and Climate
Section I. Atmosphere (16 Chapters)
Section II. Weather and Climate (16 Chapters)
Section III. Climate Change (8 Chapters)

With the challenges and uncertainties ahead, I hope that the collective wisdom, the improved science, technology and awareness and willingness of the people could lead us toward the right direction and decision in governance of natural resources and make responsible collaborative efforts in balancing the equilibrium between societal demands and the capacity of natural resources base. I hope that this *HNR* series can help facilitate the understanding about the consequences of changing resource base to the ecological integrity and the sustainability of life on the planet Earth.

Yeqiao Wang
University of Rhode Island

Acknowledgments

I am honored to have this opportunity and privilege to work on the *Handbook of Natural Resources, Second Edition (HNR)*. It would be impossible to complete such a task without the tremendous amount of support from so many individuals and groups during the process. First and foremost, I thank the 342 contributors from 28 countries around the world, namely Australia, Austria, Brazil, China, Cameroon, Canada, Czech Republic, Finland, France, Germany, Hungary, India, Israel, Japan, Nepal, New Zealand, Norway, Puerto Rico, Spain, Sweden, Switzerland, Syria, Turkey, Uganda, the United Kingdom, the United States, Uzbekistan, and Venezuela. Their expertise, insights, dedication, hard work, and professionalism ensure the quality of this important publication. I wish to express my gratitude in particular to those contributors who authored chapters for this *HNR* and those who provided revisions from their original articles published in the *Encyclopedia of Natural Resources*.

The preparation for the development of this *HNR* started in 2017. I appreciate the visionary initiation of the restructure idea and the guidance throughout the preparation of this *HNR* from Irma Shagla Britton, Senior Editor for Environmental Sciences and Engineering of the Taylor & Francis Group/CRC Press. I appreciate the professional assistance and support from Claudia Kisielewicz and Rebecca Pringle of the Taylor & Francis Group/CRC Press, which are vital toward the completion of this important reference series.

The inspiration for working on this reference series came from my over 30 years of research and teaching experiences in different stages of my professional career. I am grateful for the opportunities to work with many top-notch scholars, colleagues, staff members, administrators, and enthusiastic students, domestic and international, throughout the time. Many of my former graduate students are among and/or becoming world-class scholars, scientists, educators, resource managers, and administrators, and they are playing leadership roles in scientific exploration and in management practice. I appreciate their dedication toward the advancement of science and technology for governing the precious natural resources. I am thankful for their contributions in *HNR* chapters.

As always, the most special appreciation is due to my wife and daughters for their love, patience, understanding, and encouragement during the preparation of this publication. I wish my late parents, who were past professors of soil ecology and of climatology from the School of Geographical Sciences, Northeast Normal University, could see this set of publications.

Yeqiao Wang
University of Rhode Island

Aims and Scope

Land, water, and air are the most precious natural resources that sustain life and civilization. Maintenance of clean air and water and preservation of land resources and native biological diversity are among the challenges that we are facing for the sustainability and well-being of all on the planet Earth. Natural and anthropogenic forces have affected constantly land, water, and air resources through interactive processes such as shifting climate patterns, disturbing hydrological regimes, and alternating landscape configurations and compositions. Improvements in understanding of the complexity of land, water, and air systems and their interactions with human activities and disturbances represent priorities in scientific research, technology development, education programs, and administrative actions for conservation and management of natural resources.

The chapters of *The Handbook of Natural Resources, Second Edition (HNR)*, are authored by world-class scientists and scholars. The theme topics of the chapters reflects the state-of-the-art science and technology, and management practices and understanding. The chapters are written at the level that allows a broad scope of audience to understand. The graphical and photographic support and list of references provide the helpful information for extended understanding.

Public and private libraries, educational and research institutions, scientists, scholars, resource managers, and graduate and undergraduate students will be the primary audience of this set of reference series. The full set of the *HNR* and individual volumes and chapters can be used as the references in general environmental science and natural science courses at different level and disciplines, such as biology, geography, Earth system science, environmental and life sciences, ecology and natural resources science. The chapters can be a value-added collection of references for scientific research and management practices.

Editor

Yeqiao Wang, PhD, is a professor at the Department of Natural Resources Science, College of the Environment and Life Sciences, University of Rhode Island. He earned his BS from the Northeast Normal University in 1982 and his MS degree from the Chinese Academy of Sciences in 1987. He earned the MS and PhD degrees in natural resources management and engineering from the University of Connecticut in 1992 and 1995, respectively. From 1995 to 1999, he held the position of assistant professor in the Department of Geography and the Department of Anthropology, University of Illinois at Chicago. He has been on the faculty of the University of Rhode Island since 1999. Among his awards and recognitions, Dr. Wang was a recipient of the prestigious Presidential Early Career Award for Scientists and Engineers (PECASE) in 2000 by former U.S. President William J. Clinton, for his outstanding research in the area of land cover and land use in the Greater Chicago area in connection with the Chicago Wilderness Program.

Dr. Wang's specialties and research interests are in terrestrial remote sensing and the applications in natural resources analysis and mapping. One of his primary interests is the land change science, which includes performing repeated inventories of landscape dynamics and land-use and land-cover change from space, developing scientific understanding and models necessary to simulate the processes taking place, evaluating consequences of observed and predicted changes, and understanding the consequences of change on environmental goods and services and management of natural resources. His research and scholarships are aimed to provide scientific foundations in understanding of the sustainability, vulnerability and resilience of land and water systems, and the management and governance of their uses. His study areas include various regions in the United States, East and West Africa, and China.

Dr. Wang published over 170 refereed articles, edited *Remote Sensing of Coastal Environments* and *Remote Sensing of Protected Lands*, published by CRC Press in 2009 and 2011, respectively. He served as the editor-in-chief for the *Encyclopedia of Natural Resources* published by CRC Press in 2014, which was the first edition of *The Handbook of Natural Resources*.

Contributors

Jazuri Abdullah
Department of Civil and Environmental Engineering
Engineering Research Center
Colorado State University
Fort Collins, Colorado

Jose A. Amador
Laboratory of Soil Ecology and Microbiology
University of Rhode Island
Kingston, Rhode Island

James L. Baker
Department of Agricultural and Biosystems Engineering
Iowa State University
Ames, Iowa

Derek B. Booth
Center for Water and Watershed Studies
University of Washington
Seattle, Washington

John Borrelli
Department of Civil Engineering
Texas Tech University
Lubbock, Texas

Thomas Boving
Department of Geosciences/Department of Civil and Environmental Engineering
University of Rhode Island
Kingston, Rhode Island

John Van Brahana
Division of Geology
Department of Geosciences
University of Arkansas
Fayetteville, Arkansas

Helen Bramley
The UWA Institute of Agriculture
University of Western Australia
Crawley, Western Australia, Australia

Thomas J. Burbey
Department of Geosciences
Virginia Tech University
Blacksburg, Virginia

Tim P. Burt
Department of Geography
University of Durham
Durham, United Kingdom

Lucila Candela
Department of Geotechnical Engineering and Geo-Sciences
Technical University of Catalonia (UPC)
Barcelona, Spain

Gargi Chaudhuri
Department of Geography and Earth Science
University of Wisconsin
La Crosse, Wisconsin

Daniel L. Civco
Department of Natural Resources and the Environment
University of Connecticut
Storrs, Connecticut

John C. Clausen
Department of Natural Resources and the Environment
University of Connecticut
Storrs, Connecticut

Sharon A. Clay
Plant Science Department
South Dakota State University
Brookings, South Dakota

Ahmed M. Degu
Department of Civil and Environmental Engineering
Tennessee Technological University
Cookeville, Tennessee

Naomi Detenbeck
U.S. Environmental Protection Agency
Narragansett, Rhode Island

Michael E. Dietz
Center for Land Use Education and Research (CLEAR)
University of Connecticut
Storrs, Connecticut

Peili Duan
School of Business
Ludong University
Yantai, China

Charles W. Fetter, Jr.
C. W. Fetter, Jr. Associates
Oshkosh, Wisconsin

Thomas G. Franti
Department of Biological Systems Engineering
University of Nebraska
Lincoln, Nebraska

Dan Gao
School of Geography and Environmental Science
Ministry of Education's Key Laboratory of Poyang Lake Wetland and Watershed Research
Jiangxi Normal University
Nanchang, China

Richard W. Healy
U.S. Geological Survey (USGS)
Lakewood, Colorado

Robert W. Hill
Biological and Irrigation Engineering Department
Utah State University
Logan, Utah

Curtis H. Hinman
School of Agricultural
Human and Natural Resource Sciences
Washington State University
Tacoma, Washington

Kyle D. Hoagland
School of Natural Resources
University of Nebraska
Lincoln, Nebraska

Faisal Hossain
Department of Civil and Environmental Engineering
Tennessee Technological University
Cookeville, Tennessee

Fang Huang
Key Laboratory of Geographical Processes and Ecological Security in Changbai Mountains
Ministry of Education
School of Geographical Sciences
Northeast Normal University
Changchun, China

C. Rhett Jackson
Daniel B. Warnell School of Forest Resources
University of Georgia
Athens, Georgia

Pierre Y. Julien
Department of Civil and Environmental Engineering
Engineering Research Center
Colorado State University
Fort Collins, Colorado

Darryl J. Keith
Atlantic Ecology Division
U.S. Environmental Protection Agency (EPA)
Narragansett, Rhode Island

Jaehoon Kim
Department of Civil and Environmental Engineering, Engineering Research Center
Colorado State University
Fort Collins, Colorado

Matthias Langensiepen
Department of Modeling Plant Systems
Humboldt University of Berlin
Berlin, Germany

Lin Li
Department of Earth Sciences
Indiana University–Purdue University Indianapolis
Indianapolis, Indiana

Gregory McIsaac
Natural Resources and Environmental Sciences
University of Illinois
Urbana, Illinois

Miguel A. Medina, Jr.
Department of Civil and Environmental Engineering
Duke University
Durham, North Carolina

Cevza Melek Kazezyılmaz-Alhan
Department of Civil Engineering
Istanbul University
Istanbul, Turkey

Jessica Morgan
Department of Science
United States Coast Guard Academy
New London, Connecticut

Miranda Y. Mortlock
School of Agriculture and Food Sciences
University of Queensland
Brisbane, Queensland, Australia

John Nimmo
U.S. Geological Survey (USGS)
Menlo Park, California

Soni M Pradhananga
Department of Geosciences
University of Rhode Island
Kingston, Rhode Island

Lijie Qin
Key Laboratory of Geographical Processes and Ecological Security in Changbai Mountains
Ministry of Education
School of Geographical Sciences
Northeast Normal University
Changchun, China

Yongwei Sheng
Department of Geography
University of California, Los Angeles
Los Angeles, California

Kadambot H. M. Siddique
The UWA Institute of Agriculture
University of Western Australia
Crawley, Western Australia, Australia

Thomas R. Sinclair
Crop Genetics and Environmental Research Unit
U.S. Department of Agriculture (USDA)
University of Florida
Gainesville, Florida

Roy F. Spalding
Water Science Laboratory
University of Nebraska
Lincoln, Nebraska

Jean L. Steiner
Agricultural Research Service
U.S. Department of Agriculture (USDA-ARS)
El Reno, Oklahoma

David A. Stonestrom
U.S. Geological Survey (USGS)
Menlo Park, California

Marzia Tamanna
Department of Geosciences
University of Rhode Island
Kingston, Rhode Island

Ivan A. Walter
Ivan's Engineering, Inc.
Denver, Colorado

Jianqin Wang
School of Geographical Sciences
Northeast Normal University
Changchun, China

Yeqiao Wang
Department of Natural Resources Science
University of Rhode Island
Kingston, Rhode Island

Penny E. Widdison
Department of Geography
University of Durham
Durham, United Kingdom

Jian Xu
School of Geography and Environmental Science
Ministry of Education's Key Laboratory of Poyang Lake Wetland and Watershed Research
Jiangxi Normal University
Nanchang, China

Peng Yin
Institute of Marine Development
Ocean University of China
Qingdao, China

Yuyu Zhou
Department of Geological and Atmospheric Sciences
Iowa State University
Ames, Iowa

Carl L. Zimmerman
Department of Urban and Environmental Policy and Planning
Tufts University
Medford, Massachusetts

I

Fresh Water and Hydrology

1

Aquifers: Groundwater Storage

Thomas J. Burbey
Virginia Tech

Introduction 3
Compressibility 3
Confined Aquifer Storage 5
Unconfined Aquifer Storage 6
Confining Units 7
Three-Dimensional Storage 8
Conclusion 9
References 9
Bibliography 10

Introduction

All rocks or sediments that contain pore space have the capacity to store water. However, the concept of aquifer storage has more to do with how water is released from the rock or sediments than the fact that water is simply contained in these geologic materials. The idea that water can be "squeezed" out of a rock or sediment infers the concept that aquifers can store and release water. This idea was perhaps first formally described by O.E Meinzer,[1] who noted the occurrence of abrupt water-level fluctuations due to the passing of trains over a railroad near a well. Meinzer concluded that "aquifers are apparently all more or less compressible and elastic though they differ widely in the degree and relative importance of these properties." Shortly thereafter, Theis[2] formulated one of the most commonly used expressions today that describe the cylindrical drawdown response of a pumping well as being dependent on the aquifer transmissivity and storativity (storage coefficient) without formulating a rigorous description of the storage coefficient. Low-storage materials tend to have broader and deeper drawdown cones than materials with larger storage values. Theis[3] recognized that water released to wells from storage was a result of aquifer compression in artesian (confined) aquifers or due to the physical drainage of water in unconfined aquifers. Jacob[4] derived a mathematical formulation of the storage coefficient by assuming that the compressibility of the aquifer material under stress from a reduction in hydraulic head occurs solely in the vertical direction and that storage in artesian aquifers is dependent on both skeletal (aquifer) and fluid compressibility. Thus, to provide a complete description of aquifer storage, the concept of compressibility must first be developed.

Compressibility

Compressibility is simply defined as the ratio of strain over stress and describes how a material or fluid deforms as a result of an applied external stress. Strain refers to how a material deforms relative to an original or previous state. When Jacob[4] defined strain, he made a convenient mathematical decision to

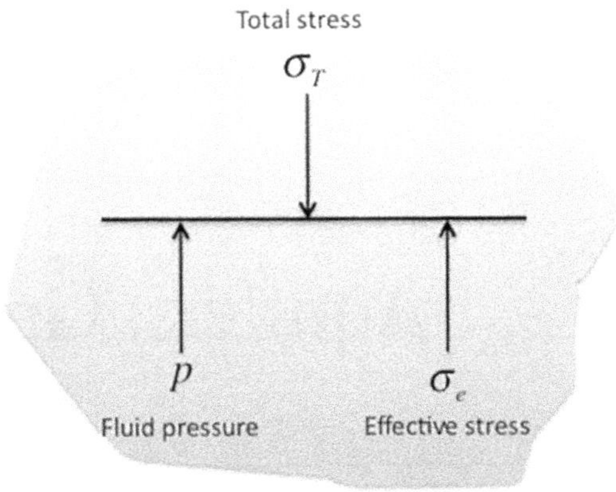

FIGURE 1.1 Depiction of Terzaghi's principle of effective stress on an arbitrary plane within a saturated aquifer.

assume that all the strain induced by an imposed stress occurs in the vertical direction only. For now, we will also make this assumption and then later qualitatively address this issue in three dimensions. Stress represents the weight per unit area acting at any given location within an aquifer system. Consider an arbitrary vertical plane through a saturated porous media as depicted in Figure 1.1. According to Terzaghi's principle of effective stress,[5] the total stress acting on the plane is balanced by the pore pressure and effective stress as

$$\sigma_T = p + \sigma_e. \tag{1.1}$$

That is, the total weight per unit area acting on the plane from above, from both the porous material and fluid, is supported by the buoyant fluid stress and the grain-to-grain stress of the matrix. The differential of each term in Equation 1.1 describes how the stress state may change in time. In confined aquifer systems, the total stress can be considered to be constant; therefore, $\boldsymbol{d\sigma_e} = -\boldsymbol{dp}$. Under such a condition, a decline in the pore pressure (say, from pumping) will result in an equal and opposite increase in effective stress. That increase in effective stress is what causes the matrix to compress. The pore pressure can be converted to hydraulic head as $\boldsymbol{p} = \rho_w \boldsymbol{g}\,(\boldsymbol{h} - z)$ where ρ_w is the fluid density, $\boldsymbol{g}$ is the gravitational constant, $\boldsymbol{h}$ is the hydraulic head, and z is the elevation head. If $\boldsymbol{z}$ is constant (negligible vertical compaction at the plane of interest relative to the change in hydraulic head), then the change in pressure is directly related to a change in hydraulic head as

$$dp = \rho_w g dh. \tag{1.2}$$

Now that the components of compressibility are defined, one can surmise that compressibility represents a material property that can be applied to the pore fluid (water), the granular matrix or aquifer material, and the individual grains that make up the aquifer system framework. This latter type of compressibility is typically ignored in groundwater systems because its contribution is typically considered to be negligible in most aquifer environments. Now a reduction in hydraulic head can both cause the fluid to expand and the matrix to contract or compress. Thus, both matrix or porous media compressibility and fluid compressibility need to be defined before the storage contributions from these constituents can be properly defined.

Compressibility of the porous media is defined as

$$\alpha = \frac{dV_T / V_T}{d\sigma_e} \tag{1.3}$$

TABLE 1.1 The Range of Compressibility of Some Common Geologic Materials and Pure Water

Material	Compressibility, α (m2/N)
Clay	10^{-6} to 10^{-8}
Sand	10^{-7} to 10^{-9}
Gravel	10^{-8} to 10^{-10}
Fractured rock	10^{-8} to 10^{-10}
Unfractured rock	10^{-9} to 10^{-11}
Water (β)	4.4×10^{-10}

Source: Modified from Freeze and Cherry.[17]

where dV_T is the change in the total volume of the medium (aquifer material) from its initial state V_T. Again, for convenience, here we assume that the volume change occurs only in the vertical direction. The total volume is the sum of the volume of solids, V and the volume of voids, V_e. That is, $V_T = V_s + V_v$. If we examine the change in volume resulting from a change in effective stress, then $dV_T = dV_s + dV_v$ but because the change in the volume of solids is considered negligible (that is, the compressibility of individual grains), $dV_s = 0$. Therefore, $dV_T = dV_v$.

Water compressibility under isothermal conditions is defined as

$$\beta = \frac{-dV_w / V_w}{dp} \tag{1.4}$$

where dV_w is the change in the volume of water from an initial state V_w. The negative sign indicates that when the pressure decreases, there is an increase in fluid volume. In other words, fluid expands when the pressure declines. This is analogous to an expanding helium balloon that ascends into the atmosphere as a result of the declining atmospheric pressure with increasing altitude. Table 1.1 lists the range of compressibility for common geologic materials and pure water.

Confined Aquifer Storage

The storage coefficient in confined aquifers is defined[6] as the volume of water released from storage (dV_w) over a unit cross-sectional area of aquifer (A) per unit decline in hydraulic head ($-dh$) and is expressed as

$$S = \frac{-dV_w}{Adh}. \tag{1.5}$$

The specific storage, S_s, is related to the storage coefficient through the aquifer thickness, $S = S_s b$, and is defined as the volume of water released by a unit volume of aquifer due to a unit decline in hydraulic head

$$S_s = \frac{-dV_w}{V_T dh}. \tag{1.6}$$

It should be stressed that the confined aquifer is assumed to remain saturated; that is, the head, even after the water-level decline, remains higher than the top of the aquifer as depicted in Figure 1.2.

The question then is what is the source of the volume of water, dV, that is, either released from storage when the water level is lowered or added to storage if the water level is increased?

We saw from Terzaghi's principle of effective stress[5] that the volume of water comes from the sum of the volumes of water released from expansion of water and compression of the aquifer matrix. Confined aquifers, by definition, are saturated so that $dV_T = dV_v = dV_w$. Substituting this relation into Equation 1.3 and rearranging to solve for dV_w yields

$$dV_w = \alpha V_T d\sigma_e = -\alpha V_T \rho_w g dh. \tag{1.7}$$

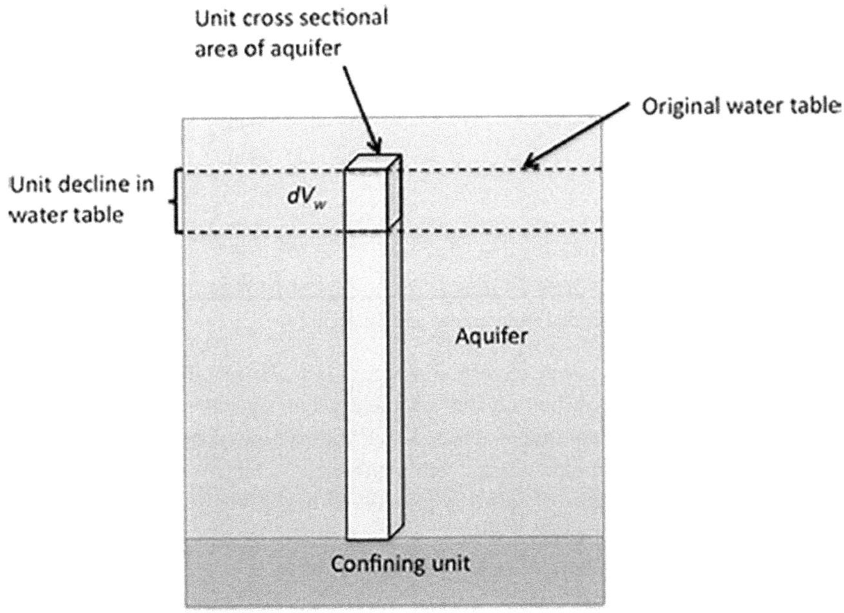

FIGURE 1.2 Diagram illustrating confined aquifer storage.

Equation 1.4 can also be rearranged to solve for $\boldsymbol{dV_w}$ and when combined with Equation 1.2, we obtain

$$dV_w = -\beta V_w \rho_w g dh. \tag{1.8}$$

Now summing Equations 1.7 and 1.8 and recognizing that $\boldsymbol{V_w = nV_T}$, yields

$$dV_w = -\rho_w g dh V_T (\alpha + n\beta).$$

Substituting the above expression into Equation 1.5 yields the expression for specific storage (1/L) used in the groundwater flow equation:

$$S_s = \rho_w g(\alpha + n\beta). \tag{1.9}$$

The related storage coefficient (dimensionless) is

$$S = \rho_w g b(\alpha + n\beta). \tag{1.10}$$

The range of storage coefficient values for confined aquifers is typically from 0.005 to 0.00005.

Unconfined Aquifer Storage

The source of water released from (or added to) storage in unconfined aquifers is largely the result of draining or dewatering of the pore space due to a water table decline and is referred to as specific yield. The specific yield is defined as the volume of water released from storage per unit surface area of aquifer per unit decline in the water table:[6]

$$S_y = \frac{dV_w}{Adh}. \tag{1.11}$$

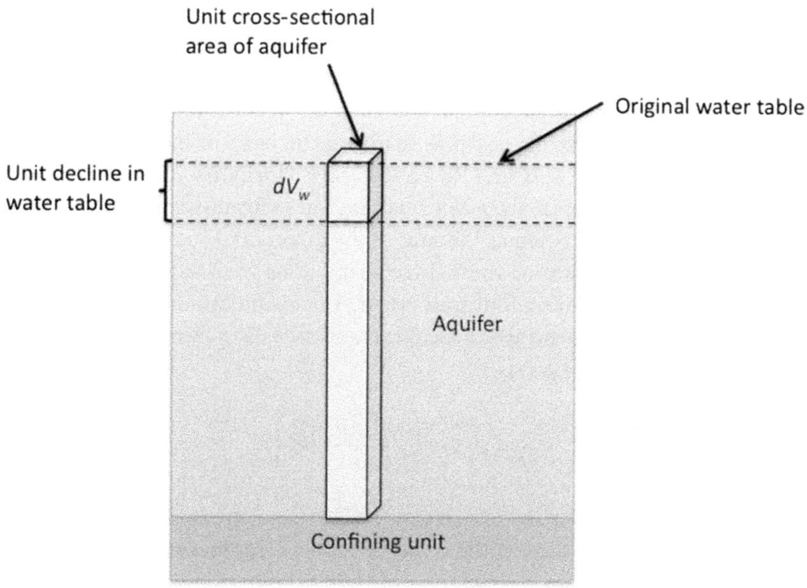

FIGURE 1.3 Diagram illustrating unconfined aquifer storage.

Unconfined storage is depicted in Figure 1.3. Because the specific yield represents a draining of the pore space, it is directly related to the porosity by the simple expression $\boldsymbol{n} = \boldsymbol{S}_y + S_r$, where $\boldsymbol{S}_r$ is the specific retention and represents the portion of the water retained in the matrix after gravity drainage. The specific yield can never be larger than the porosity.

The total storativity (equivalent to storage coefficient) of unconfined aquifers is accurately represented as

$$S = S_y + hS_s \tag{1.12}$$

where the $\boldsymbol{S}_s$ here represents the sources of storage from aquifer compression and water expansion. Because the quantity $\boldsymbol{hS}_s$ is usually 2–4 orders of magnitude less than $\boldsymbol{S}_y$, the unconfined storativity is typically considered to be the specific yield. The typical range of values for specific yield is 0.01–0.30.

Even though $\boldsymbol{S}_y$ is much larger than typical values of $\boldsymbol{S}_s$, the total storativity of confined aquifers is often much larger than that of unconfined aquifers because confined aquifers tend to be much thicker and laterally far more extensive than unconfined aquifers.

Confining Units

Confining units or aquitards often bound aquifers above and below and do not affect the storativity of the aquifer itself, but they can potentially greatly influence the quantity of water available from storage in an aquifer that is being pumped. Confining units, because they are generally composed largely of clays, have a very high compressibility and porosity (Table 1.1) and therefore a high specific storage relative to the aquifer. However, the confining units also have a very low hydraulic conductivity, and therefore the head change across the confining unit can greatly lag behind that of the aquifer. The one-dimensional diffusion equation can be used to describe how the head change propagates through the confining layer over time. As this occurs, the confining unit releases water to the aquifer in which the decline in head has occurred, thus providing additional storage to the aquifer that can be significant, especially when pumping occurs over long time periods—typically years.

Three-Dimensional Storage

In the previous analysis, the compressibility, α, is represented as a one-dimensional quantity where strain (numerator in Equation 1.3) is assumed to occur in the vertical direction only. The assumption of purely vertical strain proposed by Jacob[4] was made on the basis of mathematical expediency and intuitional reason. However, Biot[7,8] and later Rice and Cleary[9] formulated expressions for the coupling of fluid flow with three-dimensional matrix deformation. These formulations result from the fact that the strain term in α must now be written as the sum of the directional components of strain, $\varepsilon_x + \varepsilon_y + \varepsilon_z$, whose sum represents the volume strain. These three terms along with fluid pressure or hydraulic head result in a system of four equations and four unknowns. The formulation of these coupled expressions leads to a new definition of specific storage that contains two elastic parameters for characterizing volume compressibility of the aquifer matrix:[10]

$$S_s = \rho_w g\left(\frac{3(1-2\nu)}{2G(1+\nu)} + n\beta\right) \tag{1.13}$$

where $\boldsymbol{G}$ is the shear modulus and $\boldsymbol{\nu}$ is Poisson's ratio. The shear modulus is the ratio of shear stress to shear strain while the Poisson's ratio is the negative ratio of transverse to axial strain. It should be pointed out that this formulation is not compatible with the standard groundwater flow equation. The derivations of the equations leading to Equation 1.13 can be found in Burbey and Helm[11] and Burbey,[10] with Verruijt[12] providing the mathematical foundation from first principles.

The importance of horizontal strain became apparent from the deformation that resulted from oil withdrawal in the Wilmington Oil Field beginning in Long Beach, California, in the 1930s.[13] Wolff[14] was perhaps the first to conduct a controlled aquifer test to explicitly measure horizontal strain and found the horizontal strains to be significant and form a radial pattern as a function of distance from the pumped well. Burbey[15] measured the horizontal strains from a pumping test and used them to characterize aquifer properties in the absence of water-level data.

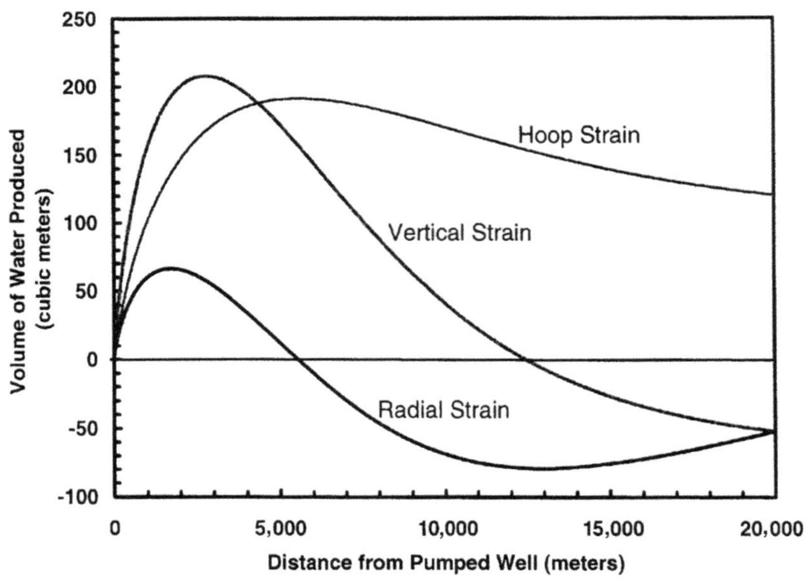

FIGURE 1.4 Simulation showing relative volumes of water removed from storage (positive) or added to storage (negative) from the three strain components as a function of distance from the pumped well for a purely homogeneous isotropic aquifer.
Source: Adapted from Burbey.[16]

Figure 1.4 illustrates the potential storage contribution of each strain component (vertical, radial, and hoop or tangential strain) from a numerical analysis of a 100-day pumping test of an infinite homogeneous and isotropic confined aquifer where the overlying and underlying confining units do not influence the simulation results.[16] The results show that radial strain provides a sizable contribution of the water released from storage near the pumping well. Furthermore, as the distance from the pumping well increases, hoop strain becomes the primary source of water released from storage. The importance of horizontal strain can vary depending on the aquifer system being evaluated.

Conclusion

Storage is one of the most important aquifer properties that influences the distribution and shape of drawdown and directly affects the quantity of water that can be extracted from pumping wells. Classical formulations of storage only consider vertical strain in the formulation of aquifer storage. However, more recent investigations show that the volume of water released from storage in confined aquifers during pumping is the result of both vertical and horizontal aquifer compression and water expansion in the vicinity of the pumping well. Unconfined storage is largely due to dewatering of the aquifer by gravity drainage during a decline in the water table.

References

1. Meinzer, O.E. Compressibility and elasticity of artesian aquifers. Econ. Geol. **1928**, *23*, 263–291.
2. Theis, C.V. The relation between the lowering of the piezometric surface and the rate and duration of discharge of a well using ground-water storage. Trans. Am. Geophys. Union **1935**, *16*, 519–524.
3. Theis, C.V. The significance and nature of the cone of depression in ground-water bodies. Econ. Geol. **1938**, *33*, 889–902.
4. Jacob, C.E. On the flow of water in an elastic artesian aquifer, EOS Trans. Am. Geophysi. Union **1940**, *21*, 574–586.
5. Terzaghi, K. Principles of soil mechanics: IV. settlement and consolidation of clay. Eng. News-Rec. **1925**, *95*, 874–878.
6. Hantush, M.S. Drawdown around a partially penetrating well. Am. Soc. Civil Eng. Trans. **1962**, *27* (part 1), 268–283.
7. Biot, M.A. General theory of three-dimensional consolidation. J. Appl. Phys. **1941**, *12* (2), 155–164.
8. Biot, M.A. Theory of elasticity and consolidation for a porous anisotropic solid. J. Appl. Phys. **1955**, *26* (2), 182–185.
9. Rice, J.R.; Cleary, M.P. Some basic stress diffusion solutions for fluid-saturated elastic porous media with compressible constituents. Rev. Geophys. Space Phys. **1976**, *14*, 227–241.
10. Burbey, T.J. Effects of horizontal strain in estimating specific storage and compaction in confined and leaky aquifer systems. Hydrogeol. J. **1999**, *7* (6), 521–532.
11. Burbey, T.J.; Helm, D.C. Modeling three-dimensional deformation in response to pumping of unconsolidated aquifers. Environ. Eng. Geosci. **1999**, *5* (2), 199–212.
12. Verruijt, A. Elastic storage of aquifers. In *Flow Through Porous Media;* DeWiest, R.J.M., Ed.; Academic Press: New York, 1969; 331–376.
13. Yerkes, R.F.; Castle, R.O. Surface deformation associated with oil and gas field operations in the United States. In *Land Subsidence;* Tison L.J., Ed., Vol 1. International Association. Sci Hydrol Pub. **1969**, *88*, 55–64.
14. Wolff, R.G. Relationship between horizontal strain near a well and reverse water level fluctuation. Water Resour. Res. **1970**, *6*, 1721–1728.
15. Burbey, T.J. Three-dimensional deformation and strain induced by municipal pumping, part 1: Analysis of field data. J. Hydrol. **2006**, *319* (1–4), 123–142.

16. Burbey, T.J. Storage coefficient revisited: Is purely vertical strain a good assumption? Ground Water **2001**, *30* (3), 458–464.
17. Freeze, R.A.; Cherry, J.A. *Groundwater;* Prentice-Hall, Inc.: Englewood Cliffs, New Jersey, 1979; 604 p.

Bibliography

Green, D.H.; Wang, H.F. Specific storage as a poroelastic coefficient. Water Resour. Res. **1990**, *26*, 1631–1637.
Lohman, S.W. Ground-water hydraulics. In *U.S. Geological Survey Professional Paper;* 1972; Vol. 708, 70 p.
Wang, H.F. *Theory of Linear Poroelasticity;* Princeton University Press: Princeton, New Jersey, 2000; 287 p.

2

Aquifers: Recharge

John Nimmo, David A. Stonestrom, and Richard W. Healy
U.S. Geological Survey (USGS)

Introduction 11
Water Budget Methods 12
Methods Based on Surface Water or Ground Water Data 13
Darcian Methods 13
Tracer Methods 13
Other Methods 14
References 14

Introduction

Aquifer recharge was defined by Meinzer[1] and Heath[2] as water that moves from the land surface or the unsaturated zone into the saturated zone. This definition excludes saturated flow between aquifers, which avoids double-accounting in large-scale studies, so it might be more precisely called "aquifer-system" or "saturated-zone" recharge. *Recharge rate* designates either a flux [L^3/T] into a specified portion of aquifer, or a flux density [L/T] into an aquifer at a point. Sources of water for recharge include precipitation that infiltrates, permanent or ephemeral surface water, irrigation, and artificial recharge ponds. Recharge may reach the aquifer directly from portions of rivers, canals, or lakes,[3] though usually it first travels by various means through the unsaturated zone.

Recharge varies considerably with time and location. Temporal variation occurs, for example, with seasonal or short-term variations in precipitation and evapotranspiration (ET). This variability is especially evident in thin unsaturated zones, where recharge may occur within a short time of infiltration. In deep unsaturated zones, recharge may be homogenized over several years so that it may occur with essentially constant flux even though fluxes at shallow depths are erratic. Spatial variation occurs with climate, topography, soils, geology, and vegetation. For example, a decrease of slope or increase of soil permeability may lead to greater infiltration and greater recharge. Many applications use a concept of recharge that is time-averaged or areally averaged.

Both the amount of infiltration and the fraction of it that becomes recharge tend to be greater with more abundant water, so the recharge process is most efficient if infiltration is concentrated in space and time. Because ET may extract most or all of the water that infiltrates, water is more likely to become recharge if it moves rapidly below the root zone. Temporal concentration occurs during storms, floods, and snowmelt, when ongoing processes such as ET are overwhelmed. Spatial concentration typically occurs in depressions and channels, where higher water contents promote rapid movement by increasing the hydraulic conductivity (K), the amount of preferential flow, and the downward driving force at a wetting front. Quantitative estimation of recharge rate contributes to the understanding of large-scale hydrologic processes. It is important for evaluating the sustainability of ground water supplies, though it does not equate with a sustainable rate of extraction.[4] Because it represents a first approximation to the rate of solute transport to the aquifer, the recharge rate is also important to estimate contaminant fluxes and travel times from sources near the land surface. Methods for obtaining a quantitative estimate of recharge mostly require a combination of various types of data which themselves may be hard to estimate, so in general it is wise to apply multiple methods and compare their results.

Water Budget Methods

The water balance for a basin can be stated as

$$P + Q_{\text{on}}^{\text{sw}} + Q_{\text{on}}^{\text{gw}} = \text{ET}^{\text{sw}} + \text{ET}^{\text{uz}} + \text{ET}^{\text{gw}} + Q_{\text{off}}^{\text{sw}} + Q_{\text{off}}^{\text{sw}} + Q_{\text{bf}} + \Delta S^{\text{snow}} + \Delta S^{\text{sw}} + \Delta S^{\text{uz}} + \Delta S^{\text{gw}} \tag{2.1}$$

where P is precipitation and irrigation; Q_{on} and Q_{off} are water flow on and off of the site, respectively; Q is runoff; Q_{bf} is baseflow (ground water discharge to streams or springs); and ΔS is change in water storage. Superscripts refer to surface water, ground water, unsaturated zone, or snow, and all parameters are in units of L/T (or volume per unit surface area per unit time). For the saturated zone only, a water balance can be written for a defined area as

$$R = \Delta Q^{\text{gw}} + Q_{\text{bf}} + \text{ET}^{\text{gw}} + \Delta S^{\text{gw}} \tag{2.2}$$

where R is recharge and ΔQ^{gw} is the difference between ground water flow off of and onto the basin. This equation implies that water arriving at the water table: 1) flows out of the basin as ground water flow; 2) discharges to the surface; 3) is evapotranspired; or 4) goes into storage. Substitution in Equation (2.1) produces a simpler form of the water balance:

$$R = P + Q_{\text{on}}^{\text{sw}} - Q_{\text{off}}^{\text{sw}} - \text{ET}^{\text{sw}} - \text{ET}^{\text{uz}} - \Delta S^{\text{snow}} - \Delta S^{\text{sw}} - \Delta S^{\text{uz}} \tag{2.3}$$

Water budget methods include all techniques based, in one form or another, on one of these water balance equations.

The most common water budget method is the "residual" approach: all other components in the water budget are measured or estimated and R is set equal to the residual. Water budget methods can be applied over a wide range of space and time scales. The major limitation of the residual approach is that the accuracy of the recharge estimate depends on the accuracy with which other components can be measured. This limitation can become significant when the magnitude of R is small relative to other variables. The time scale for applying water budget methods is important, with more frequent tabulations likely to improve accuracy. If the water budget is calculated daily, P can greatly exceed ET on a single day, even in arid settings. Averaging over longer time periods tends to dampen out extreme precipitation events and hence underestimate recharge. Annual recharge estimated with water budgets range from 23 mm in a region of India[5] to 400mm at a site in the eastern United States.[6]

Watershed, surface water flow, and ground water flow models constitute an important class of water budget methods that have been used to estimate of recharge.[7] An attractive feature of models is their predictive capability. They can be used to gauge the effects of future climate or land-use changes on recharge rates.

Methods Based on Surface Water or Ground Water Data

Fluctuations in ground water levels can be used to estimate recharge to unconfined aquifers according to

$$R = S_y \mathrm{d}h/\mathrm{d}t = S_y \; \Delta h/\Delta t \tag{2.4}$$

where S_y is specific yield, h is water table height, and t is time. The method is best applied over short time periods in regions with shallow water tables that display sharp water- level rises and declines. Analysis of water-level fluctuations can also, however, be useful for determining the magnitude of long-term change in recharge rates caused by climate or land-use change. The method is only appropriate for estimating recharge for transient events; recharge occurring under steady flow conditions cannot be estimated. Difficulties lie in determining S_y and ensuring that fluctuations are due to recharge, not to changes in pumping rates or atmospheric pressure or other phenomena. Recharge rates estimated by this technique range from 11 mm over a 26-month period in Saudi Arabia[8] to 541 mm yr^{-1} over 1 yr for a small basin in the United States.[9]

Ground water levels can also be used to estimate flow, Q, through a cross-section of an aquifer that is aligned with an equipotential line. Multiplying K by the hydraulic gradient normal to the section times the area of the section calculates Q. Recharge is determined by dividing Q by the surface area of the aquifer upgradient from the section.

Methods of estimating recharge based on surface-water data include the Channel Water Balance Method (CWBM) and determination of baseflow by hydrograph separation. The CWBM involves measuring discharge at two gauges on a stream; the difference in discharge between the upstream and downstream gauges is the transmission loss. This loss may become recharge, ET, or bank storage. Hydrograph separation involves identifying what portion of gauged stream flow is derived from ground water discharge. Rutledge and Daniel[10] developed an automated technique for this purpose and applied the method to estimate recharge at 15 sites. Drainage areas for the sites ranged from less than 5200km^2 to more than 5200 km^2; estimated annual recharge was between about 13 cm and 64 cm.

Darcian Methods

Applied in the unsaturated zone, Darcy's law gives a flux density equal to K times the driving force, which equals the recharge rate if certain conditions apply. Matric-pressure gradients must be measured or demonstrated to be negligible. Some types of preferential flow are inherently non-Darcian and if important would need to be determined separately. Accurate measurements are necessary to know K adequately under field conditions at the point of interest. For purposes requiring areal rather than point estimates, additional interpretation and calculation are necessary. In the simplest cases, in a region of constant downward flow in a deep unsaturated zone, gravity alone drives the flow. With a core sample from this zone, laboratory K measurements at the original field water content directly indicate the long-term average recharge rate.[11]

In the general case, transient water contents and matric pressures must be measured in addition to K.[12] Transient recharge computed with Darcy's law can relate to storms or other short-term events, or provide data for integration into temporal averages.

Tracer Methods

Increasing availability and precision of physical and chemical analytical techniques have led to a proliferation in applications of tracer methods for recharge estimation. Isotopic and chemical tracers include tritium, deuterium, oxygen-18, bromide, chloride, chlorine-36, carbon-14, agricultural chemicals, dyes, chlorofluorocarbons, and noble gases. In practice, concentrations measured in pore water are related to recharge by applying chemical mass-balance equations, by matching patterns inherited from infiltrating water, or by determining the age of the water. Tracer methods provide point and areal estimates of recharge. Multiple tracers used together can test assumptions and constrain estimates.

The most common tracer for estimating recharge is chloride. Chloride continually arrives at the land surface in precipitation and dust. Chloride is conservative in many environments and is non-volatile. Under suitable conditions,

$$R = P\ [Cl_p]/[Cl_r] \tag{2.5}$$

where P is precipitation and $[Cl_p]$ and $[Cl_r]$ are chloride concentrations in precipitation and pore water, respectively. Chloride mass-balance methods can be applied to unsaturated profiles[13] and entire basins.[14]

Isotopic composition of water provides a useful tracer of the hydrologic cycle. The isotopic makeup of precipitation varies with altitude, season, storm track, and other factors. Recharge estimates using isotopic varieties of water usually employ temporal or geographic trends in infiltrating water.

Non-conservative tracers can indicate the length of time that water is isolated from the atmosphere, that is, its "age." Recharge rates can be inferred from water ages if mixing is small. If ages are known along a flow line,

$$R = \theta L/(\mathrm{A}_2 - A_1) \tag{2.6}$$

where θ is volumetric water content, A_1 and A_2 are ages at two points, and L is separation length. One point is often located at the water table. Preindustrial water can be dated by decay of predominately cosmogenic radioisotopes, including carbon-14 and chlorine-36. The abundance of tritium and other radioisotopes increased greatly during atmospheric weapons testing, labeling recent precipitation. Additional compounds for dating modern recharge include chlorofluorocarbons, krypton-85, and agricultural chemicals.[15,16]

Heat is yet another tracer of ground water recharge. Daily, seasonal, and other temperature fluctuations at the land surface produce thermal signals that can be traced through shallow profiles.[17,18] Water moving through deeper profiles alters geothermal gradients, which can be used in inverse modeling to obtain recharge rates.[19]

Other Methods

Additional geophysical techniques provide recharge estimates based on the water-content dependence of gravitational, seismic, and electromagnetic properties of earth materials. Repeated high-precision gravity surveys can indicate changes in the quantity of subsurface water from recharge events.[20] Similarly, repeated surveys using seismic or ground-penetrating-radar equipment can resolve significant changes in water-table elevation associated with transient recharge.[21] In addition to surface-based techniques, cross-bore tomographic imaging can provide detailed three-dimensional reconstructions of water distribution and movement during periods of recharge.[22]

References

1. *Meinzer, O.E. The Occurrence of Ground Water in the United States, with a Discussion of Principles; U.S. Geological Survey Water-Supply Paper 489: 1923; 321 pp.*
2. Heath, R.C. *Ground-Water Hydrology;* U.S. Geological Survey Water-Supply Paper 2220: 1983; 84 pp.
3. Winter, T.C.; Harvey, J.W.; Franke, L.O.; Alley, W.M. *Ground Water and Surface Water a Single Resource*; U.S. Geological Survey Circular 1139: 1998; 79 pp.
4. Bredehoeft, J.D.; Papadopoulos, S.S.; Cooper, H.H., Jr. *Groundwater: The Water-Budget Myth, Scientific Basis of Water-Resource Management: Studies in Geophysics*; National Academy Press: Washington, DC, 1982; 51–57.

5. Narayanpethkar, A.B.; Rao, V.V.S.G.; Mallick, K. Estimation of groundwater recharge in a basaltic aquifer. Hydrol. Proc. **1994**, *8*, 211–220.
6. Steenhuis, T.S.; Jackson, C.D.; Kung, S.K.; Brutsaert, W. Measurement of groundwater recharge in eastern long Island, New York, U.S.A. J. Hydrol. **1985**, *79*, 145–169.
7. Bauer, H.H.; Mastin, M.C. *Recharge from Precipitation in Three Small Glacial-Till-Mantled Catchments in the Puget Sound Lowland*, U.S. Geological Survey Water-Resources Inv. Rep. 96–4219, Washington, 1997; 119 pp.
8. Abdulrazzak, M.J.; Sorman, A.U.; Alhames, A.S. Water balance approach under extreme arid conditions—aquifers: Recharge 57 case study of Tabalah Basin, Saudi Arabia. Hydrol. Proc. **1989**, 3, 107–122.
9. Rasmussen, W.C.; Andreasen, G.E. *Hydrologic Budget of the Beaverdam Creek Basin;* U.S. Geol. Survey. Water- Supply Paper 1472, Maryland, 1959; 106 pp.
10. Rutledge, A.T.; Daniel, C.C. Testing an automated method to estimate ground-water recharge from streamflow records. Ground Water **1994**, 32(2), 180–189.
11. Nimmo, J.R.; Stonestrom, D.A.; Akstin, K.C. The feasibility of recharge rate determinations using the steady state centrifuge method. Soil Sci. Soc. Am. J. **1994**, *58*, 49–56.
12. Freeze, A.R.; Banner, J. The mechanism of natural groundwater recharge and discharge 2. Laboratory column experiments and field measurements. Water Resour. Res. **1970**, *6* (1), 138–155.
13. Bromley, J.; Edmunds, W.M.; Fellman, E.; Brouwer, J.; Gaze, S.R.; Sudlow, J.; Taupin, J.-D. Estimation of rainfall inputs and direct recharge to the deep unsaturated zone of southern Niger using the chloride profile method. J. Hydrol. **1997**, *188–189* (1–4), 139–154.
14. Anderholm, S.K. *Mountain-Front Recharge along the Eastern Side of the Middle Rio-Grande Basin;* U.S. Geological Survey Water-Resources Investigations Report 00–4010, Central New Mexico, 2000; 36 pp.
15. Ekwurzel, B.S.P.; Smethie, W.M., Jr.; Plummer, L.N.; Busenberg, E.; Michel, R.L.; Weppernig, R.; Stute, M. Dating of shallow groundwater: comparison of the transient tracers 3H/3He, chlorofluorocarbons, and 85Kr. Water Resour. Res. **1994**, 30(6), 1693–1708.
16. Davisson, M.L.; Criss, R.E. Stable isotope and groundwater flow dynamics of agricultural irrigation recharge into groundwater resources of the central valley, California. In *Proc. Isotopes in Water Resources Management,* Vienna (Austria), Mar 20–24, 1995; International Atomic Energy Agency: Vienna, 1996; 405–418.
17. *Lapham, W.W. Use of Temperature Profiles Beneath Streams to Determine Rates of Vertical Ground-Water Flow and Vertical Hydraulic Conductivity; U.S. Geological Survey Water-Supply Paper 2337: 1989; 34 pp.*
18. Constantz, J.; Thomas, C.L.; Zellweger, G. Influence of diurnal variations in stream temperature on streamflow loss and groundwater recharge. Water Resour. Res. **1994**, *30*, 3253–3264.
19. Rousseau, J.P.; Kwicklis, E.M.; Gillies, D.C.; Eds. *Hydrogeology of the Unsaturated Zone, North Ramp Area of the Exploratory Studies Facility, Yucca Mountain,* U.S. Geological Survey Water Resources Investigations Report 98–4050, Nevada, 1999.
20. *Pool, D.R.; Schmidt, W. Measurement of Ground-Water Storage Change and Specific Yield Using the Temporal- Gravity Method Near Rillito Creek, Tucson; U.S. Geological Survey Water-Resources Investigations Report 97–4125, Arizona, 1997; 30 pp.*
21. Haeni, F.P. *Application of Seismic-Refraction Techniques to Hydrologic Studies;* U.S. Geological Survey Open File Report 84–746, 1986; 144 pp.
22. Daily, W.; Ramirez, A.; LaBrecque, D.; Nitao, J. Electrical resistivity tomography of vadose water movement. Water Resour. Res. **1992**, *28*(5), 1429–1442.

3

Artificial Reservoirs: Land Cover Change on Local Climate

Faisal Hossain and
Ahmed M. Degu
Tennessee Technological University

Introduction 17
Is the Impact of Dams Detectable from Observational Records? 20
Conclusion 24
Acknowledgments 24
References 24

Introduction

Understanding the terrestrial water cycle is important for any study that concerns future water availability. The dynamic nature of the Earth's water cycle means that the water that is available naturally in various forms (precipitation, stream flow, ground water, snow etc.) is highly variable in both space and time. Consequently, this means that the supply of water that is available for harnessing directly from nature does not always meet the rising demand to sustain our human civilization (for energy, food, water supply needs). Historically, therefore, one of the common engineering solutions to guarantee a more steady water supply against such rising demands has been to construct surface water impoundments on rivers. Such large-scale infrastructures are commonly known as dams and artificial reservoirs. These structures trap a sufficiently large amount of water from the local water cycle to make up for a shortfall when demand exceeds the variable supply from nature.

Recently published research on sustainable development stresses the need to improve our understanding of how the extremes of climate and water availability are changing. Of the many important factors, land use and land cover (LULC) change represents a major human-induced activity critical to extremes of water cycle and fresh water availability.[1–3] One example of human-induced LULC change is the construction of engineering facilities for irrigation, hydroelectric power generation, industrial and domestic water supply once a large dam is built nearby. In particular, irrigation is one of the more pervasive drivers of change in the water cycle. During the last century, irrigable land increased from 40 million hectares (Mha) to 215 Mha.[4] About 40% of the current irrigable land is supplied with surface water that is impounded by large artificial reservoirs and dams built on rivers.[5] Hereafter the term "dam" will be used interchangeably with "artificial reservoir."

The world has around 845,000 dams.[6] The water stored in these large dams amounts to about 10% (or more) of the annual river flow and covers one-third of the Earth's natural lake areas.[6] Though no accurate data are available on the volume of water impounded behind dams, estimates show that up to 10,800 cubic kilometers (km^3) of water may have been impounded. This volume is equal to a volume of the world's ocean water created to a depth of 30 mm.[7] To get an idea of the distribution of dams around the world, one can refer to the most comprehensive compilation of large dams archived by the Global Water Project and known as the GRanD database)[8] Figure 3.1 shows the global distribution of large dams. According to this GRanD database, about 34% of these large dams are engaged in irrigation.

In the United States, statistics may suggest that dam building is a thing of the past and not a current focus for the civil engineering profession (Figure 3.2).[10,11] However, for vast regions of the underdeveloped or developing world, large dam construction projects will continue in increasing numbers for tackling the rising water deficit for emerging economies. Examples of such large dam projects are the

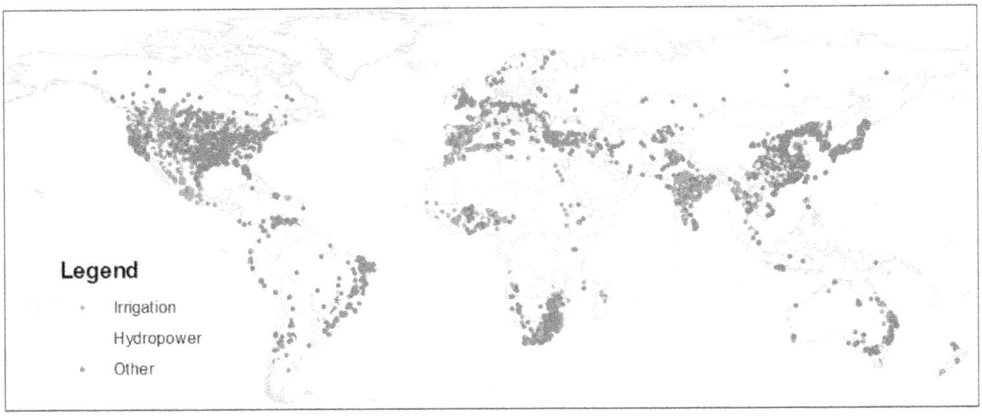

FIGURE 3.1 **(See color insert.)** The global distribution of large dams and their main purpose as archived in the GRanD database.

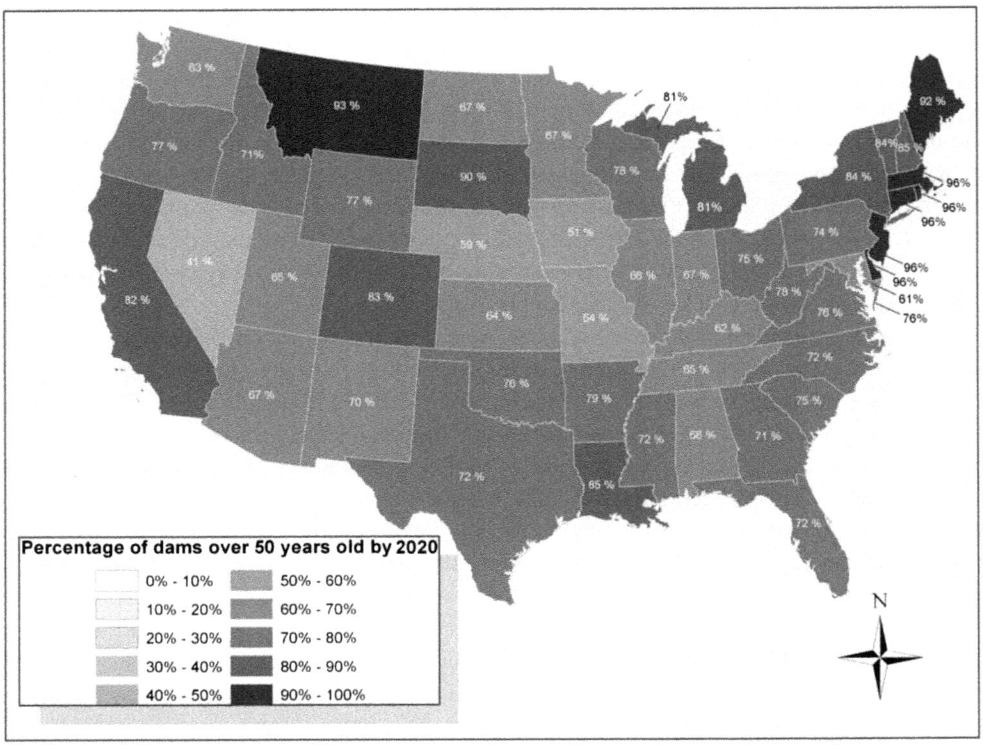

FIGURE 3.2 The percentage of dams per state that will be over 50 years old in 2020.
Source: Reproduced from U.S. Army Corps of Engineers report.[9]

Southeast Anatolia Project or GAP (Turkish acronym) project in Turkey comprising 22 dams on the Tigris and Euphrates rivers)[12] the Three Gorges Dam (TGD) in China,[13] Itaipu Dam in Brazil,[14] and the proposed Indian River Linking Project.[15] Thus, an understanding of the impact of dams on local climate and water cycle is important for sustainable development in the developing world.

In current literature, the impact of global climate change (which is usually perceived as a large-scale phenomenon based on a globally rising temperature trend) on dams and water supply has been studied at regional scales for some timed[16,17] This is also evident from a comprehensive literature synthesis published by the U.S. Bureau of Reclamation (USBR) on climate change adaptation.[18] The converse (impact of reservoirs on local climate), which is the focus of this entry, has not been explored in depth yet.[19,20]

A broader view of the change a dam typically triggers needs to be considered if we are to understand the changes a dam can initiate on the local climate. At a minimum, a dam changes the pre-existing landscape to an open body of water, which then leads to a change in surface albedo, surface roughness, and sensible and latent heat fluxes. A flood control or a hydropower dam can also accelerate the pace of urbanization in the downstream and flood-safe valley regions. Irrigation dams intensify agricultural production in the vicinity of the reservoir.

As mentioned earlier, of the various LULC changes due to dams, irrigation is the most widespread. Global data and simulation analysis by Biemans et al.[21] report that artificial reservoirs contribute significantly to irrigation water supply in many regions around the world. The additional contribution of reservoirs to irrigation has increased spectacularly from 5% (around 1900 A.D.) to almost 40% in the 21st century.[21] When such changes to the land cover and land use (LCLU) are assessed in relation to existing knowledge on their impact on climate,[22–24] it becomes apparent that a typical dam-reservoir system can, in principle, change the local climate through a gradual change in the landscape. Examples of the effect of local LULC on precipitation include Gero et al., Lei et al., Shepherd et al., and Niyogi et al.[25–28] Recently, there have been studies that report that there is global scale impact from irrigation,[29] of which a significant fraction of this irrigation is the result of water made available for vegetation from artificial reservoirs.

The change to local climate near dams is expected to be particularly noticeable on temperature and precipitation patterns during the growing season. For example, Mahmood et al.[30] showed that irrigation in the Great Plains during growing season (May through September) might have caused regional cooling by 1.41°C of growing season mean maximum temperatures at irrigated locations during the post-1945 period. A few other studies support the notion that atmospheric moisture added by irrigation can also increase rainfall, provided that the mesoscale conditions are appropriate.[31,32] De Angelis et al.[33] have reported an increase of 10–30% of rainfall downwind of the Ogallala aquifer that has been used for ground water-based irrigation in the Great Plains. A statistical analysis of observed rainfall frequency downwind of irrigated lands in South Spain (upper and lower Vegas and lower Guadalquivir) found a significant increase in the number of months with minimum precipitation after irrigation when compared with pre-irrigation records.[34] In summary, dams can, therefore, clearly cause a change of climate in their vicinity through systematic change in the land use and land cover.

While the climate effects of LULC change attributable to large dams have been explored, the hydrology impact is not understood as well yet and therefore needs to be explored. Qualitative investigation of databases archived by the National Performance of Dams Program (NPDP; http://npdp.stanford.edu) reveals that historically, the number of reported incidences of embankment overtopping far exceeds that of structural failures. According to our NPDP survey, a total of 1133 dams have overtopped hydrologically between 1889 and 2006. Of these, 625 experienced a complete "hydrologic performance failure," most likely due to increased reservoir inflow from upstream. Herein, it is important to note that the land use changes can amplify the surface runoff generation mechanism in two ways: (1) through modification of precipitation rates leading to increased infiltration-excess runoff and (2) through enhancement of rainfall partitioning as runoff due to increased imperviousness. The former cause is akin to a "strategic" cause that occurs through gradual change in the local climate in the post-dam era, while the latter is a more "tactical" (instantaneous) cause (of increasing imperviousness). Both causes may be equally important and may work together to magnify the intensification of upstream runoff, as the dam ages.

Is the Impact of Dams Detectable from Observational Records?

The previous section dwelt on a "can"-type question on how a dam–reservoir system is physically capable of changing the pre-dam climate. This section addresses the following question: *If dams can really change local climate in principle, can we detect it from observational records?* Wu et al.[35] examined the effect of the TGD of China, built in 1998, on the regional precipitation around its vicinity by analyzing satellite precipitation data. Based on their 8-year data analysis (1998–2006), they concluded that filling up of the TGD reservoir was found to correlate with an increase in precipitation in the nearby region. However, at other regions, precipitation was found to experience a decrease.

For a more in-depth and comprehensive analysis of observational data (comprising several dams and multidecadal data), a recent effort by Hossain et al.[20] is recounted here as is hereafter with kind permission granted by the American Society of Civil Engineers (ASCE). Ninety-two dams located in the United States and defined as large, according to the standards set by the International Commission on Large Dams, were recently studied in a manner similar to a recent study by Degu et al.[19] The database on dams was available from a series of world dam registers published by the Global Water Systems Project Digital Water Atlas.[36] These dams were spread across eight distinct Köppen climate zones (Figure 3.3). For the observational record of climate, reanalysis data from the National Center for Environmental Prediction, North American Regional Reanalysis (NARR),[37,38] were used. NARR provides good quality and finer-resolution climate dataset for North America. Daily NARR fields span a period from 1979 to 2009 (30 years) and focus on the daily average of the surface Convective Available Potential Energy (CAPE; J/kg) as a proxy signature of dams.[39] Among the many important ingredients required for

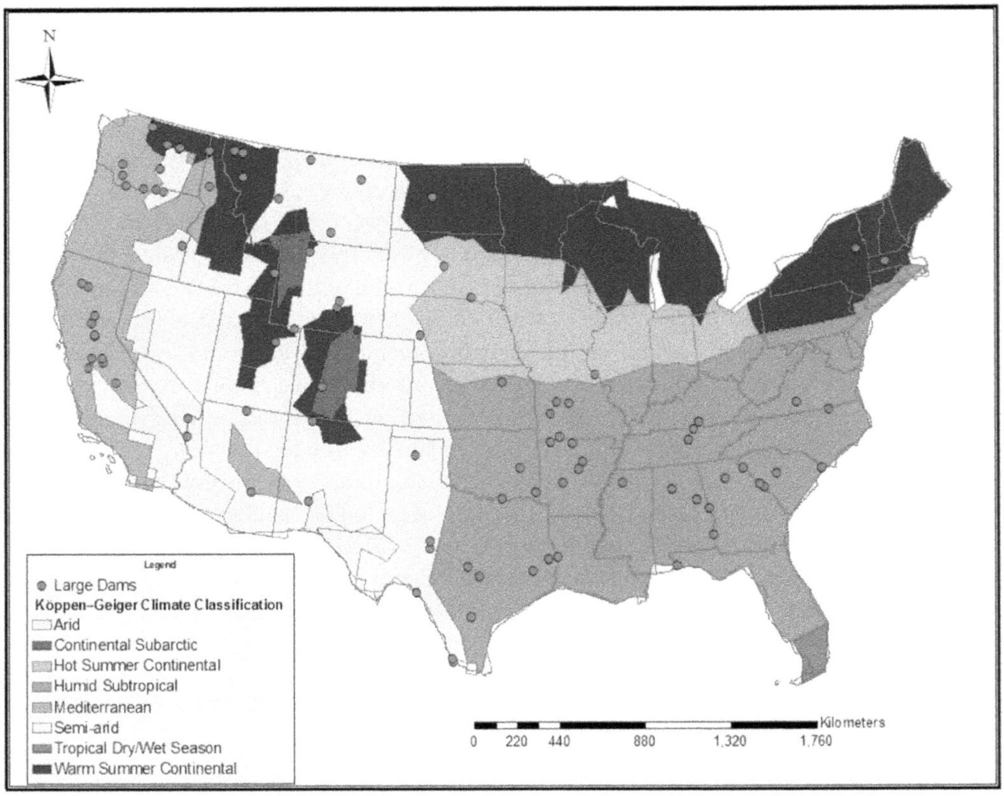

FIGURE 3.3 **(See color insert.)** Location of the 92 large dams according to the Köppen climate map. **Source:** Reprinted with kind permission from American Society of Civil Engineers.

rainfall, CAPE can be considered an important atmospheric indicator of the presence of heavy rainfall process. The objective was to identify detectable changes in the CAPE climatology (available at 32 km spatial scales) in the vicinity of the dams for a given climate zone.

Because the NARR record of CAPE does not date back before the construction of most large dams in the United States, the ergodic assumption was invoked to answer the "detection" and "have"-type question posed in this section. The spatial average of CAPE over a region far away from dams (referred to as "no-dam" region) was considered as a substitute for the temporal average of CAPE during the pre-dam era. In other words, the average CAPE over the "no-dam" region was used to represent the "pre-dam" climatology of convective atmospheric instability. Vice versa, the average CAPE over the "dam" region was used as a proxy for the post-dam climatology. Herein, the "no-dam" region was defined as the annular region 100 km outside the dam's spillway having the same area as the "dam" region (Figure 3.4). To attribute the unique and local impact due to the dam alone, a "dam" region was defined as a circular area having a 50 km radius around the dam location (Figure 3.4).

For a statistical generalization of the 92 dams, Table 3.1 shows the relative ranking of the dams in terms of the % change (increase or decrease in the climatologic average of CAPE) between the "dam" and "no-dam" region. Here, the % change in CAPE is defined as

$$\text{Percent increase} = \frac{\left(\text{mean of "Dam"} - \text{mean of "No Dam"}\right)}{\text{mean of No dam}} \times 100\%$$

Herein, the "mean" refers to the 30-year climatologic average for the period 1979–2009. It is very clear from Table 3.1 that dams in the Mediterranean climate (Koppen class name is Dry Summer Subtropical)

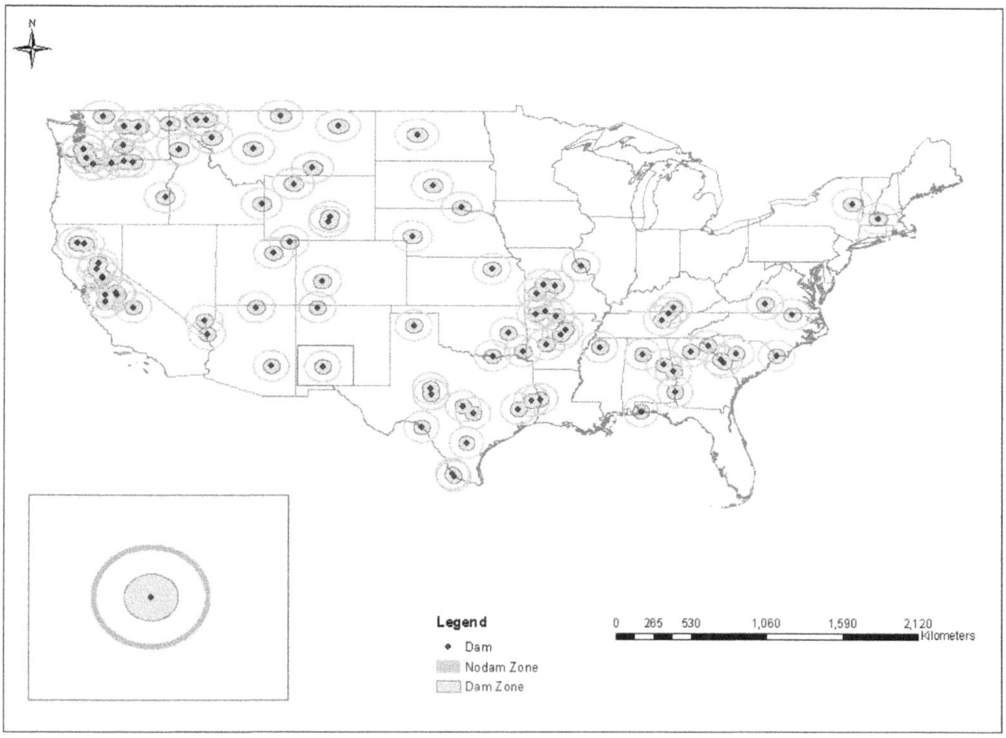

FIGURE 3.4 The 50 km buffer zone showing the "dam" and the 100 km away annular region (outer encircle) as "no-dam region."
Source: Reprinted with kind permission from American Society of Civil Engineers.

TABLE 3.1 Ranking of Dams Studied according to % Change in CAPE Climatology between Dam and No-Dam Regions. Only a % Change of up to 1% Is Reported in the Table

Köppen–Geiger Climate	Dam	State	% Increment
Dry Summer Subtropical	New Bullards Bar	California	105.01
Dry Summer Subtropical	Oroville	California	99.00
Dry Summer Subtropical	New Exchequer	California	80.91
Dry Summer Subtropical	Shasta	California	66.89
Dry Summer Subtropical	Folsom	California	56.86
Dry Summer Subtropical	Don Pedro	California	56.06
Dry Summer Subtropical	Trinity	California	42.41
Dry Summer Subtropical	Monticello	California	33.79
Dry Summer Subtropical	Pine Flat Lake	California	31.31
Dry Summer Subtropical	Swift	Washington	27.22
Warm Summer Continental	Conklingville	New York	26.36
Dry Summer Subtropical	Dworshak	Idaho	25.73
Cold Semiarid	Kingsley	Nebraska	21.27
Cold Semiarid	Yellowtail	Montana	20.61
Humid Subtropical	Hartwell	Georgia	20.42
Warm Summer Continental	Albeni Falls	Idaho	18.01
Warm Summer Continental	Hungry Horse	Montana	16.66
Warm Summer Continental	Kerr	Montana	15.35
Cold Arid	Hoover	Nevada	14.16
Humid Subtropical	Smith Mountain	Virginia	13.45
Humid Subtropical	Pensacola	Florida	13.09
Warm Summer Continental	Seminoe	Wyoming	12.13
Warm Summer Continental	Palisades	Idaho	11.19
Hot Arid	Davis	Arizona	11.17
Humid Subtropical	Center Hill	Tennessee	10.87
Hot Semiarid	Robert Lee	Texas	7.86
Humid Subtropical	Little River	South Carolina	7.49
Continental Subarctic	Blue Mesa	Colorado	7.26
Dry Summer Subtropical	Mossyrock	Washington	6.13
Humid Subtropical	Sardis	Mississippi	5.97
Cold Semiarid	Owyhee	Oregon	5.47
Cold Semiarid	Pathfinder	Wyoming	4.78
Dry Summer Subtropical	New Melones	California	4.24
Humid Subtropical	Richard B. Russell	Georgia	4.02
Humid Subtropical	Wolf Creek	Kentucky	3.97
Humid Subtropical	Eufaula	Oklahoma	3.75
Humid Subtropical	Dale Hollow	Tennessee	3.38
Continental Subarctic	Oahe	South Dakota	3.35
Cold Semiarid	Coolidge	Arizona	3.28
Cold Semiarid	Buffalo Bill	Wyoming	3.03
Warm Summer Continental	Libby	Montana	2.98
Humid Subtropical	Stockton	Missouri	2.79
Cold Semiarid	Tiber	Montana	2.23
Hot Semiarid	Twin Buttes	Texas	2.19
Cold Semiarid	Fort Peck	Montana	1.86

consistently experience the highest alteration of CAPE near the reservoir ("dam") region. Subsequently, this change is matched by dams located in hot or cold arid climates. Dams in humid or arctic climates exhibit spatially insignificant increases (<5%) in CAPE near the reservoir. An important aspect to note is the likely absence of irrigation in humid and arctic climates that may be one of the contributing factors for spatial uniformity of CAPE near and away from the reservoir. As an example, Figure 3.5 shows the time series of 30-year CAPE climatology for two dams in contrasting climates (Walter Dam in Alabama and Coolidge Dam in Arizona). Two possible reasons may be attributed for the intensification of CAPE by dams in arid and Mediterranean climates: (1) more widespread irrigation; and (2) stronger spatial gradients in humidity and surface evaporation around the reservoir–shoreline interface.

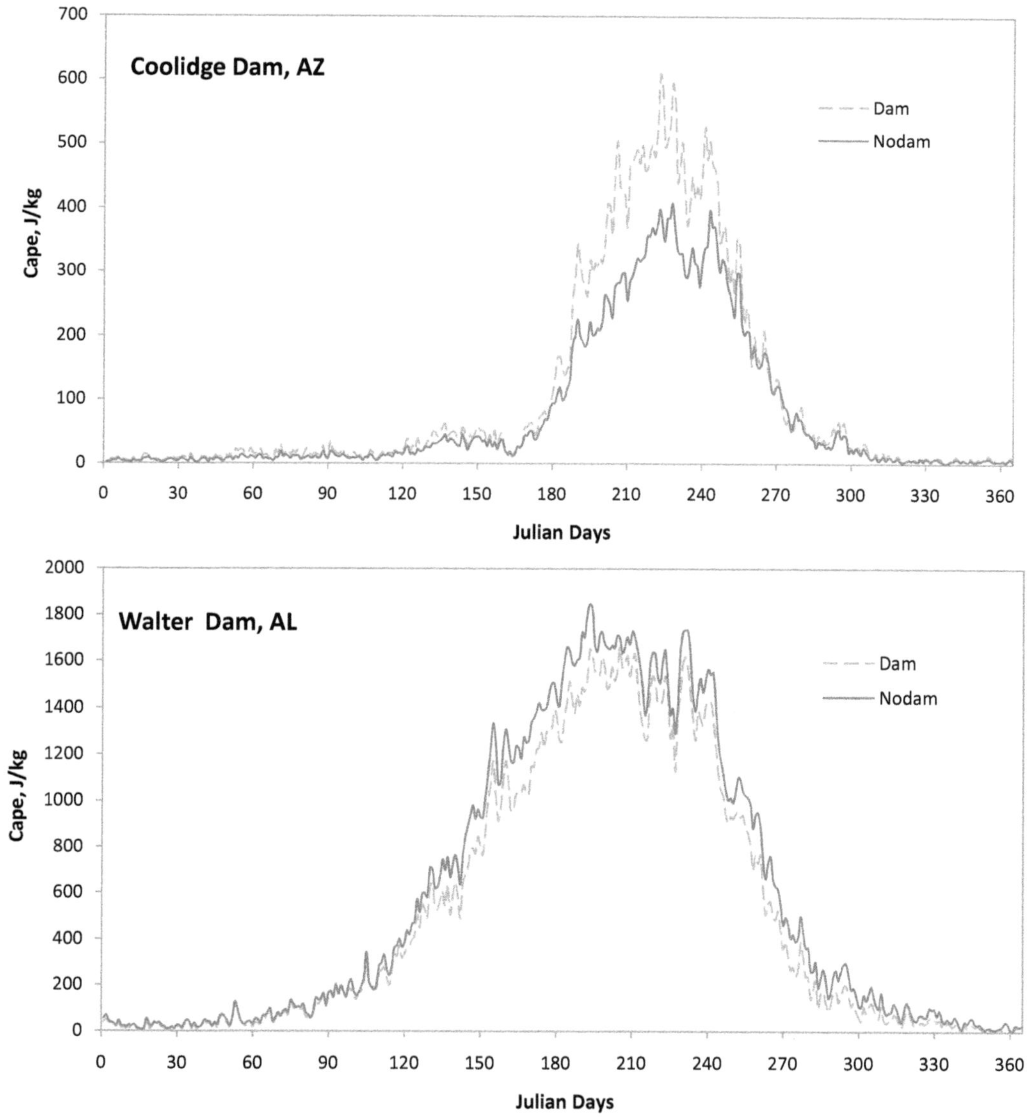

FIGURE 3.5 Comparison of Convective Available Potential Energy (CAPE) climatology (30 year) for dam- and no-dam region for Coolidge (in arid climate, Arizona) and Walter (in humid climate, Alabama) dams.
Source: Reprinted with kind permission from American Society of Civil Engineers.

Conclusion

This entry presented a review of what is currently known about the potential impact of large artificial reservoirs and dams on the local climate based on the premise of systematic change in climate-sensitive LCLU that most dams initiate. To better understand the implication of local climate changes, it is, therefore, important to have a broader view of the changes to the landscape in the post-dam era. However, in the future, other factors, such as urbanization,[40] atmospheric aerosols,[28] and global warming,[18] need to be explored as well. These factors are well known to initiate changes in local climate, particularly to precipitation. Systematic changes in precipitation patterns can also gradually alter the hydrology of the impounded river basin, which is currently not well understood. For example, more intensified storms near the reservoir shorelines or overly dense irrigated landscapes may result in more runoff, erosion, and water-logging issues. The groundwater table may rise and gradually change infiltration–excess runoff areas to a more excess saturation type. Overall, it is intuitive to expect that the hydrological impact due to changing rainfall patterns will differ upstream and downstream of the reservoir due to the hydraulic head differences and seepage issues. For better water resources management, it is now timely to investigate the changing hydrology of aging dams, which undergo a shift in local rainfall dynamics.

Acknowledgments

This entry was reproduced from an earlier forum article that appeared in ASCE Journal of Hydrologic Engineering titled "Climate Feedback-based Considerations to Dam Design, Operations and Water Management in the 21st Century" (vol. 17(8), pp. 837–850, doi:10.1061/(ASCE) HE.1943–5584.0000541).

References

1. Chase, T.N.; Pielke Sr., R.A.; Kittel, G.F.; Nemani, R.R.; Running, S. W. Simulated impacts of historical land cover changes on global climate in northern winter. Clim. Dyn. **2000**, *16*, 93–105.
2. Vörösmarty, C.J.; Sahagian, D. Anthropogenic disturbance of the terresterial water cycle. Bioscience **2000**, *50* (9), 753.
3. Hossain, F.; Niyogi, D.; Adegoke, J.; Kallos, G.; Pielke Sr., R. Making sense of the water resources that will be available in future use, EOS Forum **2011**, *92* (17).
4. Freydank, K.; Siebert, S. Towards mapping the extent of irrigation in the last century: a time series of irrigated area per country. Germany: Frankfurt Hydrology Paper 08, Institute of Physical Geography, University of Frankfurt (Main), 2008.
5. Lempérière, F. The role of dams in XXI century: Achieving a sustainable development target. International Journal of Hydropower and Dams, http://www.hydropower-dams.com/Issue-Three.php?c_id=60 (accessed May 2011).
6. Jacquot, J. Dams, from Hoover to Three Gorges to the Crumbling Ones, Discover Magazine. March 2009, http://discovermagazine.com/2009/mar/08-dams-hoover-three-gorges-crumbling-ones (accessed May 2011).
7. Chao, B.F.; Wu, Y.H.; Li, Y.S. Impact of artificial reservoir water impoundment on global sea level. Science **2008**, *320*, 212–214.
8. Lehner, B.; Döll, P. Development and validation of a global database of lakes, reservoirs and wetlands. J. Hydrol. **2004**, *296*, 1–22.
9. Hossain, F.; Jeyachandran, I.; Pielke, R., Sr. Have Large Dams altered extreme precipitation patterns?, EOS-AGU **2009**, *90* (48), 453–454.
10. Graf, W.L.; Wohl, E.; Sinha, T.; Sabo, J.L. Sedimentation and sustainability of western American reservoirs. Water Resour. Res. **2010**, *46*, W12535, (doi:10.1029/2009WR008836).
11. Graf, W.L. Downstream hydrologic and geomorphic effects of large dams on American rivers. Geomorphology **2006**, *79* (3–4), 336–360.

12. Unver, O.H.I. South-eastern Anatolia integrated development project (GAP), Turkey: an overview of issues of sustainability Int. J. Water Resour. Develop. **1977,** *13* (2), 187–208.
13. Shen, G.; Xie, Z. Three Gorges project: chance and challenge, Science **2004**, *681* (doi:10.1126/science.304.5671.681b).
14. Pierce, F. The biggest dam in the world New Sci. **1995,** *145* (1962), 25–29.
15. Misra, A.K.; Saxena, A.; Yaduvanshi, M.; Mishra, A.; Bhauduriya, Y.; Takur. A. Proposed river linking project of India: Boon or bane to nature? Env. Geol. **2007,** *51* (8), 1361–1376. (doi: 10.1007/s00254-006-0434–7).
16. Hamlet, A.F.; Lettenmaier, D.P. Effects of climate change on hydrology and water resources in the Columbia River basin J. Amer. Water Res. Assoc. **1999,** *35* (6), 1597–1623.
17. Christensen, N.S.; Wood, A.W.; Voisin, N.; Lettenmaier, D.P.; Palmer, R.N. The effects of climate change on the hydrology and water resources of the Colorado river basin. Clim. Change **2004,** *62* (1–3), 337–363 pp.
18. USBR. Literature synthesis on climate change implications for reclamation's water resources, USBR Technical Services Report (Technical Memorandum 86–68210-091), 2nd Edition, 2009, available online at http://www.usbr.gov/research/docs/climatechangelitsynthesis.pdf (last accessed May 2011).
19. Degu, A.M.; Hossain, F.; Niyogi, D.; Pielke, R., Sr.; Shepherd, J.M.; Voisin, N.; Chronis, T. The influence of large dams on surrounding climate and precipitation patterns. Geophys. Res. Lett. **2011,** *38* (4), L04405, doi:10.1029/ 2010GL046482.
20. Hossain, F.A.M. Degu; Yigzaw, W.; Niyogi, D.; Burian, S.; Shepherd, J.M.; Pielke, R., Sr. Climate feedback-based considerations to dam design, operations and water management in the 21st century. ASCE J. Hydrol. Eng. **2011,** *17* (8), 837–850.
21. Biemans, H.; Haddeland, I.; Kabat, P.; Ludwig, F.; Hutjes, R.W.A.; Heinke, J.; von Bloh, W.; Gerten, D. Impact of reservoirs on river discharge and irrigation water supply during the 20th century, Water Resour. Res. **2011,** *47,* W03509 (doi:10.1029/2009WR008929).
22. Pielke Sr., R.A. Land use and climate change. Science **2005,** *310,* 1625–1626.
23. Feddema, J.J.; Oleson, K.W.; Bonan, B.; Mearns, L.O.; Buja, L.E.; Meehl, G.A.; Washington, W.M. The importance of land cover change in simulating future climates. Science **2005,** *310,* 1674–1678.
24. Pielke, Sr., R.A.; Adegoke, Sr., J.; Beltran-Przekurat, A.; Hiemstra, C.A.; Lin, J.; Nair, U.S.; Niyogi, D.; Nobis, T.E. An overview of regional land use and land cover impacts on rainfall. Tellus B **2007,** *59,* 587–601.
25. Gero, A.F.; Pitman, A.J.; Narisma, G.T.; Jacobson, C.; Pielke, R.A., Sr. The impact of land cover change on storms in the Sydney Basin. Global Planet. Change **2006,** *54,* 57–78.
26. Lei, M.; Niyogi, D.; Kishtawal, C.; Pielke, R., Sr.; Beltrán-Przekurat, A.; Nobis, T.; Vaidya, S. Effect of explicit urban land surface representation on the simulation of the 26 July 2005 heavy rain event over Mumbai, India, Atmos. Chem. Phys. **2008,** 8, 5975–5995.
27. Shepherd, J.M.; Carter, W.M.; Manyin, M.; Messen, D.; Burian, S. The impact of urbanization on current and future coastal convection: a case study for Houston. Env. Plan. **2010,** *37,* 284–304.
28. Niyogi, P. Pyle, Lei, M.; Arya, S.; Kishtawal, C.; Shepherd, M.; Chen, F.; Wolfe, B. Urban modification of thunderstorms - an observational storm climatology and model case study for the Indianapolis urban region. J. Appl. Meteor. Climatol **2011,** *50,* 1129–1144.
29. Puma, M. J.; Cook, B.I. Effects of irrigation on global climate during the 20th century, J. Geophys. Res. **2011,** *115,* D16120 (doi:10.1029/2010JD014122).
30. Mahmood, R.; Hubbard, K.G.; Carlson, C. Modification of growing-season surface temperature records in the Northern Great Plains due to land use transformation: Verification of modeling results and implications for global climate change. Int J Climatol. **2004,** *24,* 311–327.
31. Lohar, D.; Pal, B. The effect of irrigation on pre-monsoon season precipitation over Southwest Bengal, India. J. Clim. **1995,** *8* (10), 2567–2570.
32. Barnston, A.G.; Schickedanz, P.T.; The effect of irrigation on warm season precipitation in the Southern Great Plains. J. Appl. Meterol. Clim. **1984,** *23* (6).

33. DeAngelis, A.; Dominguez, F.D.; Fan, Y.; Robock, A.; Kustu, M.D.; Robinson. D. Evidence of enhanced precipitation due to irrigation over the Great Plains of the United States. J. Geophys. Res. **2010**, *115*, D15115(doi:10.1029/2010JD013892).
34. Jodar, J.; Carrera, J.; Cruz, A. Irrigation enhances precipitation at the mountains downwind. Hydrol. Earth Syst. Sci. Discussion, **2010,** 7, 3109–3127.
35. Wu, L.G.; Zhang, Q.; Jiang, Z.H. Three Gorges Dam affects regional precipitation. Geophys. Res. Lett. **2006**, 33L13806 10.1029/2006gl026780.
36. GWSP, Digital Water Atlas, Map 41: Dams and capacity of artificial reservoirs (V1.0), Bonn, Germany, 2008 (available at http://atlas.gwsp.org/)
37. Mesinger, F. North American Regional Reanalysis, Bull. Am. Meteorol. Soc. **2006**, *87* (3), 343–360.
38. Rutledge, G.K.; Alpert, J.; Ebuisaki, W. NOMADS: A Climate and Weather Model Archive at the National Oceanic and Atmospheric Administration. Bull. Am. Meteorol. Soc. **2006**, 87, 327–341.
39. Pielke Sr., R.A. Influence of the spatial distribution of vegetation and soils on the prediction of cumulus convective rainfall. Rev. Geophys. **2001**, *39*, 151–177.
40. Shepherd, J.M.; Pierce, H.; Negri, A.J. On rainfall modification by major urban areas: Observations from spaceborne radar on TRMM. J. Appl. Meteorol. **2002**, *41*, 689–701.

4

Drainage and Water Quality

Environmental Conditions 27
Soils/Hydrology • Climate/Precipitation
Pollutant Properties 28
Persistence • Adsorption/Filtration
Management Practices 29
Tillage Systems • Cropping • Controlled Drainage
Conclusion 30
References 31

James L. Baker
Iowa State University

Environmental Conditions

Soils/Hydrology

Whether water added to the soil surface as precipitation or irrigation infiltrates or becomes surface runoff is critical to water quality. Topography/slope, soil moisture content, texture, and soil structure, including the existence of preferential flow paths or "macropores," can affect both the rate and route of water infiltration (as shown in Figure 4.1). While different drain spacings are used to provide a desired "drainage coefficient" (e.g., 0.5 in./day) based on the internal drainage characteristics of the subsoils, the conditions of the surface soil will determine what percentage of applied water (rain or irrigation) will infiltrate. In general, the existence of subsurface drainage increases the volume of infiltration and thus decreases the volume of surface runoff (and pollutant loss with that runoff), increases shallow percolation, and lowers water tables. This effect will be more pronounced for "lighter" soils that have higher infiltration rates and lower water holding capacities. These changes affect the final quality of cropland drainage because of differences in time and type of soil-water-chemical interactions. Surface runoff allows water to come in contact only with the surface soil and materials such as crop residue and surface-applied fertilizers, manures, and pesticides present on it for short periods of time. On the other hand, water that infiltrates and percolates through the soil comes in intimate contact with all the soils in the profile to at least the depth of the tile drain, and generally, for much longer periods of time. The soil residence time is shorter for that portion of infiltrated water intercepted by the tile drains than for water that must flow through the underground strata to appear as base flow.

Climate/Precipitation

The timing, amount, and intensity of water inputs, including precipitation and irrigation, relative to evapotranspiration (ET) determine the timing and amount of excess water that will leave agricultural land as surface runoff and subsurface drainage. Inputs at low intensities generally will totally infiltrate,

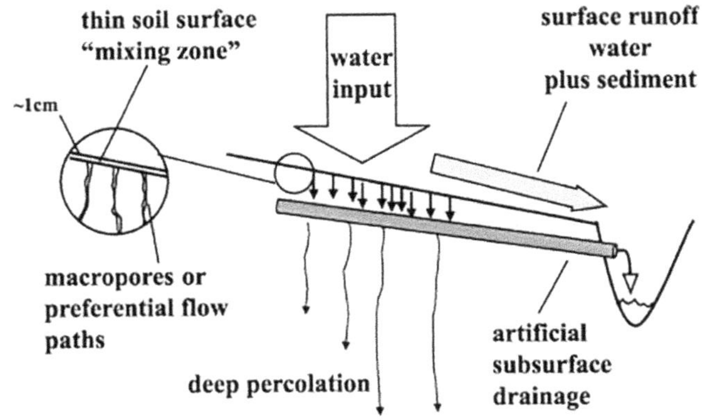

FIGURE 4.1 Schematic of transport processes.

but as intensities increase past the rate of infiltration, surface runoff begins. As just discussed, because decreased antecedent soil moisture contents, which result from the existence of artificial subsurface drainage, increase infiltration rates, the volume of surface runoff generally is decreased when subsurface drainage exists. This effect will be more pronounced for areas where input intensities often exceed the rates of infiltration.

For a given crop and climatic region, the amount of ET is relatively constant, so the volume of subsurface drainage is very much dependent on the amount of input water above that value. As an example, in a 4-year Iowa tile drainage experiment,[1] during a low rainfall year, only a trace of subsurface flow occurred; in a wet year (116 cm of precipitation vs. the average of 80 cm), there was 29 cm of flow; and the overall average flow for the four years was 15 cm.

Pollutant Properties

Persistence

Pollutants that have a limited existence in agricultural lands because of plant uptake/chemical transformation (i.e., nutrients), degradation (i.e., pesticides), and die-off (i.e., micro-organisms) have different potentials for off-site transport with water based on their persistence. Because surface runoff is much more of an immediate process than subsurface drainage, concentrations of nonpersistent pollutants in subsurface drainage are usually lower than in surface runoff. This effect will be more pronounced, the lesser the persistence is. However, the "route" of infiltration, where in some cases the drainage water moves quickly through the soil profile because of macropores (see Figure 4.1), plays a role and can somewhat negate the expected dissipation effect.

Adsorption/Filtration

Based on their chemical properties and/or their physical size, some pollutants transported down into and through the soil with subsurface drainage are removed from the flow stream by the soil. In particular, pollutants that are positively charged (e.g., ammonium-nitrogen (NH_4–N)) and larger, less soluble, organic compounds (e.g., some pesticides) can be attenuated by adsorption to soil clay and organic matter. Inorganic P ions can be removed by complexation or precipitation with soil cations. Microorganisms and sediment can be filtered out by small soil pores. Thus, in general, with the exception of soluble salts of nitrate-nitrogen (NO_3–N), sulfate (SO_4), and chloride (Cl) anions, pollutant concentrations in subsurface drainage are lower than in surface runoff. This effect will be more pronounced, the greater the interaction between the pollutant and soil is.[2]

For sediment itself, because soil erosion is dependent on the erosive ability and transport capacity of surface runoff, reducing the runoff flow rate and volume with subsurface drainage reduces sediment loss. For example, in a 6-year study in the lower Mississippi Valley, surface runoff volumes from plots on a clay loam soil with subsurface drainage were 34% less than for plots without subsurface drainage, and the corresponding soil loss of 3500 kg/ha/yr represented a decrease of 30%. For the plots with subsurface drainage, both sediment concentrations and losses in subsurface drainage were about one-tenth those in surface runoff.[3]

For nitrogen (N), loss from poorly drained soils is usually much less than that from soils with improved drainage systems.[4] While significant N can be transported with sediment (often sediment has at least 1000 ppm N), land needing subsurface drainage is usually not highly susceptible to erosion, and N loss is dominated by soluble inorganic-N loss. In a 3-year tile-drained watershed study in northeast Iowa,[5] NO_3–N losses in solution represented over 85% of the total N losses, including NH_4–N, organic-N in solution, and N associated with sediment. In a 5-year study in east-central Iowa,[6,7] where nutrient concentrations in both surface runoff and subsurface drainage from cropland were monitored, NO_3–N concentrations in subsurface drainage averaged about 12 mg/L and 2–3 times those in surface runoff. While NH_4–N concentrations were usually 2–10 times higher in surface runoff, on an absolute scale, the concentrations were overall much lower than those for NO_3–N and constituted only a fraction of the total N loss.

For phosphorus (P), transport is primarily in surface runoff with sediment (often sediment has at least 500 ppm P) and dissolved in surface runoff. For conventionally tilled cropland, about 75–90% of P transported in surface runoff is with sediment. In areas where soil erosion is minimal, soluble P in surface runoff water can dominate transport. The soluble P in surface runoff (and subsurface drainage) is regulated by adsorption/desorption characteristics of soil. Therefore, with P in surface soils generally much higher than in subsoils, subsurface drainage usually has much lower soluble P concentrations than in surface runoff.

For pesticides, because of their adsorption characteristics, like with P, concentrations in surface runoff water are usually much greater than in subsurface drainage. However, this effect is even more dramatic for pesticides because unlike P, pesticides have a limited persistence, which decreases their potential for movement with subsurface flows with long travel times. In a series of studies on herbicides in surface and subsurface drainage,[8–10] atrazine concentrations in May (the period of application) were 75 μg/L in surface runoff for plots with surface drainage only, and were 51 μg/L and 1 μg/L in surface runoff and subsurface drainage, respectively, for plots that also had subsurface drainage. Not only were there much lower atrazine concentrations in subsurface drainage water, but the presence of the subsurface drains delayed and reduced surface runoff such that concentrations in surface runoff from the plots with subsurface drainage were reduced about one-third.

For bacteria, a review by Crane et al.[11] showed that fecal coliform counts in surface runoff from manured lands often were greater than 10,000/100 mL. In a comparison of surface runoff and subsurface drainage, Culley and Phillips[12] found similar counts (>10,000/100 mL) for fecal coliform in surface runoff from both manured and fertilized plots, but with much, much lower counts (<5/100 mL) in subsurface drainage.

Management Practices

Tillage Systems

The hydrologic interactions between tillage systems and subsurface drainage that affects water quality involves how tillage affects the timing, route, and volume of infiltration (and hence the relative volumes of subsurface drainage and surface runoff). In a review of the hydrologic effects of conservation tillage, Baker[13] noted that changing the relative volumes of subsurface drainage and surface runoff also can affect chemical concentrations in those carriers. In general, increasing infiltration increases the time for beginning of surface runoff, which in turn reduces the concentrations of chemicals at the soil surface (shown as a thin mixing zone in Figure 4.1) and therefore in surface runoff. However, the effect of conservation tillage on infiltration is time dependent. For the first storm after any tillage there is usually less runoff from the tilled soil, although on an annual basis, conservation (or less) tillage often results in lower total surface runoff volumes.

Cropping

As with tillage, the hydrologic interactions between cropping systems and subsurface drainage that affects water quality involves how cropping affects the timing, route, and volume of infiltration (and hence the relative volumes of subsurface drainage and surface runoff). A major difference between perennial crops such as forages, and row-crops such as corn and soybeans, is the volume and timing of ET demands. In general, with higher, more consistent ET, perennial crops would have lower total drainage volumes. A bigger effect of cropping would be the effect of needed chemical applications and their potential losses. For example, the large amounts of N needed (added, recycled, and/or fixed) in a continuous corn or corn–soybean rotation means there is usually high NO_3–N concentrations in the soil profile, and hence in subsurface drainage when it occurs. However, for grasses and alfalfa. NO_3–N concentrations in subsurface drainage are much lower.[14,15]

Controlled Drainage

In areas where subsurface drainage exists, controlling the timing of outflows has been suggested as one method to reduce chemical losses. This controlled drainage could reduce losses by reducing both subsurface volumes and chemical concentrations. The potential for reduced concentrations is probably the greatest for NO_3–N, where the process of denitrification would reduce NO_3 to N gases in the soil profile where high water tables and the presence of organic matter drive the system anaerobic. The results summarized from 125 site-years of data from North Carolina[16] showed that controlled drainage reduced subsurface drainage volumes an average of 30% compared to uncontrolled drainage systems. Reductions in N and P lost with subsurface drainage were 45% and 35%, respectively. While almost all the reduction of P loss was due to decrease in drainage volume; for N, reductions in NO_3–N concentrations also contributed to the reduction.

Conclusion

The total effect of surface drainage on surface water resources receiving drainage from agricultural lands involves the relative volumes of surface runoff and subsurface drainage, and the relative concentrations of sediment, nutrients, pesticides, and bacteria. In general, the existence of subsurface drainage increases infiltration rates which delays and reduces the volume of surface runoff. For pollutants lost mostly with surface runoff, which include sediment, NH_4–N, P, pesticides, and bacteria, not only is the volume of the carrier reduced, but also the concentrations. This is because delayed runoff and more water moving through the surface-mixing zone reduce the amounts of contaminants at the soil surface available to interact with added water and overland flow. Thus the only real water quality negative to subsurface drainage is the increased volume of water moving thorough the soil profile carrying the soluble unadsorbed NO_3–N anion. The use of improved infield N management in the way of rate, method, and timing of N applications has some potential to reduce this problem[17] However, other practices such as controlled drainage or construction/reconstruction of wetlands may be needed to provide some NO_3–N reduction treatment. Alternatively, reducing the amount of row-crops grown on subsurface-drained lands could have a large impact although the current economics of doing that would be quite negative.

References

1. Baker, J.L.; Campbell, K.L.; Johnson, H.P.; Hanway, J.J. Nitrate, phosphorus and sulfate in subsurface drainage water. J. Environ. Qual. **1975**, *4*, 406–412.
2. Gilliam, J.W.; Baker, J.L.; Reddy, K.R. Water quality effects of drainage in humid regions. *Drainage Monograph;* Am. Soc. Agron.: Madison, WI, 1999.
3. Bengston, R.L.; Carter, C.E.; Morris, H.F.; Bartkiewicz, S.A. The influence of subsurface drainage practices on nitrogen and phosphorus losses in a warm, humid climate. Trans. ASAE **1988**, *31*, 729–733.
4. Gambrell, R.P.; Gilliam, J.W.; Weed, S.B. Nitrogen losses from soils of the North Carolina coastal plain. J. Environ. Qual. **1975**, *4*, 317–323.
5. Baker, J.L.; Melvin, S.W.; Agua, M.M.; Rodecap, J. *Collection of Water Quality Data for Modeling the Upper Maquoketa River Watershed;* Paper No. 99-2222; ASAE: St. Joseph, MI, 1999.
6. *Johnson, H.P.; Baker, J.L. Field-to-Stream Transport of Agricultural Chemicals and Sediment in an Iowa Watershed: Part I. Data Base for Model Testing (1976–1978); Rep. No. EPA-600/S3-82-032; USEPA Environ. Res. Lab.: Athens, GA, 1982.*
7. *Johnson, H.P.; Baker, J.L. Field-to-Stream Transport of Agricultural Chemicals and Sediment in an Iowa Watershed: Part II. Data Base for Model Testing (1979–1980); Rep. No. EPA-600/S3-84-055; USEPA Environ. Res. Lab.: Athens, GA, 1984.*
8. Southwick, L.M.; Willis, G.H.; Bengston, R.L.; Lormand, T.J. Effect of subsurface drainage on runoff losses of atrazine and metolachlor in Southern Louisiana. Bull. Environ. Contam. Toxicol. **1990**, *45*, 113–119.
9. Southwick, L.M.; Willis, G.A.; Bengston, R.L.; Lormand, T.J. Atrazine and metolacholor in subsurface drain water in Louisiana. J. Irrig. Drain. Eng. **1990**, *116*, 16–23.
10. Bengston, R.L.; Southwick, J.M.; Willis, G.H.; Carter, C.E. The influence of subsurface drainage practices on herbicide losses. Trans. ASAE **1990**, *33*, 415–418.
11. Crane, S.R.; Moore, J.A.; Grismer, M.E.; Miner, J.R. Bacterial pollution from agricultural source: A review. Trans. ASAE **1983**, *26*, 858–866, 872.
12. Culley, J.L.B.; Phillips, P.A. Bacteriological quality of surface and subsurface runoff from manured sandy clay loam soil. J. Environ. Qual. **1982**, *11*, 155–158.
13. Baker, J.L. Hydrologic effects of conservation tillage and their importance relative to water quality. In *Effects of Conservation Tillage and Groundwater Quality;* Logan, T.J., Davidson, J.M., Baker, J.L., Overcash, M.R., Eds.; Lewis Publishers, Inc.: Chelsea, MI, 1987; Chap. 6.
14. Baker, J.L.; Melvin, S.W. Chemical management, status, and findings. *Agricultural Drainage Well Research and Demonstration Project—Annual Rep. and Project Summary;* Iowa Dep. of Agric. and Land Stewardship and Iowa State University: Ames, 1994; 27–60.
15. Weed, D.A.J.; Kanwar, R.S. Nitrate and water present in and flowing from root-zone soil. J. Environ. Qual. **1996**, *25*, 709–719.
16. Evans, R.O.; Skaggs, R.W.; Gilliam, J.W. Management practice effects on water quality. Proc. Irrig. Drain. 1990 Natl. Conf. ASCE Irrig. Div., Durango, CO, 1990; 182–191.
17. Baker, J.L. Limitations of improved nitrogen management to reduce nitrate leaching and increase use efficiency. Optimizing nitrogen management in food and energy production and environmental protection: proceedings of the 2nd international nitrogen conference on science and policy. Sci. World **2001**, *1*(S2), 10–16.

5

Endorheic Lake Dynamics: Remote Sensing

Yongwei Sheng
University of California

Introduction 33
Global Distribution of Endorheic Lakes 34
Remote Sensing Monitoring of Endorheic Lakes 35
Lake Area Change • Lake Level Change • Lake Water Salinity Remote Sensing • Paleolake Change
Conclusion 42
Acknowledgments 42
References 42

Introduction

Atmospheric precipitation over much of the Earth's land eventually returns to the ocean through rivers, whereas precipitation falling into some interesting areas does not flow back into the ocean. These areas are endorheic basins. An endorheic basin, also known as a closed basin, is a drainage basin that retains water within it, prohibiting water outflow to external rivers or the ocean.[1] The retained water accumulates in the basin bottom, forming endorheic lakes. Endorheic lakes are water bodies whose water does not flow into the ocean. Endorheic lakes may include both through-flow lakes and terminal lakes in an endorheic basin. The water in through-flow lakes eventually flows to its final destination: terminal lakes. As endorheic lakes do not have outlets for the water to flow out of their basins, the only pathways for water to leave the drainage system are evaporation and seepage, either evaporating into the air or seeping into the ground. The water falling as precipitation in the basin dissolves and carries minerals during the course of transporting to its destinations—endorheic lakes, and the water in the lakes evaporates, leaving a high concentration of minerals and other inflow erosion products in the lakes. This process over time can render endorheic lakes rather saline and sensitive to environmental pollutions. Endorheic lakes include some of the world's large lakes, such as the largest, the Caspian Sea and the Aral Sea in Asia, the Lake Chad in Africa, the Great Salt Lake in North America, and Lake Eyre in Australia. They largely are saline or salt lakes located in dry climates.

Endorheic lakes are important natural resources. First of all, they are crucial water resources in the arid environment that they belong to, feeding people and animals and irrigating agriculture. They are also valuable industrial mineral resources. The minerals dissolved in the water are concentrated during the natural evaporation process and deposit in lakes. Zabuye Caka, for example, is a hypersaline lake on the western Tibetan Plateau with a salinity concentration of ~300 g/L as measured in 2011.[2] Lithium carbonate was discovered in this lake in the 1980s,[3] and it now is China's major source of lithium, which is a main component of lithium batteries extensively used in electronic devices.

Closed to major rivers and oceans, these lakes are particularly sensitive and vulnerable to climate change and human activities. Endorheic lakes are among the most sensitive recorders of past precipitation–evaporation balance due to changes in lake size, water level, and chemistry.[4] The size and water level of an endorheic lake are largely determined by the balance between precipitation in its basin and evaporation from the lake. Most endorheic lakes are rather sensitive to precipitation. Under global warming, precipitation, however, will likely become more unstable and variable in the future,[5] which will reinforce the vulnerability of endorheic lakes in arid regions. Water level, area, and volume of endorheic lakes fluctuate in response to changes in the evaporation and precipitation rates within their basins. Therefore, monitoring of these lake changes is not only important for hydrological and economic purposes but can also provide a climate record, a proxy indicator of climate change.[6,7]

Endorheic regions have been home to human cultures for thousands of years, and the population continues to increase in the areas.[8] However, population growth demands surface and ground water for irrigation. As a consequence, many endorheic lakes are under great pressure from human water demands, leading to their shrinkage and even desiccation. Rapid shrinkage of endorheic lakes may result in severe environmental problems. The environmental disaster caused by the Aral Sea's shrinkage presents a classic example. The Aral Sea, formerly the world's fourth largest lake in 1960 with an area of 68,000 km^2 only behind the Caspian Sea, Lake Superior, and Lake Victoria, has been dramatically shrinking since the 1960s due to the diversion of the Sya Darya and Amu Darya rivers for agriculture irrigation. Its lake level dropped nearly 13 m, its area decreased by 40%, and its salinity rose from 10 to 27 g/L between 1960 and 1987, which is the most rapid and pronounced recession in history, turning the Aral Sea into a residual brine lake.[9] The current Aral Sea comprises two separate bodies of water: the small North Aral Sea and the relatively large South Aral Sea, with the total area under 20% of its original size. The Aral is estimated to contain ~10 gigaton (GT) salts. An estimated 43 million tons of salts are annually transported by winds from the Aral's dried lakebed into adjacent areas and deposited as aerosols by rain and dew over 150,000 km^2. The wind blows salts and dust to travel a thousand kilometers to the Black Sea coast and even to the Russian Arctic shore. The biological diversity and productivity has steeply declined. Twenty of 24 native fish species disappeared, and the commercial catch is out of business.[9] In addition, desert animals have suffered from the greatly increased mineral content in their drinking water, and native plant communities have degraded, causing desertification in this region. The land around the Aral Sea is heavily polluted, and the people living in the area are suffering from a lack of fresh water and from health problems, including high rates of cancer and respiratory illnesses.

Many other endorheic lakes like the Aral Sea are under stress due to climate change and human activities. In the era of global change, it is of crucial importance to inventory and monitor endorheic lakes in the context of global warming. However, the abundance, distribution, and dynamics of these lakes are not well understood. There is a pressing need to assess endorheic lake dynamics under global warming and anthropogenic pressures and their potential impacts on water resources.

Global Distribution of Endorheic Lakes

Endorheic lakes and basins occur in arid to semihumid climates but are commonly found in semiarid regions. Figure 5.1 shows the distribution of major endorheic basins (darker regions). These basins were extracted from global watershed data sets as lake basins closed to major rivers and oceans. Approximately 20% of the Earth's land drains to endorheic lakes. Continents vary in their concentration of endorheic regions due to conditions of geography and climate. In general, endorheic basins are extensively distributed in western and central Asia, northern Africa, central Australia, the Altiplano Plateau of South America, and the Great Basin of North America. Western and Central Asia are home to the world's largest group of endorheic lakes, including the largest, the Caspian Sea, the Dead Sea, and the Aral Sea. The Tibetan Plateau in the southern part of the region also contains a large number of endorheic lakes at an altitude of ~4500 m. Africa hosts three major groups of endorheic lakes in the Sahara Desert, the East African Rift, and the Kalahari Desert. Lake Chad in the southern Sahara Desert continues to shrink due to the decreased precipitation and increased irrigation pressure. The Great Basin in the United States contains a large region of endorheic basins, in which the Great Salt Lake is the largest endorheic

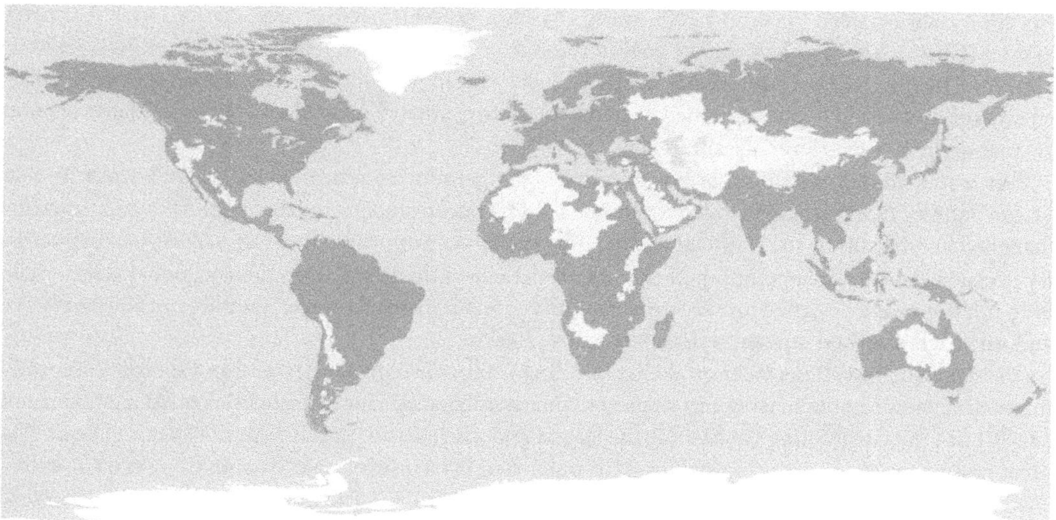

FIGURE 5.1 Global distribution of major endorheic basins. Endorheic basins (darker regions) are extensively distributed in western and central Asia, northern Africa, central Australia, the Great Basin of North America, and the Altiplano Plateau of South America.

lake in North America. The Valley of Mexico in Central America also contains a number of dried lakes. The Altiplano basin in South America includes a number of closed lakes such as Lake Poopó. Central Australia has many endorheic drainages, including the highly variant Lake Eyre.

Remote Sensing Monitoring of Endorheic Lakes

Lake area, water level, and salinity are measurable indicators of lake dynamics. However, field-based investigation of endorheic lakes is challenging due to the low accessibility and the harsh environment. These lakes are seldom observed, let alone monitored for dynamics. Spaceborne remote sensing provides a feasible tool for multitemporal lake mapping and monitoring at local, regional, and global spatial scales and monthly-to-decadal temporal scales. Harris[10] and Birkett[11] discussed the need for simultaneous measure of level and area of endorheic lakes and promoted remote sensing of areas and levels for large endorheic lakes from satellite images and radar altimeters.

Many remote sensing satellites that are currently in space can be used in mapping and monitoring lake dynamics. At high-medium resolutions, Landsat, *Satellite Pour l'Observation de la Terre* (SPOT), and Terra/Advanced Space-borne Thermal Emission and Reflection Radiometer (ASTER) images are widely used in detecting subtle lake area changes.[12–14] Coarse spatial resolution satellite sensors normally are able to monitor large lakes at much higher temporal resolutions to catch even seasonal variability. National Oceanographic and Atmospheric Administration/Advanced Very High Resolution Radiometer (NOAA/AVHRR; ~1km resolution) and Terra/Aqua Moderate Resolution Imaging Spectroradiometer (MODIS; 0.25 to 1km resolutions) are used to monitor lake area dynamics as frequently as every two days, even to catch changes within closed basins as a result of possible changes in regional rainfall patterns.[15,16]

Satellite radar and laser altimetry technology provides a precise and efficient tool for measuring and monitoring water level of large inland lakes at 0.5 dm accuracy)[11,17] The widely used radar altimetry missions include Topex/Poseidon (1992–2005), Jason-1 (2001 to present), Jason-2 (2008 to present), and Envisat (2002 to present), providing measurements since the early 1990s.[18] The only available spaceborne lidar mission Ice, Cloud, and land Elevation Satellite (ICESat, launched in January 2003) collected elevation data between 2003 and 2009 using the laser altimetry sensor Geoscience Laser Altimeter System (GLAS). Though the primary goal of the ICESat mission is to measure ice sheet mass balance and cloud heights, ICESat data have been found successful in monitoring water levels for oceans, lakes, rivers, and wetlands.[19] The combined available satellite altimetry missions have accumulated about a 20-year archive of lake level measurements.

Two water salinity satellites have been recently in operation in space, including Soil Moisture and Ocean Salinity (SMOS) launched in November 2009 by European Space Agency (ESA)[20] and Aquarius launched in June 2011 by the National Aeronautics and Space Administration (NASA).[21] As ocean salinity is crucial to the understanding of the role of the ocean in climate through the global water cycle, both missions were designed for sea surface salinity measurements; however, they are not able to resolve endorheic lakes at their current resolutions of 100–200 km.

The subsequent sections focus on the Great Siling Co (Co: lake in Tibet) basin as a specific case study to illustrate how remote sensing can be used to monitor dynamics for endorheic lakes. At a total area of 58,570 km^2, the Great Siling Co basin is the largest endorheic basin on the central Tibetan Plateau. The great basin consists of the primary Siling Co watershed (34,638 km^2), the Gyaring Co–Urru Co watershed (15,838 km^2), the Ringco Ogma Co watershed (3,364 km^2), the Pangkog Co watershed (2,658 km^2), and the Namka Co (2,072 km^2) (Figure 5.2). This region is cold all year round with an annual mean temperature of −2°C. The annual evaporation reaches 2,167 mm while the mean annual precipitation is only ~290 mm, 90% of which falls in the months of June to September.[22] Centered at (31.8°N, 89°E), Siling Co is the primary saline lake in the Great Siling Co basin with a salinity of ~18.5 g/L. The lake has been expanding dramatically over the past decades. The lake was only 1,640 km^2 in 1976 but expanded

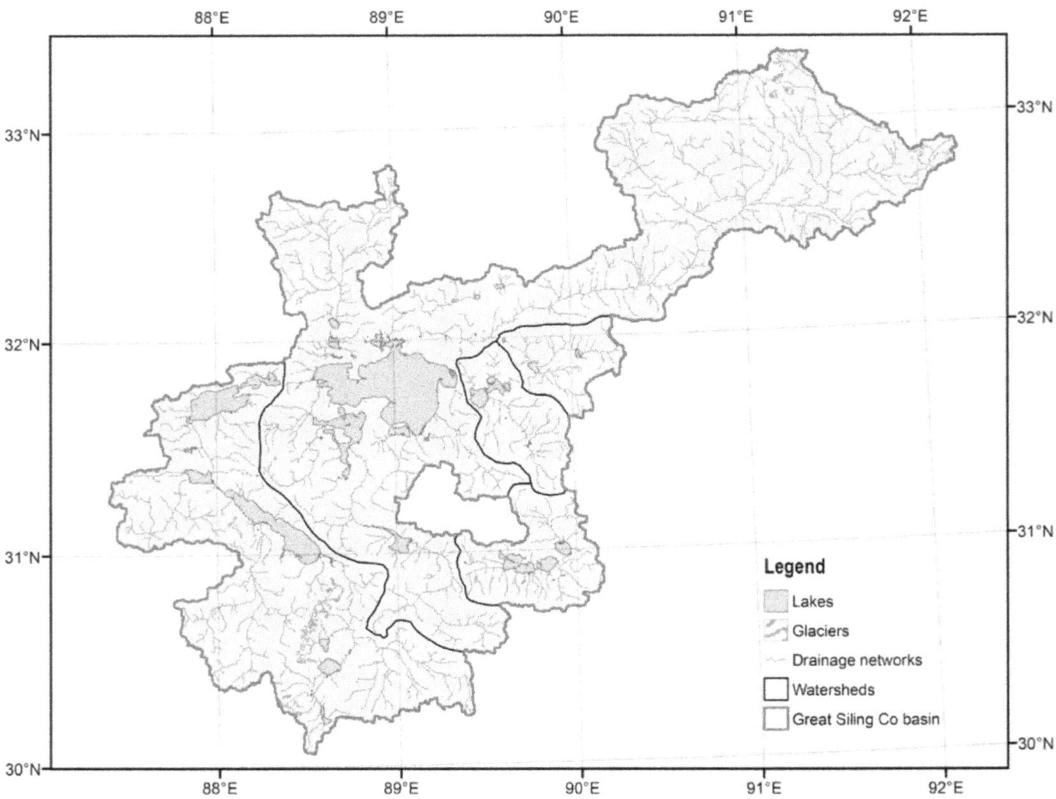

FIGURE 5.2 Great Siling Co basin and lakes. The basin consists of the Siting Co watershed, the Gyaring Co–Urru Co watershed, the Ringco Ogma Co watershed, the Pangkog Co watershed, and the Namka Co.

to its current size at 2,335 km^2, and its water level reached 4,544.5 m in 2009. In addition to atmospheric precipitation, it is fed by glacier meltwater from Mount Geladaindong in the far northeast via the Zagya Zangbo River entering the lake from the north and from Mount Gyagang in the south via the Zagen Zangbo River entering the lake from the west through Gyaring Co and Urru Co.

Lake Area Change

The Landsat program is particularly suitable and irreplaceable for detailed lake inventory and lake dynamics monitoring at regional and even global scales not only because of the recent release of the entire Landsat data archive for free public access but also due to the fact that the Landsat program provides the longest high-resolution image records of the Earth.[13] Images acquired by the early Multispectral Scanner (MSS; provided in 60 m resolution) sensors onboard Landsat 1–5, the later Thematic Mapper (TM, 30 m) onboard Landsat 4–5, and the recent Enhanced Thematic Mapper plus (ETM+, 30 m) onboard Landsat 7 have continuously monitored the Earth in the past ~40 years repeatedly every ~16 days. With the recent Landsat 8 launched in February 2013 carrying the Operational Land Imager (OLI; also in 30 m resolution) sensor, the Landsat mission is expected to continue into the next decade.

Decadal lake area changes in the Great Siling Co basin were monitored using 1976 MSS, 1990 TM, 2000 ETM+, and 2009 TM images acquired between October and December when seasonal variability is minimal. Theoretically, lakes and other surface water bodies are readily identified in Landsat images owing to their very low reflectance in near infrared channels of Landsat sensors. However, the variant mineral content in endorheic lakes complicates the lake water spectral responses and imposes challenges to remote sensing lake mapping. The normalized difference water index (NDWI), derived using green and near infrared bands, can enhance water signature and depress disturbing factors and has been widely used in water body delineation from satellite images.[23] NDWI is less sensitive to mineral content in lake water and demonstrates better capability in delineating water bodies from land than individual bands. However, lakes cannot be properly delineated in the image using a single threshold to segment an NDWI image due to the various conditions of the lakes.

Lakes in the Great Siling Co basin were mapped and monitored using an automated hierarchical lake mapping approach. The method simulates a human operator's procedures in delineating lakes and segments the image at both global and local levels.[13] In the global level of segmentation, lakes are first located by segmenting the entire NDWI image using a conservative threshold. Then, each identified lake is treated as an object and is further refined only in the area surrounding the lake (i.e., a buffer zone) for a fine adjustment at the local level. This hierarchical method can effectively map from Landsat imagery lakes under various complex water conditions in the Great Siling Co basin.

The monitoring results in Figure 5.3 show that Siling Co has experienced significant change over the past ~35 years. It used to be the second largest lake on the central plateau (just after Num Co) in 1976 with an area of ~1,641 km^2, but it expanded steadily afterward (1990 and 2000) and became the largest lake in 2009 with an area of 2,335 km^2, a 42% expansion from the 1970s. Moreover, the lake has been growing at an accelerating rate: 0.33%/year between 1976 and 1990, 0.93%/year in the 1990s, and 2.7%/year in the recent decade. The increased precipitation and the accelerated glacier melting in response to the rising temperature are responsible for the observed lake expansions.[13]

Lake Level Change

ICESat/GLAS collects elevation data with a high precision (~3 cm) every ~170 m along the spacecraft's ground track with small footprints (50–90 m), compared to Jason's ~5 km footprints.[19] As such, GLAS is able to measure more precisely water levels for smaller inland lakes in endorheic regions than the satellite radar altimeters.

The water level changes of Siling Co were monitored using ICESat/GLAS global land surface altimetry product, GLA14 delivered by National Snow and Ice Data Center (NSIDC), covering the operation

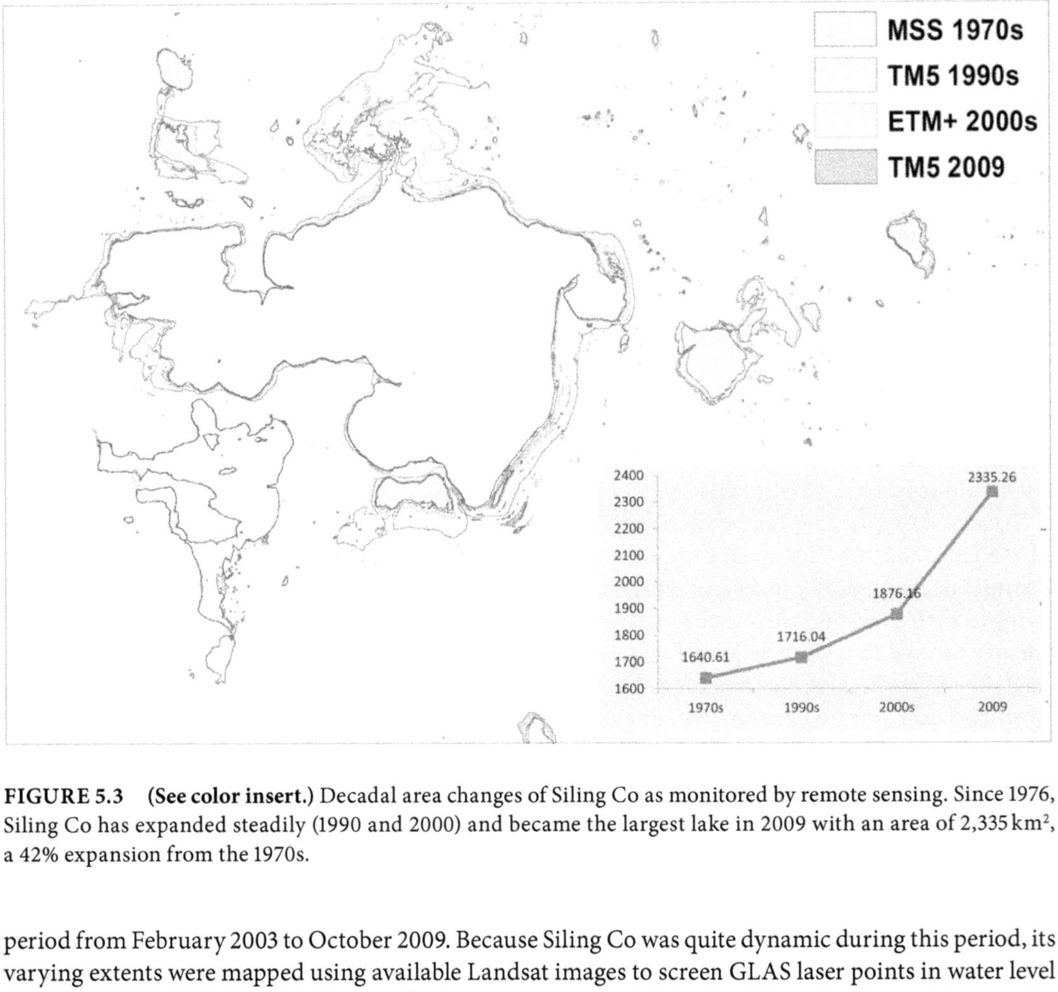

FIGURE 5.3 (See color insert.) Decadal area changes of Siling Co as monitored by remote sensing. Since 1976, Siling Co has expanded steadily (1990 and 2000) and became the largest lake in 2009 with an area of 2,335 km^2, a 42% expansion from the 1970s.

period from February 2003 to October 2009. Because Siling Co was quite dynamic during this period, its varying extents were mapped using available Landsat images to screen GLAS laser points in water level calculation. As such, a high resolution lake level series was established for Siling Co. Figure 5.4 shows the lake level changes of Siling Co derived from GLAS data and the corresponding lake area changes (black dots) extracted from Landsat. The lake area measured by Landsat in 2002, 2005, 2006, 2007, and 2009 synchronizes rather well with the GLAS-detected lake level changes. Over the monitoring period, the lake level rose steadily with three rapid ~1 m jumps in the summers of 2003, 2004, and 2005. Siling Co expanded by ~15.2% from 2009 km^2 in December 2002 to 2314 km^2 in July 2009 with a ~4.5 m level rise. Combining the area and level changes, the total volume change of Siling Co between 2003 and 2009 is estimated at 9.7 billion m^3.

A major issue with the current lake level remote sensing is that the tracks of satellite radar and laser altimeters are rather sparse. Envisat have a 35-day temporal resolution and 70 km intertrack spacing at the equator, while Topex/Poseidon, Jason-1, and Jason-2 have a 10-day orbital cycle and 350 km equatorial intertrack spacing.[18] ICESat has a 91-day repeat orbit with a 33-day subcycle. This 33-day scenario has a repeat track spacing of ~80 km at the equator and ~20 km at 75° latitude.[19] As a result, the current lake level data sets are only able to measure hundreds of large lakes (such as Siling Co) occurring under the satellite tracks. The future Surface Water Ocean Topography mission (SWOT) planned to launch in 2020 by NASA will simultaneously monitor both lake area and level changes in nearly global coverage.[24] Flying two radar antennas, the SWOT satellite will measure surface elevations using radar interferometry along a 120 km wide swath. The radar sensors can penetrate both clouds and darkness and provide all-weather lake observations at high resolutions, overcoming the cloud-free limitation of

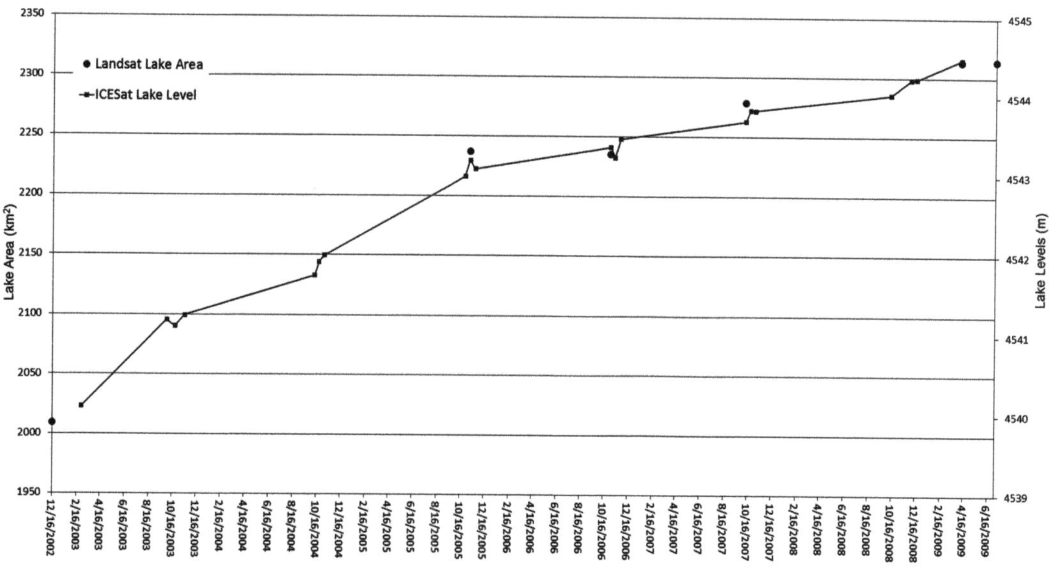

FIGURE 5.4 Level change of Siling Co detected by Ice, Cloud, and land Elevation Satellite (ICESat)/Geoscience Laser Altimeter System (GLAS). The lake level rose steadily up to 4.5 m from 2003 to 2009, with three rapid ~1 m jumps in the summers of 2003, 2004, and 2005. The level changes synchronize rather well with the lake area change (black dots) detected by Landsat images.

its optical counterparts. The SWOT mission will periodically monitor lake areas at the spatial scale of decameters and the vertical water level change in centimeter accuracy. With the measurements of both lake surface area and water level changes, the SWOT mission will be able to monitor lake water storage changes, which is revolutionary in endorheic lake dynamics monitoring.

Lake Water Salinity Remote Sensing

Salinity is an important water parameter of endorheic lakes. Water salinity content in endorheic lakes changes along with their area and level dynamics. It is desirable to measure lake salinity using remote sensing. Figure 5.5 shows the Landsat/ETM+ image acquired in the western Tibetan Plateau on October 28, 2000, on which lakes appear in different colors, largely indicating water salinity content. Taro Co in the south is a freshwater lake with a low salinity of 0.77 g/L and appears completely dark on the image. The bluish lake north of Taro Co is Zabuye Caka (the major lithium mineral site in China) in two parts: the North Zabuye Caka and the South Zabuye Caka. The dark blue north lake is a brine lake with a salinity of 393 g/L. The south lake appears blue, light blue, and white on the image from east to west. The blue part is brine waters at high salinity (~440 g/L) while the saturated salt deposits appear light blue, and the playa appears cyanish white on the image. The small playa lake in a valley north of the North Zabuye Caka is also surrounded by cyan-white lacustrine deposits. As this figure suggests, the optical image is able to separate these endorheic lakes into fresh water lakes, saline lakes, salt lakes, and playa lakes. However, quantitative retrieval of lake salinity using optical satellite remote sensing is far from robust and practical.

The only direct way to remotely assess water salinity from space seems to be through the use of passive microwaves. The dielectric constant of saltwater is a function of its salinity and temperature and directly impacts sea surface emissivity. Currently, both the SMOS and Aquarius satellite missions have only been able to derive sea surface salinity. Aquarius is an L-band radiometer and scatterometer instrument combination designed to map the salinity field at the surface of the ocean from space. The instrument is designed to provide global sea surface salinity maps on a monthly basis with a spatial

FIGURE 5.5 Appearance of lakes with various salinity content on satellite imagery. The Landsat/Enhanced Thematic Mapper plus (ETM+) image acquired on October 28, 2000 illustrates the appearance of the lakes with varying salt content. Freshwater Taro Co appears rather dark, while Zabuye Caka shows the color tones of white outline black, dark gray, and white corresponding to brine water, salt deposits, and dry playas, respectively, along the west shore. In addition, Zabuye Caka is surrounded by abandoned paleo-shorelines, attesting to the existence of larger lakes in ancient time.

resolution of 150 km.[21] SMOS carries an L-band interferometric radiometer, which receives the radiation emitted from the Earth's surface and converts it to salinity in the surface waters of the oceans at a resolution of 100–200 km.[20] At such coarse resolutions, inland lakes cannot be resolved. In addition, endorheic basins are not as uniform as the ocean surface and the salinity range of endorheic lakes is much higher and more variant than that of the ocean. All these leave satellite remote sensing of inland lake salinity a currently unsolved problem.

Paleolake Change

Paleo-changes of endorheic lakes with paleo-shoreline relics in their basins can be recovered through integration of high-resolution imagery and digital elevation models (DEMs). Many endorheic lakes (e.g., Zabuye Caka in Figure 5.5) are surrounded by paleo-shorelines, attesting to a substantially moister condition in the past. The paleolake extent is critical for us to understand the paleoclimatic and paleo-hydrologic environments in the basin. Aerial photographs, satellite images, and terrain data have been used to reconstruct paleolake extents for Lake Chad in Africa[25] and Lake Eyre in Australia.[26] Integrating high-resolution satellite imagery and DEMs, Sheng (2009) developed a semiautomated tool for paleolake extent recovery at regional scales.[27]

Paleo-shorelines are found extensively on Landsat imagery in the Great Siling Co basin. With the help of the interactive paleolake recovery tool, paleolake extent was recovered in the basin using Landsat ETM+ images and Shuttle Radar Topography Mission (SRTM) DEMs (Figure 5.6). Four paleo-lakes were recovered in the basin, occupying at least a total area of 10,221 km². These paleolakes, after shrinkage,

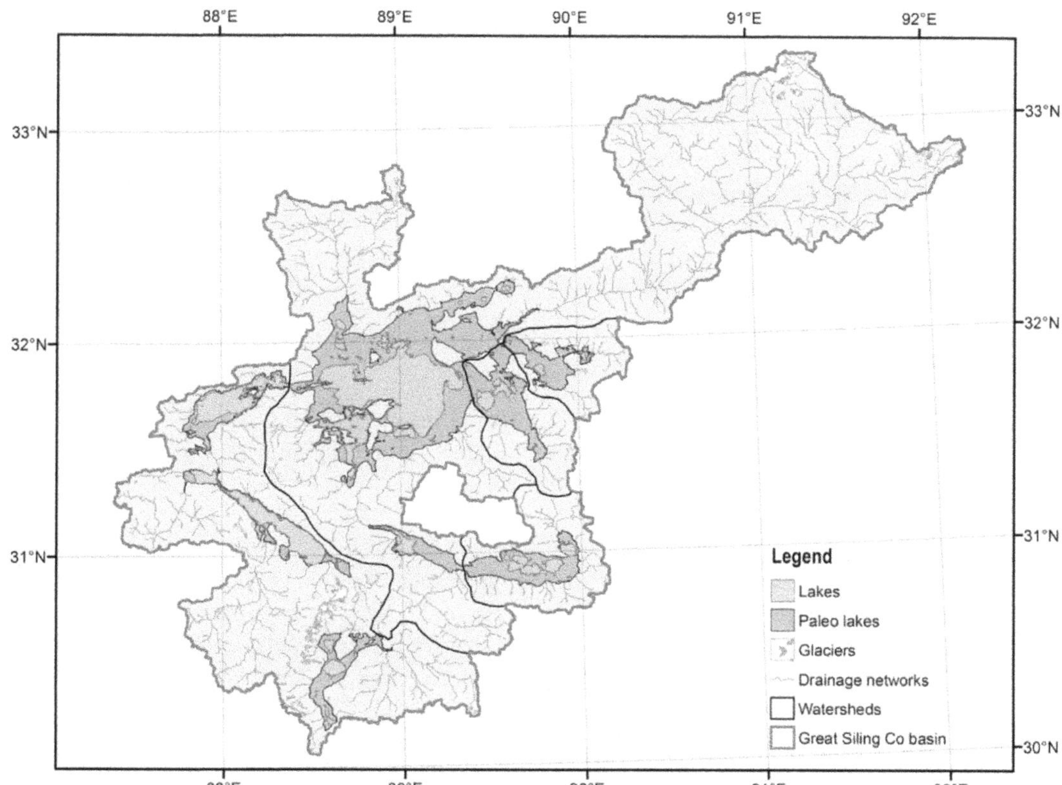

FIGURE 5.6 **(See color insert.)** Paleolake recovery in Great Siling Co basin. The extents of four paleolakes (in dark gray) were recovered in the basin using current Landsat images and digital elevation data. They had shrunk in Holocene up to ~79% in area with an up-to-56 m level drop in the basin.

broke into 95 contemporary lakes with a total area of 3,759 km^2. The total area shrinkage and water loss are estimated at 6,462 km^2 and 372 billion m^3, respectively. Nearly two-thirds of the lake area has disappeared since the great lake period, which occurred ~11 ka BP in the basin according to optically stimulated luminescence (OSL) dating results from paleo-shore sand samples collected in the field.

Paleo Siling Co alone used to occupy an area of 7,665 km^2 in early Holocene and subsequently shrank as the climate became drier, evolving into 56 contemporary lakes including the current Siling Co and the sizable Pangkog Co, Urru Co, and Qiagui Co. The lake was recovered at the water level of 4,600 m on satellite images with field validation in 2009, which was ~56 m above the present-day water level. The recovery reveals that the lake shrinkage that occurred in Holocene led to a loss of ~310 billion m^3 water and a disappearance of ~64% of the original lake extent. The water level of Paleo Gyaring Co is found to have dropped by ~14 m with a loss of 11.8 billion m^3 of water and 33% of lake area. Paleo Yueqia Co and Paleo Ringco Ogma, the two paleolakes on highlands in the basin, suffered even more significant shrinkage. Paleo Yueqia Co experienced a 34 m water level drop, losing 18.6 billion m^3 of water and 79% of lake area, while Paleo Ringco Ogma experienced a 26 m water level drop, losing 31.5 billion m^3 of water and 77% of lake area. In summary, paleolakes in the Great Siling Co basin had shrunk in Holocene up to ~79% in area with an up-to-56 m level drop.

Conclusion

Endorheic lakes have been changing dramatically due to the climate change and human activities, for example, the rapid shrinkage of the Aral Sea since the 1960s and the accelerated expansion of Siling Co in the past four decades. Spaceborne remote sensing provides diversified means to monitor lake area changes, level changes, and thus the lake volume changes at regional and global scales. Future satellite missions are expected to simultaneously monitor lake area and level dynamics for a majority of endorheic lakes at high resolutions. Paleolake extents can be recovered through the integration of remote sensing and digital elevation data to reveal paleo-endorheic lake changes. Though water salinity satellite missions have recently been in operation, lake salinity remote sensing from space is still not practical in its current stage. Various remote sensing tools have demonstrated their success in monitoring lake dynamics in the case study of the Great Siling Co basin. Siling Co and other endorheic lakes in the basin had dramatically shrunk in Holocene but have been significantly expanding in recent decades in response to the increased precipitation and the accelerated melting of glaciers in the basin.

Acknowledgments

This work was supported in part by NASA New Investigator Program under Grant NNX06AE58G and in part by the U.S. Geological Survey (USGS) Landsat Science Team Program Grant G12PC00071.

References

1. Hammer, U.T. *Saline Lake Ecosystems of the World*; Springer: Boston, 1986.
2. Yuan, F.; Sheng, Y.; Yao, T.; Fan, C.; Li, J.; Zhao, H.; Lei, Y. Evaporative enrichment of oxygen-18 and deuterium in lake waters on the Tibetan Plateau. J. Paleolimnol. **2011**, *46* (2), 291–307.
3. Zheng, M.; Liu, X. Hydrochemistry of salt lakes of the Qing-hai-Tibet Plateau, China. Aquat. Geochem. **2009**, *15*, 293–320.
4. Smol, J.P. *Pollution of Lakes and Rivers: A Paleoenviron-mental Perspective;* 2nd Ed.; Blackwell: Malden, MA, USA, 2008.
5. IPCC. *Climate Change 2007: The Physical Science Basis;* Cambridge University Press: Cambridge, 2007.
6. Street-Perrott, E.A.; Harrison, S.P. Lake levels and climate reconstruction. In *Paleoclimate Analysis and Modelling*; Hecht, A.D., Ed.; Wiley: New York, 1985; 291–340.

7. Mason, I.M.; Guzkowska, M.A.J.; Rapley C.G.; Street-Perrott, F.A. The response of lake levels and areas to climatic change. Clim. Change **1994**, *27* (2), 161–197.
8. Shen, Y.; Chen, Y. Global perspective on hydrology, water balance, and water resources management in arid basins. Hydrol. Process. **2010**, *24*, 129–135.
9. Micklin, P.P. Disiccation of the Aral Sea: A water management disaster in the Soviet Union. Science **1988**, *241* (4870), 1170–1176.
10. Harris, A.R. Time series remote sensing of a climatically sensitive lake. Remote Sensing Environ. **1994**, *50*, 83–94.
11. Birkett, C.M. The contribution of Topex/Poseidon to the- global monitoring of climatically sensitive lakes. J. Geophys. Research-Oceans, **1995**, *100* (C2), 25179–25204.
12. Millington, A.C.; Drake, N.A.; Townshend, J.R.G.; Quarmby, N.A.; Settle, J.J.; Reading, A.J. Monitoring salt playa dynamics using Thematic Mapper data. IEEE Trans. Geosci. Remote Sensing **1989**, *27* (6), 754–761.
13. Sheng, Y.; Li, J. Satellite-observed endorheic lake dynamics across the Tibetan Plateau between circa 1976 and 2000. In *Remote Sensing of Protected Lands;* Wang, Y., Ed.; CRC Press: New York, 2011; 305–319.
14. Sivanpillai, R.; Miller, S.N. Improvements in mapping water bodies using ASTER data. Ecol. Inform. **2010**, *5* (1), 73–78.
15. Bryant, R.G. Application of AVHRR to monitoring a climatically sensitive playa, case study: Chott el Djerid, Southern Tunisia. Earth Surf. Process. Landforms **1999**, *24*, 283–302.
16. McCullough, I.M.; Loftin, C.S.; Sader, S.A. High-frequency remote monitoring of large lakes with MODIS 500 m imagery. Remote Sensing Environ. **2012**, *124*, 234–241.
17. Shum, C.; Yi, Y.; Cheng, K.; Kuo, C.; Braun, A.; Calmant, S.; Chambers, D. Calibration of Jason-1 Altimeter over Lake Erie. Mar. Geod. **2003**, *26* (3–4), 335–354.
18. Cretaux, J.F.; Jelinski, W.; Calmant, S.; Kouraev, A.; Vuglinski, V.; Bergé-Nguyen, M.; Gennero, M.C.; Nino, F.; Abarca Del Rio, R.; Cazenave, A.; Maisongrande, P. SOLS: A lake database to monitor in the near real time water level and storage variations from remote sensing data. Adv. Space Res. **2011**, *47*, 1497–1507.
19. Urban, T.J.; Schutz, B.E.; Neuenschwander, A.L. A survey of ICESat coastal altimetry applications: Continental coast, open ocean island, and inland river. Terr. Atmos. Ocean Sci. **2008**, *19* (1–2), 1–19.
20. Kerr, Y.H.; Waldteufel, P.; Wigneron, J.P.; Delwart, S.; Cabot, F.; Boutin, J.; Escorihuela, M.J.; Font, J.; Reul, N.; Gruhier, C.; Juglea, S.E.; Drinkwater, M.R.; Hahne, A.; Martin-Neira, M.; Mecklenburg, S. The SMOS mission: New tool for monitoring key elements of the global water cycle. Proc. IEEE **2010**, *98* (5), 666–687.
21. Le Vine, D.M.; Lagerloef, G.S.; Torrusio, S.E. Aquarius and remote sensing of sea surface salinity from space. Proc. IEEE **2010**, *98* (5), 688–703.
22. Wang, S.; Dou, H. *Catalog of China's Lakes;* Science Press: Beijing, China, 1998.
23. McFeeters, S.K. The use of the normalized difference water index (NDWI) in the delineation of open water features. Int. J. Remote Sensing **1996**, *17* (7), 1425–1432.
24. Lee, H.; Durand, M.; Jungab, H.C.; Alsdorf, D.; Shum, C.K.; Sheng, Y. Characterization of surface water storage changes in Arctic lakes using simulated SWOT measurements. Int. J. Remote Sensing **2010**, *31* (14), 3931–3953.
25. Leblanc, M.; Favreau, G.; Maley, J.; Nazoumou, Y.; Leduc, C.; Stagnitti, F.; van Oevelen, PJ.; Delclaux, F.; Lemoalle, J. Reconstruction of Megalake Chad using shuttle radar topographic mission data. Palaeogeogr. Palaeoclimatol. Pal-aeoecol. **2006**, *239* (1/2), 16–27.
26. DeVogel, S.B.; Magee, J.W.; Manley, W.F.; Miller, G.H. A GIS-based reconstruction of late quaternary paleohydrol-ogy: Lake Eyre, arid central Australia. Palaeogeogr. Palaeo-climatol. Palaeoecol. **2004**, *204* (1/2), 1–13.
27. Sheng, Y. PaleoLakeR: A semiautomated tool for regionalscale paleolake recovery using geospatial information technologies. IEEE Geosci. Remote Sensing Lett. **2009**, *6* (4), 797–801.

6

Eutrophication

Kyle D. Hoagland
and Thomas
G. Franti
University of Nebraska

Introduction....45
Cultural Eutrophication....45
Rate of Eutrophication....46
Effects of Eutrophication....46
Reduction and Management....47
References....47

Introduction

Eutrophication is the nutrient enrichment of surface water and the subsequent impacts on water quality and the aquatic ecosystem. The overabundance of plant nutrients, usually nitrogen and phosphorus—but sometimes silicon, potassium, calcium, iron, or manganese—creates the conditions for excessive plant growth.[1] Algal blooms are an example of excessive growth caused by oversupply of nutrients. A body of water is classified by its trophic state based on the amount of nutrients supplied to it. An oligotrophic state is low in nutrients, a mesotrophic state is intermediate, and a eutrophic or hypereutrophic state is high in nutrients.[2] Eutrophication is a natural process. However, as a natural process it takes generations, or even thousands of years, for eutrophication to cause significant changes. It is the acceleration of the process, known as cultural eutrophication, that is of the greatest concern to water quality and the health of aquatic systems.

Cultural Eutrophication

Cultural eutrophication is the result of excess nutrients—primarily nitrogen and phosphorus—delivered to rivers, lakes, and estuaries by the activities of humans. Nutrient loading can come from both point sources and diffuse sources. Point sources of nutrients are generally urban sewage treatment or industrial water treatment. Diffuse sources (or non-point sources) of nutrients include agriculture, deforestation, and urban lawn runoff. The use of commercial fertilizer in crop production can lead to losses of nutrients to surface water. Concentrated animal feeding operations are also a potential source of excess nutrients.[2,3] Finally, deforestation and subsequent soil erosion can deliver excess nutrients, sediment, and organic matter to surface water, creating the conditions for eutrophication.

The natural aging process of lakes in some climates can lead from an oligotrophic state, with clean open water supporting usually cold water fish species, to a eutrophic state of warmer, shallow water supporting different species of plants and fish, and ultimately, to a filled lake closed over by a bog or fen. This process usually takes thousands of years. In contrast, cultural eutrophication can occur in a matter of one or two generations.

The last 50 years of the twentieth century saw the greatest impact of eutrophication on the world's waters. Cultural eutrophication has accelerated the lake aging process, reduced water quality, and impacted aquatic plant and animal populations, at economic costs. These impacts are being felt worldwide.[1]

Modern society's use of phosphorus additives in detergents has led to excessive phosphorus loading to surface waters.[1] Modern agriculture's use of commercial fertilizer and fertilizer's relative abundance, low cost, and over use, has led to excessive nitrogen loading to surface water. Agriculture also contributes excessive phosphorus from fertilizer and from intensive livestock operations and land application of manure.[3,4] Urban sources, such as sewage treatment plants, storm water runoff, industrial water treatment facilities, and food processing contribute excess nutrients. Sewage treatment loading to waters, including treated gray water, is generally considered the greater source of phosphorus. The greater contributor of nitrogen is agricultural crop production. Other significant agricultural sources are animal confinement operations, forestry, and atmospheric inputs.

Rate of Eutrophication

The rate of eutrophication is controlled by the rate at which nitrogen and phosphorus are delivered to a body of water. It is generally accepted that phosphorous is the limiting nutrient for algal growth in lakes.[1,5] Another limiting factor for algal growth is light. As excess phosphorus enters a lake it can trigger high levels of algal growth. Excessive growth, or blooms, can reduce water clarity. Aquatic macrophytes may also thrive under these conditions. This process can increase the productivity of a lake to an extent. When it becomes excessive, the process of cultural eutrophication has begun.

The nutrients supplied to an aquatic system is the most important factor that determines the species and amount of plant material, which in turn controls available oxygen and the animal species that thrive.[1] As plants and macrophytes grow and die, organic matter accumulates on the lake bottom. Decomposition of excessive amounts of organic matter can consume available oxygen and create anoxic conditions. An anoxic condition drives away plant and animal species dependent on oxygen. Slowly a lake's original community changes.

This anoxic state creates another problem if a lake becomes stratified. Temperature differences in the water at the lake's surface and at depth can create thermal stratification. In this condition the cold, low oxygen water in the hypolimnion, or lower layer, does not mix with the epilimnion, the warmer surface water. In this condition the anoxic conditions in the hypolimnion creates a condition where the sediment releases phosphorus, which is transported upward to the epilimnion and contributes to increased algal growth. This internal cycling of phosphorus continues the eutrophication process even if no additional phosphorus enters the lake.[5,6]

Effects of Eutrophication

There are various biological effects of eutrophication. Besides the problems of algal blooms and anoxic conditions previously described, there are changes in water temperature, reduction in water clarity, increased macrophyte production, population shifts in both plants and animals, and accelerated aging of lakes. Fish kills and the coastal "red tides" and "brown tides" are also potential environmental impacts from eutrophication.[5,7]

There are many impacts on human use of water from eutrophication. Potable water—water used for human consumption—can be significantly degraded by eutrophication. Besides the excess nutrients themselves, there is excess algae and other plant growth which can contribute to unwanted odors and tastes. Thus, additional water treatment is needed to make the water drinkable.[1,4]

The decline of commercial and recreational fisheries is also an indirect result of eutrophication. Fish populations may move away, or over time shift to less desirable fish species as water temperature, clarity, and quality change.[1] There are also negative impacts on recreational use of surface waters such as boating and swimming as a result of floating plant growth, smell, and the overall loss of aesthetic quality.[4] Finally, the general decline of an aquatic ecosystem and the loss of biodiversity are impacts of eutrophication.[1,3,7]

Some of these impacts can cause significant economic losses as well. Water treatment costs rise as water quality decreases. Algal blooms can contribute to shutting down water treatment plants. Advanced water treatment of municipal water to remove nutrients (such as alum addition)—hence removing a contributing factor to eutrophication—adds another expense to water treatment. Loss of commercial fisheries is an economic impact to a region. Finally, loss of recreation income from tourists, boaters, and recreational fisherman can have significant impacts on a local economy.[4,7]

Reduction and Management

Two approaches can be taken to the reduction and management of eutrophication: 1) reduction of nutrient loads; and 2) managing the existing high-nutrient state.[1] Reduction of loading is clearly the more robust approach. Reduced loads are necessary if long-term improvement is expected. However, because of the ecosystem changes brought on by eutrophication and the potential problem of phosphorus cycling previously described, water quality and ecosystem improvements may not respond quickly to reduced loading.[3] A combination of reduced loads and in-lake controls may be needed.

Reducing nutrient loads requires reduction of both point and diffuse sources. To reduce point sources, sewage and industrial water treatment plants need advanced water treatment to remove nutrients.[1,7] Advanced water treatment is expensive, but is the only way to reduce these point source nutrients. Most diffuse sources contributing nutrients to surface water come from agricultural operations. Loading is dependent upon the type of crops grown, soil type, climate, cultural practices, fertilizer use, and whether animal waste management practices are sufficient to reduce loading. Nutrient reduction from these sources will require improved management of fertilizer and animal waste, and continued reduction of soil erosion throughout the watershed.[1]

There are several methods used to manage nutrient levels in lakes. These fall into two categories: those that 1) remove nutrients; or 2) manage nutrient levels without nutrient removal. Methods to remove nutrients include: 1) lake flushing; 2) hypolimnetic water withdrawal; 3) sediment removal (e.g., dredging); and 4) nutrient inactivation by precipitation (e.g., alum treatment). Methods to manage nutrients without removing them include: 1) artificial mixing and/or aeration; 2) dilution by the addition of water lower in nutrients; 3) bottom sealing to prevent internal nutrient cycling; 4) manipulation of lake biological communities (e.g., selective fish harvesting or the introduction of fish predators such as largemouth bass, walleye, or brown trout); 5) introduction of biological controls for unwanted macrophyte growth (e.g., weevils or grass carp); and 6) herbicide or other chemical treatments (e.g., copper sulfate) of excessive algal or macrophyte growth.[1,6]

Clearly the best way to reduce or reverse eutrophication is to reduce nutrient loading, that is, targeting the source of the problem.[1] This long-term solution involves participation and management by people throughout the entire watershed. In-lake nutrient management can be done, but may require annual inputs and regular management.

References

1. *Harper, D. Eutrophication of Freshwaters: Principles, Problems and Restoration; Chapman & Hall: London, 1992.*
2. Smith, V.H.; Tilman, G.D.; Nekola, J.C. Eutrophication: impacts of excess nutrient inputs on freshwater, marine, and terrestrial ecosystems. Environ. Pollut. **1999**, *100*, 179–196.
3. Carpenter, S.R.; Caraco, N.F.; Correll, D.L.; Howarth, R.W.; Sharpley, A.N.; Smith, V.H. Nonpoint pollution of surface waters with phosphorus and nitrogen. Ecol. Appl. **1998**, *8* (3), 559–568.

4. Daniel, T.C.; Sharpley, A.N.; Lemunyon, J.L. Agricultural phosphorus and eutrophication: a symposium overview. J. Environ. Qual. **1998**, *27* (4), 251– 257.
5. Correll, D.L. The role of phosphorus in the eutrophication of receiving waters: a review. J. Environ. Qual. **1998**, *27* (2), 261–266.
6. Cooke, G.D.; Welch, E.B.; Peterson, S.A.; Newroth, P.R. *Lake and Reservoir Restoration*; Butterworths Publishers: Stoneham, MA, 1986.
7. Klapper, H. *Control of Eutrophication in Inland Waters*; Ellis Horwood Limited: Chichester, West Sussex, England, 1991.

7

Evaporation: Lakes and Large Bodies of Water

John Borrelli
Texas Tech University

Introduction 49
Techniques for Measuring Lake Evaporation 50
Water Budget • Energy Budget • Eddy Correlation
Estimation of Evaporation 51
References 52

Introduction

The conversion of water from liquid state to vapor state is called evaporation. Evaporation requires energy—approximately 540 cal/cm^3 of water (≈2.45 MJ/kg). Research has shown that the rate of evaporation is primarily a function of temperature, solar energy, wind velocity, vapor pressure deficit, and advected energy. The energy for evaporation comes predominately from solar radiation and wind. Evaporation is a major component of the hydrologic cycle, second only to precipitation. As such, precise documentation of evaporation from lakes and other water bodies is required for wise management of our water resources.

Annual lake evaporation across the United States has been estimated to range from 60 cm/yr to over 200 cm/yr.[1] The annual evaporation rates for several typical lakes vary from 51 cm/yr for Hungary Horse Reservoir (cool northern climate) to 223 cm/yr for Lake Mead (desert southwest) (Table 7.1). Consequently, there is a necessity to accurately measure evaporation rates and provide numerical models for estimating evaporation from numerous lakes and reservoirs where direct measurement is too costly to undertake.

TABLE 7.1 Annual Evaporation from Lakes

Lake	Annual evaporation (cm)	Longitude	Latitude	Area (ha)	Average depth (m)
Pyramid Lake[2]	128	119°40′	40°00′	46,640	61
Salton Sea[2]	179	116°10′	33°05′	88,100	8
Lake Ontario[2]	73	77°00′	44°00′	1,940,000	86
Hyco Lake[2]	94	79°05′	36°15′	1,760	6
Hungary Horse Reservoir[2]	51	113°55′	46°00′	9,700	15
Lake Kerr[2]	118	81°50′	29°20′	1,040	5
Lake Mead[2]	223	114°30′	36°05′	51,400	54
Lake Okeechobee[9]	147	80°55′	27°00′	182,130	3
Amistad Reservoir[2]	203	101°20′	29°20′	27,900	16
Great Salt Lake[2]	101	112°30′	41°00′	388,900	10

Techniques for Measuring Lake Evaporation

There are three widely accepted methods for measuring the evaporation rates of lakes: a water budget, an energy budget, and the eddy correlation method. The water budget and the energy budget require a considerable amount of investment in personnel, instruments, and time. As a result, these methods are applied sparingly to calibrate numerical models.[2] With today's dependable computer technology, the eddy correlation method has become widely used in recent years. Most studies will employ two or all of the methods.

Water Budget

If all components of the water budget could be measured accurately, it is the only method that directly measures evaporation. The water budget for a lake is as follows:

$$Evap = \left[\left(SW_{in} - SW_{\mathrm{out}} + GW_{\mathrm{in}} - GW_{\mathrm{out}}\right]\right.$$
$$\left. + S_{\mathrm{b}} - S_{e}\right) / \mathrm{Area} + \mathrm{PPT} / \mathrm{Time}$$

where *Evap* [LT^{-1}] is evaporation, SW_{in} [L^3] is surface water inflow, SW_{out}. [L^3] is surface water outflow, GW_{in} [L^3] is groundwater inflow, GW_{out} [L^3] is groundwater outflow, S_{b} [L^3] is lake storage at the beginning of the time period, S_{e} [L^3] is the lake storage at the end of the time period, Area [L^2] is the surface area of the lake, PPT [L] is precipitation, and Time [T] is the time period over which the measurements are made.[3] Evaporation is the residual of several measured terms and contains the errors included in the measurement of all those terms. Precipitation, for example, can have a bias error of up to 20% due to wind currents around the orifice of a rain gauge.[4] To use the water budget to measure evaporation, the inflow and outflow from the lake must be relatively small compared to the storage; otherwise, the errors in measurement will dominate the determination of evaporation. Overall, the error of measurement is ±5–10%.

Energy Budget

The energy budget uses the conservation of energy principle to determine net transfer of energy into and out of a lake. Like the water budget, the evaporation rate is computed as the residual of all other terms; thus, it will contain residual measurement errors. Sturrock, Winter, and Rosenberry[5] used the following energy budget equation in the study of Williams Lake:

$$Q_x = Q_s - Q_r + Q_a - Q_{ar} - Q_{bs} + Q_v - Q_e - Q_h - Q_w + Q_b$$

where Q_x is the change in energy content of the body of water, Q_s is incoming short-wave radiation, Q_r is reflected short-wave radiation, Q_a is incoming long-wave radiation, Q_{ar} is reflected long-wave radiation, Q_{bs} is long-wave radiation emitted from the body of water, Q_v is net energy advected to the body of water, Q_e is energy used for evaporation, Q_h is energy conducted from the water as sensible heat, Q_w is energy advected from the body of water by the evaporated water, and Q_b is heat transfer to the water from the bottom sediments. All terms are expressed in W/m^2.

Eddy Correlation

At the surface of the water, the water vapor in the air is nearly saturated. As air moves across the surface, small eddies transport the water vapor vertically at a net air movement of zero in the vertical direction. With current instrumentation, it is now possible to measure the vertical flux of water vapor or evaporation above the surface of a lake. The eddy correlation method directly measures the evaporative flux as presented by Shuttleworth[6] in the following formula:

$$E = 86.4\overline{\rho_a w' q'}$$

where E is the evaporation rate (mm/day), ρ_a is the air density (g/m^3), w' is the vertical wind velocity (m/sec^1), and q' is the specific humidity (g of water/g of air). The overbar denotes a mean value over a specific interval and the prime denotes an instantaneous deviation from the mean. Kizer and Elliot[7] provide a complete procedure to measure and calculate all terms needed to use the eddy correlation method. The accuracy for the eddy correlation measurements is 5–10%.[6] This compares favorably with the energy and water budget methods, which have the same range of accuracy. Measurements are taken at a point but are used to represent a large area of a lake. This causes some error because there are different microclimates over a large lake.

Estimation of Evaporation

Evaporation cannot be measured at all lakes and reservoirs by using the methods described above. Thus, researchers have developed several equations that use climatological data for estimating evaporation. The most widely used equation is the modified Penman equation that was originally developed for evaporation as well as to estimate evapotranspiration from vegetation.[8] The modified Penman equation requires data on wind, net solar radiation, humidity, and temperature. There are many equations called modified Penman. The following is a good example of a modified Penman equation:[6]

$$E_P = \frac{\Delta}{\Delta+\gamma}(R_n + A_h) + \frac{\gamma}{\Delta+\gamma}\frac{6.43(1+0.536\,U_2)D}{\gamma}$$

where E_p is estimated potential evaporation (mm/d), R_n is net radiation exchange for the free water surface (mm/d), A_h is significant energy advected to the water body (mm/d), U_2 is wind speed at 2 m (m/sec), D is vapor pressure deficit (kPa), λ is latent heat of vaporization (MJ/kg), Δ the gradient of the saturation vapor–temperature curve (kPa/°C), and γ is the psychrometric constant (kPa/°C). Please refer to Shuttleworth[6] for details on the calculation of different variables.

Investigators have found that the modified Penman equation (not necessarily the same modifications as above) has estimated evaporation within the accuracy of measured evaporation rates.[9–11] The modified Penman equation does not take into account the heat stored in a lake, which can be significant. The Penman equation will overpredict evaporation during warmer months and underpredict evaporation during the colder months.[11] On an annual basis, the modified Penman has proven reliable over a wide range of locations and climatic conditions.

Pan evaporation rates have been widely used to estimate lake evaporation. Kohler, Nordenson, and Fox[12] reported on an extensive study at Lake Hefner in Oklahoma comparing lake evaporation and pan evaporation. They reported that the annual ratio for a U.S. Weather Bureau Class A pan evaporation to lake evaporation was 0.7. This proportional constant is called the pan coefficient. The USGS[13] reported that monthly pan coefficients varied from 0.13 in February to 1.32 in November. Annual pan coefficients have been reported as low as 0.51 at Lake Mead[14] to 0.75 at Lake Okeechobee.[9] Evaporation pans provide reliable results if several stations are used. However, pan evaporation records are often erratic and often trend downward with time because of environmental changes of surroundings and poor maintenance of the pan.

There are many other equations that have been developed to estimate evaporation. They include mass-transfer equations,[5] radiation equation,[9] temperature equations,[10] etc. The applicability of these equations is generally limited to their use in environments similar to those in which the equations were calibrated.

References

1. Farnsworth, R.K.; Thompson, E.S.; Peck, E.L. *Evaporation Atlas for the Contiguous 48 United States,* NOAA Technical Report NWS-33; National Weather Service Reports: Washington, DC, 1982; 1–28.
2. Anderson, M.E.; Jobson, H.E. Comparison of techniques for estimating annual lake evaporation using climatological data. Water Resour. Res. **1982**, 18(3), 630–636.
3. *Rose, W.J.; Robertson, D.M. Hydrology, Water Quality, and Phosphorus Loading of Kirby Lake, Barron County,* Wisconsin; Fact Sheet FS-066–98; U.S. Geological Survey, U.S. Department of the Interior: Washington, DC, 1998; 1–4.
4. Wanielista, M.; Kersten, R.; Eaglin, R. *Hydrology—Water Quantity and Quality Control;* John Wiley & Sons, Inc.: New York, 1997; 68–117.
5. Sturrock, A.M.; Winter, T.C.; Rosenberry, D.O. Energy budget evaporation from Williams Lake: a closed lake in North Central Minnesota. Water Resour. Res. **1992**, 28(6), 1605–1617.
6. Shuttleworth, W.J. Evaporation. In *Handbook of Hydrology;* David, R.M., Ed.; McGraw-Hill, Inc.: New York, 1993; 4.1–4.53.
7. Kizer, M.A.; Elliott, R.L. *Eddy Correlation Measurement in Crop Water Use.* The 1987 International Winter Meeting of the American Society of Agricultural Engineers, Chicago, IL, Dec 15–18, 1987; American Society of Agricultural Engineers: St. Joseph, MI; Paper No. 87–2504, 1–14.
8. Penman, H.L. Natural evaporation from open water, bare soil, and grass. Proc. R. Soc., Ser. **A 1948,** *193,* 120–145.
9. Abtew, W. Evaporation estimation for Lake Okeechobee in South Florida. J. Irrig. Drain. Eng., Am. Soc. Civil Eng. **2001**, 127(3), 140–147.
10. Hill, R.W. Consumptive Use of Irrigated Crops in Utah. Research Report 145; Utah Agricultural Experiment Station: Utah State University: Logan, Utah, 1994; 1–361.
11. Vardavas, I.M.; Fountoulakis, A. Estimation of lake evaporation from standard meteorological measurements: application to four Australian lakes in different climatic regions. Ecol. Model. **1996,** *84,* 139–150.
12. Kohler, M.A.; Nordenson, T.J.; Fox, W.W. *Evaporation from Pans and Lakes,* Research Paper No. 38; Weather Bureau, U.S. Department of Commerce: Washington, DC, 1955–20; 1–20.
13. U.S. Geological Survery (USGS). *Water-Loss Investigations: Lake Hefner Studies,* USGS Professional Paper 269; U.S. Geological Survey, Department of the Interior: Washington, DC, 1954; 1–158.
14. Hughes, G.H. *Analysis of Techniques Used to Measure Evaporation from Salton Sea, California.* Geological Survey Professional Paper 272 H. U.S. Geological Survey: Washington, DC, 1967; 151–176.

8

Evaporation and Energy Balance

Matthias Langensiepen
Humboldt University of Berlin

Introduction 53
Energy Balance and Water Availability 54
Unlimited Water Availability • Snow Covered and Frozen Surface Layers • Water Scarcity • Vegetation Control of Water Availability
Determination of the Surface Energy Balance 56
Conclusion 56
Bibliography 57

Introduction

Various forms of energy drive water transport through the hydrological cycle. Radiant energy, originating from the sun, provides the input energy for the cycle. Once matter absorbs this energy, it is converted into sensible heat that elevates the temperature of the air and the ground, and latent heat that causes evaporation, driving thereby the cycle against the pull of gravity. Further transport is generated by kinetic energy and pressure energy of the moving air masses. Translocation of vapor is accompanied by continuous interchanges among radiant, thermal, kinetic, and pressure energy. Large amounts of latent heat are released when water condenses in the clouds and falls as precipitation on the Earth's surface. It carries kinetic energy while flowing through watersheds. Vertical movement and percolation through the Earth's crust finally causes changes in potential and pressure energies.

The first law of thermodynamics states that energy is neither created nor destroyed only converted from one form into another. This effectively means that the input and output energies of a completely defined system must balance. Storage effects may temporarily disturb this equilibrium condition. The energy balance must thus be expressed in its most general form as

$$\text{Energy Input} = \text{Energy Output} + \text{Energy Storage}$$

The water balance of the Earth–atmosphere system can be treated analogically as the mass of water is conserved at all times. Evaporation is the connecting link between the system's water and energy balances. It is a surface process, which takes place at the lower boundary of the atmosphere and is an important component of the surface energy balance (see Figure 8.1):

$$R_n = LE + H + G - A + S - L_p F_p$$

where R_n is the flux of net all wave-radiation, L the latent heat, E the evaporation rate, H the flux of sensible heat, G the heat flux at the lower boundary of the surface, and A the energy advected to the surface when the ground properties have horizontal discontinuities. The energy balance is

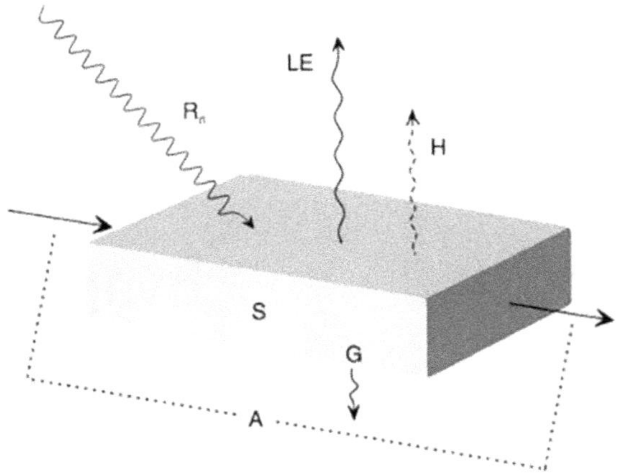

FIGURE 8.1 Schematic illustration of the surface energy balance. (R_n = net radiation, LE = latent heat flux, H = sensible heat flux, G = ground heat flux, S = heat storage, A = advection. All symbols are expressed in W m^{-2}.)

sometimes parameterized for a volume of surface material (for example water body, soil, or canopy volume). As the solar energy input undergoes diurnal and annual fluctuations, heat storage S may become an important component of the balance when it is applied at time intervals shorter than the fluctuation period. When the layer includes vegetation, biochemical energy storage due to photosynthesis can also be considered. L_p is then the thermal conversion factor of carbon dioxide, and F_p is the flux of CO_2.

Shortwave radiation from the sun is the sole energy input of the Earth–atmosphere system. Its net amount available for heat conversion is related to geographical location, time, atmospheric transparency, atmospheric path length, geometrical distribution of the surface elements, and their optical properties. Complementary long wave radiation exchange is governed by surface to air temperature differences and cloudiness. Net shortwave and long wave radiation form the net all wave radiation R_n.

The partitioning of R_n into the remaining terms of the surface energy balance determines the rate of surface evaporation and depends on the availability of surface water.

Energy Balance and Water Availability

Unlimited Water Availability

When water availability is unlimited on a large scale, such as in oceans, vertical temperature gradients within the atmosphere tend to be very close to the adiabatic value, and most of the available energy (R_n) is diverted to latent heat (L_vE) from moisture flux at the surface. Wind gradients near the surface are typically very steep under such conditions and quickly approach values that remain nearly constant throughout the convective boundary layer. Vertical motion is damped out by strong subsidence inversion at the upper boundary of this well-mixed layer. Heat Storage (S) has a dominant effect on the diurnal course of the ocean's energy balance leaving only little energy for transport ($L_vE + H$) into the air until late in the afternoon. This situation is reversed during the night, where heat released from the ocean surface becomes the major source of energy. Large-scale advection (A) partly disturbs the thermal inertia of oceans, which has considerable effects on the global weather systems (Gulfstream, Humboldt-Stream, El Nino). The smaller the water volume becomes the more it is likely that its thermal inertia is disturbed by local advection due to horizontal discontinuities of thermal surface properties. Since radiation can penetrate into water bodies it is also possible that the underlying floor becomes a source of heat.

Snow Covered and Frozen Surface Layers

When water is bound in snow covered or frozen surface layers, energy partitioning is affected by the penetration of shortwave radiation, phase changes, and internal distribution of water. Net radiation gain (R_n) is commonly restricted by a high surface albedo. The optical depth of snow and ice also affect radiation absorption and penetration. The available energy is mainly partitioned between storage energy (S) and energy required to allow water to change between frozen and liquid states (L_f). Only little energy is consumed by atmospheric transport ($LE + H$). Phase changes of water within the layer (freezing, melting, condensation, evaporation, sublimation) are accompanied by the continuous consumption or liberation of energy. The internal partitioning of available energy is thus influenced by the physical states of water: Water has a high specific heat of 4216 J kg^{-1} K^{-1} at 0°C due to a strong intermolecular bonding force. Fifteen percent of the hydrogen bonds break when water changes from a solid to liquid state. The energy required to effect this change is 0.334 MJ kg^{-1} and is called the latent heat of fusion L_f. Nearly 7.5 times as much energy is required at this temperature level to cause water to further change from a liquid to a gaseous state. The corresponding energy is called latent heat of vaporization L_v and is temperature dependent (2.5 MJ kg^{-1} at 0°C, 2.45 MJ kg^{-1} at 20°C, and 2.41 MJ kg^{-1} at 40°C). In the event that water changes directly from a solid to a gaseous state (sublimation) the required latent heat of sublimation L_s is the algebraic sum of L_f and L_v. Freezing or condensation liberates energy, the amount depending on the corresponding phase shift. When the surface layer is below the freezing point and the sky is clear, net radiation can become negative under conditions of decreased radiation availability (high latitudes). It becomes positive, however, when the sky window is obstructed by clouds or surface emission is exceeded by incoming radiation. When the surface melts, both radiation and convection act as energy sources, sometimes accompanied by additional heat input from rainfall. Surface temperatures change only little during this process, because most energy is stored as latent heat of fusion.

Water Scarcity

When water is scarce, as is the case in deserts, most of the available energy (R_n) is consumed by surface heating, which can be sensed as a rise in surface temperature. Sensible heat (H) dissipation from dry surfaces lowers the density of air increasing its instability and tendency to rise. The instable air parcels form plumes (thermals) that progressively cool down as they mix with the surrounding air and are finally capped off by the inversion layer. Additional air is entrained from the top of the capping inversion layer and dragged to the ground by sinking motion of the cooling air masses. The height of the inversion layer is dependent on the amount of energy available for surface heating. At night and during early morning, winds in deserts are light, turbulence is low, air is stable or neutral, and the inversion layer is close to the ground. Net radiation (R_n) is partitioned into surface heat-flux (G) and heat storage (S) under such conditions. However, low thermal admittance of the barren dry soil diverts a major portion of the available energy to sensible heat (H), increasing air instability and turbulence. They promote the buildup of miniature whirlwinds known as dust devils. The situation is reversed during afternoons, where sinking radiation energy input stabilizes air masses. High differences between day and night temperatures are a consequence of lacking water, the magnitude depending on the diurnal evolution of net radiation (R_n). Sloping terrain and thermal surface heterogeneities induce horizontal heat transport, known as advection (A). It causes the buildup of wind gusts and turbulence, which act as kinetic energy sources in soil erosion. Advection also plays a significant role in the energy balance of wet surface islands in dry areas (Oasis effect). The evaporative demand of the atmosphere is generally high under conditions of elevated air temperature and limited water availability.

Vegetation Control of Water Availability

When vegetation cover controls water availability, energy partitioning is affected by the physiological state of the plants. The sites of regulation are *stomata* (from Greek "mouth"), tiny pores serving as pathways between the plant interior and the atmosphere. Each pore is surrounded by a pair of specialized cells (guard cells), which control its aperture and respond to plant internal and external signals. Light, vapor pressure deficit, and water potential are the principal controlling signals. Carbon dioxide, hormones (abscisic acid and cytokinins), and photosynthetic assimilation capacity have so far been detected as additional regulating factors. Signals and plant responses are acting in an integrated manner and form the canopy resistance against water loss. Development and growth determine the evolution of plant stand architecture and hence the spatial distribution of exchange surfaces. The more the surfaces are vertically exposed against airflow, the higher is their capacity to absorb momentum. In neutral transport conditions, the logarithmic portion of the wind-profile above a canopy extrapolates downward to a height where wind speed becomes zero. This level is called *zero plane displacement* and is defined as the average height of mass and heat exchange within a canopy volume. This height changes in accordance with foliage density distribution, form drag, and wind speed. The type of surface vegetation cover thus influences the magnitude of heat and mass exchange. When determining the energy balance of a plant stand, two sources of water have to be considered, canopy and soil. If the vegetation cover is sparse or is at an early development stage, significant portions of the available energy (R_n) can reach the soil level. In this case, the availability of water depends on biological factors and the soil hydraulic properties (water retention, hydraulic conductivity, and soil water diffusivity). The partitioning of available energy (R_n) into the heat terms of the energy balance is largely determined by the water status of the soil–plant system. Latent heat (L_vE) from transpiration is the major energy sink when soil water is abundantly available. In case radiation reaches the canopy floor, latent heat (L_vE) from soil evaporation as well as soil heat flux (G) are additional sinks of energy. Advection (A) may become an additional source of energy in hot climates. Energy storage due to photosynthesis (L_pF_p) is very small in comparison with the other components of the energy balance and is therefore often neglected. Heat storage (S) becomes important in massive canopies like forests. When water becomes limited, surface regulation restricts latent heat loss (L_vE) and sensible heat (H) becomes the principal sink of energy causing rises in surface temperature. Plants have flexible capabilities to optimize production in response to such conditions.

Determination of the Surface Energy Balance

Model determinations of the surface energy balance are commonly carried out with the combination equation, which emphasizes the mutual relation between latent and sensible heat fluxes. Practical methods assume equality, either between scalars and momentum (aerodynamic method) or between the eddy diffusivities for heat and vapor. The ratio of sensible to latent heat is then proportional to the ratio of air temperature over vapor concentration (Bowen ratio $\beta = H/L_vE$). Instrumentation can be categorized in accordance to their application. Surface parameters are commonly measured with net-radiometers (R_n), heat flow sensors (G), and lysimeters (L_vE). Gradient measurements above exchange surfaces involve determinations of wind speed (anemometers), air temperature (thermometers), air humidity (hygrometers, psychrometers), and CO_2 (infrared gas analyzers). Sonic anemometers, quartz thermometers, Lyman-alpha, and Krypton hygrometers are applied with the eddy correlation method. Remote sensors can be used to deduce turbulence parameters, heat and momentum fluxes from backscattered or forward-propagated signals (sodars, radars, and lidars).

Conclusion

The first law of thermodynamics states that the input and output energies of any given system must balance. Solar radiation is the sole energy input of the Earth-atmosphere system. Its partitioning into surface fluxes of latent and sensible heat is determined by the physical properties and availability of surface water, the "evaporative demand" of the atmosphere, and the nature of the surface. Evaporation is the connecting link between the system's energy and water balances. The quantification of a system's energy balance requires a definition of its boundary conditions.

They consist of the spatial and temporal dimensions of the system and its exchange surfaces, their physical transport properties, the energy states across the system boundaries, and possible modes of energy transfer.

Bibliography

Brutsaert, W. *Evaporation into the Atmosphere,* 1st Ed.; Kluwer: Dordrecht, 1982; 299 pp.

Campbell, G.S.; Norman, J. *Introduction to Environmental Biophysics,* 2nd Ed.; Springer: Berlin, 1998; 286 pp.

Hillel, D. *Environmental Soil Physics,* 1st Ed.; Academic Press: San Diego, 1998; 771 pp.

Jones, H.G. *Plants and Microclimate,* 2nd Ed.; Cambridge University Press: Cambridge, 1992; 428 pp.

Kaimal, J.C.; Finnigan, J.J. *Atmospheric Boundary Layer Flows: Their Structure and Measurement,* 1st Ed.; Oxford University Press: Oxford, 1994; 289 pp.

Oke, T.R. *Boundary Layer Climates*, 2nd Ed.; Routledge: London, 1987; 435 pp.

Ross, J. The Radiation Regime and Architecture of Plant Stands, 1st Ed.; Junk: Den Haag, 1982; 391 pp.

Zeiger, E.; Farquhar, G.D.; Cowan, I.R. *Stomatal Function,* 1st Ed.; Stanford University Press: Stanford, 1987.

9

Field Water Supply and Balance

Introduction 59
Water Balance Components 60
Soil Water Content • Precipitation and Irrigation • Horizontal Movement of Water • Evaporation and Transpiration • Energy Balance • Radiation Balance and Soil Heat Flux • Drainage of Water across the Lower Boundary
Applications 62
Irrigation Scheduling • Rainfed Cropping • Plant Growth and Natural Resource Modeling
Conclusion 63
References 63

Jean L. Steiner
U.S. Department of Agriculture (USDA-ARS)

Introduction

The soil water balance (Figure 9.1) can be given as:

$$SW_t = SW_i + P + I - R - E - T - D \tag{9.1}$$

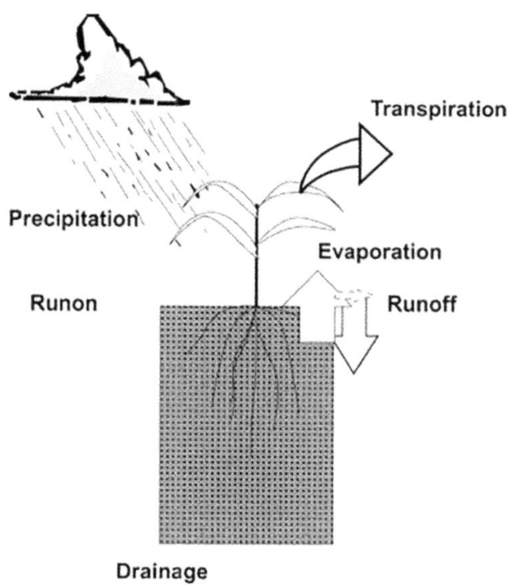

FIGURE 9.1 Soil water balance for an agricultural plant.

where SW is soil water content within a defined root zone, t and i subscripts represent the end and beginning of a time period, respectively, P is precipitation, I is irrigation, R is runoff, D is drainage below the root zone, E is evaporation from the soil surface, and T is transpiration, with all terms in the same units and over the time period defined by the t and i subscripts. The R term might be modified to include any horizontal movement of surface water or shallow water table flow, which can be either imported to or exported from the defined soil volume. In most circumstances, the D term is downward flux below the root zone, but can be defined to include vertical flux across the bottom of the root zone that could include upward movement from a shallow water table to deep rooted plants. The E term can be considered to include water evaporated from any wetted surface (e.g., ponded water, wetted plants, evaporation or sublimation of accumulated snow), as well as evaporation of water from the soil profile. Frequently, the soil water balance is used to determine terms of Equation (9.1) (e.g., E+T, $SW_t - SW_i$,) by measuring or estimating the remaining terms. Infiltration is often estimated by measuring P and R, if the other terms can be considered negligible during the precipitation event. Gardner[1] provides additional detail about the soil water balance.

Water Balance Components

Soil Water Content

Many different methods have been used to measure soil water content. A direct method that has been used since the early days of soil and agricultural research and remains common today is the gravimetric method. For the gravimetric method, soil samples, often cores, are collected and the water content is determined by weighing the sample before and after oven drying to determine the quantity of water lost by evaporation. Gravimetric sampling offers the benefit of providing a direct measurement of soil water content using simple equipment. However, it is time consuming and cannot be used to provide repeated measurement at the same location because it is destructive sampling. Given the high degree of spatial variability in most field soils, this limits the ability to determine temporal changes in soil moisture precisely, limiting the application of the soil water balance to relatively longer time periods. Methods that provide repeated measures of soil water content at the same location, such as using neutron probe or other technologies, reduce the problem of spatial variability and allow the soil water balance to be applied over shorter time periods.

Precipitation and Irrigation

Water is added to the system through precipitation or irrigation. Since precipitation is often highly variable, the rain gauge should be as near the site of the investigation as possible and should use standard weather gauges, properly sited away from tall buildings or vegetation that can distort rainfall catch. At field scales, irrigation applications are not totally uniform, so the gross irrigation amount may have to be adjusted by efficiency and uniformity factors to determine the net input to the soil water balance. However, well-managed, modern irrigation techniques such as low pressure applicators on center pivots and drip or subsurface irrigation methods, can provide very uniform distributions of water in a field with high efficiency.

Horizontal Movement of Water

The horizontal surface movement of water is largely runoff. In some situations, there may be a net gain through run-on of water from another portion of the landscape. However, there is normally a net loss to the water balance of a particular field when horizontal movement occurs. The proportion of precipitation that runs off depends on soil properties, vegetation, and topography, as well as the intensity and duration of rainfall. In some soil water balance calculations, runoff is assumed to be negligible because of level topography, or in some cases berms are constructed around small experimental areas to restrict horizontal movement. When a large amount of precipitation falls as snow, there can be considerable movement caused by the wind and drifting. In some field situations, there is a substantial horizontal movement of water below the surface, making it difficult to use the water balance method.

Evaporation and Transpiration

Because it is difficult in field studies to separate loss of water by evaporation from soil and transpiration from plants, these terms are often linked into a single term, usually called evapotranspiration, E_t. From a management perspective, options for influencing soil evaporation and transpiration are different, so the terms are presented separately in Equation (9.1).

Energy Balance

The energy balance approach, often used to estimate E_t, was developed for a plant–soil system under conditions where transpiration dominated the total loss. This approach has been successful because a large portion of energy that enters the Earth's atmosphere is required to transform water from liquid to vapor (the evaporation process). The combination evaporation equation derived by Penman[2] and Monteith[3] can be expressed as

$$E_t = \left\{\Delta\left(R_n + G\right) + \left[\rho C_p\left(e_s - e_a\right)/r_{av}\right]\right\} \Big/ L\left[\Delta + \gamma\left(1 + r_s/r_{av}\right)\right] \tag{9.2}$$

where Δ is the slope of the saturation vapor pressure curve, R_n is net radiation, G is soil heat flux, ρ is air density, C_p is the specific heat of dry air, e_s and e_a are the vapor pressures at the evaporating surface (assumed to be the saturation vapor pressure at the surface temperature) and of the atmosphere, respectively; r_{av} is the aerodynamic resistance, L is the latent heat of vaporization, γ is the psychrometric constant, and r_s is surface resistance. Because it is difficult to measure some of these terms, particularly the surface resistance term, methods have been developed to evaluate the equation for potential evaporation conditions using standard weather data, applying assumptions about a well-watered, vegetated surface; and then relating the potential evaporation to actual evaporation for a particular vegetative surface (such as early or late in the season with low vegetative cover or under water-stressed conditions) using crop coefficients and other adjustments. These methods are discussed thoroughly in Allen et al.[4] and Allen.[5]

Radiation Balance and Soil Heat Flux

Net radiation is the balance resulting from the incoming and outgoing fluxes of short and long-wave radiation and is affected by several aspects of the soil and plant cover at the surface.[6] Short-wave radiation is strongly affected by surface roughness, color, soil water content, and solar angle. For soils and plants, emitted long-wave radiation is largely determined by surface temperature, so wetter soils (which are generally cooler) have less outgoing long-wave radiation than drier, warmer soils. Incident long-wave radiation is influenced by sky conditions (cloudiness, cloud type, etc.), and it will generally decline with increased sky cover. As vegetative cover increases, the effect of soil conditions on net radiation decreases to negligible levels. For calculation of short-term evaporation, soil heat flux is sometimes taken as a fraction of net radiation (~10%). For bare soils, such as early or late in a growing season, the fraction can be much higher. For longer time periods, changes in air temperature can be used to estimate the flux of heat into or out of the soil over the period.[4] As vegetative cover or crop residue cover increases, the flow of heat in the soil is reduced because temperature gradients in the soil decrease as the soil is shaded.

Drainage of Water across the Lower Boundary

In some cropping systems, the flux across the lower boundary of the root zone can be considered to be negligible. This is true primarily in semiarid or arid climatic regimes and in soils with a high water holding capacity and/or a low hydraulic conductivity, such as clay loams, silty clay loams, or in some cases silt loams and loams. It may also be true where there is a restrictive layer in the soil profile that prevents or slows the flow of water below the root zone. However, when considering year round water balances, soil water is often lost below the root zone during at least part of the annual cycle in periods of high precipitation or low evapotranspiration. Water losses below the root zone can be measured using lysimeters, or can be calculated using measurements of soil water tension or content over time along with knowledge of soil hydraulic properties. In some situations, water can move upward from a shallow water table into the root zone of deep rooted plants.

Applications

Irrigation Scheduling

Irrigated agriculture is one of the most intensive forms of agriculture. Having control of the water supply to ensure adequate water for crop growth allows a producer to invest more in other inputs that ensure a high yield and quality of the crop. However, irrigation applications can be expensive and excessive water application can result in loss of nutrients and other production inputs as well as causing environmental problems. Therefore, it is important to apply enough, but not too much, irrigation water. One way to do this is to monitor soil water content during the growing season and apply knowledge of the soil water balance to guide timing and amount of irrigation applications. Some irrigation scheduling models use weather data to simulate the evapotranspiration and maintain a soil water budget to predict changes in the soil water content. Depending on the rate of water use by the crop and the amount of soil water storage capacity, the producer can project the upcoming needs for irrigation applications.

Rainfed Cropping

Rainfed cropping is subject to great risks because of the high variability of rainfall in most agricultural regions. In many regions, water stored in the soil at planting time is an important component of the seasonal water supply. If there is a large amount of water stored at planting time, that stored water provides a buffer against dry periods during the growing season, and the producer might plan for an average or good yield level and invest in inputs to support those yield levels. If the water storage at planting is low, then the risk of crop losses due to growing season drought is high and the producer may decide to reduce or delay investment in some inputs until later in the season when more is known about growing season precipitation and forecasts. Analysis of long-term climatological records using a soil water balance approach can be used to evaluate alternative crops or rotations for a region. In some cases, high soil water levels at the end of the growing season will result in a low faction of off-season precipitation being stored for the next crop with greater losses to percolation (with some possible nutrient leaching) or runoff (with possible greater erosion).

Plant Growth and Natural Resource Modeling

Soil water balance calculations are an integral part of plant growth and hydrologic models. Many of these models have been developed to operate at a daily time step and have been applied to a wide range of analyses. Some examples of such models that are available for downloading from the internet include crop growth, erosion, and hydrology models developed at the Grassland Soil and Water Research Laboratory at Temple, Texas, http://arsserv0.tamu.edu/intro.htm; the soil organic matter model, CENTURY, developed at the Natural Resource Ecology Laboratory at Colorado State University, http://www.nrel.colostate.edu/projects/century5/; water balance, irrigation management, and soils models developed by Dr. J. T. Ritchie and colleagues, http://nowlin.css.msu.edu/; the Decision Support System for Agrotechnology Transfer, DSSAT, which is a series of crop models and associated weather, crop, and soil data bases, http://icasanet.org/dssat/ and a suite of models, ranging from nutrient management and water quality models to an operational tool for whole farm/ranch strategic planning developed by the Great Plains System Research Unit at Fort Collins, Colorado, http://gpsr. ars.usda.gov/products/. These models, and many more, include a soil water balance as an integral part of the system.

Conclusion

The soil water balance approach has played an important role in improving our understanding and management of plant, water, and soil resources. The field soil water balance involves accounting for inputs of water to the system, such as precipitation and irrigation, as well as water leaving the system via evapotranspiration, runoff, and drainage below the root zone. The way a field is managed can have a large impact on the magnitude of the components of the water balance, such as runoff and drainage, as well as patterns of evapotranspiration and partitioning of the water loss into soil evaporation and transpiration. In agriculture, increasing the amount of transpiration increases productivity of the system. Soil water balance approaches can be applied to irrigated and rainfed agriculture and are an integral part of all plant growth and natural resource model, so understanding the basic concepts and principles is important for sound water management.

References

1. Gardner, W.R. Soil properties and efficient water use: an overview. In *Limitations to Efficient Water Use in Crop Production;* Taylor, H.M., Jordon, W.R., Sinclair, T.R., Eds.; Am. Soc. Agron., Crop Sci. Soc. Am., Soil Sci. Soc. Am.: Madison, WI, 1983; 45–71.
2. Penman, H.L. Natural evaporation from open water, bare soil, and grass. Proc. R. Soc. Lond. **1948**, 120–145.
3. Monteith, J.L. *Evaporation and Environment,* Symp. Soc. Exp. Biol.; Academic Press: New York, 1965; Vol. 19, 205–234.
4. *Allen, R.G.; Pereira, L.S.; Raes, R.; Smith, M. Crop Evapotranspiration—Guidelines for Computing Crop Water Requirements—FAO Irrigation and Drainage Paper 56;* Food and Agriculture Organization of the United Nations: Rome, 1998; 300.
5. Allen, R.G. Using the FAO-56 dual crop coefficient method over an irrigated region as part of an evapotranspiration intercomparison study. J. Hydrol. **2000,** *229,* 27–41.
6. Steiner, J.L. Crop residue effects on water conservation. In *Managing Agricultural Residues;* Unger, P.W., Ed.; Lewis Publishers: Boca Raton, Florida, 1994; 41–76.

10

Hydrologic Cycle

John Van Brahana
University of Arkansas

Processes and Pathways 65
Energy Sources 66
Hydrologic Reservoirs and Implications for Humans 66
References 67

Processes and Pathways

The dominant processes of the hydrologic cycle, and the pathways along which we can trace water movement, include the following (Figure 10.1): evaporation from oceans and open bodies of water on the Earth's surface into the atmosphere; evapotranspiration by plants of soil water into atmospheric vapor; condensation of water vapor into liquid or solid particles (clouds); precipitation as the condensed water or ice falls from the atmosphere back to the surface; infiltration of the water into subsurface (soil and groundwater) reservoirs; baseflow contribution to streams from groundwater; streamflow recharging of groundwater; streamflow runoff if the precipitation rate exceeds the infiltration rate; surface and subsurface flow back to oceans or intermediate reservoirs; and storage.[1–10]

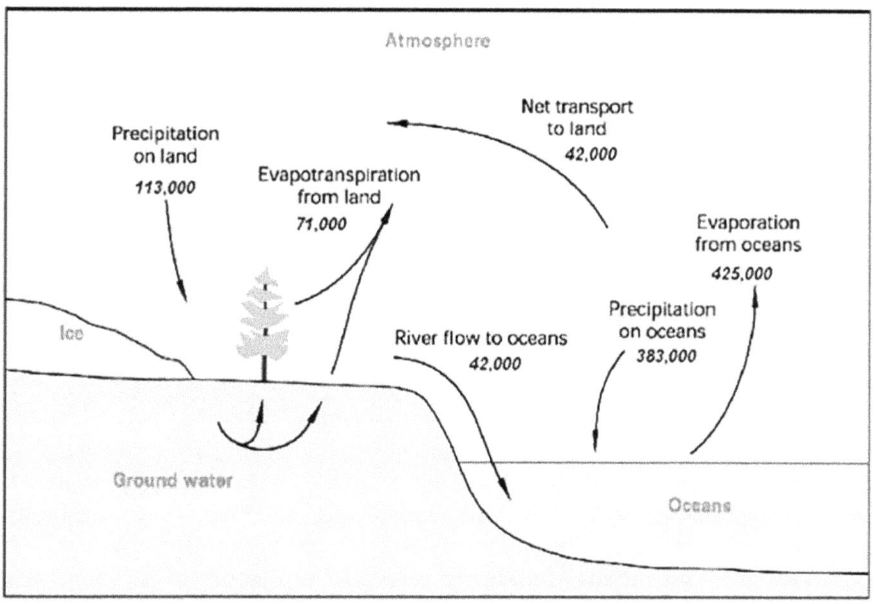

FIGURE 10.1 Quantitative representation of processes in the hydrologic cycle showing transfer rates between reservoirs in units of cubic kilometers per year.
Source: Adapted from Winter et al.[5]

In one sense, the hydrologic cycle is one of the most basic concepts of water science, yet in detail the concept is complex because it involves all forms of water of the hydrosphere, and it is affected by many influencing factors that are not always obvious. Because this circulation of water is intimately tied to energy transfer, it is helpful to start with the basic physics of the forces that drive this seemingly endless flow of water on our planet.

Energy Sources

The underlying source of energy that drives the movement of water throughout the hydrosphere of the Earth is solar radiation—thermal energy from our sun. Solar energy heats water, causes it to evaporate and change state from liquid to a gas, and in so doing, facilitates its movement through the atmosphere in response to wind and pressure changes. Every gram of liquid water at its point of vaporization requires an input of 540 cal of thermal energy to convert it to a gas. Worldwide, water vapor represents a huge source of energy storage and transport in a hydrologic link between atmosphere, oceans, and continents, which we call "climate." The residence time of water vapor in the atmosphere is short, usually no more than several days or weeks, until it condenses and falls as precipitation. In undergoing condensation to a liquid, the energy stored in the vapor is released.

As a liquid, water is controlled by gravity. If it can move, it will, always move downhill to a lower potential energy state, and always along the path of least resistance, down the steepest gradient. As water moves, it expends energy. Fast-flowing runoff, particularly in streams, is the single most dominant agent of erosion of the surface of our planet. Glaciers likewise are effective at sculpting the land surface, but owing to their limited occurrence on only 10% of the continents, their impact is not nearly as widespread as that of flowing streams. Thus, erosion and the Earth's landforms are intimately tied to the hydrologic cycle. Ultimately, water reaches the lowest accessible level possible, which for most places on the Earth is sea level. Internally drained basins that are isolated from the oceans by mountain ranges and other high divides may exist below sea level (i.e., Death Valley in California; Caspian Sea in Kazakhstan; Dead Sea in Jordan), but these represent local base-level conditions rather than regional or global conditions. These areas of internal drainage are typically formed by tensional tectonics, where blocks of rock are down-faulted (grabens) due to forces that tend to pull the continents apart. Water drains into these depressions under the force of gravity from the surrounding highlands, and escapes only by evaporation.

Hydrologic Reservoirs and Implications for Humans

Oceans form the largest of our hydrologic reservoirs (Table 10.1), covering about 70% of the Earth's surface, and including about 96% of all of its water.[8,11,12] Ocean water unfortunately is saline and non-potable (undrinkable), containing about 35,000 mg L^{-1} of dissolved solids,[13] much too salty for human consumption. Of the remaining 3–4% of the Earth's water found in reservoirs on the continents, approximately 1% of this (principally saline lakes or deep, saline groundwater) also is non-potable.[6,8,11–13] Thus, about 97% of all the water on this Earth is too mineralized for humans to drink without expensive desalinization. Of the approximately 3% of the total water that is fresh, the largest percentage, estimated as 1.7–2.97% by different experts, is stored in icecaps and glaciers, far removed from most of the Earth's population and its water needs. Thus, the freshwater needs of the world are served by a fraction of 1% of the total hydrologic budget, primarily water in storage and transit as shallow groundwater, freshwater lakes, soil moisture, water in manmade reservoirs, and rivers. These data are synthesized from several comprehensive studies, and although the values do not match exactly, they generally do not vary by more than 1% or 2%.[8,12]

As an integrated Earth system, the hydrologic cycle has no discrete beginning or ending point, but from a consumptive human point of view, the oceans are the major source of water, the atmosphere is the deliverer, and the land is the user. In this system, no water is lost or gained, but the amount of water available to the user may fluctuate mostly because of variations in the delivering agent. In the geologic past, large alterations in the cyclic roles of the atmosphere and the oceans produced deserts

TABLE 10.1 Comparisons of Quantities and Percentages of Earth's Water in the Major Storage Reservoirs of the Hydrologic Cycle, Based on Estimates from UNESCO and NRC

	UNESCO		NRC	
Storage reservoirs of the hydrologic cycle	Volume, in 106 km^3	%	Volume, in 106 km^3	%
Oceans	1,338	96.5	1,400	95.96
Icecaps and glaciers	24.3	1.73	43.4	2.97
Groundwater	23.4	1.69	15.3	1.05
Lakes	0.176	0.013	0.125	0.009
Atmospheric water	0.0129	0.001	0.0155	0.001
Soil moisture	0.0165	0.0012	0.065	0.004
Rivers	0.0021	0.0002	0.0017	0.00012
Biologic water	0.001	0.0001	0.002	0.00013
Total	1,386		1,459	

Source: From U.S.S.R.[8,12]

and glaciation across entire continents. During the last ice age, the colder climate resulted in a greater percentage of water being stored as snow and ice, with a decreased percentage of water being stored in the oceans. Scientists see evidence that major sea-level declines corresponded with maximum glacial development, and in fact, point to "drowned" valleys (e.g., Chesapeake Bay in the United States, and the fiords in the Scandinavian countries) that were eroded and formed when sea level was much lower, and have since been inundated and flooded by rising sea level as the glacial ice melted.

Historically, freshwater in the hydrologic cycle has been enough to serve human needs, but exponential population growth and water usage in regions with little freshwater are posing ever-increasing political and planning problems. For these areas, freshwater has to be imported from great distances, at great expense, and to the detriment of other regions that need the water for their own use. Many of the most pressing problems of the 21st century will be related to obtaining freshwater for the world's expanding population.

References

1. http://observe.arc.nasa.gov/nasa/earth/hydrocycle/hydro2.html (accessed July 2002).
2. http://www.uwsp.edu/geo/faculty/ritter/geog101/modules/hydrosphere/hydrosphere_title_page.html (accessed July 2002).
3. http://ww2010.atmos.uiuc.edu/(Gh)/guides/mtr/hyd/home.rxml (accessed July 2002).
4. http://ga.water.usgs.gov/edu/mearth.html (accessed July 2002).
5. Winter, T.C.; Harvey, J.W.; Franke, O.L.; Alley, W.M. *Ground Water and Surface Water—A Single Resource;* U.S. Geological Survey Circular 1139: Denver, Colorado, 1999; 1–79.
6. Maidment, D.R.; Ed. *Handbook of Hydrology;* McGraw-Hill, Inc.; 1993; 1-1–1-15.
7. Driscoll, F.G. *Groundwater and Wells*, 2nd Ed.; Johnson Division: St. Paul, MN, 1986, Chapter 1.
8. U.S.S.R. Committee for the International Hydrological Decade, World Water Balance and Water Resources of the Earth, English translation. *Studies and Reports in Hydrology*; UNESCO: Paris, 1978; Vol. 25.
9. World Meteorological Organization–UNESCO. *International Glossary of Hydrology,* 2nd Ed.; WMO Report No. 385; Geneva, Switzerland, 1992.
10. Wilson, W.E.; Moore, J.E.; Eds. *Glossary of Hydrology*; American Geological Institute, 1998.
11. Bras, R.L. *Hydrology*; Addison-Wesley: Reading, MA, 1990.
12. National Research Council (NRC). *Global Change in the Geosphere-Biosphere*; National Academy Press: Washington, DC, 1986.
13. Hem, J.D. *Study and Interpretation of the Chemical Characteristics of Natural Water*; U.S. Geological Survey Water Supply Paper 2254; Washington, DC, 1992.

11

Hydrologic Modeling of Extreme Events

Introduction 69
Trex Model 69
Upland Processes • Overland Flow • Channel Flow
Site Description and Model Calibration 71
Site Description • Model Organization and Parameterization • Calibration and Validation
Extreme Event Simulation 76
Conclusion 78
Acknowledgments 78
References 79

Jazuri Abdullah,
Jaehoon Kim, and
Pierre Y. Julien
Colorado State University

Introduction

The simulation of extreme floods is one of the main concerns for public safety and hydrologic engineering.[1] In the analysis of extreme floods, the concepts of Probable Maximum Flood and paleo-hydrology are very useful.[2] In the United States, the models HEC-1 and HEC-HMS are used by the U.S. Army Corps of Engineers, whereas the Flood Hydrograph and Runoff (FHAR) model is promoted by the U.S. Bureau of Reclamation. The major advances in Geographical Information System (GIS)-based computer models in the past two decades offer a new perspective on the simulation of extreme flood events. We explore the use of a physically based and distributed model to simulate floods on a watershed under monsoon precipitation in South-East Asia.

The objectives of this research are to test the potential of GIS-based rainfall-runoff modeling to simulate extreme floods; and demonstrate modeling applicability through calibration, validation, and simulation of extreme floods. The Two-dimensional Runoff, Erosion and EXport (TREX) model was selected for the analysis and is tested at Sungai Semenyih in Malaysia where the 236 km^2 watershed can be subjected to some of the highest precipitation levels ever recorded world-wide.

Trex Model

The physically based TREX model is fully distributed and simulates rainfall-runoff, sediment transport, and chemical fate and transport at the watershed scale.[3] TREX uses a GIS-based raster format combining the CASC2D surface runoff model,[4–7] and the water quality WASP/IPX model.[8] TREX is classified as an event model as it simulates the Hortonian overland flow and the watershed responses from a single storm with no soil infiltration capacity recovery between events. The selection of the computational time step was done by satisfying the Courant Condition.

Upland Processes

The hydrologic processes in the model are as follows: 1) rainfall, interception, and surface storage; 2) infiltration and transmission loss; and 3) overland and channel flow. The model state variables are water depth in the overland plane and stream channels. Rainfall can be uniform or distributed in both time and space and can also be specified using several grid-based formats for point rain gauges or radar rainfall data.[9] Rainfall can be representing as the gross volume of water that reached the near surface as:

$$\frac{\partial V_g}{\partial t} = i_g A_s \tag{11.1}$$

where V_g = gross rainfall water volume [L^3]; i_g = gross rainfall rate [L/T]; A_s = surface area over which rainfall occurs [L^3]; and t = time [T]. The model calculates the volume of interception as:

$$V_i = (S_i + Et_r)A_s \tag{11.2}$$

where V_i = interception volume [L^3]; S_i = interception capacity of projected canopy per unit area [L^3/L^2]; E = evaporation rate [L/T]; and t_r = precipitation event duration [T].

Infiltration and transmission loss rates are simulated using the Green and Ampt relationship[10]:

$$f = K_h \left(1 + \frac{(H_w + H_c)(1 - S_e)\theta_e}{F}\right) \tag{11.3}$$

where: f = infiltration or transmission loss rate [L/T]; K_h = effective hydraulic conductivity [L/T]; H_w = hydrostatic pressure head (pond water depth) [L]; H_c = capillary pressure (suction) head at the wetting front [L]; $\theta_e = (\varphi - \theta_r)$ effective soil porosity [-]; φ = total soil porosity [-]; θ_r = residual moisture content [-]; S_e = effective saturation [-]; and F = cumulative infiltrated depth (depth to wetting front) [L].

Overland Flow

Overland flow is two dimensional and simulated using the diffusive wave approximation. Flow occurs when the water depth exceeds the storage depth:[4,10]

$$\frac{\partial h}{\partial t} + \frac{\partial q_x}{\partial x} + \frac{\partial q_y}{\partial y} = i_n - f + W = i_e \tag{11.4}$$

$$q_x = \alpha_x h^\beta$$
$$q_y = \alpha_y h^\beta \tag{11.4a}$$

where: h = surface water depth [L]; q_x, q_y = unit flow discharge in the x- or y-direction = Q_x/B_x, Q_y/B_y [L^2/T]; Q_x, Q_y = flow in the x- or y-direction [L^3/T]; B_x, B_y = flow width in the x- or y-direction [L]; i_n = net precipitation rate (gross rainfall minus interception) [L/T]; W = flow point source (unit discharge) [L/T]; i_e = excess precipitation rate [L/T]; α_x, α_y = resistance coefficient for flow in the x- or y-direction = $|S_{fx}|^{1/2}/n_M$, $|S_{fy}|^{1/2}/n_M$ [$L^{1/3}/T$]; n_M = Manning roughness coefficient [$T/L^{1/3}$]; S_{fx}, S_{fy} = $S_{0x} - dh/dx$, $S_{0y} - dh/dy$ friction slope (energy grade line) in the x- or y-direction [-]; S_{0x}, S_{0y} = ground surface slope in the x- or y-direction [-] and β = exponent of the stage-discharge relationship. To solve the resistance coefficient equations for α_x and α_y, the absolute value of S_f is used and the sign of S_f indicates the flow direction.

Channel Flow

Channel flow is one dimensional and simulated using the diffusive wave approximation. Flow occurs when the water depth exceeds the storage depth and the friction slope is not zero:[4,10]

$$\frac{\partial A_c}{\partial t}+\frac{\partial Q_x}{\partial x}=q_l+W \tag{11.5}$$

$$Q_x=\frac{1}{n_M}A_c R_h^{2/3}\left|S_{fx}\right|^{1/2} \tag{11.6}$$

where A_c = cross-sectional area of flow [L^2]; Q = total discharge [L^3/T]; q_l = lateral unit flow (into or out of the channel) [L^2/T]; W = unit discharge from/to a point source/sink (including direct rainfall to the channel) [L^2/T]; R_h = hydraulic radius = A_c/P_c [L]; P_c = wetted perimeter of channel [L]. To solve Equation (11.5), the absolute value of S_f is used and the sign indicates the flow direction.

In floodplain areas, water and any transported constituents are transferred between the overland plane and channel network based on the difference in water surface elevations. Floodplain transfers are bi-directional. Water and transported constituents move into stream channels by overland flow and can return to the overland plane when water levels in the stream exceed bank height. Similarly, materials can be moved from the sediment bed and can be delivered to the land surface by floodwaters.[11,12]

Site Description and Model Calibration

Site Description

Located in the state of Selangor in Malaysia, the Semenyih watershed covers 236 km² (Figure 11.1). The lowest elevation at the outlet is 40 m above sea level, whereas the highest point reaches 1100 m at the upstream end of the watershed. The average terrain slope is about 45% and ranges between 4% and 85% with very steep mountains overhanging flat and wide valleys (Figure 11.2a). The length and width of the watershed are 17.7 km and 23.9 km, respectively. The average normal depth of the main channel in Sungai Semenyih ranges between 0.8 m and 2.49 m. The watershed is partly used for agriculture and the urbanization through residential and industrial development has rapidly transformed the area in the past 20 yr.

Located near the equator, the watershed climate is categorized as equatorial, being hot and humid throughout the year. The average rainfall precipitation reaches 2500 mm/yr and the average temperature is 27°C. Influenced by the southwest and northwest monsoons, the study area falls into the west coast rainfall region, where June and July are the driest months and November is the wettest.

Model Organization and Parameterization

The Semenyih watershed was simulated using the TREX watershed model. To resolve surface topography, the watershed was discretized at a 90 m × 90 m grid scale. This should yield satisfactory results according to the grid size analysis of Molnar and Julien.[13] The Digital Elevation Model (DEM) data for the site (Figure 11.2a) were obtained from the Department of Surveying and Mapping Malaysia (DSMM). The resulting rectangular raster grid has 265 rows and 197 columns. Within this raster grid, the watershed area is defined by 29,139 cells. The DEM also allowed a delineation of the channel network with the watershed. The channel network includes seven links (reaches) and 399 nodes for a total stream length of approximately 36 km. Four soil types (Figure 11.2b) and six land uses (Figure 11.2c) were incorporated into the raster-based GIS representation of the watershed for the determination of hydraulic conductivity and surface roughness.

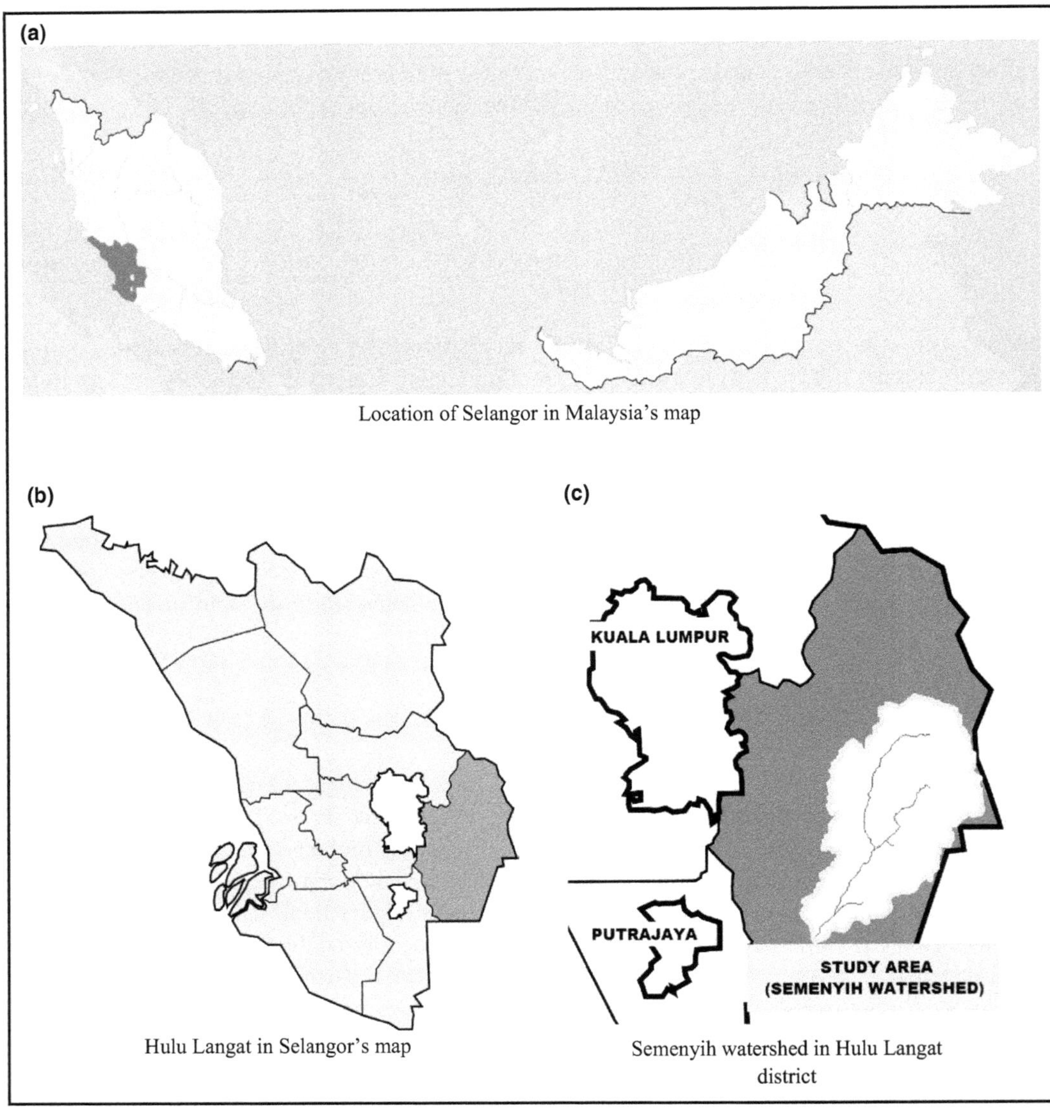

FIGURE 11.1 Semenyih study watershed location map.

Calibration and Validation

The April 13, 2003, storm was used for calibration using the precipitation and flow records collected by Department of Irrigation and Drainage (DID). The calibration procedure focused on properly simulating peak flow, discharge volume, and time to peak at the outlet. Model parameters subject to calibration were simply the effective hydraulic conductivity (K_h), and roughness (Manning n). The values of the calibrated parameters are summarized in Table 11.1. The values of hydraulic conductivity K_h were adjusted during calibration to achieve good agreement between measured and simulated runoff. The antecedent moisture condition for the watershed was assumed to be fully dry.

The Nash-Sutcliffe Efficiency Coefficient (NSEC) and Percent BIAS (PBIAS) were used to evaluate model performance. The value of discharges and volumes from the simulated result were evaluated using these methods, respectively. The value of NSEC can be between $-\infty$ and 1.0, with NSEC = 1.0 being the optimal value. Generally, NSEC values between 0.0 and 1.0 are acceptable levels of performance,

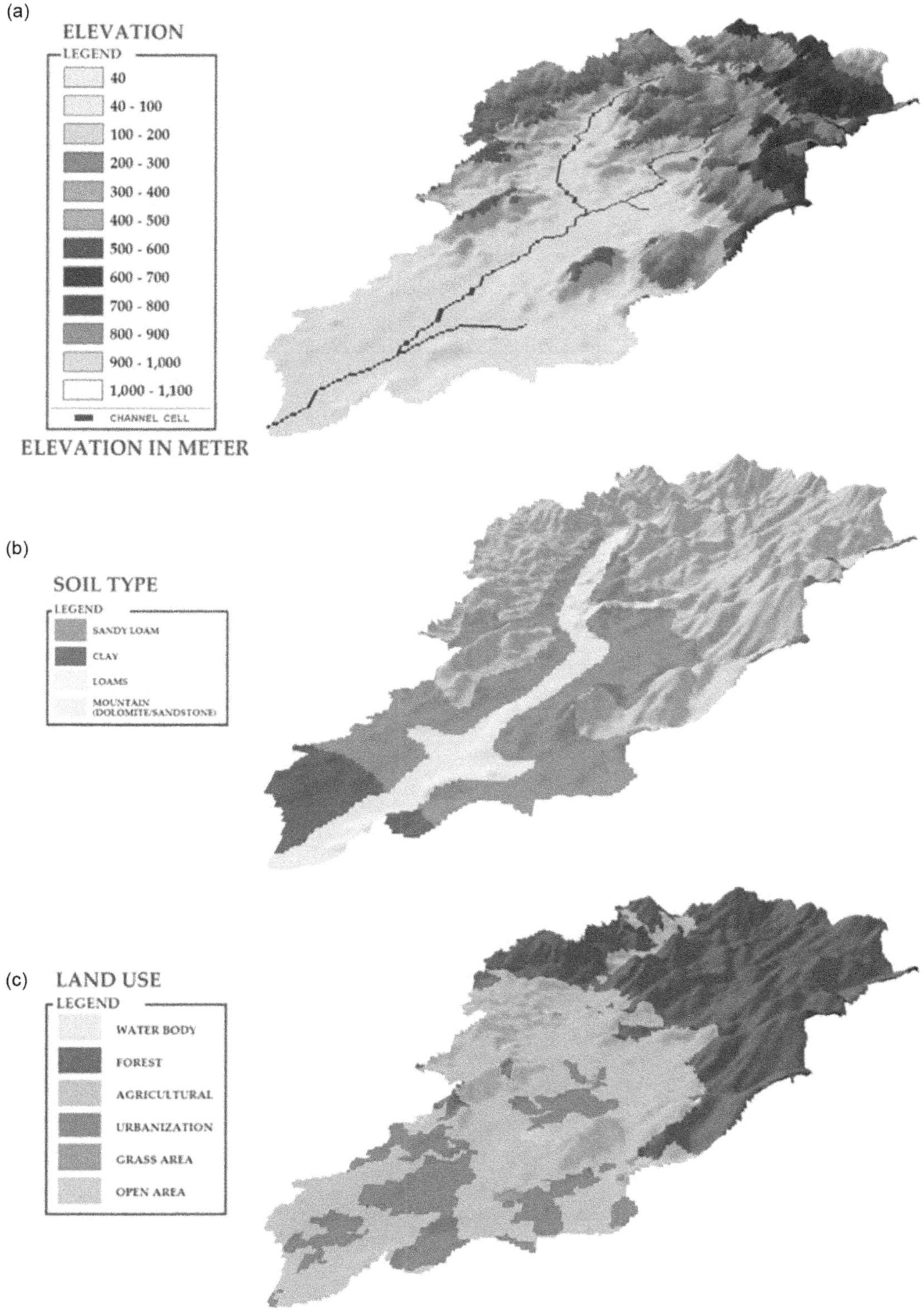

FIGURE 11.2 **(See color insert.)** Sungai Semenyih 90 m TREX GIS data layers: (**a**) elevation grid in meter and channels cells; (**b**) soil-type classes; (**c**) land use classes.

TABLE 11.1 Summary of Model Parameter Values for Semenyih Watershed

Parameter	Range	Application
Hydraulic	$1.5 \times 10^{-8} - 6.1 \times 10^{-7}$	Sandy Loam
Conductivity	$3.7 \times 10^{-8} - 6.9 \times 10^{-8}$	Loam
K_h (m/s)	$1.7 \times 10^{-9} - 1.3 \times 10^{-8}$	Clay
	$3.2 \times 10^{-8} - 3.2 \times 10^{-10}$	Mountain-Limestone
	0.05 – 0.18	Agricultural
	0.05 – 0.15	Urbanization
Manning's n (s/m$^{1/3}$)	0.20 – 0.40	Forest
	0.05 – 0.20	Grass area
	0.05 – 0.15	Open area
	0.02 – 0.04	Channel bed

whereas values less than 0.0 indicate unacceptable performance. PBIAS measures the tendency of the simulated data to be larger or smaller than the observed data.[14] Positive and negative values indicate a model underestimated and overestimated, respectively. The optimal value of PBIAS is 0.0. In this study, the classifications for NSEC and PBIAS defined by Moriasi et al.[15] were used to determine whether the simulation results were good or very good.

The TREX model parameterization for the calibration on April 13, 2003, shows that peak flow, flow volume, and time to peak were all accurately simulated at the outlet (Figure 11.3). A calibrated channel Manning n (0.04) was within the range proposed by Zakaria et al.[16] The Relative Percent Difference (RPD) for peak flow, total volume, and time to peak was 0.4, 19.1, and 1.5, respectively. The NSEC and PBIAS value peak discharge and total volume were 0.8% and –19.3%, respectively, and indicate a very high simulation performance. For the validation events, the peak discharge found to be

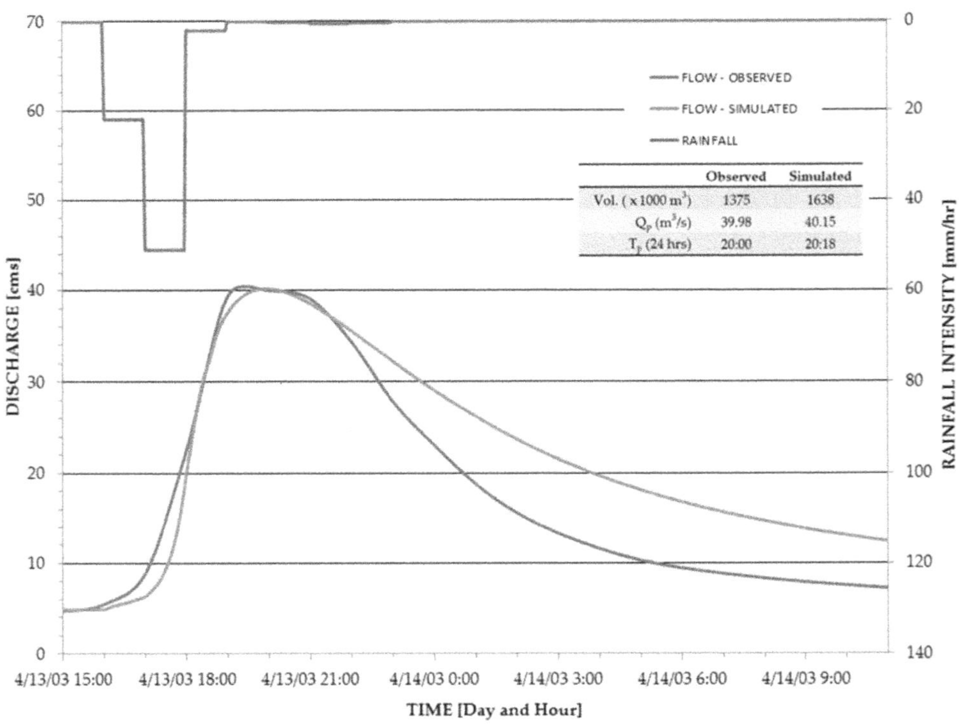

FIGURE 11.3 April 13, 2003, hydrograph and TREX calibration.

near-perfectly simulated with an average RPD value as low as 0.28%. The simulated hydrographs generally had the same shape as the observed hydrographs with an average total flow volume RPD of 32%. The peak discharge from measured and simulated for calibrated and validated model is presented in Figure 11.4. The model is numerically very stable and matches the peak discharge and time to peak. Performance evaluations for these simulations are presented in Table 11.2.

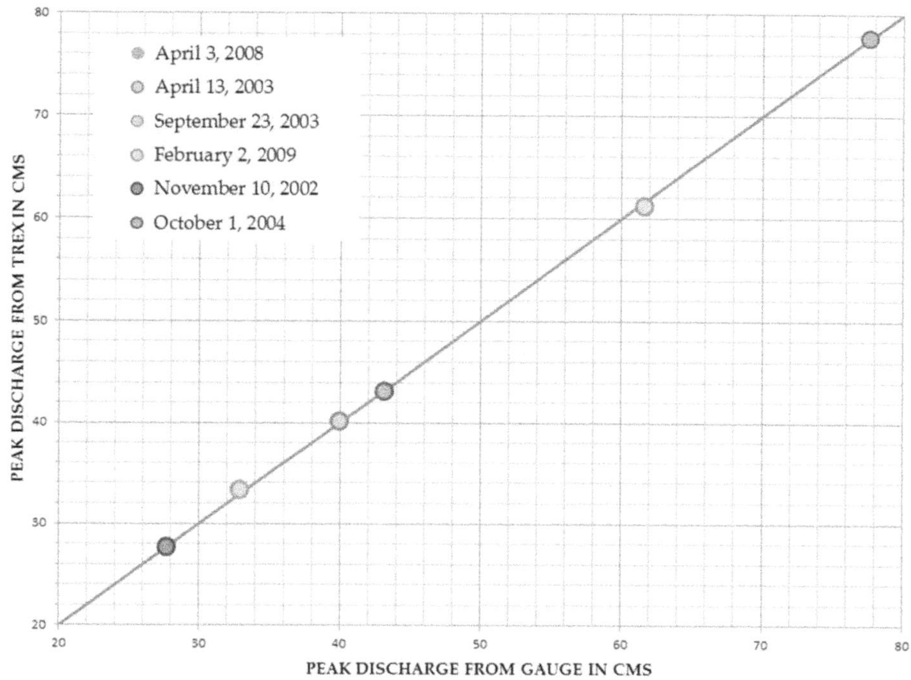

FIGURE 11.4 Simulated and measured water discharge for the model calibration and validation events.

TABLE 11.2 Hydrologic Model Performance Evaluation Summary

	Peak flow (m^3/sec)			Total volume (× 1000 m^3)			Time to peak (hr)			Performance evaluation	
Event	Observation	Simulated	RPD (%)	Observation	Simulated	RPD (%)	Observation	Simulated	RPD (%)	NSEC	PBIAS(%)
					Calibration						
April 13, 2003	39.98	40.15	0.4	1375	1638	19.1	20:00	20:18	1.5	0.8	−19.3
					Validation						
April 3, 2008	77.58	77.77	0.2	2939	3052	3.9	23:00	23:54	3.9	1.0	−7.6
September 23, 2003	32.83	33.37	1.6	590	950	61.2	7:00	7:42	10.0	0.1	−57.7
February 23, 2003	61.59	61.23	−0.6	1924	2530	31.5	22:00	22:45	3.4	0.4	−31.7
November 10, 2002	27.71	27.74	0.1	947	1277	34.9	0:00	0:42	2.9	0.8	−25.9
October 1, 2004	43.12	43.18	0.1	1236	1590	28.7	19:00	19:21	1.8	0.8	−28.9

Abbreviations: RPD, relative percent different; NSEC, Nash Sutcliffe Efficiency Coefficient; PBIAS, percent BIAS.

Extreme Event Simulation

The importance of Probable Maximum Precipitation (PMP) is always highlighted and emphasized in terms of public safety and hazard downstream of any major river regulating structures, especially if it is located upstream of populated areas. In the case presented in this analysis, the world's greatest precipitation (WGP) events are used as input to the calibrated TREX model to determine the magnitude of extreme floods. A 16 hr precipitation duration obtained from Jennings[17] with an intensity of 68.24 mm/hr has been applied (Figure 11.5). The precipitation duration was selected at 16 hr because precipitation records for shorter duration can be determined with a sufficient level of certainty in this monsoon climate according to MSMA[18] (Urban Stormwater Management Manual). Table 11.3 shows the summary of the input for TREX. The 2 yr, the 100 yr[18] and also the PMP[19] conditions for Malaysia have also been applied in the model for a comparison of the peak discharge between the four simulated storms. The watershed runoff coefficient, $C = Q/iA$ (where Q = peak discharge [L^3T^{-1}]; i = rainfall intensity [LT^{-1}]; and A = watershed area [L^2]), increases in magnitude as intensity and peak discharge increased. The C values for the 2 yr and 100 yr storm events were lower because the infiltration and flow characteristics in the flat areas were dominant compared with steep areas. The C values for the PMP and WGP storm events were closer to unity because the flat areas became fully saturated and almost impervious.

Figure 11.6 shows the results from the WGP event. Figure 11.6a represents the watershed starting to receive precipitation at time 01:39 hours. The WGP simulation result indicates that the lowlands would

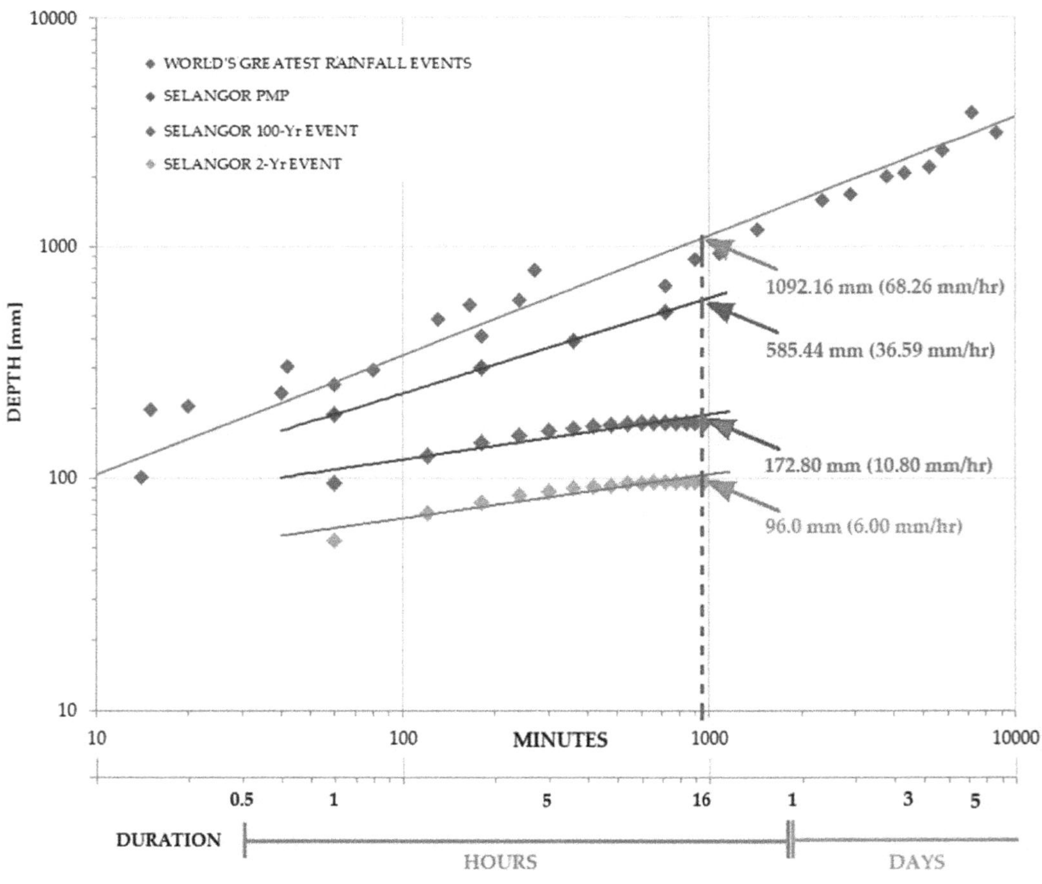

FIGURE 11.5 Comparison between world's greatest rainfall events and Selangor extreme events.

TABLE 11.3 Summary of Intensity and Peak Discharge at Semenyih Using 2 Yr, 100 Yr, PMP, and World's Greatest Precipitation

Events		Input		Results		
		Depth (mm)	Intensity (mm/hr)	Peak discharge (cm)	Time to peak (24-hr)	Runoff coefficient
	2-Yr	96	6	136	18:12	0.3
Semenyih	100-Yr	173	11	223	23:36	0.3
	PMP	585	37	1484	19:06	0.6
World's Precipitation		1092	68	3686	17:30	0.8

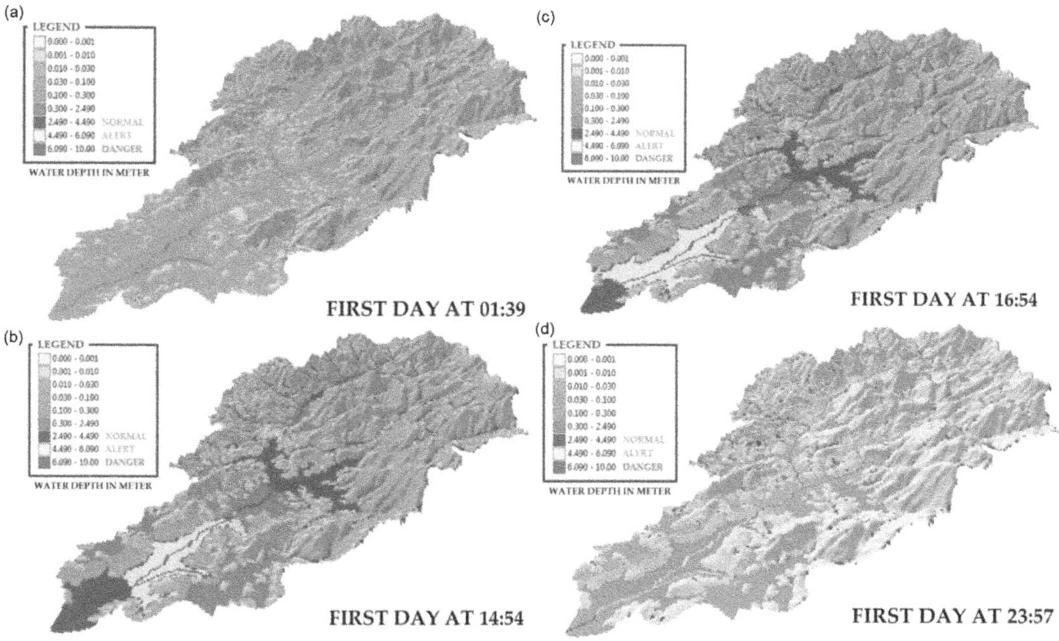

FIGURE 11.6 **(See color insert.)** Spatial distribution of water depth (m) with world's greatest precipitation event (68.24 mm/hr at 16 hr duration).

be flooded with a water depth as high a 6 m (Figure 11.6b). The DID website defines three water levels in relation to floods in Malaysia: 1) the "normal" water level corresponds to water depths less than 4.49 m; 2) the "alert" level corresponds to water depths between 4.49 m and 6.09 m; and 3) the "danger" level is reached when the flow depth exceeds 6.09 m. This "danger" condition is reached during our simulation as indicated in red color on Figure 11.6c. Flooding is expected to be widespread in the low valley area. In this simulation, the "alert" condition (yellow in color) started around 11:00 hours and submerged about 50% of lowest area and lasted approximately 4 hr until 14:54 hours during the first day of the storm event. The flood took about 6 hr to recede and drain most of the surface water as shown in Figure 11.6d.

The hydrograph (Figure 11.7) for Semenyih watershed demonstrates that the TREX model can be effectively used to simulate and estimate the magnitude of flooding during extreme storms. The TREX result based on WGP input data are compared with floods of lesser magnitude, including the 2 yr, the 100 yr flood, as well as the PMP condition for Malaysia. A summary of the conditions and simulation results for these floods is presented in Table 11.3.

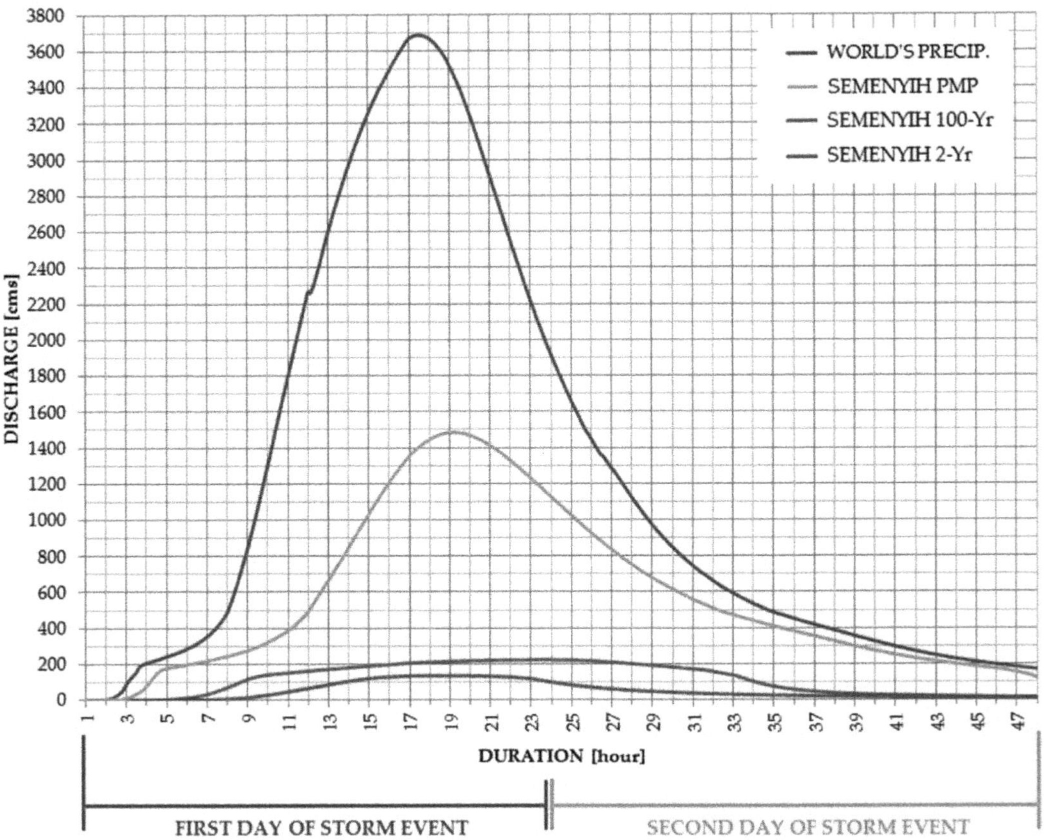

FIGURE 11.7 TREX hydrograph with world's greatest precipitation event (68.24 mm/hr at16hr duration).

Conclusion

The model TREX has been successfully applied for the simulation of extreme floods on the 236 km^2 of Semenyih watershed in Malaysia. The April 13, 2003, storm was used to calibrate the model, whereas several other storm events served for the model validation. The calibrated TREX model can then simulate the hydrograph using the world's greatest precipitation (68.24 mm/hr for 16 hr duration). TREX successfully simulates extreme hydrographs for comparison with 2 yr and 100 yr flood events. TREX also provides visual flood hazard maps for this site.

Acknowledgments

Financial support to the first author was granted through the Ministry of Higher Education (MOHE), Malaysia. Data for the Semenyih watershed application were provided by Mohd Rozi Talib (Ex-DSMM); Nor Haslinda Mohamed Yusop and Arshad Mohd Isa from DSMM; and Lizawati Turi, Abu Salim Abd. Aziz, Khairul Fadzilah Mohd Omar, Mohd Shawal Abd. Wahid, and Azmi Md. Jafri from DID. We also appreciate the assistance of Prof. Junaidah Ariffin, Joe Nyuin, Azmi Ibrahim, and Norizan Ismail from Universiti Teknologi MARA, Nur Shazwani Muhammad and Othman Jaafar from National University of Malaysia to provide us some useful information regarding Semenyih watershed. Additional support from Mark Velleux (HydroQual, New Jersey) and John England (U.S. Bureau of Reclamation) in using TREX software is also gratefully acknowledged.

References

1. England, J.F. Jr.; Velleux, M.L; Julien, P.Y. Two-dimensional simulations of extreme floods on a large watershed. J. Hydrol. **2007**, *347* (12), 229241.
2. England, J.F. Jr.; Godaire, J.E.; Klinger, R.E; Bauer, T.R.; Julien, P.Y. Paleohydrologic bounds and extreme flood frequency of the upper Arkansas river, J. Geomorphol. **2010**, *124*, 1–16.
3. Velleux, M.L.; England, J.F. Jr.; Julien, P.Y. *TREX Watershed Modeling Framework User's Manual: Model Theory and Description;* Department of Civil Engineering; Colorado State University, Fort Collins, Colorado, 2006a; 83.
4. Julien, P.Y.; Saghafian, B.; Ogden, F.L. Raster-based hydrologic modeling of spatially-varied surface runoff. Water Resour. Bull. AWRA **1995**, *31* (3), 523-536.
5. Johnson, B.E.; Julien, P.Y.; Molnar, D.K.; Watson, C.C. The two-dimensional upland erosion model CASC2D-SED. J. Am. Water Resour. Assoc. AWRA **2000**, *36* (1), 31–42.
6. Ogden, F.L.; Julien, P.Y. CASC2D: A two-dimensional, physically-based, Hortonian hydrologic model. In *Mathematical Models of Small Watershed Hydrology and Applications;* Singh, V.P.; Frevert, D. Eds.; Water Resources Publications: Littleton, CO, 2002; 69–112.
7. Julien, P.Y.; Rojas, R. Upland erosion modeling with CASC2D-SED. Int. J. Sediment Res. **2002**, *17* (4), 265–274.
8. Velleux, M.L.; Julien, P.Y.; Rojas-Sanchez, R.; Clements, W.H.; England J.F. Simulation of metals transport and toxicity at a mine-impacted watershed: California Gulch, Colorado. Environ. Sci. Technol. **2006b**, *40* (22), 6996–7004.
9. Jorgeson, J.; Julien, P.Y. Peak flow forecasting with CASC2D and radar data, Water International. IWRA, **2005**, *30* (1), 40–49.
10. Julien, P.Y. *River Mechanics;* Cambridge University Press: Cambridge, 2002; 434.
11. Julien, P.Y. *Erosion and Sedimentation;* Cambridge University Press: Cambridge, 2nd Ed., 2010; 371.
12. Velleux, M.L.; England, J.F. Jr.; Julien, P.Y. TREX: Spatially distributed model to assess watershed contaminant transport and fate. J. Sci. Total Envir. **2008**, *404*, 113–128.
13. Molnar, D.K.; Julien, P.Y. Grid size effects on surface runoff modeling. J. Hydrol. Eng., ASCE, **2000**, *5* (1), 8–16.
14. Gupta, H.V.; Sorooshian, S.; Yapo, P.O. Status of automatic calibration for hydrologic models: comparison with multilevel expert calibration. J. Hydrol. Eng., ASCE, **1999**, *4* (2), 135–143.
15. Moriasi, D.N.; Arnold, J.G.; Van-Liew, M.W.; Bingner, R.L.; Harmel, R.D.; Veith, T.L. Model evaluation guidelines for systematic quantification of accuracy in watershed simulations. Am. Soc. Agric. Biol. Eng. **2007**, *50* (3), 885–900.
16. Zakaria, N.A.; Azamathulla, H.M.; Chang, C.K.; Ab. Ghani, A. Gene expression programming for total bed material load estimation-a case study, Sci. Total Env. **2010**, *408*, 5078–5085.
17. Jennings, A. H. *World's Greatest Observed Point Rainfalls.* Hydrometeorological Section, U. S. Weather Bureau: Washington D.C., 1950; 4–5.
18. DID. Urban Stormwater Management Manual (MSMA Manual), 2000, 13 (3).
19. Hii, C.P.; Heng, H.H. Probable maximum precipitation derivation in Malaysia: Review and Comparison. Int. J. Hydro-Clim. Eng. **2010**, 37–72.

12

Hydrology: Environmental

Introduction 81
History 81
Environmental Hydrology and Water Quality 82
Environmental Hydrology and Water Quantity 83
Conclusion 84
Acknowledgments 85
References 85

Thomas Boving
University of Rhode Island

Introduction

Life on Earth depends on water that is both sufficient in quantity and quality. Anthropogenic activity, however, has degraded or depleted many water resources and humanity is now faced with remediating and restoring this essential natural resource. This endeavor can only be successful if scientists and engineers understand the complex connections between the various compartments of the environment that influence the availability and quality of water. This entry is about environmental hydrology. It describes how this branch of science stands at the nexus of competing interest for Earth's limited water resources and what role environmental hydrology can play in alleviating current and future ecohydrological stresses.

History

The science of hydrology, in general, focuses on the water on Earth, where it occurs, its circulation and distribution in biotic and abiotic systems, including its chemical and physical properties and interactions with the environment. Because water is vital for sustaining life on Earth, its study must be conducted in an environmental continuum[1] in which water interacts with the solid earth and the atmosphere in a system of complex and continuous reactions. This environmental continuum approach is emphasized in what is referred to as environmental hydrology.

The term environmental hydrology has been widely used since the early 1970s when hydrogeologists began debating and systematically investigating the effects of widespread pollution of the aqueous environment.[2] In those days, the public was confronted with massively polluted sites, such as the Cuyahoga River in Ohio, which became known as "the river that caught fire" or the notorious Love Canal site in Niagara Falls, NY, where an inadequate cover on a chemical company's dumpsite allowed a deadly brew of toxins to leak into a suburban neighborhood. The Love Canal became the site of one of the worst environmental disasters in American history, forcing 900 families to vacate their homes and causing damage worth over hundred million dollars.

As it became obvious that Love Canal was not an isolated case and that various types of toxic substances were dumped all over the environment, President Carter declared that uncovering these dumpsites was "one of the grimmest discoveries of the modern era."[3] The legacy of these sites eventually gave rise to the field of Environmental Hydrogeology, which represents a branch of hydrology that is primarily concerned about water quality. Environmental hydrology may be defined as the science dealing with the spatial and temporal variability of water quality and its evolution in the hydrosphere, lithosphere, and atmosphere. Atmospheric water must be considered a part of environmental hydrology because water cycling through the atmosphere impacts the water quality as demonstrated by the deposition of air-borne methyl mercury into surface water[4] or acid precipitation.[5] Hence, environmental hydrology encompasses the occurrence, distribution, and variability of water quality in ground water, surface water, the unsaturated (vadose) zone, and the atmosphere.[1] Further, there is mounting evidence that climate change is having a dramatic impact on the health of the ecohydrological environment,[6] which is why the hydrological consequences of climate change should also be covered under the umbrella of environmental hydrology.

Environmental hydrology is a multidisciplinary science, drawing knowledge from several branches of engineering and science, including chemistry, physics, biology, meteorology, geography, ecology, computer science, and mathematics, as well as agriculture and forestry. In addition, practicing environmental hydrologists are expected to have a working knowledge of the applicable environmental laws as well as an awareness of the potential sociopolitical implications of their work.

Many countries around the world have adopted environmental standards and regulations originally formulated by the U.S. Environmental Protection Agency (EPA). The EPA was proposed by President Nixon and charged by U.S. Congress on December 3, 1970 with protecting human health and the environment. The EPA enforces environmental regulations passed into law by congress. One of the earliest environmental laws is the Clean Water Act (CWA) of 1972. The CWA is the cornerstone of surface water quality protection in the United States and employs a variety of regulatory and nonregulatory tools to sharply reduce direct pollutant discharges into waterways, finance municipal wastewater treatment facilities, and manage polluted runoff.[7] It is noteworthy that CWA does not deal directly with ground water or with water quantity issues. In 1974, the Safe Drinking Water Act was passed to protect the drinking water supply by establishing enforceable national drinking water standards (e.g., maximum contaminant level). Probably the most important sets of environmental rules and regulations pertaining to the practicing Environmental Hydrologist are the Resources Conservation and Recovery Act (RCRA) and the Comprehensive Environmental Response Compensation and Liability Act (CERCLA), which were passed into law in 1976 and 1980, respectively. Under RCRA, the EPA has the authority to control hazardous waste from the "cradle-to-grave," including supervision of the generation, transportation, treatment, storage, and disposal of hazardous waste. CERCLA, also known as "Superfund law," provides broad federal authority to respond directly to releases or potential releases of hazardous substances that may endanger public health or the environment. Under CERLA, the EPA determines which hazardous waste sites warrant further investigation and are eligible for cleanup funding from the Superfund program. These sites are then placed on the National Priority List (NPL). As of February 2012, there were 1293 NPL sites.[8]

Environmental Hydrology and Water Quality

Environmental hydrology utilizes tools such as geographic information systems, hydraulic and transport/fate models, and global circulation models to solve complex environmental and ecological problems. Environmental hydrology research relies on studies and data generated in the field, including geophysical surveys, water and soil samples analysis, or tracer tests; in the laboratory, such as bench scale and column tests or isotherm experiments; or three dimensional pilot-scale test setups at dedicated research centers such as the Borden site in Canada[9] or the VEGAS installation in Germany.[10]

The information and data gathered in the field and laboratory have been used very successfully to predict the transport and fate of contaminants and to aid in the remediation of hazardous waste sites. The theory and concepts of contaminant transport and fate, including their modeling, are summarized in a number of noteworthy textbooks.[11–15] Chemical and physical properties of environmental pollutants are readily available in published format, such as[16–19] and on the internet.[20] Further, there are several textbooks specifically aimed at environmental hydrology, for example.[1,21]

A principal challenge of environmental hydrology science is to describe in quantitative terms the complex chemical, physical, and microbiological processes that influence the solute transport in surface and subsurface. The aqueous phase is the principal carrier of dissolved and suspended contaminants. In the absence of any interactions between the solute and the porous matrix, the advection-dispersion equation[11,22] can be utilized to describe solute transport in saturated media in terms of mixing and dilution. However, many contaminants interact with the porous matrix (sorption, reaction) or other compartments of the environment, including the atmosphere (volatilization) or biosphere (mineralization) or simply decay abio-logically. Also, if nonaqueous phase liquids, such as petroleum products or chlorinated solvents are present, multiphase flow phenomena have to be considered as well.[11,23] To account for the various processes that influence the fate and transport of solutes in the environment, the advection-dispersion equation has to be expanded for the appropriate processes and solved numerically using a variety of powerful computer models for either saturated or unsaturated flow conditions. In a review of existing modeling software, over 250 codes were identified and rated for their ability to model the fate and transport of organic contaminants in the environment.[24] Environmental hydrologists commonly use these models as aides in designing remediation schemes or for predicting the response of hydrologic systems to changing environmental conditions. However, great care has to be taken to correctly set up these computer models and interpret their results.

Environmental Hydrology and Water Quantity

While environmental hydrology primarily focuses on water quality, its study cannot be isolated from understanding the quantity of water interacting with the environment. Many places on Earth already face severe water shortages; as the world's population continues to increase, the demand for water increases accordingly. Water scarcity is among the main problems many societies face in the 21st century.[25] For instance, as of 2004 it was estimated that 1.1 billion people lacked access to improved water supplies and 2.6 billion people lack adequate sanitation.[26] In many locations, trying to satisfy the demand for water resulted in falling water tables, depleted aquifers, or sea-water intrusion into over-pumped coastal aquifers.[27] The ineffective use of water for irrigation has contributed to water depletion and, locally, has led to salinization of previously fertile farmland[28] and desertification.[29] In addition, the densely populated urban areas in the developing world will make up an estimated 81% of all urban inhabitants by 2030.[30] Supplying the growing number of urban and rural population with clean water and ensuring the effective treatment and disposal of wastewater will remain an enormous challenge for this and following generations of environmental hydrologists.

The consequences of climate change further stress the world's water resources. In 2007, the Intergovernmental Panel on Climate Change came to the conclusion that "all regions of the world show an overall net negative impact of climate change on water resources and freshwater ecosystems. Areas in which runoff is projected to decline are likely to face a reduction in the value of the services provided by water resources. The beneficial impacts of increased annual runoff in other areas are likely to be tempered in some areas by negative effects of increased precipitation variability and seasonal runoff shifts on water supply, water quality and flood risks."[6] Changes in the timing, intensity, and duration of precipitation together with higher water temperatures can affect water quality. Higher temperatures reduce dissolved oxygen levels, which can have an effect on aquatic life. Where stream flow and lake levels fall, there will be less dilution of pollutants; however, increased frequency and intensity of rainfall will produce more pollution and sedimentation due to runoff.[6] Today, environmental hydrologists are faced with trying to understand how a changing climate is affecting the water quality and availability and how the negative consequences of climate change can be mitigated.

Environmental problems have forced people to either vacate water-depleted, drought-stricken areas or pushed them into consuming polluted water. Besides pollution caused by the discharge of chemicals, water borne diseases like diarrhea pose a major threat to human health. Waterborne diseases are caused by pathogenic microorganisms (viruses, bacteria, protozoa), which can be transmitted when contaminated water is consumed. The World Health Organization estimates that diarrheal disease alone is responsible for the deaths of 1.8 million people every year and is especially affecting children in developing countries.[26] The socioeconomic impacts of limited access to clean water and basic sanitation are substantial, estimated at annual losses of 6.4%, 5.2%, and 7.2% of GDP in India, Ghana, and Cambodia, respectively.[31] In 2000, the international community committed to the Millennium Development Goals (MDGs) with the aim to halve the proportion of people without access to clean water and basic sanitation by 2015. Currently, the world is on track to meet the water supply MDG but there are many regions and countries still lagging behind, particularly in sub-Saharan Africa.

Only about 3% of all the water on Earth is fresh water. Two-thirds of it is frozen, one-third is groundwater and a mere 0.3% remains as accessible surface water. Of the available renewable freshwater resources, about 54% is already appropriated by humanity for various uses: irrigated agriculture (70%), industry (22%), and domestic use (8%). By 2025, water withdrawals are projected to increase by 50% and 18% in the developing and developed world, respectively.[31] Water scarcity, commonly defined as water supplies of less than $1000\,m^3$ per person per year, is already a reality in much of North Africa and Arabia. By 2025, the United Nations expects absolute water scarcity to extend into Pakistan and India, affecting more than 1.8 billion people.[32] These statistics and predictions underline the need for integrated, cooperative solutions and adaptive water governance and management. This is particularly urgent in the 263 river basins, which are shared by two or more states, and in which nearly half the territory and population of the world are located.[33] In the absence of strong institutions and agreements, changes within a basin, such as dam construction or water diversion, can lead to transboundary tensions. When major water projects proceed without regional collaboration, they can become a point of conflict, heightening regional instability. Some of the world's largest watersheds and rivers, such as the Nile, the Jordan, the Colorado, and the Danube, may serve as examples. Given their expertise in dealing with complex water quantity and quality issues, environmental hydrologists are well placed to serve society as scientific advisors to political decision makers and water resources managers.

Conclusion

Environmental hydrology became a branch of hydrology in the early 1970s in response to ubiquitous environmental pollution problems. Since then it has evolved into a multidisciplinary science concerned not only with water quality but also the quantity of water on Earth. The underlying theoretical framework that describes the fate and transport of pollutants in surface and subsurface water is well developed, although there is plenty of room for future breakthroughs. Practicing environmental hydrologists perform their services within a legal framework, which grew out of the need for remediating, at great cost, a large number of hazardous waste sites. The prospect for employment in this area is likely to remain high for years to come.

The future of environmental hydrology is directly connected to how our society decides to deal with water scarcity and climate change. For billions of humans, safe and accessible water already is or shortly will be the natural resource that limits their standard of living and dictates where they can or cannot live. Now and in the future, environmental hydrologists have to play an important role in alleviating this global water crisis by making available to society their unique expertise regarding the sustainable management of the world's stressed water resources. This not only puts a great responsibility on the shoulders of currently practicing environmental hydrologists but also poses a tremendous economic opportunity for the next generation of young water scientists and professionals.

Acknowledgments

Over the past few decades, many people have contributed to the field of environmental hydrology and a great number of excellent scientific articles, textbooks, and other materials have been published. Unfortunately, only a limited few could be cited in this entry. The selection of these sources, including emphasizing specific areas of research, is admittedly subject to this author's background and research interest. Many thanks to Ms. Laura Schifman for her critical review.

References

1. Singh, V.P. What is environmental hydrology? In: *Environmental Hydrology* (Chap. 1). Singh, V.P., Ed.; Springer: 1995, Reprinted 2010.
2. Pettyjohn, W.A. *Water Quality in a Stressed Environment: Readings in Environmental Hydrology;* Burgess Pub. Co.: Minneapolis, 1972.
3. Beck, E.C. 1979. The Love Canal tragedy. EPA Journal -January 1979. http://www.epa.gov/aboutepa/history/topics/lovecanal/01.html (accessed September 2011).
4. Ullrich, S.; Tan ton, T.; Abdrashitova, S. Mercury in the aquatic environment: a review of factors affecting methylation. Crit. Rev. Environ. Sci. Technol. **2001**, *31,* 241–293.
5. Likens, G. E.; Bormann F. H. Acid rain: a serious regional environmental problem. Science **1974**, *184* (4142), 1176–1179.
6. Intergovernmental Panel on Climate Change (IPCC). Climate Change. Impacts, Adaptation, and Vulnerability. Contribution of Working Group II to the Fourth Assessment Report of the Intergovernmental Panel on Climate Change; Parry, M.L., Canziani, O.F., Palutikof, J.P., van der Linden, P.J., and Hanson, C.E. Eds.; Cambridge University Press: Cambridge, United Kingdom, 2007.
7. United States Environmental Protection Agency, 2008. Introduction to the Clean Water Act. http://www.epa.gov/owow/watershed/wacademy/acad2000/cwa/ (accessed September 2011).
8. United States Environmental Protection Agency. National Priorities List (NPL), http://www.epa.gov/superfund/sites/npl/ (accessed September 2011).
9. Sudicky, E.A.; Illman, W.A. Lessons learned from a suite of CFB Borden experiments. Rev. paper Ground Water **2011**, *49* (5), 630–648.
10. Koschitzky, H.P.; Braun, J. *VEGAS: Versuchseinrich- tung zur Grund wasser-und Altlastensanierung.* Wasser und Abfall, Boden, Altlasten, Umweltschutz, **2008**, *10,* 54–55.
11. Bear, J., *Dynamics of Fluids in Porous Media;* Dover Publications: 1988.
12. Domenico, P.A.; Schwartz, F.W. *Physical and Chemical Hydrogeology;* 2nd Ed., Wiley: 1997.
13. Schwarzenbach, R.P.; Gschwend, P.M.; Imboden, D.M. *Environmental Organic Chemistry*; 2nd Ed., Wiley-Inter-sciences: 2002.
14. Fetter, C.W. *Contaminant Hydrogeology;* 2nd Ed., Waveland Press: 2008.
15. Bear, J.; Cheng, A.H.-D. *Modeling Groundwater Flow and Contaminant Transport: Theory and Applications of Transport in Porous Media,* Springer: 2010.
16. Yaws, C.L. *Chemical Properties Handbook: Physical, Thermodynamics, Environmental Transport, Safety & Health Related Properties for Organic & Inorganic Chemicals;* McGraw-Hill Professional: 1998.
17. Mackay, D.; Shiu, W.Y.; Ma, K.C. *Illustrated Handbook of Physical-Chemical Properties and Environmental Fate of Organic Chemicals;* 2nd Ed.; Vol. 1–4, CRC Press; 2006.
18. Montgomery, J.H. *Groundwater Chemicals Desk Reference;* 4th Ed., CRC Press: 2007.
19. Verschueren, K. *Handbook of Environmental Data on Organic Chemicals;* 5th Ed., Vol. 1–4, Wiley: 2008.
20. United States Environmental Protection Agency. Environmental Databases. http://www.epa.gov/opp00001/science/efed_databasesdescription.htm (accessed September 2011).

21. Ward, A.D.; Trimble, S.W. *Environmental Hydrology;* 2nd Ed., Lewis Publishers: 2003.
22. Freeze, A.R.; Cherry, J.A. *Groundwater;* Prentice Hall: 1979.
23. Brennen, C.E. *Fundamentals of Multiphase Flow;* Cambridge University Press: 2005.
24. MDH Engineering solutions corporation. Evaluation of computer models for predicting the fate and transport of hydrocarbons in soil and groundwater. The Water Research Users Group, Alberta Environment, Canada, 2005.
25. United Nation Development Program (UNDP). Human Development Report: Water Scarcity. Chapter 4. http://hdr.undp.org/en/media/HDR_2006_Chapter_4.pdf (accessed 09/12/2011).
26. United Nations Children's Fund (UNICEF) & World Health Organization (WHO). Meeting the MDG Drinking Waterand Sanitation Target: A Mid-Term Assessment of Progress. UNICEF/ WHO, Geneva, Switzerland, 2004.
27. Barlow, P.M.; Wild, E.C. *Bibliography on the Occurrence and Intrusion of Saltwater in Aquifers along the Atlantic Coast of the United States;* U.S. Geological Survey: 2002.
28. United States Department of Agriculture (USDA). Soil quality resources concerns: Salinization. United States Natural Resources Conservation Service - Soil quality information sheet. http:// soils.usda.gov/sqi/publications/files/salinzation.pdf (accessed September 2011).
29. United Nations International Strategy for Disaster Reduction (UNISDR). Drought, desertification and water scarcity. http://www.unisdr.org/we/inform/publications/3815 (accessed October 2013).
30. World Water Council. World water day 2011. http://www.wodd-watercouncil.org/index.php?id=32 (accessed September 2011).
31. United Nation Development Program (UNDP). Water and Ocean Governance. http://www.undp. org/water (accessed September 2011).
32. United Nation. Coping with water scarcity. ftp://ftp.fao. org/agl/aglw/docs/waterscarcity.pdf (accessed September 2011).
33. United Nations Educational, Scientific and Cultural Organization (UNESCO). From potential conflict to co-operation potential: Water for peace. Prevention and resolution of water related conflicts. http://www.unesco. org/water/wwap/pccp/pubs/brochures.shtml (accessed September 2011).

13

Hydrology: Urban

Introduction 87
Alteration of Hydrologic Processes 87
Hydrograph Effects 89
Mitigation 90
Conclusions 91
References 91

C. Rhett Jackson
University of Georgia

Introduction

Since the passage of the 1972 Clean Water Act, the water quality of most United States rivers and streams has improved markedly, but urban streams have been an exception. Given the importance of flow to channel geomorphology and aquatic biota, the permanent and severe hydrologic changes wrought by urbanization make it difficult to restore good water quality conditions to previously urbanized streams.

Alteration of Hydrologic Processes

Urbanization—broadly defined as all aspects of converting natural or agricultural land to residential, commercial, and institutional land—radically alters the hydrology of local streams, principally by reducing infiltration rates over most of the landscape. The infiltration rate is the depth of water per unit of time that can enter the soil from the surface. When infiltration rates are lowered, rainfall that formerly infiltrated into the ground and reached streams by relatively slow subsurface pathways instead runs over the ground surface, picks up pollutants spilled or applied to the ground or paved surfaces, typically enters engineered conveyance systems, and rapidly reaches streams during storms.

The influence of urbanization on local hydrologic systems can only be understood in comparison to the hydrology of natural watersheds. In a forest or grassland, soil structure and hydrologic behavior are strongly influenced by biological activity, the presence of leaf litter, and carbon accumulation from plant matter. Root growth, root decay, cracking due to freeze/thaw and wetting/drying processes, animal burrowing, the windthrow of weak trees, subsurface erosion, and other natural processes all increase soil porosity (the ratio of void space to total soil volume), the number and size of macropores, and the conductivity of the soil to water. Leaf litter on the soil surface dissipates raindrop energy and allows rainfall to drip into the soil. Relatively high organic contents of natural soils increase the stability of soil aggregates, and stable aggregates prevent soil crusting during rainfall, reduce detachment of small soil particles, and maintain high surface infiltration rates. For all of these reasons, almost all rainfall infiltrates into the ground surface except during extremely intense rainfall events. When rainfall rates exceed infiltration rates, excess precipitation runs downhill over the ground surface. This type of surface runoff is called Horton overland flow and is rare on natural upland soils.

During the process of urbanization, lands are cleared of native vegetation; slopes and soils are graded to improve the topography for building; impervious surfaces (rooftops, parking lots, roads) are constructed; and typically curbs, gutters, and storm drains are created to hasten the flow of surface water off the landscape (Figure 13.1). The process of clearing and grading compacts the soil, reducing porosity and macropore density and thus decreasing infiltration rates. Urban soils typically lack litter layers and have low carbon contents, so they are susceptible to surface sealing during rainfall. Horton overland flow occurs at lower rainfall rates on these urban soils. Impervious surfaces have infiltration rates approaching zero and thus convert nearly 100% of incident rainfall into overland flow.

Forests and grasslands have high leaf area indices—the ratio of vegetative surface area to the underlying ground area—and therefore a significant portion of rainfall is captured by the canopy and evaporated before ever reaching the ground. Urbanization typically reduces leaf area indices, increasing the amount of rainfall reaching the ground. Additionally, vegetative canopies provide temporary storage of rainfall, reducing precipitation intensities at the ground surface. Between storms, some water held by soils in the natural landscape is used by plants for transpiration and the rest moves by subsurface pathways to the water table and to streams, lakes, or wetlands at the base of the slope. Depending on hillslope position and the structure of the soils, travel times through these subsurface flow paths range from days to years, as opposed to minutes and hours for Horton overland flow. Because of interception and transpiration, a significant portion of rainfall on natural watersheds evaporates before reaching surface waters. The fraction of rainfall that becomes runoff is called the yield of a watershed, and yields increase in urbanized areas because of increased Horton overland flow and decreased transpiration. Furthermore, because of reduced evaporative cooling, reductions in soil water storage and canopy cover also increase urban heat island effects.

Where water tables occur near the ground surface, typically at the base of slopes and around the margins of streams, wetlands, and lakes, water tables may rise to the surface during rainfall. When the soil saturates from below, additional rainfall runs over the ground surface as saturation excess flow.[1] The areas that produce saturation excess flow are called variable source areas because they expand during wet periods and contract during dry periods.[2] These areas, along with direct precipitation on channels, are responsible for stormflow response in natural basins. Urbanization has relatively little effect on variable source areas but adds impervious surfaces and compacted soils to these rapidly responding areas. Thus, in relatively small basins, stormflow volumes and peak flow rates increase almost linearly with increasing impervious surface coverage and also increase strongly with increasing coverage of compacted soil.

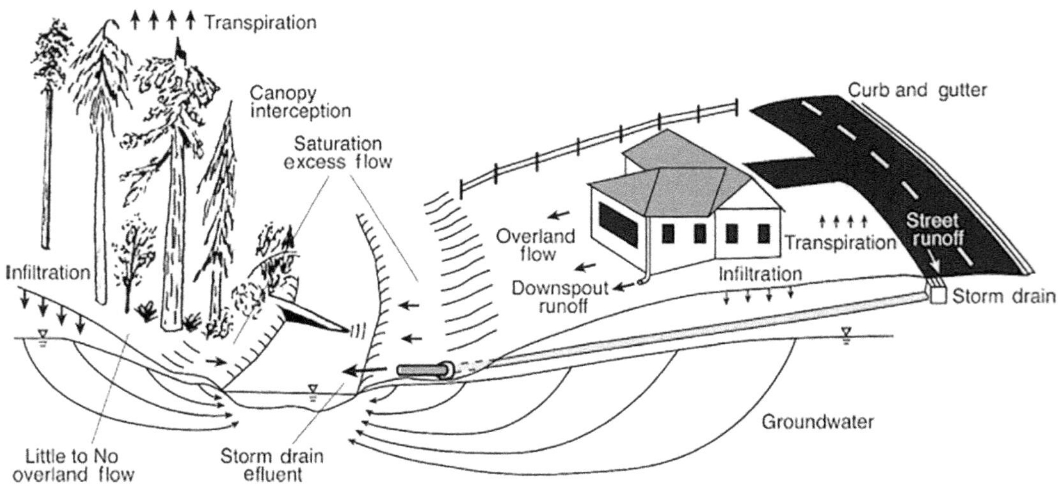

FIGURE 13.1 Schematic illustrating multiple hydrologic changes wrought by urbanization including loss of forest cover, grading of soils, construction of impervious surfaces and stormwater conveyance facilities, occurrence of frequent overland flow, and reduced interception, infiltration, and evaporation.

Hydrograph Effects

Within relatively small basins, the hydrologic alteration caused by urbanization can increase peak flow rates more than five times over their pre-development levels—with greater increases for frequent storm events and smaller increases for low frequency events[3]—and therefore increase flooding costs and problems. Stormflow volumes and cumulative durations of high flows also increase.[4] The additional erosive power leads to cycles of channel erosion and degradation[5,6] and to channel widening and simplification.[4,7] Channel erosion associated with flow increases in urban basins can contribute two-thirds of the sediment yield of an urban watershed.[8] These hydrologic and geomorphic alterations, along with changes in water chemistry associated with urbanization, are manifested in reduced diversity of aquatic organisms.[9,10] In addition to its direct hydrologic effects, increased percentages of impervious surfaces have been associated with increased nutrient loads and bacteria concentrations, higher summer stream temperatures, and reduced plant diversity in wetlands receiving urban runoff.[10]

Urbanization alters streamflow response to larger storms much less than the response to smaller storms because, in undeveloped watersheds, variable source areas expand and produce surface runoff over large areas during very large storms. The additive effect of impervious surfaces becomes less important during such storms (Figure 13.2). Furthermore, the effects of peak flow alteration diminish going downstream as hydrograph phasing and channel routing progressively dampen flood flows as basin area increases. A city located on a large river has little effect on the large river flows; rather, it affects the tributaries draining the city into the river and also affects the river's chemical and biological quality.

Stormflows in urban streams also increase in frequency. In natural basins, when soils are dry, rainfall events are absorbed by the soils and streams respond very little. In most of the humid United States, natural streams experience very few large flows in late summer and early fall when soils are dry. Impervious surfaces provide no such buffering and produce stormflows during all rainfall events regardless of soil moisture conditions. Cumulative hydrologic alterations of an urban basin can also dramatically affect wetland hydropatterns (the time series of water levels), causing shifts in the composition of vegetative and amphibian communities. Urbanization reduces recharge rates over much of the landscape,

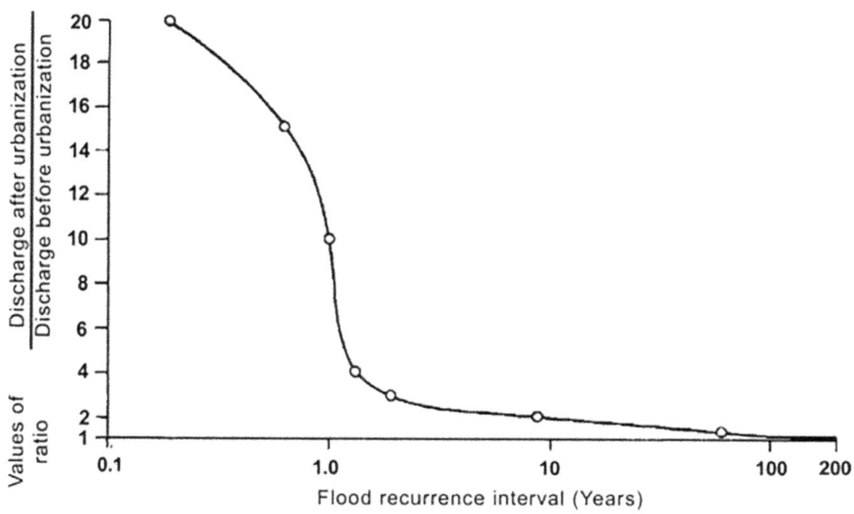

FIGURE 13.2 Relationship between the relative increase in flood discharge due to urbanization (for watersheds with 20% pavement) and the flood recurrence interval. Urbanization has a very large effect on frequent floods and progressively less effect on large, low-frequency floods.
Source: Graph taken from Hollis.[3]

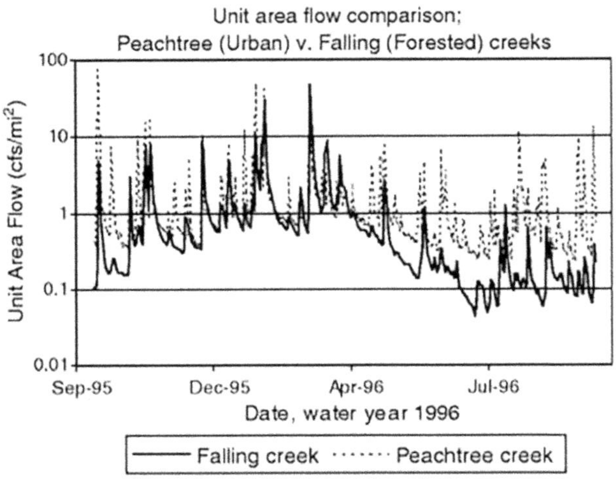

FIGURE 13.3 Comparison of unit area runoff (cubic feet per second per square mile) between a highly urbanized stream (Peachtree Creek in Atlanta, Georgia) and a nearby forested stream (Falling Creek near Juliette, Georgia). Data are United States Geological Survey daily flow measurements shown for water year 1996 (October 1, 1995 through September 30, 1996).

thus reducing local sources of stream baseflow; but urban areas often import water from other basins, and groundwater and baseflows in urban streams may be supplemented by leakage from the water and sewage distribution system, lawn watering, and car washing. Thus, different studies have found both increases and decreases in base- flows as a result of urbanization.

The hydrograph of Peachtree Creek, Atlanta, when compared to the hydrograph of a nearby forested stream, clearly illustrates the multiple hydrograph effects of urbanization (Figure 13.3). Peachtree Creek begins in the highly developed eastern suburbs of Atlanta and flows through Atlanta's urban core, draining 86.8 mi^2. Compared to a nearby forested stream on a unit area basis, Peachtree Creek has larger peak flows, greater stormflow volumes, more frequent high flow events, and greater total streamflow.

Mitigation

Mitigating the potential hydrologic effects of urbanization involves maintaining infiltration rates and soil storage to the degree possible and creating artificial storage for runoff from impervious surfaces and compacted soils. The complete suite of stormwater mitigation strategies includes clustering development to reduce impervious surfaces and maintain forest cover, using pervious pavements where feasible,[11] employing low-impact development principles at the site scale (e.g., downspout infiltration, rain gardens, infiltration/biofiltration, and swales), incorporating green roof technology into commercial developments, constructing infiltration or detention basins, and maintaining riparian buffers around streams.[12,13] Detention facilities collect water from developed areas and release it at slower rates than it enters, whereas infiltration facilities are sited over soils that allow infiltration of a substantial portion of the collected waters. Riparian buffers provide little hydrologic benefit, but they increase the physical and biological resiliency of channels to hydrologic alteration. Structural solutions alone have been inadequate to mitigate urbanizations hydrologic alterations,[12] and protecting urban streams requires full integration of structural, non-structural, and site design strategies.[12,14]

In conventional structural stormwater mitigation, simple hydrologic models of runoff from single rain events are used to estimate stormflow peaks and volumes for pre- and post-development conditions. Then, a hydraulic routing model is used to design detention or infiltration ponds to reduce post-development peak flow rates to pre-development levels for various storms of interest, typically the 2-year, 10-year, and 25-year 24-hour rainstorms. There are two main problems with this design protocol: (1) matching only the peak flow rates still allows the duration of high flows to increase, thus the so-called peak matching design standards do not protect receiving streams from increased erosion[4] and (2) single rainfall event models are inherently incapable of matching peak flow rates because they do not account for problems associated with successive storms.[15] To protect receiving channels, detention and infiltration ponds should be sized to match durations of erosive flows. Development and acceptance of continuous rainfall/runoff models for stormwater design will result in better stormwater mitigation design and analysis.[15]

Because it takes years of post-development monitoring to evaluate hydrologic and water quality responses of streams to urbanization and because developments incorporating all modern concepts of stormwater mitigation have only recently been built, the cumulative effectiveness of modern stormwater management techniques for protecting the hydrology and water quality of urbanizing watersheds is unknown.

Conclusions

Urbanization reduces the volume of water storage provided by soils and vegetation, increases the fraction of rainfall that becomes surface runoff during storms, hastens the movement of runoff to streams, reduces evaporation, and increases watershed yield. Protecting the hydrology and water quality of streams in newly urbanizing basins requires comprehensive implementation of structural and non-structural mitigation, including but not limited to the retention of natural vegetation, the minimization of impervious surfaces, low-impact site designs, rain gardens, detention and infiltration facilities, and water quality treatments. However, the maximum possible extent of urbanization that can be accommodated while protecting the hydrology, geomorphology, and biota of receiving waters is uncertain.

References

1. Dunne, T.; Moore, T.R.; Taylor, C.H. Recognition and prediction of runoff producing zones in humid regions. Hydrol. Sci. Bull. **1975**, 20(3), 305–327.
2. Hewlett, J.D.; Hibbert, A.R. Factors affecting the response of small watersheds to precipitation in humid areas. In *Forest Hydrology*; Sopper, W.E., Lull, H.W., Eds.; Pergamon Press: New York, 1967; 275–290.
3. Hollis, G.E. The effects of urbanization on floods of different recurrence intervals. Water Resour. Res. **1975**, *11*, 431–435.
4. Booth, D.B.; Jackson, C.R. Urbanization of aquatic systems—degradation thresholds, stormwater detention, and the limits of mitigation. J. Am. Water Resour. Assoc. **1997**, *33*(5), 1077–1090.
5. Wolman, M.G. A cycle of sedimentation and erosion in urban river channels. Geogr. Ann. **1967**, *49A*, 385–395.
6. Wolman, M.G.; Schick, A. Effects of construction on fluvial sediment, urban and suburban areas of Maryland. Water Resour. Res. **1967**, 3, 451–464.
7. Hammer, T.R. Stream and channel enlargement due to urbanization. Water Resour. Res. **1972**, *8*, 1530–1540.
8. Trimble, S.W. Contribution of stream channel erosion to sediment yield from an urbanizing watershed. Science **1997**, *278*, 1442–1444.
9. Klein, R. Urbanization and stream quality impairment. Water Resour. Bull. **1979**, *15*, 948–963.
10. Schueler, T. The importance of imperviousness. Watershed Prot. Tech. **1995**, *1*(3), 100–111.
11. Ferguson, Bruce, K. *Porous Pavements*; CRC Press: Boca Raton, FL, 2005.
12. Booth, D.B.; Hartley, D.; Jackson, C.R. Forest cover, impervious surface area, and mitigation of stormwater impacts in King County, WA. J. Am. Water Resour. Assoc. **2002**, *38*(3), 835–846.

13. McCuen, R.H. Smart growth: hydrologic perspective. ASCE J. Prof. Issues Eng. Edu. Pract. **2003,** *129,* 151–154.
14. Walsh, C.J.; Roy, A.H.; Feminella, J.W.; Cottingham, P.D.; Groffman, P.M.; Morgan, R.P., II. The urban stream syndrome: current knowledge and the search for a cure. J. N. Am. Benthol. Soc. **2005**, *24*, 706–723.
15. Jackson, C.R.; Burges, S.J.; Liang, X.; et al. Development and application of simplified continuous hydrologic modeling for drainage design and analysis. In *Land Use and Watersheds: Human Influence on Hydrology and Geomorphology in Urban and Forest Areas*; Wigmosta, M.S., Burges, S.J. et al., Eds.; Water Science and Application; American Geophysical Union: Washington, DC, 2001; Vol. 2, 39–58.

14

Impervious Surface Area: Effects

Introduction 93
Land Cover Change and Impervious Cover 93
Environmental Impacts • Measuring and Mapping IC • Extending the Utility of IC
Conclusion 97
Acknowledgments 98
References 98
Bibliography 99

Carl L. Zimmerman
Tufts University

Daniel L. Civco
University of Connecticut

Introduction

In this section, the concept of impervious cover (IC) as it relates to human activities will be introduced. The reader will learn how IC impacts and alters the environment, and an introduction on the measurement of IC at different scales will be provided. Finally, the entry will conclude with a discussion on a few methods that may ameliorate the impacts of IC on the environment.

Land Cover Change and Impervious Cover

The influence and impact of humans on the world has become an important area of research for scientists and planners. Some scientists are suggesting that we have entered a new geological epoch called the Anthropocene where human activities have become the dominant controller of the planet's geologic, climate, and biological systems.[1] An important component of our influence on the world is how we use and manipulate the landscape. Humans have managed and manipulated the environments in which we live for thousands of years to improve access to food and improve living conditions; however, the scale of anthropogenic land-use change has accelerated and altered in the past decade both in North America and in many developing regions undergoing rapid economic and population growth, such as India and China.

One of the most obvious changes in our environment is the rapid growth of urban areas. Accompanying the land development process is the engineered infrastructure such as roads, highways, and parking, and buildings with rooftops. Typically, these anthropogenic materials are made of materials such as concrete, tar, or asphalt and interfere with the natural infiltration process because of their lack of permeability. Collectively, these areas are called impervious cover (IC).

Surprisingly, the relationship between population growth and IC coverage is not proportional. The expansion of IC in metropolitan areas of the United States such as Washington D.C., Houston, Texas, or Las Vegas, Nevada, has rapidly outstripped population growth.[2] In rapidly developing countries such as China and India, IC has expanded explosively in cities because of rapid industrialization and immigration. Even outside urban centers, IC has become more widespread and dispersed in areas that previously were dominated by other land cover types such as agriculture, forest, or grassland.[3]

As early as the 1930s,[4] aerial photography and mapping of land use and environmental variables was occurring at county or state scales. Later, multiple methods for understanding and mapping changes to the Earth's surface from anthropogenic activities such as urban growth, sprawl, and economic development emerged after the launch of Earth resource satellites such as Landsat in the 1970s and early 1980s. These images changed the way both citizens and scientists perceived the world. The medium-resolution, multispectral sensors on remote sensing satellites allowed for continental-scale classification of land cover such as the biological (e.g., forest) and physical (e.g., water) materials on the surface of the Earth, which can be captured within the field of view of a sensor.[5] As early as 1961, a definition of land cover included human influences on the planet's surface. Land cover was defined as "the vegetational and *artificial construction* covering the land surface."[6] These areas of "artificial construction" are made of man-made materials and are collectively known as IC. IC impacts the infiltration and runoff components of the water cycle.

Environmental Impacts

The process of land development places buildings and roads on areas that were typically utilized for agricultural land uses or covered by pervious land cover types such as grassland or forest. The alteration of a predevelopment area without IC to an area that has been developed adds more civil engineering infrastructure, such as stormwater drainage systems, and greater amounts of IC in the form or buildings and roads, one parcel or project at a time.[7] While a plethora of environmental impacts occurs in these areas, the sources of the environmental impacts are generally related to the alteration of the local hydrologic cycle. IC disrupts the natural infiltration of precipitation process in most landscape types and greatly increases the rate and amount of runoff in watersheds, which causes a number of cascading impacts (Figure 14.1). The conversion of pervious natural land cover to IC in urbanizing areas generates the greatest runoff depth of any land cover type and introduces affiliated hydraulically efficient

FIGURE 14.1 Impervious cover without the infiltration of water.

stormwater drainage systems found in roads and other transportation systems that rapidly remove water and interrupt infiltration. The flow pathway of water is simplified and shortened, and water is routed quickly away from homes and expensive land development projects.[8]

The resulting impacts of IC and development on the environment begin to occur at very low percentages of IC, typically in the 5% to 15% range of IC within a watershed.[9] "Vertical" hydrologic processes like infiltration and groundwater storage are reduced, which decrease year-round base flow in local streams and rivers. Water that previously infiltrated, stays at the surface that can, in some cases, exponentially increase runoff depth and volume with very high flow rates into receiving waters.[10] The hydrologic changes are related to structural changes in the local stream channels. They become rapidly incised with disproportionate deepening and increases in the cross-sectional area,[11] and the local riparian corridors and stream channels of the local receiving water become unstable because of the alteration of flow rate and volume. The channels erode and release a great deal of sediment and nutrients into the waters. Changes in the water quality impact stream ecology.

Fish populations of streams are negatively impacted and altered by the hydrologic, geomorphological, and water quality changes associated with IC especially when they are directly connected to receiving waters. Storm sewers and IC areas adjacent to streams play a critical role in delivering a wide variety of unattenuated pollutants into streams and rivers that degrade the quality of local habitat.[12] In the state of Connecticut, one study[13] found a wide number of nutrients and processes related to runoff from IC that degraded the quality of macroinvertebrate habitat when the IC percentage within watersheds reached 12%. Other possible impacts include those on coastal waters from fast-growing urban areas such as Tampa Bay, Florida, that could impact coastal water quality[14] and extreme heat events that have been related to the IC associated with sprawling urbanizing areas in the United States.[15] Finally, IC plays a role in fragmenting forested and other natural areas. One study found that approximately 62% of forest patches in the lower 48 states are located within 164 yards (150 m) of the nearest edge, and this edge is often related to some sort of IC-containing land-use area such as roads or developments.[16]

Measuring and Mapping IC

Professionals in land planning and development are using the percentage of IC within watersheds as a metric for evaluating the environmental impact of urbanization upon local receiving waters.[17] This metric is typically called percentage of total impervious area (%IC). Rather than representing one specific environmental impact or process, the percentage of IC within a watershed represents a whole collection of covarying pollutants associated with IC. Methods of determining %IC are derived from planning and geospatial data and are typically determined at either coarse planning scales or much finer site scales. Methods of determining %IC are implemented in a Geographic Information System (GIS) software package. These methods include:[18]

1. Deriving percent impervious surface coefficients from land cover (Figure 14.2)
2. Modeling IC at watershed or census tract levels
3. Use of "subpixel" modeling to create spatially explicit per-pixel predictions of IC
4. Spatially explicit digitizing of IC into vector layers (Figure 14.3)

The two primary data types used for mapping of IC are vector and raster data, and these data types are typically processed in a GIS software package. The vector data type is more commonly utilized in urban planning and is usually derived from engineering or surveying sources. It uses vertices located in a geographic coordinate system to define the perimeter of polygons of uniform IC surfaces. Typically, this data type has been utilized to define IC spatially at a site or local scale. This information can be acquired from construction documents, imported from computer-assisted design (CAD) maps, or created using different forms of digitizing. Accuracy is determined by the density of digitized points and by the quality of the source maps.

FIGURE 14.2 (See color insert.) Impervious surfaces derived through on-screen digitizing using high-resolution aerial imagery.

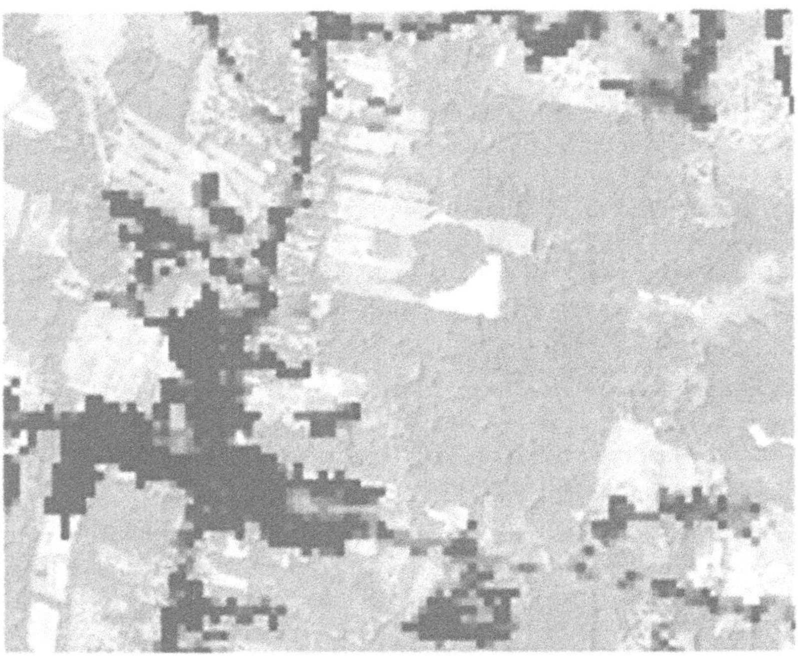

FIGURE 14.3 (See color insert.) Percent impervious cover derived from processing of 30 m resolution Landsat satellite remote sensing data.

Source: Adapted from the United States Geological Survey National Land Cover Dataset, USGS NLCD.

The other data type that is typically used to map and measure IC is the raster data type which is typically derived from remote sensing imagery. A raster data type is organized similar to the pixels within a digital photograph and each pixel, also called a cell, is the same size. In the case of a Landsat image, the cell is 30 m×30 m and incorporates reflectance information of the land cover acquired from the satellite for a specific location. The pixel does not explicitly define the IC. The analyst utilizes remote sensing software and interprets IC land use for specific locations based on the unique spectral characteristics of pavement, rooftops, roads, and other anthropogenic land cover. There are three advantages of the raster data sets over other types of environmental and geospatial data. The first advantage is that every location has data; so it is easy to get an overview of a larger area such as a watershed, and no gaps exist in the data. The second advantage is that satellite imagery data is in the raster format; so multidate comparisons of land cover change are simplified. The third advantage is that raster data is particularly well suited for integration with other environmental data and for environmental modeling of dynamic systems.

Extending the Utility of IC

Significant research efforts by geographers, ecologists, and hydrologists have gone toward extending the utility of land cover in modeling biological, hydrological, and ecological aspects of the environment. Typically run in a GIS, the goal is often to integrate different types and sources of collateral information with land cover data. As the resources for these professions are limited, the ability to scale up environmental models using land cover gives them potential to look at larger areas and problems that could not be investigated by field work alone. Many types of scientific modeling, planning, and environmental research have utilized IC in hydrologic modeling,[9] research on climate change, investigations into urban sprawl,[3] fragmentation of the habitat,[16] and land-use change and assessment research.[2] Landscape ecologists[19] and urban planning professionals have utilized IC to create spatial metrics that evaluate relationships between different types of land cover.[3]

One example of the extension of IC land cover data type is the application to some research related to water quality and stream ecology. By itself, a raster grid cell identified as being IC lacks the knowledge of its context in the urban or residential landscape, does not identify critical stormwater infrastructure, and does not identify flow paths to local rivers, streams, and ponds. By adding information about the connection or linkages and the location of stormwater infrastructure or flow patterns, a new type of IC assessment model can be created called directly connected impervious cover.[17] These hydrologically linked IC areas in this new type of model identify those areas that run off directly into local water resources, and hence are the most likely to carry nutrients directly into a water body without attenuation. Those impervious land cover areas that drain toward pervious areas and allow runoff to infiltrate are called unconnected impervious areas. Directly connected impervious areas have been found to be a better predictor of water quality, impacts on stream ecology, and fish assemblages than IC alone. Other researchers are extending the utility of land cover data by creating integrated landplanning models that evaluate environmental and hydrologic impacts within urban watersheds under different planning scenarios.[20]

Conclusion

Urbanization and growth of cities will continue to alter the environment both within and beyond urban centers as rapid economic growth occurs in many previously undeveloped nations throughout the world. Measuring and mapping of IC are improving and are becoming scalable, thanks to the availability of satellite data, the use of GIS software, and rapid improvements in computing hardware. Areas of IC growth can be measured, and potential environmental changes can be analyzed. New integrated landscape models using land cover such as IC are being explored to extend the utility of this data type for understanding the environment.

Fortunately, many new "greener" development methods are becoming practical and low-impact development (LID) methods are beginning to be widely used. These methods utilize pervious materials

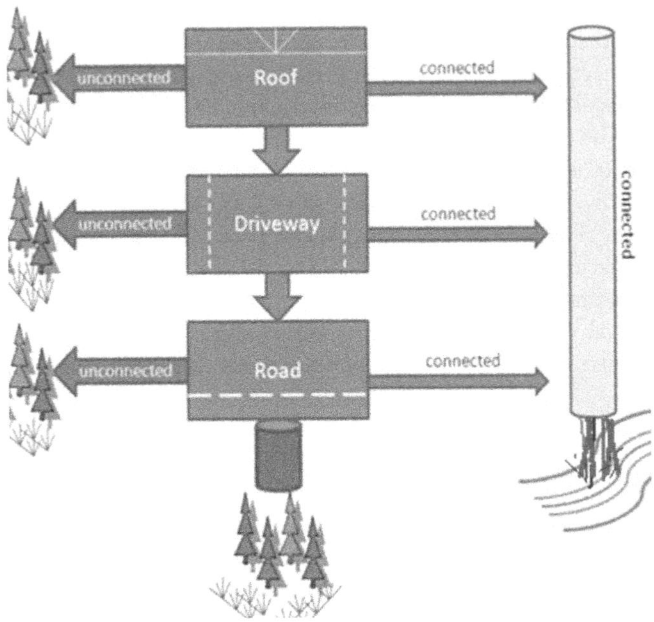

FIGURE 14.4 Model of IC connection in residential area.

such as pavers, reroute stormwater runoff from IC to locally distributed infiltration areas such as rain gardens, and employ innovative approaches of routing stormwater runoff (Figure 14.4). All these methods are intended to lessen or ameliorate the impacts of IC and its impact on the local hydrologic cycle.

Acknowledgments

The authors would like to thank the University of Connecticut NEMO program (Nonpoint Education for Municipal Officials) for the utilization of their figures.

References

1. Kolbert, E. Enter the anthropocene: The age of man. Nat. Geogr. Mag. **2011**, 60–85.
2. Jantz, P.; Goetz, S.; Jantz, C. Urbanization and the loss of resource lands within the Chesapeake Bay watershed. Environ. Manag. **2005**, *36* (6), 808–825.
3. Angel, S.; Parent, J.; Civco, D.; Biel, A. The dimensions of global urban expansion: Estimates and projections for all countries, 2000-2050. Prog. Plan. **2011**, *75* (2), 53–108.
4. Rice, S. MAGIC Connecticut Historical Aerial Photography. Found at the University of Connecticut Libraries, Magic. http:// magic.lib.uconn.edu/historical_aerial_indexes.html (accessed December 2013).
5. Arnoff, S. *Remote Sensing for GIS Managers;* ESRI Press: Redlands, CA, 2005; 1–524.
6. Anderson, J.; Hardy, E.; Roach, J.; Witmer, R. *A land use and land cover classification system for use with remote sensor data,* Paper 964; United States Geological Survey: Washington DC, 2006; 1–41.
7. Endreny, T. Land use and land cover effects on runoff processes: urban and suburban development. In *The Encyclopedia of Hydrologic Science;* Anderson, M., McDonnel, J., Eds.; John Wiley and Sons, 2005; Vol. 126, 1–29.
8. Endreny, T.; Thomas, K. Improving estimates of simulated runoff quality and quantity using road enhance road-enhanced land cover data. J. Hydraulic Eng. **2009**, *14* (4), 346–351.

9. Schueler, T.; Fraley-McNeal, L.; Cappiella, K. Is impervious cover still important? Review of recent research. J. Hydrol. Eng. **2009**, *14* (4), 309–315.
10. Dietz, M.; Clausen, J. Stormwater runoff and export changes with development in a traditional and low impact subdivision. J. Environ. Manag. **2007**, *87* (4), 560–566.
11. Booth, D. Stream channel incision following drainage-basin urbanization. Water Resour. Bull. **1990**, *26* (3), 407–417.
12. Walsh, C.; Papas, P; Crowther, P.; Sim, T.; Yoo, J. Stormwater drainage pipes as a threat to a stream dwelling amphipod of conservation significance, *Austrogammarus australis,* in southeastern Australia. Biodivers. Conserv. **2004**, *13,* 781–793.
13. Bellucci, C. Stormwater and aquatic life: Making the connection between impervious cover and aquatic life impairments. In *Water Environment Federation,* TMDL 2007 Conference Proceedings, 2007; 1003–1018.
14. Xian, G.; Crane, M. Assessment of urban growth in the Tampa Bay watershed using remotely sensing data. Remote Sensing Environ. **2005**, *97,* 203–215.
15. Stone, B.; Hess, J.; Frumkin, H. Urban form and extreme heat events: Are sprawling cities more vulnerable to climate change than compact cities? Environ. Health Perspect. **2010**, *118* (10), 1425–1428.
16. Riitters, K.; Wickham, J.; O'Neill, R.; Jones, K.; Smith, E.; Coulston, J.; Wade, T.; Smith, J. Fragmentation of continental United States forests. Ecosystems **2002**, *5,* 815–822.
17. Brabec, E.; Schulte, S.; Richards, P. Impervious surfaces and water quality: A review of current literature and its implications for watershed planning. J. Plan. Lit. **2002**, *16* (4), 1–17.
18. Chabaeva, A.; Civco, D.; Hurd, J. An assessment of impervious surface estimation techniques. J. Hydrol. Eng. **2009**, *14* (4), 377–387.
19. Moilanen, A.; Hanski, I. On the use of connectivity measures in spatial ecology. Oikos **2001**, *95* (1), 147–151.
20. Hong, B.; Limburg, K.; Hall, M.; Mountrakis, G.; Groffman, P.; Hyde, K.; Luo, L.; Kelly, V.; Myers, S. An integrated monitoring/modeling framework for assessing human-nature interactions in urbanizing watersheds: Wap-pinger and Onondaga Creek watersheds, New York, USA. Environ. Model. Software **2011**, *32,* 1–15.

Bibliography

Skidmore, E. Environmental Modeling with GIS and Remote Sensing; Taylor and Francis: New York, 2002: 268 p.

Weng, Q. (Ed.). Remote sensing of impervious surfaces; CRC Press/Taylor and Francis Group: New York, 2008: 454 p.

15

Infiltration Systems and Nitrate Removal

Introduction 101
Microbial Removal Processes 102
Denitrification • Nitrification • Uptake
Onsite Wastewater Treatment Systems 103
Conventional Systems • Advanced Systems
Permeable Reactive Barriers 104
Conclusion 105
References 105

Jose A. Amador
University of Rhode Island

Introduction

Nitrate NO_3^- is an inorganic form of nitrogen commonly found in soil, ground, and surface water and precipitation. It is highly water soluble and anionic, properties that make it very mobile in the environment. Along with ammonia (NH_3), NO_3^- is the main form of N available to plants, and it is the preferred form of N for many terrestrial and aquatic plant species. Nitrate is also a source of N for microbial biomass and can serve as a terminal electron acceptor for denitrifying microorganisms that reduce NO_3^- to N_2O (nitrous oxide) and N_2 (nitrogen gas), resulting in a loss of N to the atmosphere.

Its function as an N source for organisms and as a terminal electron acceptor, and its mobility in the landscape, make NO_3^- an important component of the nitrogen cycle of terrestrial, freshwater, and marine ecosystems. These properties also translate into a high potential for pollution when NO_3^- is found in excess quantities. For example, nitrogen is often the *limiting nutrient* for primary production in coastal marine ecosystems. Because of its high mobility, NO_3^- can be the main form of inorganic N reaching these areas. Elevated NO_3^- levels can cause *eutrophication*, resulting in excessive algal growth, depletion of dissolved oxygen associated with decomposition of algal biomass, and fish kills. Excess NO_3^- levels in drinking water $\left(10 \text{ mg } NO_3^- N/L\right)$ from ground and surface sources are considered problematic because of negative effects on human and animal health.

The large-scale industrial synthesis of NH_3 from N_2 gas to manufacture fertilizers has resulted in a human-induced alteration of the global N cycle and the enrichment of natural and man-made ecosystems with nitrogen. Elevated NO_3^- levels in terrestrial and aquatic ecosystems can result from direct inputs, with major sources of NO_3^- including inorganic fertilizers and runoff. Some sources of N, such as wastewater, animal manure, and organic and inorganic fertilizers, may not contain much NO_3^-, but the nitrogen found in them can be readily converted to it. Thus, organic N present in amino acids, amino sugars, nucleic acids, and proteins can be released as ammonia (NH_3) by microbial processes known as *ammonification*. Whether ammonia is present in the original input or results from ammonification, it can be oxidized to NO_3 via a process known as *nitrification*, by specific groups of autotrophic aerobic bacteria and archaea.

The ubiquity, mobility, and potential negative effects on the health of ecosystems, animals, and humans often require the removal of NO_3^- from water. One approach to removing nitrate present in water involves the infiltration of contaminated water into soil or other porous media with adequate metabolic capacity to transform this contaminant to gaseous forms that are readily lost to the atmosphere. We present two common examples of this approach to NO_3^- removal using infiltrations systems: 1) onsite wastewater treatment systems (OWTS); and 2) permeable reactive barriers (PRB). Because these systems rely on microbial processes for the removal of NO_3^-, a brief introduction to these processes is given first.

Microbial Removal Processes

Denitrification

Most approaches for the removal of nitrate in infiltration systems take advantage of the fact that NO_3^- can be used as an electron acceptor by microorganisms through a process known as *denitrification*. It results in net loss of N from the system through the production of a gaseous form of nitrogen, namely, N_2 and N_2O. The general equation for the process is:

$$\mathrm{NO_3^-} \rightarrow \mathrm{No} \rightarrow N_2\mathrm{O}\uparrow \rightarrow N_2\uparrow$$

The ability to reduce NO_3^- to N_2O and N_2 is mainly associated with bacteria, although fungi have also been shown to carry out this process. Denitrifying bacteria are facultative anaerobes that use NO_3^- as an alternate electron acceptor when O_2 is not available. The majority are heterotrophs that use organic carbon compounds as electron donors.

Nitrification

As mentioned previously, oxidation of ammonia to nitrate, or nitrification, is carried out by both bacteria and archaea.

The general equation for the process is:

$$\mathrm{NH_3 + O_2 \rightarrow NO_2^- + O_2 \rightarrow NO_3^-}$$

Nitrifying organisms can produce nitric oxide (NO) and N_2O when the levels of O_2 are low, resulting in net loss of N from the system. The relative importance of nitrification to gaseous N losses in environmental settings is still under debate. The amount of N lost as N gases via nitrification at low O_2 levels can be substantial relative to losses via denitrification,[1] and thus nitrification may also be important in N removal in infiltration systems.

Uptake

Microorganisms can reduce NO_3^- through a series of biochemical reactions to NH_3, which is then used in the synthesis of amino acids and nucleotides (*assimilatory nitrate reduction*). The resulting microbial biomass results in the storage of N in organic forms (e.g., proteins, nucleic acids) in microbial cells. If the microbial biomass is not removed from the system, the N associated with dead biomass can be released as NH_3 by ammonification, with the resulting NH_3 subject to oxidation to NO_3^-. Thus, although microorganisms take up NO_3^-, this process does not result in net NO_3^- removal.

Onsite Wastewater Treatment Systems

Conventional Systems

Nearly a quarter of the households in the U.S.A. rely on OWTS for treatment and disposal of domestic wastewater.[2] In conventional OWTS, wastewater from the household flows by gravity to a septic tank, where the settling of solids and some removal of carbon and nutrients takes place (Figure 15.1). The high biochemical oxygen demand in the septic tank results in anaerobic conditions. The anaerobic septic tank effluent (STE) flows by gravity to a distribution box and subsequently to a series of underground trenches (known as the drainfield, leachfield, or soil absorption area) where STE infiltrates into the soil. The infiltrative surface of the drainfield is generally at least 150 cm below the ground surface, and the separation distance between the infiltrative surface and the seasonal high water table generally ranges from 30 to 120 cm.[3] Nitrogen removal in a conventional OWTS is relatively low (0–30%), because these systems are not designed for N removal.[3,4] Because the anaerobic conditions prevalent in the septic tank preclude the oxidation of ammonia to nitrate, inorganic N in STE is present only as NH_3. However, as STE flows and infiltrates into the soil, O_2 present in the atmosphere of the trench can support some level of nitrification. The resulting NO_3^- may be removed via denitrification in the leachfield; however, a considerable amount of N still leaches from these systems, mostly as NH_3. This ammonia can be readily oxidized to NO_3^- once it reaches the vadose zone, posing a threat to groundwater quality.

Advanced Systems

The conventional OWTS design can be modified to address the poor N-removal efficiency and minimize NO_3^- leaching. These modified systems, often referred to as advanced treatment systems (ATS), are designed to promote the nitrification of ammonia in the STE, followed by denitrification before water is disposed of in the drainfield (Figure 15.2). They are often installed in areas sensitive to nitrogen pollution, on lots that present site constraints, or in places where consistent system performance is desirable.[5] Although ATS designs vary, they involve the same approach: nitrification of ammonia, followed by denitrification. In order to promote nitrification, O_2 is introduced into the STE. This can be accomplished by passing STE through a filter filled with sand or other inert porous media, where O_2 can be entrained in the STE as it moves through the filter. Media filters provide habitat for microorganisms, in the form of particle surfaces, as well as aerobic conditions, since they are not saturated. Alternatively,

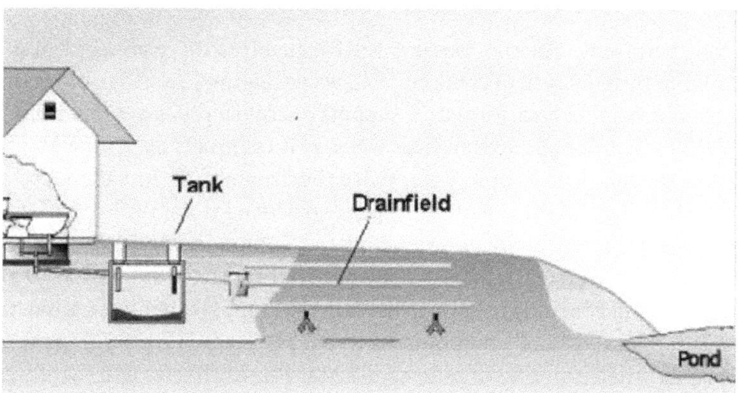

FIGURE 15.1 Schematic diagram of a conventional onsite wastewater treatment system, or OWTS. Wastewater drains by gravity from home to the septic tank, where the removal of solids and some nutrients takes place, and on to the drainfield, where the removal of pathogenic organisms is the main function. Very little nitrogen is removed in the drainfield of a conventional OWTS, and NO_3^- leaching to groundwater can be a significant problem.
Source: Adapted from Joubert et al.[5]

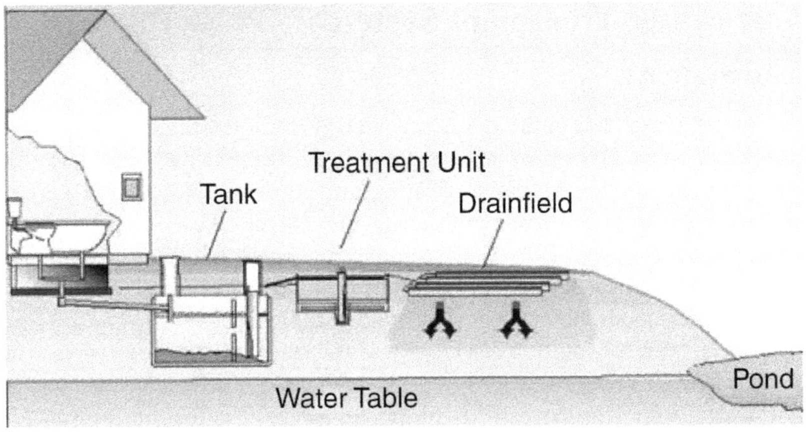

FIGURE 15.2 Schematic diagram of an advanced onsite wastewater treatment system, or ATS. In systems designed for N removal, the additional treatment unit is used to convert NH_3 in STE to NO_3^-, which can be subsequently denitrified to N_2O and N_2 in the septic tank. This results in net loss of N from the ATS to the atmosphere, preventing NO_3^- contamination of groundwater.
Source: Adapted from Joubert et al.[5]

aeration can be accomplished by bubbling air into a separate chamber through which STE flows. To promote denitrification of NO_3^- resulting from the aeration step, the nitrified STE is returned to the septic tank, where the anaerobic conditions and high levels of organic carbon support denitrification. A somewhat different ATS design relies on these two microbial processes taking place directly in the drainfield.[6] The first step involves accumulating STE in the septic tank or pump chamber, which is followed by aeration of the drainfield soil for a short period of time to promote nitrification. STE is then applied to the drainfield soil to displace oxygen and provide a carbon source, promoting denitrification.

Nitrogen removal efficiencies greater than 50% can be accomplished using ATS,[5] making them a highly effective technology for the removal of N from infiltrating wastewater, and thus the prevention of groundwater contamination with NO_3^-.

Permeable Reactive Barriers

The construction of permeable reactive barriers (PRBs) involves the placement of a layer of porous media perpendicular to the flow path of contaminated water (Figure 15.3). The porous medium can react directly with the contaminant to transform it, or support microbial processes that result in contaminant transformation. PRBs may be configured to treat water as it infiltrates the soil, with the barrier placed parallel to and below the soil surface, or as water moves horizontally in the vadose and saturated zones, in which case a trench is dug and filled with appropriate reactive media. PRBs have been employed successfully to remove a wide variety of organic and inorganic compounds, primarily from groundwater.[7]

PRBs have been used to effectively remove nitrate present in surface, drainage, and groundwater.[8–10] PRBs used for NO_3^- removal generally contain a medium that supports denitrification, the main mechanism by which NO_3^- is removed from contaminated water. The reactive medium in the trench is usually an inexpensive, easily available, porous organic material that is relatively rigid (to prevent collapse) and has a sufficiently high hydraulic conductivity to allow water to flow relatively unimpeded. Commonly used barrier materials include wood chips, peat, and straw. The porous, organic material serves as a source of organic carbon for denitrification, promotes the establishment of anaerobic conditions for denitrification, and serves as a physical support for the bacteria involved in denitrification. The trench may be placed near the soil surface, to allow the infiltration of NO_3^- containing runoff, or at an appropriate depth to allow for the interception of drainage water or groundwater flow.

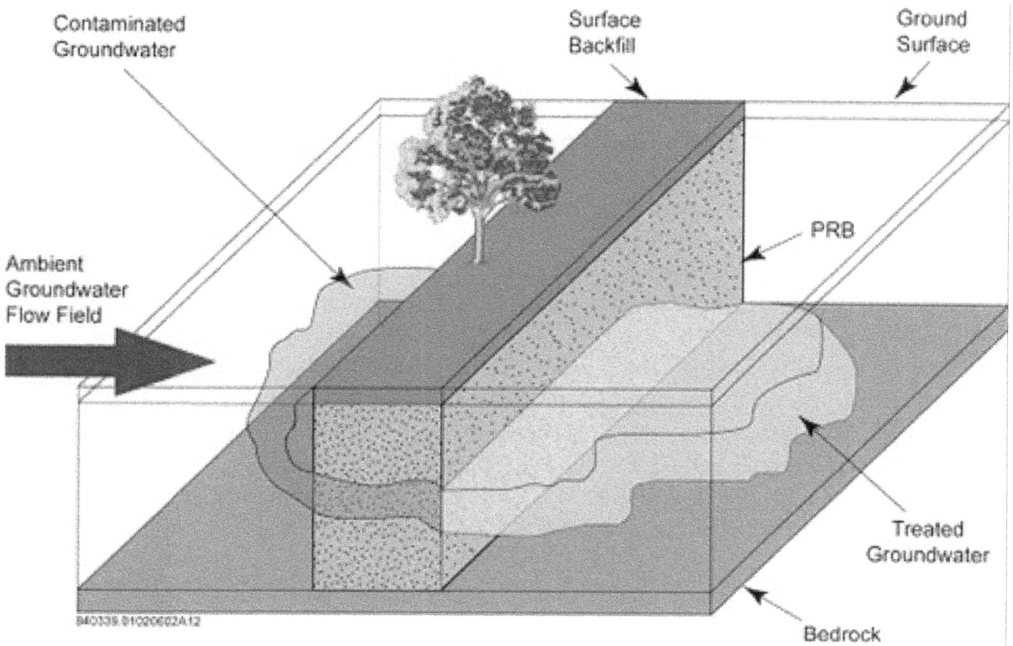

FIGURE 15.3 Schematic diagram of a permeable reactive barrier, or PRB. For the removal of NO_3^- from groundwater, the PRB usually consists of a porous organic material that serves as habitat and a source of carbon for denitrifying bacteria, which convert NO_3^- to N_2O and N_2.

Source: From Los Alamos National Laboratory (© 2011 Los Alamos National Security, LLC All rights reserved) as included in Environment at LANL: Monitoring, Compliance and Risk Reduction. http://www.lanl.gov/environment/all/overview.shtml, accessed May 2012.

Removal rates between 50% and nearly 100% of initial NO_3^- concentration[10] have been reported for these systems. PRBs represent a relatively inexpensive, passive approach to the remediation of NO_3^--contaminated water that can provide long-term treatment with minimal need for maintenance or replenishment of reactive media.

Conclusion

Soil infiltration systems rely on the microbial reduction of nitrate to gaseous forms of N—denitrification—for the removal of this contaminant from ground, surface, and wastewater. Advanced OWTS involve a two-step approach, promoting the nitrification of ammonia to nitrate in wastewater prior to removal via denitrification. PRB rely on denitrification to remediate ground and surface water contaminated with nitrate. The barrier material supports nitrate removal by providing habitat and organic carbon compounds for denitrifying microorganisms.

References

1. Kool, D.M.; Dolfing, J.; Wrage, N.; Van Groenigen, J.W. Nitrifier denitrification as a distinct and significant source of nitrous oxide from soil. Soil Biol. Biochem. **2010**, *43* (1), 174–178.
2. United States Census Bureau. American housing survey for the United States, Washington, DC, http://www.census.gov/hhes/ www/housing/ahs/ahs99/tab1a4.html (accessed May 2012).
3. U.S. Environmental Protection Agency. *Onsite Wastewater Treatment Systems Manual*; U.S. government: Washington, DC, 2002.

4. Kaplan, O.B. *Septic Systems Handbook*; Lewis Publishers: Chelsea, MI, 1987.
5. Joubert, L.; Loomis, G; Dow, D.; Gold, A.; Brennan, D.; Jobin, J. *Choosing a Wastewater Treatment System*; University of Rhode Island: Cooperative Extension, Kingston, RI, 2005.
6. Amador, J.A.; Loomis, G.W.; Kalen, D.; Patenaude, E.L.; Gorres, J.H.; Potts, D.A. Evaluation of Leachfield Aeration Technology for Improvement of Water Quality and Hydraulic Functions in Onside Wastewater Treatment Systems. Final Report to the Cooperative Institute for Coastal and Estuarine Environmental Technology (CICEET); NOAA/University of New Hampshire: Durham, NH, 2007.
7. Blowes, D.W.; Ptacek, C.J.; Brenner, S.G.; McRaea, C.W.T.; Bennetta, T.A.; Puls, R.W. Treatment of inorganic contaminants using permeable reactive barriers. J. Contam. Hydrol. **2000**, *45* (1–2), 123–137.
8. Ergas, S.J.; Sengupta, S.; Siegel, R.; Pandit, A.; Yao, Y.; Yuan, X. Performance of nitrogen removing bioretention systems for control of agricultural runoff. J. Environ. Eng. **2010**, *136* (10), 1105–1112.
9. van Driel, P.W.; Robertson, W.D.; Merkley, L.C. Denitrification of agricultural drainage using wood-based reactors. Trans. ASABE **2006**, *49* (2), 565–573.
10. Robertson, W.D.; Blowes, D.W.; Ptacek, C.J.; Cherry, J.A. Long-term performance of in situ reactive barriers for nitrate remediation. Ground Water **2000**, *38* (5), 689–695.

16

Low-Impact Development

Curtis H. Hinman
Washington State University

Derek B. Booth
University of Washington

Introduction 107
Typical Characteristics of Low-Impact Development 107
Limitations 108
Site Design and Management Objectives for Low-Impact Development 109
Conservation Measures • Minimization Techniques • Flow Attenuation • Distributed Integrated Management Practices • Pollution Prevention Measures
Conclusion 111
References 111

Introduction

Low-Impact Development (LID) is a strategy for stormwater management that uses on-site natural features integrated with engineered, small-scale hydrologic controls to manage runoff by maintaining or closely mimicking predevelopment watershed hydrologic functions.[1] Planning for LID is most effective at the scale of an entire subdivision or watershed; engineering and site design elements, however, are implemented at the scale of individual parcels, lots, or structures. In combination, these actions seek to store, infiltrate, evaporate, or otherwise slowly release storm-water runoff in a close approximation of the rates and processes of the predevelopment hydrologic regime.

Typical Characteristics of Low-Impact Development

Most applications of LID have several common components (Figure 16.1):

- Preserving elements of the natural hydrologic system that are already achieving effective storm-water management, recognized by assessment of a site's watercourses and soils; channels and wetlands, particularly with areas of overbank inundation; highly infiltrative soils with undisturbed vegetative cover; and intact mature forest canopy.
- Minimizing the generation of overland flow by limiting areas of vegetation clearing and soil compaction; incorporating elements of urban design such as narrowed streets, structures with small footprints (and greater height, as needed), use of permeable pavements as a substitution for asphalt/concrete sur- faces for vehicles or pedestrians;[3] and using soil amendments in disturbed areas to increase infiltration capacity.
- Storing runoff with slow or delayed release, such as in cisterns or distributed bioretention cells, across intentionally roughened landscaped areas or on vegetated roofing systems ("green roofs").[4,5] Runoff storage in LID differs from traditional storm-water management, notably the latter's use of detention ponds, primarily by its scale—small and distributed in LID, large and centralized in traditional approaches.

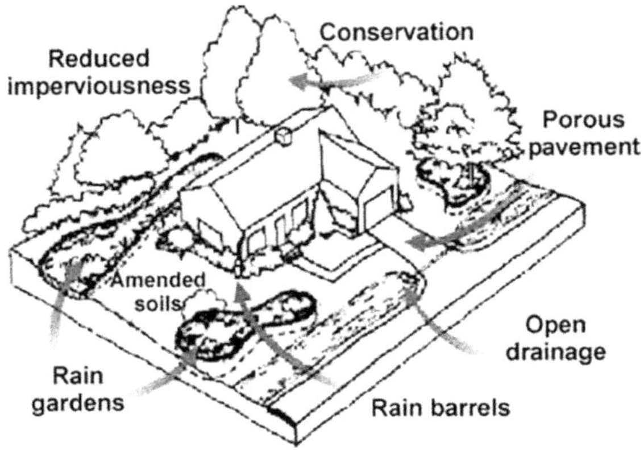

FIGURE 16.1 Schematic of typical LID applications at the scale of a residential building lot.
Source: Modified from Puget Sound Action Team.[2]

Native soils play a critical role in storage and conveyance of runoff, particularly in humid regions. In such regions, one to several meters of soil, generally high in organic material and relatively permeable, overlay less permeable substrates of largely unweathered geologic materials. While water is held in this soil layer, solar radiation and air movement provide energy to evaporate surface soil moisture and contribute to the overall evapotranspiration component of the water balance. Water not evaporated, transpired, or held interstitially moves slowly downslope or down gradient as shallow subsurface flow (also called interflow) over many hours, days, or weeks before discharging to streams or other surface-water bodies. In arid regions with relatively lower organic content soils and vegetation cover, precipitation events can produce rapid overland flow response naturally; however, the principles of LID remain—retain native soils, vegetation, topography, and hydrologic regime to preserve aquatic ecosystem structure and function.

The transition from a native landscape to a built environment increases the coverage of impervious surface from roads, parking areas, sidewalk, rooftops, and landscaping.

The upper soil layers that evaporate, store, or infiltrate storm water are compacted or covered altogether. As a result, the watershed area contributing direct overland flow to streams, lakes, and wetlands increases; hence, precipitation will reach receiving waters much more rapidly and in greater volumes.[6]

Typical storm-water management focuses on flood control and thus emphasizes the efficient collection and rapid conveyance of precipitation away from residential and commercial development, commonly to central control ponds. Several factors have led to this approach: Storm water has been perceived as a liability and applications have evolved from wastewater technology, hard conveyance structures and central control ponds are considered reliable and relatively simple to maintain, and the conveyance and collection approach is relatively simple to model for regulatory requirements. Although newer conveyance and pond strategies, if properly designed and maintained, can manage predevelopment peak flows under some conditions, a number of problems (e.g., increased runoff volume and extended flow durations, conversion of dispersed flow to point discharges) will continue to challenge this traditional strategy of stormwater management.[7]

Limitations

Although the goal of LID to mimic the predevelopment hydrologic regime is laudable, it cannot feasibly be achieved everywhere or at all times. The hydrologic system evolved from, and is dependent on, the characteristics of undisturbed watersheds—mature vegetative cover, uncompacted soils, ungullied hillslopes—whose function cannot be expected to remain unchanged where half or more of the landscape has been appropriated for human uses. Thus the objectives of any given LID must be strategically chosen, recognizing both the opportunities and the limitations of any given site. The limitations are not simply those of subdivision design and development density; they are also limitations of site topography, soil permeability and depth, and groundwater movement. They are likely to be most prominent during periods of extended rainfall, where the distributed on-site infiltration reservoirs common to most LID designs will experience their highest water levels and soil layers approach or reach full saturation. Under such conditions, the downstream impacts of uncontrolled runoff—be it flooding, channel erosion, or aquatic habitat disruption—could be as severe as with conventional stormwater control. Regulatory requirements, typical zoning and housing type, and costs of sophisticated control technology required on sites with higher development density and soils with low infiltration rates also create significant challenges for reducing or eliminating hydrologic impacts from development sites.

The potential failure of LID to control all flows from all storms, no matter how severe, is not an indictment of the approach as a whole. Indeed, LID probably represents the best opportunity to maintain a "natural," or at least minimally disrupted, flow regime in an urban watershed. Where downstream flooding or channel erosion are concerns, however, a comprehensive drainage design will need to consider more traditional storm-water management approaches (such as detention ponds or bypass pipelines) in addition to LID applications. Although such ponds would be smaller, they may still be required to achieve human or ecosystem protection at storm recurrence intervals that may be relatively long but still fall within regulatory thresholds.

Site Design and Management Objectives for Low-Impact Development

The goals for LID are achieved through the following site objectives:

- Maximize retention of mature vegetation cover and restore disturbed vegetation to intercept, evaporate, and transpire precipitation.
- Preserve permeable, native soil and enhance disturbed soils to store and infiltrate storm flows.
- Retain and incorporate topographic site features that slow, store, and infiltrate storm water.
- Retain and incorporate natural drainage features and patterns.
- Locate buildings and roads away from critical areas and soils that provide effective infiltration.
- Minimize total impervious area and eliminate effective impervious surfaces ("effective" impervious surfaces are the subset of the total imperviousness that has a direct hydraulic connection to the stream or wetland).
- Manage storm water as close to its origin as possible by utilizing small-scale, distributed hydrologic controls.
- Create a hydrologically rough landscape that slows storm flows and increases the time of concentration.
- Increase reliability of the storm-water management system by providing multiple or redundant points of control.
- Integrate storm-water controls into the development design and utilize the controls as amenities, creating a multifunctional landscape.
- Utilize an interdisciplinary approach that incorporates planners, engineers, landscape architects, and architects working together from the initial phases of the project.
- Reduce the reliance on traditional conveyance and pond technologies.
- Provide community education and promote community participation in the protection of LID systems and receiving waters.

These objectives can be grouped into five basic elements that constitute a "complete" LID design:[8]

1. Conservation measures.
2. Minimization techniques.
3. Flow attenuation.
4. Distributed integrated management practices.
5. Pollution prevention measures.

Although these five elements can be applied to any development, the manner in which they are used must be determined by the local climate and soils.

Conservation Measures

Conservation measures maintain as much of the natural landscape as possible. This includes retaining forests and other native vegetation, not filling wetlands, and providing buffers around wetlands and streams. These natural areas can then provide passive storm-water management opportunities, and they also double as open space.

Minimization Techniques

Minimization techniques reduce impacts of development on the hydrologic regime by reducing the amount of disturbance when preparing a site for development. Instead of grading an entire development site, only the lots and roads are graded while the rest of the ground is left undisturbed. Impervious surfaces are limited to areas where they are absolutely required. Cluster design is used to decrease the amount of the site developed and increase the amount of open space. Graded soils with high infiltration-capacity soils are stockpiled and reused.

Flow Attenuation

By slowing runoff velocity, the opportunity for storm-water infiltration increases and the magnitude of peak discharges decline. Whereas traditional storm-water management directs water from a site as quickly as possible, LID holds runoff on-site as long as possible without causing flooding or other potential problems.

Distributed Integrated Management Practices

LID incorporates a range of integrated best management practices throughout a site, commonly in sequence. An example of this is connecting a bioretention area to a natural area by conveyance through a grass swale. During high flow events, excess storm water is given the opportunity to continue to infiltrate while flowing from the site.

Pollution Prevention Measures

Pollution prevention measures are accomplished through a variety of source control, rather than treatment, approaches. For example, community outreach activities, such as the publication of educational materials, can control pollution by not allowing contaminants to enter the watershed or to be released in the first place. This strategy complements other elements of LID design that also reduce the downstream delivery of pollutants by minimizing storm-water runoff volumes and by promoting filtration through soil.

FIGURE 16.2 Retrofitted residential block in north Seattle, displaying a variety of LID techniques: limited impervious area, amended soils, rain gardens, and bio swales. (Photo courtesy of Seattle Public Utilities.)
Source: Adapted from Seattle Public Utilities Natural Drainage Systems Program.[9]

Conclusion

Low-impact development uses on-site natural features and small-scale hydrologic controls to manage runoff, closely mimicking predevelopment watershed hydrologic functions. Challenges for more widespread use of this approach include uncertainty in its application on relatively noninfiltrative soils, its role in mitigating high-intensity and (or) large-volume storms, and its construction in new or previously developed areas where high urban densities are desired or already present (Figure 16.2). Although additional research will be needed to characterize fully the performance of LID practices across different physical settings and development scenarios, available information indicates that LID can fully control flows associated with low- intensity storms, improve water-quality treatment by increasing storm-water contact with soils and vegetation, and significantly reduce flows from larger-volume storms on sites with relatively permeable soils. LID is a significant conceptual shift from conventional storm-water management; its broader application should encourage designers to incorporate native hydrologic processes as important organizing principles in the development of the urban and suburban landscape.

References

1. Environmental Protection Agency. *Low Impact Development—A Literature Review*: EPA-841-B-00-005; 2000; 35 pp. (available at http://www.epa.gov/owow/ nps/lid/lid.pdf).
2. Puget Sound Action Team. *Low Impact Development (Brochure);* Olympia: Washington, 2001; 4 pp. (available at http://www.psat.wa.gov/Programs/LID/lid_cd/ brochure.pdf).
3. Brattebo, B.O.; Booth, D.B. Long-term stormwater quantity and quality performance of permeable pavement systems. Water Res. **2003**, *37*, 4369–4376.
4. Konrad, C.P.; Burges, S.J. Hydrologic mitigation using on-site residential storm-water detention. J. Water Resour. Plan. Manage. **2001**, *127*, 99–107.
5. *Penn State Center for Green Roof Research;* College of Agriculture, Pennsylvania State University: State College, PA, 2002. (available at http://hortweb.cas.psu. edu/research/greenroofcenter/).

6. Booth, D.B. Urbanization and the natural drainage system—impacts, solutions, and prognoses. Northwest Environ. J. 1991, *7*, 93–118.
7. Booth, D.B.; Jackson, C.R. Urbanization of aquatic systems—degradation thresholds, stormwater detention, and the limits of mitigation. Water Resour. Bull. 1997, *33*, 1077–1090.
8. Coffman, L.S. Using low-impact development in stormwater management. Water Environ. Res. Foundation, Progress Newslett. Winter 2001, 12(1). (available at http://www.werf.org/press/winter01/01w_low.cfm).
9. *Seattle Public Utilities Natural Drainage Systems Program.* (available at http://www.ci.seattle.wa.us/util/ NaturalSystems/default.htm).

II

Water Management

17

Groundwater Contamination

Introduction 115
Types of Contaminants 115
Sources of Contamination 116
Effect of Climate 116
Transport of Contaminants 117
Fate of Contaminants 117
References 118

Charles W. Fetter
Oshkosh

Introduction

Contamination can be defined as the presence of a biological or chemical agent in groundwater in such a concentration that it renders water unfit for a particular use.[1] Agricultural uses of water include domestic drinking water, stock watering, and irrigation. Water that is contaminated for purposes of drinking might be perfectly suitable for use in irrigation.

Contaminants can be from both anthropogenic and natural sources, for example, arsenic. Arsenic found in groundwater in northeastern Wisconsin comes from a naturally occurring mineral, arsenopyrite, present in aquifer. Arsenic has also become a contaminant in groundwater due to use its use in agriculture as a pesticide as well as industrial sites where arsenic was used as a wood preservative.[1] The drinking water standard for arsenic in the United States for many years was 50 μg/L (micrograms per liter). However, in May 2000 the U.S. Environmental Protection Agency reviewed the standard and it was lowered to 10 μg/L.

Types of Contaminants

Groundwater contaminants fall into two broad categories, biological and chemical. Biological contaminants include bacteria, viruses, and protozoa. Chemical contaminants can be classified as organic or inorganic. Organic chemicals are based on a framework of carbon and hydrogen atoms. Inorganic compounds include all other chemicals, although some will have carbon present in an inorganic form, such as carbonate (CO_3^{2-}) and bicarbonate (HCO_3^-).

Organic chemicals include fuels and most pesticides. Fuels such as gasoline and diesel are composed of hundreds of different organic chemicals in varying proportions depending upon the source, and their composition will vary depending upon the season. Fuels do not mix with groundwater, rather if present in the ground they will float on the water table. They are sometimes referred to as Light Non-Aqueous Phase Liquid (LNAPL) as they are less dense than water. However, some of the chemicals that comprise gasoline and diesel will separate from fuel into a dissolved form in the groundwater. The most soluble of these chemicals are benzene, toluene, ethylbenzene, and xylenes. They are referred to by acronym BTEX.[2] Some organic pesticides may be soluble in water as they may be mixed with water prior to application to a field.

Inorganic chemicals found in groundwater are salts that dissociate into cations and anions when in contact with water. The cations include heavy metals such as iron, lead, manganese, cadmium, chromium, zinc, and mercury. The anions include nitrate (NO_3^-), nitrite (NO_2^-), sulfate (SO_4^{2-}), fluoride (F^-), chloride (Cl^-), arsenate (AsO_4^{3-}) and arsenite (AsO_3^{3-}).

Sources of Contamination

Sources of contamination can be divided into point sources and non-point sources. As the name implies, point sources can be traced to a very specific location. An example of a point source might be a septic tank, a landfill, or a pesticide mixing area. Non-point sources are dispersed across the landscape. Fertilizer and pesticides applied to fields are examples of non-point sources.

Human and animal wastes are sources of potential groundwater contamination due to the presence of bacteria and viruses as well as nitrogen compounds. One chemical compound frequently found in groundwater in rural areas is nitrate. This can come from cesspools and septic tanks, barnyards, manure spread as fertilizer, and chemical fertilizers. Nitrate and nitrite in drinking water in excess of 10 mg/L (milligrams per liter) as nitrogen have been implicated in infant methemoglobinemia or "blue baby syndrome." Another salt found in animal waste is chloride. This will impart a salty taste in drinking water if present in amounts in excess of 250 mg/L.

Pesticides are also a potential source of groundwater contamination. They can be found concentrated in areas where pesticides are mixed or equipment is washed. Likewise pesticides can be a non-point source of contamination when they are spread on a field. For example, atrazine has been found in groundwater in Wisconsin as a result of use on corn crops. Not only can pesticides occur in the environment, but breakdown products called metabolites can also occur in groundwater.

When water is used for irrigation, some will evaporate. This will concentrate the soluble salts in the remaining water, which will drain down to the water table. As a result, toxic salts may build up in the soil and groundwater. This situation has developed in some areas of California with selenium.

Fuels used on the farm can leak from underground storage tanks resulting in the formation of a pool of LNAPL on the water table below the tank and dissolved BTEX chemicals in the groundwater. The federal drinking water standard for benzene in the United States is 5 μg/L. In some states the groundwater standard is even lower, 1 μg/L.

Chemicals used for degreasing equipment can also contaminate groundwater if improperly disposed.[3] Many degreasers contain chlorinated organic compounds such as trichloroethylene (TCE) and 1,1,1-trichlorethane (TCA). These liquids are denser than water and mix poorly with water. They are referred to by acronym DNAPL. If disposed into the environment, for example by spilling on soil, they can migrate vertically to the water table and then sink below the water table into the underlying aquifers. These compounds are sparingly soluble in water, but even small amounts are dangerous. The federal drinking water standard for trichloroethylene in the United States is 5 μg/L. In some states the groundwater standard is even lower, 1 μg/L.

Chemicals used in wood preservatives are also potential groundwater contaminants. These include creosote and CCA (copper, chromium, arsenic). Treated wood itself would most likely not contaminate groundwater, but spilled or improperly disposed wood-treating chemicals could contaminate the groundwater.

Effect of Climate

In humid climates, the water table may be close to the surface and frequent rains can leach contaminants from the soil and transport them down to the water table. If the climate is more arid, contaminants in the soil zone are less likely to be transported to the water table, which itself is likely to be deeper than in a corresponding area that is more humid. However, evaporation of irrigation water in arid climates may result in a build up of soluble salts in the soil and the excess irrigation water that may eventually reach the water table.

Transport of Contaminants

Dissolved contaminants are carried by flowing groundwater through a process called *advection.* If the contaminant is *conservative,* it will move at the same rate as the groundwater in which it is dissolved. An example of a dissolved salt that is conservative is chloride. Water flowing through an aquifer will not all be moving at the same rate. Groundwater moves through pores and cracks in the ground. Some of these openings in the ground are larger than others, and water in the larger openings will be moving faster than water in the smaller openings. As a result, the faster moving water will spread out in front of the rest of the mass of the water. If a contaminant is present in a low concentration, the closer the contaminant gets to the moving front the faster moving water mixes with uncontaminated water. This process is called *longitudinal dispersion.* Through dispersion and diffusion a *plume* of groundwater contamination is formed. This plume is nothing more than a contiguous zone where the contaminant is present in the groundwater. If there is an ongoing source of contamination at the start of the plume, the greatest concentration of the contaminant will be found there and the concentration will decrease in the direction of the groundwater flow. The contaminant plume will extend along the direction of groundwater flow, but also spread sideways through a process called *lateral dispersion.* This is due to the flowing groundwater taking branching pathways.[4]

Non-aqueous phase liquids also have the potential to move through the soil and underlying aquifers. Their movement is dependent upon the ability of the non-aqueous phase liquid to overcome capillary forces and displace air in the pores above the water table and water in the pores of the earth below the water table.

Fate of Contaminants

Biological agents are particles of protoplasm. As such they can travel through large pores and cracks in the earth, but not small ones. Fine-grained soils can remove bacteria and viruses by *filtration,* usually within a few hundred meters or less of the source. Some aquifers such as coarse gravel, fractured rock, and carbonate rock have larger openings. Bacteria and viruses can travel for significant distances in such aquifers.

Ionic substances can be removed from groundwater by *ion exchange.* In this phenomenon, ions such as sodium and calcium, which are loosely bound to clay particles can be exchanged for other cations, such as lead, mercury, cadmium, and manganese. The heavy metal contaminants will thus be removed from the groundwater. The ability of a soil to remove contaminants by ion exchange is measured as the ion-exchange capacity of the soil.

Dissolved organic compounds can be removed from groundwater by *adsorption* onto organic matter contained in the soil or rock. The rate of adsorption is inversely proportional to the water solubility of the organic compound. Those that have a low solubility are more tightly bound to the soil organic matter than those that are more soluble. This propensity to be absorbed is measured by a property known as the octanol–water partition coefficient. The other important factor is the percentage of organic matter in the soil. Obviously, the greater the percentage of organic matter, the more of dissolved organic compounds it can absorb.

Finally, many of the dissolved organic compounds can potentially be broken down into simpler compounds by the action of microbes in the soil and aquifer. This process is known as *biodegradation.* The components of petroleum based fuels can be degraded by soil bacteria. The end result is either carbon dioxide or methane, depending upon the presence or absence of dissolved oxygen in the aquifer. BTEX compounds are most readily degraded under aerobic conditions, i.e., with dissolved oxygen present.

However, under certain geochemical conditions in the aquifer they can also be degraded in the absence of oxygen, but at a slower rate. Many other organic chemicals dissolved in groundwater, such as the chlorinated solvents, can be degraded either biologically or abiotically under the right geochemical conditions.[5]

References

1. Fetter, C.W., Jr. *Contaminant Hydrogeology,* 2nd Ed.; Prentice-Hall, Inc.: Upper Saddle River, NJ, 1999; 500 pp.
2. Bedient, P.B.; Rifai, H.S.; Newell, C.J. *Ground Water Contamination, Transport and Remediation,* 2nd Ed.; Prentice-Hall PTR: Upper Saddle River, NJ, 1999; 604 pp.
3. Pankow, J.F.; Cherry, J.A. *Dense Chlorinated Solvents;* Waterloo Press: Portland, Oregon, 1996; 522 pp.
4. Charbeneau, R.J. *Groundwater Hydraulics and Pollutant Transport;* Prentice-Hall, Inc.: Upper Saddle River, NJ, 2000; 593 pp.
5. Chapelle, F.H. *Ground-Water Microbiology and Geochemistry,* 2nd Ed.; Wiley and Sons, Inc.: New York, 2000; 475 pp.

18

Groundwater: World Resources

Lucila Candela
Technical University of Catalonia (UPC)

Introduction 119
Aquifer Types 120
Unconsolidated Sand and Gravel Aquifers • Consolidated/Fractured Sedimentary Aquifers • Carbonate Rock Aquifers • Basaltic and Other Volcanic-Rock Aquifers
Availability and Use 121
Groundwater Resources Distribution 121
Europe • Africa • South America, Central America, and the Caribbean • North America • Oceania • Asia and Middle East
References 124

Introduction

Groundwater constitutes an important source of water supply for domestic, industrial, and irrigation uses in different countries due to its availability, which is not subject to multiannual and seasonal fluctuations. At present it is the main source of domestic water supply in most European countries, the United States, Australia, and some countries of Asia and Africa both in large and small towns and in rural areas.[1,2] Groundwater is used for irrigation in about one-third of all irrigated lands.[3]

Natural groundwater resources are understood to be the total amount of recharge (replenishment) of groundwater under natural conditions as a result of infiltration of precipitation, seepage from rivers and lakes, leakage from overlying and underlying aquifers, and inflow from adjacent areas. In some cases, the average annual recharge of aquifers, evaluated from average annual precipitation, equals groundwater runoff. Natural groundwater resources may be equated to groundwater discharge (runoff) when the evaporation from the water table may be ignored or estimated separately.[4] Under this assumption, groundwater runoff data are widely used to characterize regional groundwater resources and are an important component of the hydrologic cycle and environment.[5–7]

The role of groundwater in the water balance and water resources of regions is quantitatively characterized by the groundwater runoff/precipitation ratio or groundwater recharge. The runoff/precipitation ratio is extremely variable depending on meteorological factors, composition of rocks, etc. Distribution of groundwater recharge to river/total river runoff ratios shows the effect of geographical and altitudinal zonality. The quantity of recharge ranges widely. Analysis of conditions of generation of groundwater resources within continents shows that this global process depends on a complex combination of various natural factors. Principal amongst these are precipitation, vegetation, soil type and geology, and the hydrogeological features of the area.

In regions where aquifers are mainly composed of sands, specific groundwater discharge values are twice as large as in regions where the percentage of sands in aquifers is small. In this regard distribution of specific values of groundwater discharge on a global scale is subject to latitudinal zonality. Values generally increase from subartic regions to medium-latitude zones, in humid tropics and tropics, and decrease in semiarid and arid regions. Large groundwater discharge values may be found in karst limestones (up to 20L $sec^{-1}km^{-2}$), sand quaternary deposits (up to 18L $sec^{-1}km^{-2}$) or highly fractured rocks (up to 10L $sec^{-1}km^{-2}$), although values are normally dependent on topographic elevations and annual precipitation. Marine sandy and clayey sediments show minimal discharge values (0.1L $sec^{-1}km^{-2}$ and smaller).[4]

The main task of areal hydrogeological subdivision when compiling groundwater runoff and resources maps is to distinguish territories which are sufficiently uniform in terms of groundwater distribution and particularities of groundwater generation.[8]

Aquifer Types

The principal aquifers are found in six types of permeable geologic materials: unconsolidated deposits of sand and gravel; semiconsolidated sand; sandstone; carbonate rocks interbedded sandstone and carbonate rocks; and basalt and other types of volcanic rocks.[8] Large areas of the world are underlain by crystalline rocks permeable only where they are fractured or weathered, and generally yield only small amounts of water to wells. In many places, they are the only source of water supply. However, because these rocks extend over large areas, important volumes of groundwater are withdrawn from them.

Unconsolidated Sand and Gravel Aquifers

Unconsolidated sand and gravel aquifers are characterized by intergranular porosity and all contain water primarily under unconfined or water-table conditions, but locally confined conditions may exist where aquifers contain beds of low permeability. Different categories can be distinguished, which occupy different geologic settings. The sediments are mostly alluvial deposits, but locally may include windblown sand, coarse-grained glacial outwash, and fluvial sediments deposited by streams recharge. Large areas of the world are covered with sediments deposited during several advances and retreats of continental glaciers. The glacial sand and gravel deposits form numerous local but productive aquifers.

Aquifers commonly receive direct recharge from precipitation and streamflow infiltration. Regional movement is down the valley in the direction of stream flow, lake or playa (located in the center of the basin). Basins in arid regions might contain deposits of salt, anhydrite, gypsum or borate produced by evaporation or mineralized water in their central parts. Also, much of the infiltrating water is lost by transpiration by riparian vegetation.

Consolidated/Fractured Sedimentary Aquifers

Aquifers in sandstones are more widespread than those in all other kinds of consolidated rocks. Sandstone retain some primary porosity unless cementation has filled all the pores, but most of the porosity in these consolidated rocks consists of secondary openings such as joints, fractures, and bedding planes. The water is not highly mineralized in areas where the aquifer outcrops or are buried to shallow depths, but mineralization generally increases as the water moves downgrading toward the structural basin.

Carbonate Rock Aquifers

The water-yielding properties of carbonate rocks are highly variable; some yield almost no water and are considered to be confining units, whereas others are among the most productive aquifers known. The original texture and porosity of carbonate deposits can range from 1% to more than 50%. Recharge water enters the aquifer through sinkholes, swallow holes, and sinking streams, some of which terminate at large depressions called blind valleys.

Basaltic and Other Volcanic-Rock Aquifers

Volcanic rocks have a wide range of chemical, mineralogic structural, and hydraulic properties due largely to rock type. Unaltered pyroclastic deposits have porosity and permeability characteristics like those of poorly sorted sediments; rhyolites have low permeability except where they are fractured.

Availability and Use

Except for widely scattered places, existing data are not uniformly distributed in space and time because hydrologic investigations have been mostly conducted in areas where water supply or water quality problems existed, or where large quantities of groundwater were withdrawn. Long term hydrologic records are rare and usually collected only during the course of a study or perhaps for a few years after the study has ended. No systematic investigation on groundwater resources and exploitation in many regions of the world have been conducted.[3,9]

The annual groundwater use for the world as a whole can be placed at 750–800 × $10^9 m^3$, a modest value when compared to overall water availability (Tables 18.1 and 18.2). But an overwhelming majority of the world's cities and towns depend on groundwater for municipal water supplies. Over 35 countries of the world use more than 1 × $10^9 m^3$ of groundwater annually.[10] Because of spatial imbalances in the occurrence of groundwater and the pattern of demand, massive problems of groundwater overexploitation are found in areas where high population exist or under intensive agriculture development.

Groundwater Resources Distribution

Europe

All types of aquifers are currently exploited: large well fields in artesian basins of platform type, such as Paris and London; river valleys (France, Volga region); cones and intermontane depressions (Italy, Switzerland).[11,12] In many cases their exploitation is accompanied by the generation of large and deep cones of depression.

Groundwater runoff in Europe is quite irregular, depending on the geostructural, climatic, and orographic conditions and the flow media generation: karst, porous fractured, and porous. Specific discharge values distribution is governed by the geological structure.

According to the available data (Table 18.1), groundwater use estimation in Europe is 78 $km^3 yr^{-1}$, which constitutes 21% of the total water consumption. Urban and rural population constitute the most important water consumers, accounting for 56% of the total water consumption. Groundwater is the

TABLE 18.1 Annual Groundwater Recharge and Withdrawals in the World

World Region	Average Annual Recharge ($km^3 yr^{-1}$)[a]	Annual Groundwater Withdrawals ($km^3 yr^{-1}$)[b]
Asia	2505	352
Europe	1368	78
Middle East and N. Africa	137	75
Sub-Saharan Africa	1548	9
North America	1884	110
C. America and Caribbean	344	29
South America	3693	14
Oceania	270	2

[a] Amount of water that is estimated to annually infiltrate into aquifers. It would represent the amount of water that could be annually withdrawn.
[b] Abstractions from aquifers. These data are scarce and not currently available for all countries in each region.
Source: WRI,[24] Dzhamalov & Zektser.[8]

TABLE 18.2 Groundwater Use in Selected Areas of the World

Country	Annual Recharge ($km^3\ yr^{-1}$)	Groundwater Use (%)
Russian Federation	900	<1
China	800	10
India	450	30

Source: From http://www.fao.org.[3]

main source for public water supply (more than 70% of total resources), especially on islands and some European countries like Denmark. More than 90% of big cities and towns are exclusively supplied by groundwater (among them Berlin, Rome). Although groundwater is mainly used for irrigation in southern countries like Spain with values ranging between $0.7\,km^3\ yr^{-1}$ and $5\,km^3\ yr^{-1}$, other European countries, like the Netherlands, may also use it during dry years.

Africa

Africa is one of the regions of the world facing serious water shortages because of greater disparities in water availability and use, and because water resources are unevenly distributed. Groundwater, first considered as a main resource for water in urban, rural areas, and mining, especially in coastal areas and arid regions, is now tending to be extended to the most isolated desert and tropical regions. In Libya, groundwater accounts for 95% of country's freshwater withdrawals, while in some areas of North Africa it is a significant source for irrigated agriculture. In many parts of the continent, groundwater resources have not yet been fully explored and tapped. According to the geographic and climatic homogeneity, which has a direct influence on water resources, Africa can be divided into several regions: Northern, Sudano-Sahelian, Gulf of Guinea, Central, Eastern, Indian Ocean Islands and Southern. This vast territory can be subdivided into a number of large aquifer systems subject to very varied climatic conditions.[13,14]

Basement rocks cover most of the central territory and aquifers are not very productive except in few cases. Sedimentary formations of sandstones overlaying the basement areas may constitute good aquifers, such the Karoo basin. The coastal sedimentary basins are the most productive aquifers, being intensely exploited along the shoreline. Alluvials are among the most important and also serve large populations, especially in Northern Africa. Karstified limestones of North-West Africa and Madagascar can yield flow rates up to $100\,m^3\ hr^{-1}$. Also large fossil aquifers are present in the Saharan and Nubian deserts made of sedimentary basins and being largely exploited.

South America, Central America, and the Caribbean

This area extending from the Central America Isthmus to South America has the most abundant river flow. Groundwater is unevenly distributed in quantity, but quality is usually good for domestic and industrial supply, presently the highest priority. Total water withdrawal from the aquifers is difficult to estimate because most comes from uncontrolled private and public wells. Based on UN estimates, 50–60% of total population domestic and industrial supply is from groundwater. Water withdrawn can be estimated at between $12\,km^3\ yr^{-1}$ and $14\,km^3\ yr^{-1}$, very low in comparison with the estimated renewable resources. Groundwater reserves estimation is $238{,}000\,km^3$, discharge to rivers being $3898\,km^3\ yr^{-1}$. Discharge values are high in the humid equatorial zone and minimum in the Atlantic Andean Cordillera and northeast Brazil.

According to geologic and tectonic features, four major water-bearing domains can be distinguished:[15,16] superficial deposits; deep aquifers in sedimentary basins; folded mountain chains; and precambrian basement bedrocks. Vast areas of South America are composed of Precambrian crystalline rocks which are not highly productive unless weathered or intensively fractured. The hydrogeologic map of South America[16] shows 16 hydrogeological provinces with similar characteristics including the previously-mentioned water-bearing domains. Some of the formations' resources are considered as the most important water-bearing formations, such as the Amazon Sedimentary Basin (32,500 km^3), Parnaíba-Maranhao (17,000 km^3), and the Paraná Sedimentary basin, where the Guarani aquifer extending over 1,500,000 km^2 has 50,000 km^3 of storage.

North America

Groundwater is an important source of water in the United States and Mexico, but it represents less than 5% of Canada's total water use. About 22% of the total water use in the United States ($290 \times 10^6 m^3 day^{-1}$) is supplied by groundwater; about 50% of the U.S. population depends on groundwater for domestic uses and also major cities and metropolitan areas and irrigation has made the High Plains one of the most important agricultural areas. Half of the U.S. population draws its domestic water supply from groundwater.[1] In Mexico, where desert and semiarid conditions prevail over two-thirds of the country, groundwater is widely used. Urban areas of Mexico use groundwater as their sole or principal source.

Unconsolidated sand and gravel are the most widespread aquifers, with intergranular porosity, and water primarily under water-table conditions).[17,18] Some unconsolidated aquifers have supplied large amounts of water for irrigation, like the High Plains aquifer ($56 \times 10^6 m^3 day^{-1}$ withdrawn from the aquifer for irrigation in 1990); in the United States, about 20% of the groundwater withdrawn is derived from the High Plains aquifer.

Carbonate rock aquifers are most extensive in eastern United States and in the Bahamas, western Canada and Yucatan (Mexico), and some of them are considered among the most productive aquifers known. Most of them consist of limestone but dolomite and marble locally yield water. More than $13 \times 10^6 m^3 day^{-1}$ (1990 data) were withdrawn from the Floridan aquifer system, the sole source of water supply for the city of Miami.

Oceania

While groundwater resources in New Zealand, Pacific Islands, and New Guinea are difficult to quantify due to the limited information, the aridity of much of the Australian continent is a significant factor in the occurrence and assessment of groundwater resources.[19] A large part of western and central Australia is arid, with a mean annual rainfall below 250mm. Total amount of groundwater used in Australia is estimated at $2460 \times 10^6 m^3$ in 1983 from more than 500,000 tube-wells, 14% of the total amount of water used.[20] The greatest concentrations are near Perth, Adelaide, South Australia, western Victoria, and on the central Queensland coast. Surficial aquifers are the most important sources for irrigation, urban, and industrial supply. Fractured rock aquifers of igneous and metamorphic rocks are of relative importance, although they may locally provide high groundwater yields.

Most highly productive aquifers are the surficial sedimentary aquifers associated with inland or eastern coastal rivers, up to 100m thick. Also sand dunes, coastal, and deltaic alluvium sediments form important aquifers along the east coast and in central Queensland. Australia's main arid-zone irrigation scheme is based on groundwater extracted from sands and gravel of Central Australia. Several large deep sedimentary basin aquifers (Amadeus, Canning, Great Artesian, Murray, Otway, Perth, Eucla, Officer), extending over more than 24,000 km^2, constitute a reliable source of old, good quality groundwater. The Great Artesian basin covers $1.7 \times 10^6 km^2$, is up to 300m thick and is one of the largest basins in the world.[21] More than 20,000 non-flowing and more than 4000 flowing artesian wells have been drilled. Individual well flows exceeding 100 L sec^{-1} have been recorded. Diffuse natural discharge from the Great Artesian Basin has been calculated to be about $1.4 \times 10^6 m^3 day^{-1}$.

Asia and Middle East

Continental Asia is an immense and complex geographical area of great extremes. Some parts of China and India are among the most populated in the world while the deserts of central Asia and the interior high plateau are extremely thinly populated. Most of the land of the Arabian Peninsula and in central and eastern Iran is a desert, reflecting different patterns of groundwater use.

In Asia and Middle East groundwater has been developed since ancient times, especially in arid regions where no other source of water supply is permanently available.[22,23] Large-scale developments are found in northern and coastal areas of China, where artesian aquifers and the loess and karst areas of the central south are tapped for urban and industrial supply. Groundwater irrigation distribution by subregions is, according to AQUASTAT,[3] in the Arabian peninsula 96.6%; Middle East 18.2%; and Central Asia 34% (although for Bangladesh it represents 69%).

Groundwater exists in the area as semiconfined, unconfined shallow, and deep aquifers. Recharge is faster in the Middle East countries although the aquifers of the Arabian Peninsula contain much larger reserves. The interior arid regions of the Middle East countries may include geological formations which can be considered as aquicludes or aquitards. Among the groundwater-bearing zones, the most important are alluvial of large rivers, vast, complex sedimentary formations holding artesian water and sedimentary basins in coastal areas (Israel). Some carbonate basins of importance are also present in the Mediterranean area (Lebanon, Syria) and Pakistan. Weathered crystalline rocks and lava flows constitute important aquifers in peninsular India and Northern Syria.

Although groundwater quality is suitable for irrigation and domestic uses, salinization of groundwater has occurred in several areas of the Indus Plain and Pakistan.

References

1. http://www.iwmi.cgiar.org/pubs/WWVisn/GrWater.htm.
2. http://www.grida.no/geo2000/english.
3. http://www.fao.org/ag/agl/aglw/Aquastatweb/Main/html/aquastat.htm.
4. Zekster, I.S.; Dzhamalov, R.G. *Role of Ground Water in the Hydrological Cycle and in Continental Water Balance;* IHPUNESCO: Paris, 1988; 133 pp.
5. Van der Leeden, F.; Troise, F.L.; Todd, K.D. *Water Encyclopedia*; Lewis Publishers: Chelsea, Michigan, 1990; 808 pp.
6. Gleick, P. *Water in Crisis: A Guide to the World's Fresh Water Resources*; Oxford University Press: New York, 1993; 1–493.
7. UN. *Comprehensive Assessment of the Freshwater Resources of the World: Report of the Secretary General*; United Nations, Commission on Sustainable Development, April 7–15, 1997; E/CN/17/1997/9, 1997; 1–35.
8. Dzhamalov, R.G.; Zektser, I. *Digital World Map of Hydrogeological Conditions and Groundwater Flow,* CDromVersion 2.0; HydroScience Press: 1999.
9. http://www.wri.org/water/index.html.
10. Llamas, R.; Back, W.; Margat, J. Groundwater use: equilibrium between social benefits and potential environment costs. Appl. Hydrol. **1992**, *1*, 3–14.
11. UN. *Groundwater in the Western Hemisphere*; Natural Resources/Water Series No. 4; United Nations, Department for Technical Co-operation for Development: New York, 1976; 337 pp.
12. UN. *Groundwater in Western and Central Europe;* Natural Resources/Water Series No. 27; United Nations, Department for Technical Co-operation for Development: New York, 1991; 363 pp.
13. UN. *Les Eaux souterraines de l'Afrique Orientale, Centrale et Australe,* Ressources naturelles/Serie eau No. 19; United Nations, Department for Technical Co-operation for Development: New York, 1988; 341 pp.

14. UN. *Groundwater in North and West Africa;* Natural Resources/Water Series No. 18; United Nations, Department for Technical Co-operation for Development: New York, 1988; 405 pp.
15. Rebouças, A.C. Desarrollo y tendencias de la hidrogeologı' a en America Latina. In *Hidrogeología, Estado Actual y Perspectivas;* Anguita, Aparicio, Candela, Zurbano, Eds.; CIMNE: Barcelona, 1991; 429–453.
16. UNESCO-IHP/DNPM/CPRM. *Hydrogeological Map of South America 1: 5,000,000. Exploratory tex;* Rio de Janeiro, 1996; 212 pp.
17. http://www.water.usgs.gov/ogw.
18. http://capp.water.usgs.gov/gwa/ch_h/H-text1.html.
19. UN. *Groundwater in the Pacific Region;* Natural Resources/Water Series No. 12; United Nations, Department for Technical Co-operation for Development: New York, 1983; 289 pp.
20. DPIE. *Review of Australia's Water Resources and Water Use. Vol. 1. Water Resources Data Set*; Dep. of Primary Industries and Energy, Australian Government Publishing Service: Canberra, 1987; 158 pp.
21. Habermehl, M.A. The great artesian basin Australia. BRM J. Aust. Geol. Geophys. **1980**, 5, 9–38.
22. UN. *Groundwater in the Mediterranean and Western Asia*; Natural Resources/Water Series No. 9; United Nations, Department for Technical Co-operation for Development: New York, 1982; 230 pp.
23. UN. *Groundwater in Continental Asia (Central, Eastern, Southern, South-Eastern Asia)*; Natural Resources/Water Series No. 15; United Nations, Department for Technical Co-operation for Development: New York, 1986; 391 pp.
24. WRI. *Environmental Data Tables*; World Resources 2000–2001: 2000.

19

Irrigation: River Flow Impact

Robert W. Hill
Utah State University

Ivan A. Walter
Ivan's Engineering, Inc.

Introduction 127
Depletion
Hydrograph Modification 128
Reservoir Storage • Irrigation Return Flows • Irrigation Methods • Irrigation Efficiencies
Basin Irrigation Efficiency 131
Environmental Concerns 131
Water Quality Implications for Agriculture • Salt Loading Pick-Up • In-Stream Flow Requirements
References 132

Introduction

In hydrologic studies it is common engineering practice to quantify the impact upon the stream(s) from which the irrigation water is diverted. The impact upon the stream is actually of two kinds: 1) diversions that decrease the streamflow and 2) return flows that increase the streamflow. The engineering term used to describe the overall impact is "streamflow depletion" which means the net reduction in streamflow resulting from diversion to irrigation uses. Actual stream depletions are a function of many factors including the amount and timing of diversions, the type of diversion structure (well vs. ditch), crops grown, soil type, depth to groundwater, irrigation method, irrigation efficiency, properties of the alluvial aquifer, area irrigated, and evapotranspiration of precipitation, groundwater, and irrigation water.

Depletion

Depletion, in this context, is the consumptive abstraction of water from the hydrologic system as a result of irrigation. It is in addition to consumptive water use that would have occurred in the unmodified natural situation. As an example, waters of the Bear River Basin of southern Idaho, northern Utah, and western Wyoming, because they are an interstate system, are administered by a federally established commission under the authority of the Bear River Compact.[1] Depletion is the basis, in the compact, for allocating Bear River water use among the three states. It is defined by a "Commission Approved Procedure" which includes consideration of land use and incorporates an equation for estimating depletion based on evapotranspiration. In a study for the commission, Hill[2] defined crop depletion as:

$$\text{Dpl} = \text{Et} - \text{SMco} - \text{Pef} \tag{19.1}$$

where Dpl is estimated depletion for a given site or subbasin; Et is calculated crop water use; SMco is moisture which is "carried over" from the previous non-growing season (October 1–April 30) as stored soil water in the root zone available for crop water use subsequent to May 1; and Pef is an estimate of that portion of precipitation measured at an NWS station during May–September, which could be used by crops.

The carry-over soil moisture (SMco) was estimated by assuming that 67% of adjusted precipitation from October through April could be stored in the root zone. If this exceeded 75% of the available soil water-holding capacity of the average root zone in the sub-basin, the excess was considered as lost to drainage or runoff and not available for crop use. Growing season precipitation was considered to be 80% effective in contributing to crop water use. The effectiveness factor of 80% allowed for precipitation depths throughout a sub-basin that might differ from NWS rain-gage amounts. It also included a reduction for mismatches in timing between rainfall events and irrigation scheduling.

Hydrograph Modification

Diversion of significant amounts of water from rivers and streams for irrigation and subsequent return flows alters the shape and timing of downstream hydrographs. In watersheds where mountain snowmelt provides the irrigation supply, such as in the western United States, diversion during the spring runoff attenuates the peak flow rate while later return flows extend the flow duration into late summer and early fall.

Reservoir Storage

Storage of water in reservoirs can significantly modify the natural stream hydrograph depending on the timing and quantity of the storage right. Irrigators with junior rights may only be able to store during time periods with low irrigation demand, such as during the winter, or during peak flow periods. Reductions of stream flow during the winter time may have considerable impact on downstream in-stream flows. Whereas, storage during periods of peak runoff may not affect minimum in-stream flow needs, but could deposit considerable amounts of sediment in the reservoir.

Irrigation Return Flows

Irrigation return flows are comprised of surface runoff and/or subsurface drainage that becomes available for subsequent rediversion from either a surface stream or a groundwater aquifer downstream (hydrologically) of the initial use. Reusable return flow can be estimated as irrigation diversion minus crop-related depletions minus additional abstractions. Additional abstractions include incidental consumptive use from water surfaces as in open drains, along with non-crop vegetation. The timing of return flow varies from nearly instantaneous (recaptured tailwater) to delays of weeks and months or perhaps longer with deep percolation subsurface drainage. In a hydrologic model study of the Bear River Basin[3] delay times between diversion and subsequent appearance of the return flow at the next downstream river gage varied from 1.5 months to as long as 6 months. The delay appeared to be related to subbasin shape and size.

Irrigation Methods

Four general irrigation methods are used: surface, subsurface, sprinkler, and trickle (also known as low flow or drip). Surface methods include wild or controlled flooding, furrow, border-strip, and ponded water (basin, paddy, or low-head bubbler). Hand move, wheel move, and center pivot are examples of sprinkler irrigation. Trickle irrigation includes point source emitters, microspray, bubbler, and line-source drip tape (above or below ground). Whereas the efficiency of surface irrigation is dependent upon the skills and experience of the irrigator, the performance of trickle and sprinkler systems is more dependent on the design. Generally, the more control that the system design (hardware) has on the irrigation system performance, the higher the application efficiency (E_a) can be. Thus, typical wheel move sprinklers have higher E_a values than surface irrigation, but lower values than for center pivots or trickle, assuming better than average management practices for each method.

The impact on river flows can be quite different among the various irrigation methods. The nature of furrow and border surface irrigation generally produces tail water runoff, which can be immediately recaptured and reused, as well as deep percolation, which may not be available for reuse until after a period of time. Tailwater is essentially eliminated and deep percolation reduced with sprinklers (Figure 19.1) compared to conventional surface irrigation. Whereas, with drip methods, deep

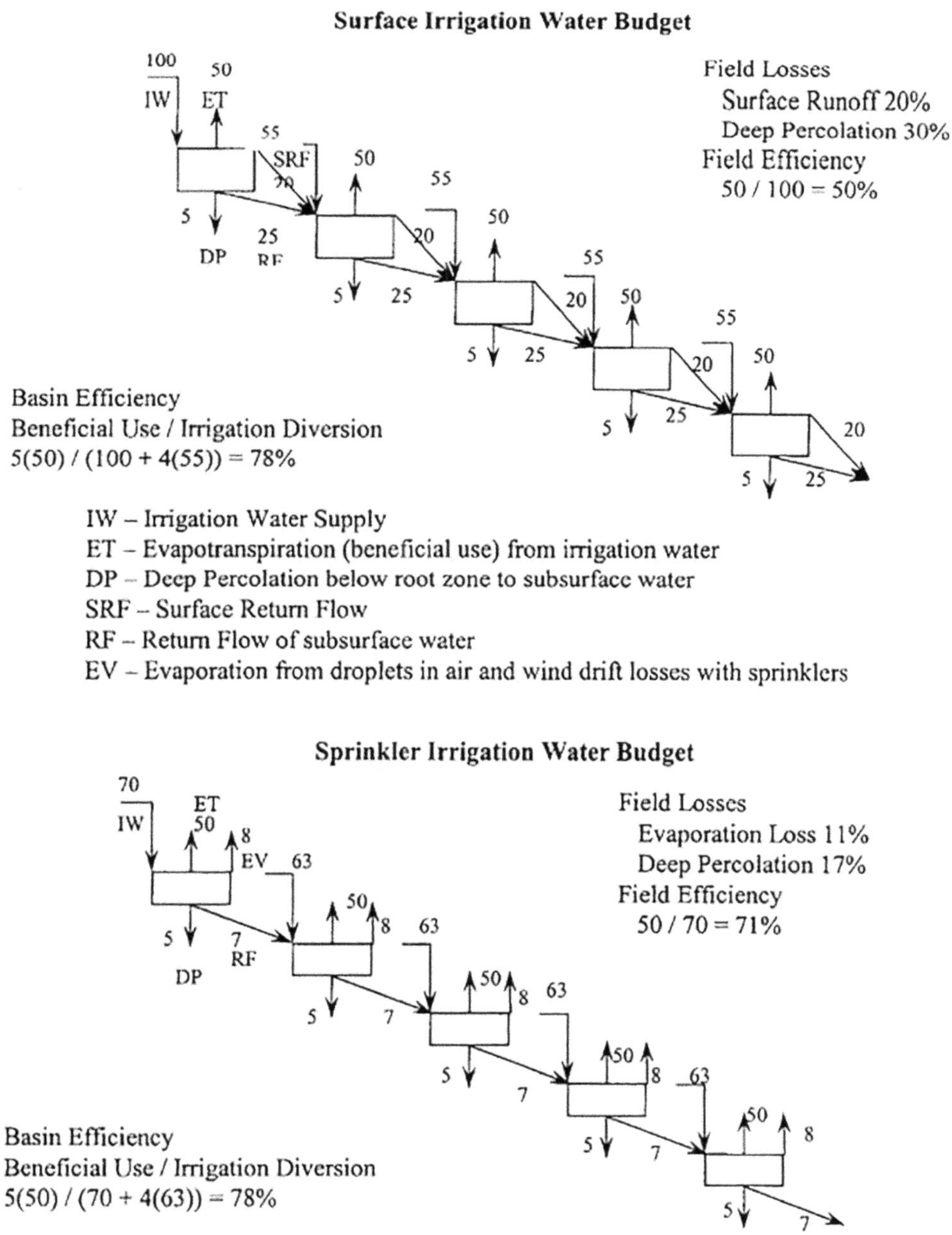

FIGURE 19.1 Comparison of basin efficiencies between surface and sprinkler irrigation methods with four return flow reuse cycles.

percolation can be further reduced. The reduction of deep percolation implies increased salt concentration in the root zone leachate, but, perhaps significant reduction in salt pick-up potential from geologic conditions.

Irrigation Efficiencies

Although a full discussion of the several variations of irrigation efficiency is beyond the scope herein, two terms will be defined and discussed. More complete discussions relating to irrigation efficiencies and water requirements are given elsewhere.[4–8] Keller and Bliesner[7] give a particularly thorough presentation of distribution uniformity and efficiencies.

Application efficiency (E_a):

$$E_a = 100 \times \frac{\text{Volume of water stored in the root zone}\left(V_s\right)}{\text{Volume of water delivered to farm or field}\left(V_f\right)}$$

Distribution uniformity:

The distribution uniformity is a measure of how evenly the on-farm irrigation system distributes the water across the field. The definition of DU is:

$$\text{DU} = 100 \times \frac{\text{Average of the lowest 25\% of infiltrated water depth}}{\text{Average of all infiltrated water depths across the field}}$$

On-farm or field application efficiencies can be affected by the distribution uniformity and vary widely for both surface and sprinkle irrigation methods. This is largely due to difference in management practices, appropriateness of design in matching the site conditions (slope, soils, and wind), and the degree of maintenance. In addition, for a given system uniformity, the higher the proportion of the field that is adequately irrigated (i.e., infiltrated water refills the soil water deficit) the lower will be the application efficiency. This is due to greater deep percolation losses in the over-irrigated portions of the distribution pattern. See Table 19.1 for an example of this phenomenon.

The E_a for a particular field may vary greatly during the season. Cultivation practices, microconsolidation of the soil surface and vegetation will alter surface irrigation efficiency both up and down from the seasonal average. Seasonal and diurnal variations in wind, humidity, and temperature will also affect sprinkle application efficiencies.

TABLE 19.1 Values Determined in Utah Field Evaluations

	Observed		
Method	High (%)	Low (%)	Typical (%)
Surface irrigation			
E_a	72	24	50
Tailwater	55	5	20
Deep percolation	65	20	30
Sprinkler irrigation			
E_a	84	52	70
Evaporation	45	8	12
Deep percolation	37	8	18

Basin Irrigation Efficiency

The actual irrigation efficiency realized for several successive downstream fields where capture and reuse of return flows is experienced is higher than the E_a of an individual field. This notion of "Basin Irrigation Efficiency"[8,9] is illustrated in Figure 19.1. This simple example comparison of surface and sprinkle methods assumes four reuse cycles. In each of the five "fields" Et is assumed to be 50 units. The surface runoff is captured for reuse on the next field. All of the irrigation-related evaporation is assumed "lost" as well as 5 units of deep percolation. After the fifth field, all surface and subsurface flows are lost. The basin efficiency for surface is 78%, which is the same as for sprinkle. The surface irrigation basin efficiency increase is dependent upon the surface return flow reuse, which is 20 units in this example. However, the depletion is greater for sprinkler due to the extra evaporation. In a Colorado field study, Walter and Altenhofen[10] found a progressive increase in irrigation efficiencies from field (average E_a of 45%), to farm, to efficiency of ditch or sectors (average of 83%). This was due to the reuse of tailwater (10–20% of delivery) and deep percolation (46% of delivery).

Environmental Concerns

The process of evapotranspiration, or crop water use, extracts pure water from the soil water reservoir, which leaves behind the dissolved solids (salts) contained in the applied irrigation water. The "evapo-concentration" of salts is an inevitable result of irrigation for crop production. As stated by Bishop and Peterson:[11]

"...Other uses add something to the water, but irrigation basically takes some of the water away, concentrating the residual salts. Irrigation may also add substances by leaching natural salts or other materials from the soil or washing them from the surface. Irrigation return flow is a process by which the concentrated salts and other substances are conveyed from agricultural lands to the common stream or the underground water supply...".

Water Quality Implications for Agriculture

Irrigated agriculture is dependent upon adequate, reasonably good quality water supplies. As the level of salt increases in an irrigation source, the quality of water for plant growth decreases. Since all irrigation waters contain a mixture of natural salt, irrigated soils will contain a similar mix to that in the applied water, but generally at a higher concentration. This necessitates applying extra irrigation water, or taking advantage of non-growing season precipitation, to leach the salts below the root zone.

Salt Loading Pick-Up

Water percolating below the root zone or leaking from canals and ditches may "pick-up" additional salts from mineral weathering or from salt-bearing geologic formations (such as the Mancos shale of western Colorado and eastern Utah). This salt pick-up will increase the salt load of return flows and consequently increase the salinity of receiving waters.

In the Colorado River Basin in the United States and Mexico salinity is a concern because of its adverse effects on agricultural, municipal, and industrial users.[12] The Salinity Control Act of 1974 (Public Law 93-320) created the Colorado River Basin Salinity Control Program to develop projects to reduce salt loading to the Colorado River. Salinity control projects include lining open canals and laterals (or replacing with pipe) and installing sprinklers in place of surface irrigation for the purpose of decreasing salt loading caused by canal leakage and irrigated crop deep percolation. Selenium in irrigation return flow has become a concern[13] and may also be reduced by salinity reduction projects.

In-Stream Flow Requirements

Diversions in some reaches in some western United States streams are "dried up" immediately downstream of diversion structures during times of peak irrigation demand. This condition eliminates any use of the reach for fisheries and other uses which depend on in-stream flow. In some instances, negotiated agreements with senior water rights users have allowed for bypass of minimal amounts of water to sustain the fishery or habitat, and for control of tailwater runoff to reduce agricultural related chemicals in the receiving water.

References

1. Bear River Commission. *Amended Bear River Compact*, S.B. 255, Utah Legislature Session, 1979.
2. Hill, R.W.; Brockway, C.E.; Burman, R.D.; Allen, L.N.; Robison, C.W. *Duty of Water under the Bear River Compact: Field Verification of Empirical Methods for Estimating Depletion. Final Report*; Utah Agriculture Experiment Station Research Report No. 125, Utah State University: Logan, Utah, January 1989.
3. Hill, R.W.; Israelsen, E.K.; Huber, A.L.; Riley, J.P. *A Hydrologic Model of the Bear River Basin*; PRWG72-1, Utah Water Resource Laboratory, Utah State University: Logan, Utah, 1970.
4. Hoffman, G.J.; Howell, T.; Solomon, K. *Management of Farm Irrigation Systems*; The American Society of Agricultural Engineers: St. Joseph, MI, 1990.
5. Jensen, M.E. *Design and Operation of Farm Irrigation Systems*; ASAE Monograph, American Society of Agricultural Engineers: St. Joseph, MI, 1983.
6. Jensen, M.E.; Burman, R.D.; Allen, R.G.; Eds. *Evapotranspiration and Irrigation Water Requirements*; ASCE Manual No. 70, American Society of Civil Engineers: New York, NY, 1990.
7. Keller, J.; Bliesner, R.D. *Sprinkle and Trickle Irrigation*; Van Nostrand Reinhold: New York, NY, 1990.
8. Willardson, L.S.; Allen, R.G.; Frederiksen, H.D. *Elimination of Irrigation Efficiencies; Proceedings 13th Technical Conference*, USCID: Denver, CO, 1994.
9. Willardson, L.S.; Wagenet, R.J. *Basin-Wide Impacts of Irrigation Efficiency;* Proceedings of an ASCE Specialty Conference on Advances in Irrigation and Drainage, Jackson, WY, 1983.
10. Walter, I.A.; Altenhofen, J. Irrigation efficiency studies— northern colorado. Proceedings USCID Water Management *Seminar. Sacramento, CA Oct 5–7; USCID: Denver, CO, 1995.*
11. Bishop, A.A.; Peterson, H.B. (team leaders). Characteristics and Pollution Problems of Irrigation Return Flow. Final Report Project 14-12-408, Fed. Water Pollution Control Adm. U.S. Dept. of Interior (Ada, Oklahoma). Utah State University Foundation, Logan, Utah. May 1969.
12. U.S. Department of the Interior. *Quality of Water— Colorado River Basin: Bureau of Reclamation, Upper Colorado Region*; Progress Report no 19; Salt Lake City, Utah, 1999.
13. Butler, D.L. *Effects of Piping Irrigation Laterals on Selenium and Salt Loads, Montrose Arroyo Basin, Western Colorado*; U.S. Geological Survey Water-Resources Investigations Report 01-4204 (in cooperation with the U.S. Bureau of Reclamation): Denver, Colorado, 2001.

20

Land Change and Water Resource Vulnerability

Gargi Chaudhuri
University of Wisconsin

Introduction 133
Agriculture 134
Urban 136
Conclusion 136
References 138

Introduction

Since the beginning of human civilization, mankind has been dependent on land and water for sustenance. Over the years, human activities have modified the natural environment to the extent that at present environmental sustainability is at danger. The term "land cover" refers to the biophysical cover of the land, and "land use" refers to the socioeconomic use of land. Land use and land cover change (LULCC), in general, means conversion of land to feed, provide shelter and other essentials to the growing human population (Defries and Eshleman 2004). Among different natural resources that are threatened by the direct and indirect impacts of LULCC, the vulnerability of fresh water resources is critical (Foley et al. 2005). Land change science has contributed significantly to the understanding of land use dynamics, yet the human use of land continues to be at the center of the most complicated and pressing problems faced by policy makers around the world today (Reid et al. 2006). The principal effects of land use change include consumption of fresh water, alteration in the amount of evaporation, groundwater infiltration, surface runoff, increasing sedimentation, and changes in the streamflow regime, which result in increased risk of flood or drought (Bullard 1966). In a nutshell, land use change affects both the quantity and the quality of fresh water.

Water quantity means the presence of adequate supply of water to sustain human and natural systems and is measured in terms of total water withdrawal and consumption from both surface water and groundwater sources. On the other hand, water quality refers to the suitability of the water supplied for its intended use (Mustard and Fisher 2012). Both water quality and amount of water use can be heavily affected by the types of land cover and land use activity and its geographical region and are important elements to consider for any watershed management program. Mustard and Fisher (2012) defined water vulnerability as *human supply and demand problem, where climate and the landscape establish the supply and human systems make the demands on the supply* (p. 265). Thus, decrease in quantity and quality of water resource due to the modification of the landscape and increase in demand due to population growth lead to increase in its vulnerability.

The relationship between land use and hydrology is well known but is complex and nonlinear (Bullard 1966). According to United Nation Environment Program (UNEP) (UNEP 2002), of the total 35 million km^3 of freshwater resources, the total usable freshwater supply for ecosystems and humans is about 200,000 km^3 or less than 1% of all freshwater resources. In the 20th century, land use and land

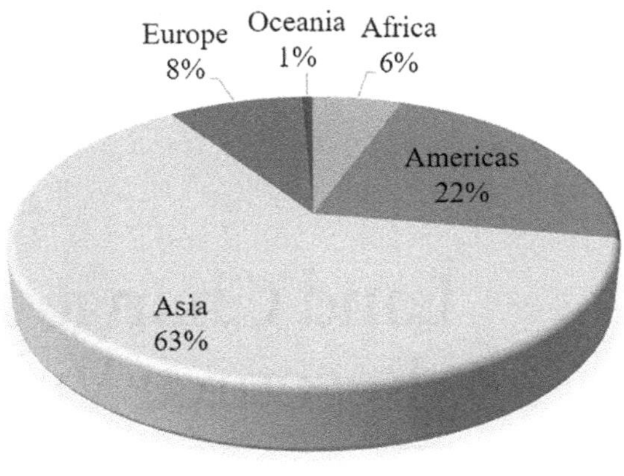

FIGURE 20.1 Total freshwater withdrawal (km³/year).
Source: AQUASTAT, www.fao.org/nr/aquastat. Update: November 2016.

cover (LULC) have changed rapidly both in terms of extent and intensity of use (Ramankutty et al. 2006), which have made its impact on water resources more intense. Global reports suggest that in the last 50 years, the world's agricultural production has increased between 2.5 and 3 times while the cultivated area has increased only by 12%. Thus, more than 40% of the increased food production came from irrigated areas, which have doubled in area (FAO 2011). Forest cover lost 135 million hectares between 1990 and 2010 (UN-FAO 2010) and at present only covers 31% of the land. In 30 years' time period (1970–2000), the world's urban area increased by 58,000 km^2 (Seto et al. 2011), where 52.1% of the global population lives at present (UNDESA Population Division 2014).

According to the World Water Development Report 4 (WWDR4) (WWDR4 2012), the world population is predicted to grow from 6.9 billion in 2010 to 8.3 billion in 2030, resulting in an expected increase of 50% in food demand (Bruinsma 2009). Thus, the major drivers of increased water demand are change in food, energy (biofuels), industry, and human settlement. In terms of present water use, agriculture accounts for 70% (approx.), industry accounts for 20% (approx.), and municipal water demand accounts for the rest 10% of the global freshwater withdrawals (WWDR4 2012). Of the total groundwater withdrawal (1,000 km^3/year), agriculture (including irrigation, livestock and aquaculture) accounts for 69% (approx.), municipal sector (including domestic) accounts for 12%, and industrial purposes account for 19% (WWDR4 2012).Thus, with the projected increase in food demand, irrigation will continue to play an important role in food production and is expected to result in 11% increase in irrigation water consumption from 2008 to 2050 in water scarce regions (FAO 2011). Among all the continents, Asia accounts for the largest freshwater withdrawal (includes agricultural, municipal, and industrial water withdrawal) (Figure 20.1). Among different sectors (Figure 20.2), agricultural activity accounts for the biggest water withdrawal in most of the world except Europe and Northern America. The impacts of agricultural and urban land use/land cover have been discussed further.

Agriculture

Agriculture (croplands and pasturelands) has transformed the surface of the Earth at the expense of forests and grasslands (Bullard 1966; Scanlon et al. 2007). Agricultural intensification and expansion have both contributed to the increasing vulnerability of the water resources, both in terms of quality and quantity. Worldwide, 60% of the crop production comes from agriculture that occupies 80% of the cultivated land, and the remaining 20% of the irrigated land accounts for 40% of the global crop production (WWDR4 2012). WWDR4 (2012) also projects an average increase of 0.6% a year in irrigated land until 2030 that will result in 36% more food being produced with 13% more water. Irrigation and intensive use

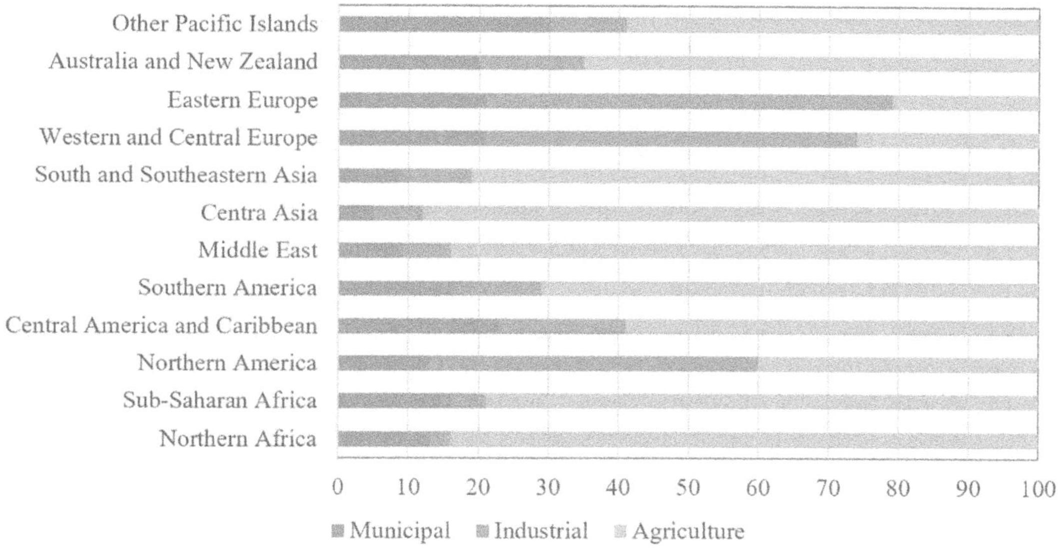

FIGURE 20.2 Proportion of water withdrawal by each sector in each region.
Source: From FAO AQUASTATS (FAO 2016).

of agricultural land may be an answer to the problem of global food security for the growing population, but it means increased risk on already stressed water resources. In addition to the food crops, the biofuel mandates in the developed countries of the world in 2007 which encourage the increasing use of biofuels and are projected to increase the demand for fresh water, too (US-EPA 2007; Congressional Research Service 2007). The International Energy Agency (IEA) indicates that if only 5% of road transport is powered by biofuel, it will amount to 20% of the water used for agriculture globally (WWDR4 2012; Comprehensive Assessment of Water Management in Agriculture 2007).

Generally, natural forested areas are being replaced by agriculture; thus, the impact of this change varies depending on the type of fauna being replaced, the nature of agricultural crop, the type of agricultural practice, and the climatic condition of the geographical region (Scanlon et al. 2007). For example, Muñoz-Arriola et al. (2009) showed that in the Yaqui River Basin, an important wheat-producing region in Mexico, the streamflow is more sensitive to the climatic variability rather than land use change, to the extent that the streamflow is higher by 76% in winter and 16% in summer during El Nino years than La Nina years. In terms of water quantity, conversion to agricultural land increases water consumption and the amount of water consumption varies highly between rain-fed and irrigated agricultural systems. Agriculture also results in a decreased amount of evapotranspiration from the land (Gordon et al. 2005 showed that deforestation resulted in 4% increase in evapotranspiration globally), increase in groundwater recharge, and increase in streamflow (Mustard and Fisher 2012; Scanlon et al. 2007). In terms of water quality, an increased intensity of agricultural land use has led to the contamination of groundwater and runoff water with nitrate and other toxic agrochemicals and increased salinization (Mustard and Fisher 2012; Scanlon et al. 2007; Scanlon et al. 2005; Schlesinger et al. 2006; Schlesinger 2008; Stonestrom et al. 2009). For example, conversion of the semiarid naturally forested land in Australia increased the groundwater recharge, but increased water levels in the shallow water tables also led to soil salinization and water logging (Allison et al. 1990). Another example of negative impact of irrigation is the fate of the Aral Sea, the size of the water body reduced from 68,000 km^2 in 1960 to 14,280 km^2 in 2010 (Gaybullaev et al. 2012) due to irrigation of the neighboring arid plains of Kazakhstan, Uzbekistan, and Turkmenistan for cotton plantation. An enormous reduction of water level in the lake, increased salinization, and a degradation of the water quality due to agricultural chemicals affected the economy of the whole region. Agriculture is the predominant user of fresh water worldwide, and the spatial scale of its impact on water resources varies from local catchment scale to regional basin scale.

Urban

From the LULC classification perspective, urban areas consist of residential, commercial, industrial, and mining areas (Anderson et al. 1976). Increase in urbanization involves the conversion of land surface into impervious surface and high intensity of water use due to dense urban population. In terms of water quantity, an impervious cover leads to the direct modification and destruction of water bodies, increased storm flows, decreased base flows during dry periods, and a decrease in infiltration and increase in runoff, thus obstructing groundwater recharge (Dreher and Price 1992). Decreased infiltration and increased peak flows of storm water runoff also increase the magnitude of urban floods. In terms of water quality, urbanization affects the water quality of both surface and groundwater from various point sources and nonpoint sources of pollution such as industrial and municipal waste discharges, leaky underground storage facilities, and spills of organic or inorganic contaminants, solvents, and fuels (Mustard and Fisher 2012).

In 2011, the world urban population is recorded to be 3.6 billion, and it is expected to increase by 72% by 2050, i.e., around 6.3 billion and much of the growth will be in Asia (UNDESA Population Division 2014). Globally, at present, the municipal water withdrawal is 12% and industrial withdrawal is 19% (FAO 2016). There are two schools of thoughts that exist regarding the relationship between urbanization and water vulnerability (Srinivasan et al. 2013). One group believes that (Lundqvist et al. 2003; Meinzen-Dick and Appasamy 2002), in the developing world, the conversion of agricultural land to urban land leads is economically efficient due to transfer water from low-value agricultural uses to high-value urban uses (Meinzen-Dick and Appasamy 2002). Moreover, improvements in agricultural water use efficiency will be adequate to increase urban water supply which is relatively less intensive than agriculture (Kendy et al. 2007; Molle and Berkoff 2009; Rogers et al. 2000). The other school of thought believes that the cities will be unable to manage the growing urban demand due to lack of proper infrastructure and poor governance (Vo 2007) and will lead to the unsustainable extraction of ground water (Güneralp and Seto 2008; Srinivasan et al. 2013).

In developing regions, urban and economic development are concomitant. Therefore, better socioeconomic status of urban population results in changing diets, which involves an increased consumption of protein (Economist 2019; Satterthwaite et al. 2010). Especially, countries like China and India with the highest population density, growing urban population and increasing purchasing power results in higher demand for meat and dairy (Oh and See 2012; Kumar et al. 2014; Ohlan 2016; Liu et al. 2009). Studies showed that, in industrialized countries, for an average person, 3,400 kcal/day (FAO 2011) of meat diet uses 3,600 L of water per day compared to 3,400 kcal/day of vegetarian diet (includes dairy) that uses 2,300 L of water per day (Hoekstra 2012; Falkenmark and Rockström 2006). Additionally, meat diets also entail the production of animal fodder in the form of crops, and during the period 2001–2007, on an average, 37% of the cereals produced in the world were reportedly used for animal feed (Hoekstra 2012; FAO 2011).

Conclusion

Empirical studies on the impact assessment of LULCC on water resources have greatly enriched the body of literature by identifying and delineating regional variations, causes, and consequences of land use change on water resources and temporal dynamics of the relationship between the two. In terms of spatial scale of effect, in humid and temperate regions with rain-fed agricultural system, the LULCC effect is limited to the drainage basin, whereas in arid and semiarid regions, which are mostly dependent on the irrigated agricultural system, the impacts are more spatially extensive (Mustard and Fisher 2012).

The temporal scales of the impact of LULCC on water resources are variable, too. At catchment scale, changes occur at an irregular timescale, whereas at plot scale for crop plantation and deforestation, changes are quite regular (Calder 1998; Chhabra et al. 2008). Moreover, there exists a time lag between LULCC and its effect on water resources that varies depending on the ecosystem, spatial extent, and climatic condition (Scanlon et al. 2007). There are a number of water vulnerability indices (Plummer et al. 2012) that show a global variation of water resource vulnerability measured on the basis of relative demand and supply of water. Figure 20.3, a water risk map produced by World Resource Institute, shows an overall global water risk (Gassert et al. 2014) and is calculated and mapped using indicators such as water quantity, water variability, water quality, public awareness of water issues, access to water, and ecosystem vulnerability. The regions that are mostly at "high to extremely high risk" class, such as western North America, northern part of Africa bordering the Saharas, Arabian Peninsula, Eurasian borders, northern China, and the Indian subcontinent, are areas of arid and semiarid regions with high human demands and a high density of human population, respectively.

It is difficult to isolate the impacts of LULCC on water resources amidst influences of other biophysical forces, sociopolitical factors, and climatic variability, but the drivers of change are mostly region specific (Chhabra et al. 2008). At present, the densely inhabited parts of the world are already heavily water stressed, and these regions are also expected to experience a high population growth in the future. To support such a growing population, both land use activities and water resources are immensely important for human sustenance, thus an integrated approach to land use and water management is important for a sustainable future. Recently, there have been an increasing number of studies where scientists are using integrated assessment models to improve the understanding of the feedback mechanism between land use changes and water resources under multiple scenarios of climate change

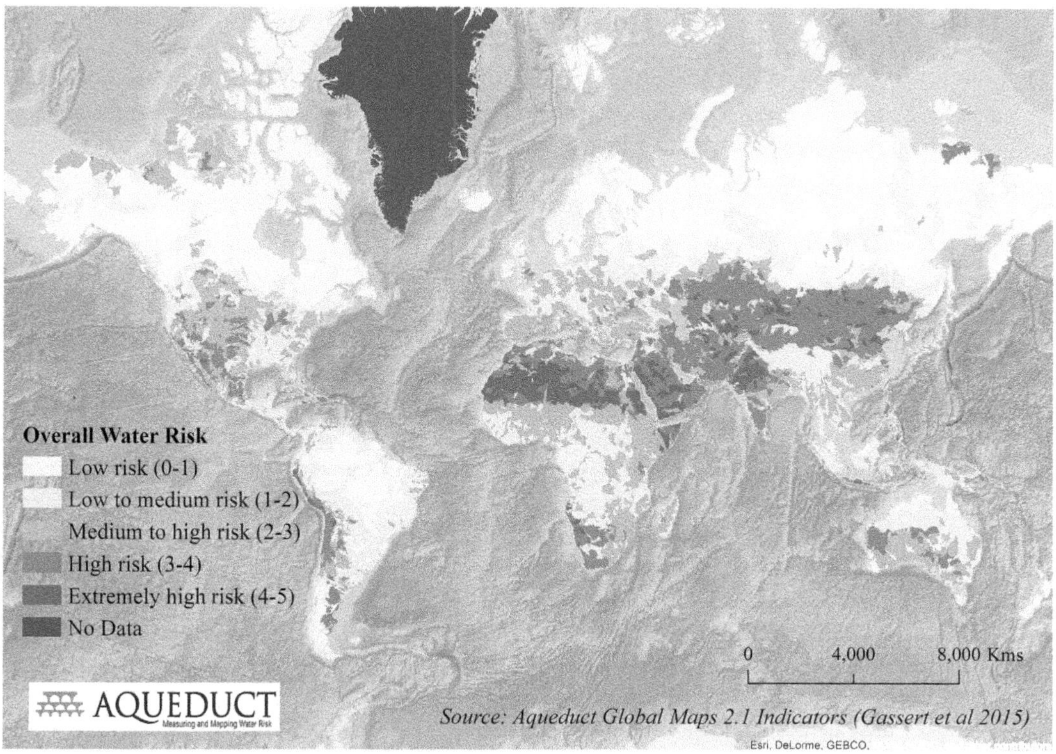

FIGURE 20.3 **(See color insert.)** Overall water risk map.

Source: WRI Aqueduct 2014 (Gassert et al. 2014); www.wri.org/applications/maps/aqueduct-atlas (accessed July 22, 2019).

(Van De Giesen et al. 2002; van Vuuren et al. 2012). These integrative approaches explore linkages between the different components of Earth and human systems to capture its dynamic relationship and assess the need for social, economic, and environmental coherence in the management of natural resources. Knowledge gained from such applications not only aids to better address policy issues but also helps to develop a scientifically sound decision support system.

References

Allison, G.B., Cook, P.G., Barnett, S.R., Walker, G.R., Jolly, I.D., Hughes, M.W. 1990. "Land Clearance and River Salinisation in the Western Murray Basin, Australia." *Journal of Hydrology* 119 (1–4): 1–20. doi:10.1016/0022-1694(90)90030-2.

Anderson, J.R., Hardy, E.E., Roach, J.T., Witmer, R.E. 1976. "A Land Use and Land Cover Classification System for Use with Remote Sensor Data." *Geological Survey Professional Paper* 964. https://pubs.usgs.gov/pp/0964/report.pdf.

Bruinsma, J. 2009. "The Resource Outlook to 2050: By How Much Do Land, Water, and Crop Yields Need to Increase by 2050?" In *FAO, Expert Meeting on How to Feed the World in 2050*. Rome. www.fao.org/3/a-ak971e.pdf.

Bullard, W.E. 1966. "Effects of Land Use on Water Resources." *Journal (Water Pollution Control Federation)* 38 (4): 645–59. www.jstor.org/stable/pdf/25035540.pdf.

Calder, I.R. 1998. "Water-Resource and Land-Use Issues." 3. SWIM Paper. Colombo, Sri Lanka. https://cgspace.cgiar.org/handle/10568/36524.

Chhabra, A., Geist, H., Houghton, R., Haberl, H., Braimoh, A., Vlek, P., Patz, J., et al. 2008. "Multiple Impacts of Land-Use/Cover Change." In *Land-Use and Land-Cover Change: Local Processes and Global Impacts*, edited by Lambin, E.F., Geist, H. pp. 71–116. Berlin/Heidelberg: Spring-Verlag. doi:10.1007/3-540-32202-7_4.

Comprehensive Assessment of Water Management in Agriculture. 2007. *Water for Food, Water for Life: A Comprehensive Assessment of Water Management in Agriculture*. www.iwmi.cgiar.org/assessment/files_new/synthesis/Summary_SynthesisBook.pdf.

Congressional Research Service. 2007. *CRS Report for Congress: Energy Independence and Security Act of 2007: A Summary of Major Provisions*. www.eere.energy.gov/manufacturing/tech_assistance/pdfs/crs_report_energy_act_2007.pdf.

Defries, R., Eshleman, K.N. 2004. "Land-Use Change and Hydrologic Processes: A Major Focus for the Future." *Hydrological Processes* 18: 2183–86. doi:10.1002/hyp.5584.

Dreher, D.W., Price, T.H. 1992. *Best Management Practice Guidebook for Urban Development*. Chicago, IL: Northeastern Illinois Planning Commission.

Falkenmark, M., Rockström, J. 2006. "The New Blue and Green Water Paradigm: Breaking New Ground for Water Resources Planning and Management Water for Food and Hunger Alleviation." *Journal of Water Resources Planning and Management* no. May/June. https://appgeodb.nancy.inra.fr/biljou/pdf/falkenmark-and-rockstrom.pdf.

FAO. 2011. *The State of the World's Land and Water Resources for Food and Agriculture (SOLAW) – Managing Systems at Risk*. Rome and Earthscan, London: Food and Agriculture Organization of the United Nations. doi:978-1-84971-326-9.

FAO. 2016. "AQUASTAT Website." Food and Agriculture Organization of the United Nations (FAO). www.fao.org/nr/water/aquastat/water_use/index.stm.

Foley, J.A., Defries, R., Asner, G.P., Barford, C., Bonan, G., Carpenter, S.R., Chapin, F.S., et al. 2005. "Global Consequences of Land Use." *Science* 309 (5734): 570–74. doi:10.1126/science.1111772.

Gassert, F., Landis, M., Luck, M., Reig, P., Shiao, T. 2014. "Aqueduct Global Maps 2.1: Constructing Decision-Relevant Global Water Risk Indicators." *World Resources Institute* 31.

Gaybullaev, B., Chen, S.C., Gaybullaev, D. 2012. "Changes in Water Volume of the Aral Sea after 1960." *Applied Water Science* 2: 285–91. doi:10.1007/s13201-012-0048-z.

Van De Giesen, N., Kunstmann, H., Jung, G., Liebe, J., Andreini, M., Vlek, P.L.G. 2002. "The GLOWA Volta Project: Integrated Assessment of Feedback Mechanisms between Climate, Landuse, and Hydrology." In *Climatic Change: Implications for the Hydrological Cycle and for Water Management*, edited by Beniston, M. pp. 151–70. Dordrecht: Springer Netherlands. doi:10.1007/0-306-47983-4_9.

Gordon, L.J., Steffen, W., Jönsson, B.F., Folke, C., Falkenmark, M., Johannessen, A. 2005. "Human Modification of Global Water Vapor Flows from the Land Surface." *Proceedings of the National Academy of Sciences of the United States of America* 102 (21): 7612–17. doi:10.1073/pnas.0500208102.

Güneralp, B., Seto, K.C. 2008. "Environmental Impacts of Urban Growth from an Integrated Dynamic Perspective: A Case Study of Shenzhen, South China." *Global Environmental Change* 18 (4): 720–35. doi:10.1016/J.GLOENVCHA.2008.07.004.

Hoekstra, A.Y. 2012. "The Hidden Water Resource Use behind Meat and Dairy." https://waterfootprint.org/media/downloads/Hoekstra-2012-Water-Meat-Dairy_2.pdf. doi:10.2527/af.2012–0038.

Kendy, E., Wang, J., Molden, D.J., Zheng, C., Liu, C., Steenhuis, T.S. 2007. "Can Urbanization Solve Inter-Sector Water Conflicts? Insight from a Case Study in Hebei Province, North China Plain." *Water Policy* 9 (S1): 75–93. doi:10.2166/wp.2007.046.

Kumar, A., Joshi, P.K., Kumar, P., Parappurathu, S. 2014. "Trends in the Consumption of Milk and Milk Products in India: Implications for Self-Sufficiency in Milk Production." *Food Security* 6 (6): 719.

Liu, H., Parton, K.A., Zhou, Z.Y., Cox, R. 2009. "At-Home Meat Consumption in China: An Empirical Study." *Australian Journal of Agricultural and Resource Economics* 53 (4): 485–501. doi:10.1111/j.1467-8489.2009.00463.x.

Lundqvist, J., Appasamy, P., Nelliyat, P. 2003. "Dimensions and Approaches for Third World City Water Security." doi:10.1098/rstb.2003.1382.

Meinzen-Dick, R., Appasamy, P.P. 2002. *Urbanization and Intersectoral Competition for Water IFPRI Gender, Agriculture, and Assets Project View Project Social Exclusion in India View Project.* Washington D.C.: The Woodrow Wilson Institute. www.researchgate.net/publication/254415389.

Molle, F., Berkoff, J. 2009. "Cities vs. Agriculture: A Review of Intersectoral Water Re-Allocation." *Natural Resources Forum* 33: 6–18. doi:10.1111/j.1477-8947.2009.01204.x.

Muñoz-Arriola, F., Avissar, R., Zhu, C., Lettenmaier, D.P. 2009. "Sensitivity of the Water Resources of Rio Yaqui Basin, Mexico, to Agriculture Extensification under Multiscale Climate Conditions." *Water Resources Research* 45. doi:10.1029/2007WR006783.

Mustard, J.F., Fisher, T.R. 2012. "Land Use and Hydrology." In *Land Change Science*, edited by Gutman, G. et al. pp. 257–76. Dordrecht: Springer. doi:10.1007/978-1-4020-2562-4_15.

Oh, S.H., See, M.T. 2012. "Pork Preference for Consumers in China, Japan and South Korea." *Asian-Australasian Journal of Animal Sciences* 25 (1): 143–50. doi:10.5713/ajas.2011.11368.

Ohlan, R. 2016. "South Asia Research Dairy Economy of India: Structural Changes in Consumption and Production." *South Asia Research* 36 (2): 241–60. doi:10.1177/0262728016638731.

Plummer, R., de Loë, R., Armitage, D. 2012. "A Systematic Review of Water Vulnerability Assessment Tools." *Water Resources Management* 26 (15): 4327–46. doi:10.1007/s11269-012-0147-5.

Ramankutty, N., Graumlich, L., Achard, F., Alves, D., Chhabra, A., DeFries, R.S., Foley, J.A., et al. 2006. "Global Land-Cover Change: Recent Progress, Remaining Challenges Navin." In *Land-Use and Land-Cover Change*, edited by Lambin, E.F., Geist, H. pp. 9–39. Berlin, Heidelberg: Springer. doi:10.1017/cbo9780511805530.028.

Reid, R.S., Tomich, T.P., Xu, J., Geist, H., Mather, A., DeFries, R.S., Liu, J., et al.. 2006. "Linking Land-Change Science and Policy: Current Lessons and Future Integration." In *Land-Use and Land-Cover Change: Local Processes and Global Impacts*, edited by Lambin E., Geist, H.J. pp. 157–71. Berlin: Springer-Verlag.

Rogers, P., Bouhia, H., Kalbermatten, J. 2000. "Water for Big Cities: Big Problems, Easy Solutions." *Urbanization, Population, Environment, and Security: A Report of the Comparative Urban Studies Project. Washington, DC*: Woodrow Wilson International Center for Scholars.

Satterthwaite, D., McGranahan, G., Tacoli, C. 2010. "Urbanization and Its Implications for Food and Farming." *Philosophical Transactions of the Royal Society of London. Series B, Biological Sciences* 365 (1554): 2809–20. doi:10.1098/rstb.2010.0136.

Scanlon, B.R., Reedy, R.C. Stonestrom, D.A., Prudic, D.E., Dennehy, K. F. 2005. "Impact of Land Use and Land Cover Change on Groundwater Recharge and Quality in the Southwestern US." *Global Change Biology* 11: 1577–93. doi:10.1111/j.1365-2486.2005.01026.x.

Scanlon, B.R., Jolly, I., Sophocleous, M., Zhang, L. 2007. "Global Impacts of Conversions from Natural to Agricultural Ecosystems on Water Resources: Quantity versus Quality." *Water Resources Research* 43 (3). doi:10.1029/2006WR005486.

Schlesinger, W.H., Reckhow, K.H., Bernhardt, E.S. 2006. "Global Change: The Nitrogen Cycle and Rivers." *Water Resources Research* 42 (3). doi:10.1029/2005WR004300.

Schlesinger, W.H. 2008. "On the Fate of Anthropogenic Nitrogen." *Proceedings of National Academy of Sciences* 106 (1): 203–8. www.pnas.org/cgi/content/full/.

Seto, K.C., Fragkias, M., Guneralp, B., Reilly, M.K. 2011. "A Meta-Analysis of Global Urban Land Expansion." *PLoS ONE* 6 (8): e23777. doi:10.1371/journal.pone.0023777.

Srinivasan, V., Seto, K.C., Emerson, R., Gorelick, S.M. 2013. "The Impact of Urbanization on Water Vulnerability: A Coupled Human-Environment System Approach for Chennai, India." *Global Environmental Change* 23 (1): 229–39. doi:10.1016/j.gloenvcha.2012.10.002.

Stonestrom, D.A, Scanlon, B.R., Zhang, L. 2009. "Introduction to Special Section on Impacts of Land Use Change on Water Resources." *Water Resource Research* 45 (7): W00A00. doi:10.1029/2009WR007937.

The Economist. 2019. *Global Meat-Eating Is on the Rise, Bringing Surprising Benefits - The Way of More Flesh*, May 4, 2019. www.economist.com/international/2019/05/04/global-meat-eating-is-on-the-rise-bringing-surprising-benefits.

UN-FAO, Food and Agriculture Organization of the United Nations. 2010. "Global Forest Resources Assessment - Main Report." Rome. doi:ISBN 978-92-5-106654-6.

UNDESA Population Division. 2014. *World Urbanization Prospects, the 2014 Revision*. doi:10.4054/DemRes.2005.12.9.

UNEP, United Nations Environment Programme. 2002. *Vital Water Graphics an Overview of the State of the World's Fresh and Marine Waters*. Nairobi, Kenya. ISBN: 92-807-2236-0.

US-EPA. 2007. *Energy Independence and Security Act (EISA) of 2007: Summary of Provisions. Public Law 110–140*. www.epa.gov/laws-regulations/summary-energy-independence-and-security-act.

Vo, P.L. 2007. "Urbanization and Water Management in Ho Chi Minh City, Vietnam-Issues, Challenges and Perspectives." *GeoJournal* 70 (1): 75–89. doi:10.1007/s10708-008-9115-2.

van Vuuren, D.P., Bayer, L.B., Chuwah, C., Ganzeveld, L., Hazeleger, W., van den Hurk, B., van Noije, T., et al.. 2012. "A Comprehensive View on Climate Change: Coupling of Earth System and Integrated Assessment Models." *Environmental Research Letters* 7 (2): 024012. doi:10.1088/1748-9326/7/2/024012.

WWDR4. 2012. *UN World Water Development Report: Managing Water under Uncertainty and Risk*. UNESCO. unesdoc.unesco.org/ark:/48223/pf0000215644.

21

Pesticide Contamination: Groundwater

Introduction 141
Pesticide Use 141
Associated Pesticide Behavior in Soils and Water 142
Insecticides • Fungicides and Fumigants • Herbicides
Groundwater Contamination 144
Management of Point Sources of Groundwater Contamination 144
Management of Non-Point Sources of Groundwater Contamination ... 144
Irrigation Management
Future Research 145
References 145

Roy F. Spalding
University of Nebraska

Introduction

Trace concentrations of most of the commonly used pesticides have been confirmed in groundwaters of the United States. Since groundwater is the source of 53% of the potable water, the more toxic pesticides and their transformation products are a concern from the standpoint of human health. Others are a risk to the environment in areas where contaminated groundwater enters surface water. Through toxicological testing, the USEPA has established Maximum Contaminant Levels (MCLs) or lifetime Health Advisory Levels (HALs) for several pesticides (Table 21.1).

The EPA also has a separate list of unregulated compounds, including newly registered pesticides and their transformation products, such as acetochlor and alachlor-ESA, that are presently being evaluated or being considered for toxicological evaluation. Based on the results of the EPAs National Pesticide Assessment,[2] 10.4% of 94,600 community systems contained detectable concentrations of at least one pesticide. Evaluation of these results led to an estimated 0.6% of rural domestic wells containing one or more pesticides above the MCL.

Pesticide Use

In the United States about 80% of pesticide usage is in agriculture. The remainder is used by industry, homeowners, and gardeners. About 500 million pounds of herbicide, 180 million pounds of insecticide, and 70 million pounds of fungicide were applied for agricultural purposes in 1993.[3] Several maps of the United States delineate usage patterns of several pesticides.[4] The majority of the triazine and amide herbicides are applied to fields in the north central corn belt states of Michigan, Wisconsin, Minnesota, Nebraska, Iowa, Illinois, Indiana, and Ohio. Commonly used organophosphorus insecticides are more heavily applied to fields in California and along the southeastern seaboard than in the northern Corn Belt. Carbamate and thiocarbamate pesticides are heavily used in potato growing areas of northern

TABLE 21.1 U.S. Maximum Contaminant Levels for Drinking Water

Organic Chemical Name	MCL (mg/L)	Organic Chemical Name	MCL (mg/L)	Organic Chemical Name	MCL (mg/L)
2,4,5-TP (Silvex)	0.05	Chlordane	0.002	Heptachlor	0.0004
2,4-D	0.07	Dalapon	0.2	Heptachlor Epoxide	0.0002
Alachlor	0.002	Dinoseb	0.007	Lindane	0.0002
Aldicarb	0.007	Diquat	0.02	Methoxychlor	0.04
Aldicarb sulfone	0.007	Endothall	0.1	Oxamyl (Vydate)	0.2
Aldicarb sulfoxide	0.004	Endrin	0.002	Picloram	0.5
Atrazine	0.003	Ethylene dibromide	0.00005	Simazine	0.004
Carbofuran	0.04	Glyphosate	0.7	Toxaphene	0.003
Carbon tetrachloride	0.005				

Source: Adapted from U.S. Environmental Protection Agency.[1]

Maine, Idaho, the Delmarva Peninsula, and vegetable fields of California and the southeastern coastal states. Fungicide use is concentrated in high humidity and irrigated areas of the coastal states and to some extent along the Great Lakes and Mississippi River Valley. The fumigants carbon tetrachloride and ethylene dibromide (EDB) were used heavily in the past at grain storage elevators throughout the Midwest and elsewhere in the United States.

Associated Pesticide Behavior in Soils and Water

Although pesticide use is a dominant factor in groundwater contamination, leaching variability among pesticides exhibiting similar behaviors is striking and explains why several heavily used pesticides seldom if ever are detected in groundwater. In general, pesticides within a class have similar chemical characteristics upon which soil leaching predictions can be made based on persistence, solubility, and mobility. Pesticide class relationships with soils and water transport described in the following text are detailed in Weber.[5] Individual frequencies of groundwater pesticide detection, in parenthesis next to commonly used products, are calculated from the Pesticide Groundwater Data Base (PGWDB)[4] and the National Water Quality Assessment (NAWQA) database.[6] High frequencies of detection identify those pesticides with a disposition to leach.

Insecticides

Chlorinated hydrocarbons are one of the oldest chemical classes of insecticides. Some of the best-known compounds include aldrin, dieldrin, DDE, DDT, endrin, and toxaphene. Although banned since the 1960s, their extremely persistent nature precludes their detection in very trace quantities in groundwater of the upper Midwest. On the other hand, heavily used organophosphates like malathion, methylparathion, disulfoton, and others have been extensively surveyed during several groundwater monitoring studies and have not been detected. The organophosphate insecticides, parathion (not reported (NR), <1; % occurrence from PGWDB and % occurrence from NAWQA data), terbufos (<1, <1), fonofos (<1, <1), and chlorpyrifos (<1, <1), which are heavily used on corn and sorghum, were also seldom detected. Diazinon (1.1, 1.3), the common garden insecticide, is occasionally detected in groundwater. Generally, the organic phosphates are rapid degraders and are strongly retained on soils.

For the most part, carbamates and thiocarbamates are very sparingly soluble and exhibit low to moderate soil retention; however, a small number have high solubility and low soil retention. Most carbamates are characterized as having short longevities. Generally, pesticides in this group having half-lives of 30 days or more have the potential to leach. The thiocarbamates butylate (<1, <1) and EPTC (2.6, <1) are extensively used in agriculture and have relatively short half-lives. Aldicarb (<1, <1) and carbofuran (14.7, <1) are at the high end for solubility and longevity in their class. Their metabolites have been frequently detected beneath high use crops, such as potatoes in the potato growing regions of the United States.

The pyrethroid insecticides have low solubilities, short half-lives, and high soil retentions that make them unlikely to leach. Yet, permethrin (<1, <1) is occasionally detected in very trace quantities in groundwater.

Fungicides and Fumigants

Fungicides are non-volatile organometallic compounds with low aqueous solubility that inhibit growth of actinomycetes and many fungi. The best-known fungicides zineb (not detected (ND), NR) and captan are zinc-based, and maneb (ND, NR) is manganese-based. Some, like bordeaux, are copper sulfate-based. Although their detection frequency is very low, fungicides have not been analyzed in many surveys.

Fumigants are very volatile halogenated compounds that generally are knifed below the soil surface. These compounds have high aqueous solubility and very low soil retention. The fumigants EDB and 1,3-dichloropropene have been frequently detected in the subsurface and in groundwater in high-use regions, such as California.[7] Ethylene dibromide and carbon tetrachloride were also used in grain storage facilities during the 1950s and 1960s. Spills, leaks, and improper handling resulted in 400 reported groundwater contamination sites in Kansas and Nebraska.

Herbicides

There are at least eight major chemical classes of herbicides. These include: quaternary N, basic, acidic, carboxylic acid, hydroxy and aminosulfonyl, amide and anilide, dinitroaniline, and phenylurea herbicides.[5] Several herbicide classes have similar behaviors with respect to soil and water.

Both quaternary N and dinitroaniline herbicides are very highly retained by soils and are not expected to be detected in groundwater. However, paraquat, pendimethalin, and trifluralin have been reported several times in groundwater. Their presence indicates that transport is dependent on factors not directly related to compound longevity, solubility, and mobility. Vertical transport by preferential flow through macropores is a commonly accepted mechanism used to explain these detections. In some instances, compounds have been described as preferentially transported attached to colloidal material.

Carboxylic, hydroxy, and aminosulfonyl acids, and thiocarbamate herbicides have very low to low soil retention and very short to moderate longevity. Thus, the more heavily used and persistent pesticides in these groups are the ones most generally detected in groundwater. They include the acids, dicamba (2.0, NR), picloram (2.5, <1), bromacil (1.8, 1.0), and dinoseb (1.4, <1).

Phenylurea herbicides have low to high soil retentions and short to moderate longevity. Linuron (16.7, <1) and diuron (<1, 1.9) are the most frequently detected in groundwater and both have moderately long half-lives ranging from 60 to 90 days.

Amide and anilide herbicides have low soil retention and short to moderate longevity. Several amide herbicides and their transformation products have been detected in groundwater. The commonly used amides in the Midwestern Corn Belt, namely alachlor (1.7, 2.7), metolachlor (<1, 12), propachlor (1.2, <1), and acetochlor (NR, <1), are the most frequent offenders because they are relatively persistent.

As suggested by the name, basic herbicides behave as bases. The group contains several subclasses including aniline, formamidine, imidazole, pyrimidine, thiadiazole, triazines, and triazole. Basic herbicides have low to high soil retention and very short to moderate longevity. Again, it is generally the most persistent and heavily used pesticides that are more frequently found in groundwater. The most frequently detected compounds in the group are the triazines, namely atrazine (5.6, 30), metribuzin (4.2, 1.9), cyanazine (2.0, 1.4), simazine (2.0, 14.8), and prometon (2.1, 11.6).

Groundwater Contamination

It stands to reason that there are generally good associations between pesticide use and their detection in groundwater. Since groundwater flows very slowly at rates normally ranging from 0.1 ft/day to 3 ft/day, pesticide sources are generally very near the monitored well. Thus, high frequencies of triazine and acetamide detections are reported in the states of the northern Corn Belt. More fungicides and fumigants were detected in warm humid states of California and Florida where vegetable and fruit crops dominate the landscape. In an analysis of the 20 NAWQAs for pesticides, frequencies of pesticide detection in groundwater were significantly related to the estimated amount of agricultural use within a 1 km radius of the sampled site.[6] They also emphasized that pesticides were detected beneath both agricultural (60.4%) and urban areas (48.5%). Discontinued used pesticides have been detected numerous times in shallow aquifers.

In general, families of pesticides have similar chemical characteristics from which predictions have been made as to the product's potential for contamination of groundwater; however, differences in the leaching behavior of pesticides exhibiting similar chemistry can be appreciable and is the reason several heavily used pesticides are seldom, if ever, detected in groundwater.

Management of Point Sources of Groundwater Contamination

Important steps are being taken to reduce water quality pollution by pesticides occurring from spills and back siphoning events (point sources). Since it is easier to resolve point than non-point sources, laws have been enacted to eliminate contamination of surface water bodies, which may be in hydraulic contact with groundwater, from used pesticide containers and rinseate from chemical wash downs. Check valves are mandatory when pesticides are mixed and/or diluted and prevent backflow to groundwater. Soils at and adjacent to agrichemical supply facilities have been surveyed in several states and found to be highly contaminated with pesticide residues. The herbicides atrazine, alachlor, metolachlor, cyanazine, and metribuzin are the worst offenders from the standpoint of pesticide mass in the soils at sites in Wisconsin and Illinois.[4] Many of these sites and those in other states are now involved in soil cleanups, which are designed to protect underlying groundwater from further pollution.

Management of Non-Point Sources of Groundwater Contamination

Normal farm chemical applications of pesticides are generally considered potential non-point sources of groundwater contamination because they are dispersed over large areas ranging from fields to watersheds. Management strategies are in place to reduce leaching of field applied chemicals.[8] These strategies vary from regulatory restrictions to outright bans on application in areas deemed more vulnerable to leaching. Integrated pest management, fostered by the office of pesticide management at the USEPA, is designed to reduce chemical applications. The practice of banding applications has reduced amounts applied. Both target more efficient pesticides and genetically engineered plants sensitive only to specific herbicidal action have been and are being developed. These new pesticides and pesticide–plant combinations require less chemical than in the past, and the altered plants allow for pest control with more environmentally sensitive chemicals. The USEPA has announced a plan to reduce the mass of applied chemicals from commonly used triazines and amides that are frequently detected in groundwater.

Irrigation Management

Irrigation practices can influence pesticide leaching. Atrazine was vertically transported deeper and faster when using flood rather than sprinkler irrigation.[9] Sprinkler systems allow for much more uniform and efficient water management practices than furrow irrigation, and recent studies have shown that they reduce chemical leaching.[10] In Nebraska's Platte Valley[11] and in the Walnut Creek watershed in Iowa,[12] peak herbicide concentrations were strongly related to rapid flushing beneath drainage areas where surface water ponds during heavy rainfall events on the cropped fields. Application of excess irrigation water was also reported to increase herbicide leaching.[9,11]

Future Research

More research is necessary to evaluate the health risks of transformation products from heavily used pesticides that are frequently detected in groundwater. Research needs to focus on precision application of pesticides to specific field problem areas as a potential mechanism to reduce chemical application.

There is a need to evaluate the environmental cost/benefit of safer product replacements used in conjunction with genetically altered crops. As new products are registered to replace more persistent and mobile pesticides, long-term fate studies, including the monitoring of the transformation product impact, on groundwater quality are necessary.

References

1. U.S. Environmental Protection Agency. *Drinking Water Standards and Health Advisories,* EPA 822-B-00-001; USEPA Office of Water: Washington, DC, Summer, 2000.
2. U.S. Environmental Protection Agency. *National Survey of Pesticides in Drinking Water Wells: Phase I Report,* EPA 570/9-90-015; USEPA Office of Pesticide and Toxic Substance: Washington, DC, November, 1990.
3. Aspelin, A.L. *Pesticide Industry Sales and Usage, 1992 and 1993 Estimates*; Economic Analysis Branch Report 733-K-92-001; USEPA Office of Pesticide Programs, Biological and Economic Analysis Division, 1994; 1–37.
4. Barbash, J.E.; Resek, E.A. *Pesticides in Ground Water. Distribution, Trends, and Governing Factors*; Ann Arbor Press, Inc.: Chelsea, MI, 1996; 1–588.
5. Weber, J. Properties and behavior of pesticides in soils. In *Mechanisms of Pesticide Movement into Ground Water*; Honeycutt, R.C., Schabacker, D.J., Eds.; Lewis Publishers: London, 1994; 15–41.
6. Kolpin, D.; Barbash, J.E.; Gilliom, R.J. Pesticides in ground water of the United States, 1992–1996. Ground Water **2000,** *38*(6), 858–863.
7. Troiano, J.; Weaver, D.; Marade, J.; Spurlock, F.; Pepple, M.; Nordmark, C.; Bartkowiak, D. Summary of well water sampling in California to detect pesticide residues resulting from nonpoint-source applications. J. Environ. Qual. **2001,** *30*(2), 448–459.
8. Guyot, C. Strategies to minimize the pollution of water by *pesticides. In Pesticides in Ground and Surface Water*; Borner, H., Ed.; Springer: Berlin, 1994; 87–148.
9. Troiano, J.; Garretson, C.; Krauter, C.; Brownwell, J.; Huston, J. Influence of amount and method of irrigation water application on leaching of atrazine. J. Environ. Qual. **1993,** *22*, 290–298.
10. Spalding, R.F.; Watts, D.G.; Schepers, J.S.; Burbach, M.E.; Exner, M.E.; Poreda, R.J.; Martin, G.E. Controlling nitrate leaching in irrigated agriculture. J. Environ. Qual. **2001,** *30*(4), 1184–1194.
11. Spalding, R.F.; Watts, D.G.; Snow, D.D.; Cassada, D.A.; Exner, M.E.; Schepers, J.S. Herbicide loading to shallow ground water beneath Nebraska's management systems evaluation area. J. Environ. Qual. **2003,** *32*, 84–91.
12. Moorman, T.B.; Jaynes, D.B.; Cambardella, C.A.; Hatfield, J.L.; Pfeiffer, R.L.; Morrow, A.J. Water quality in walnut creek watershed: herbicides in soils, subsurface drainage and groundwater. J. Environ. Qual. **1999,** 28(1), 35–45.

22

Pesticide Contamination: Surface Water

Introduction 147
Detection Frequency, Concentrations, and Seasonal Cycles of Pesticides Found in Surface Water 147
Pesticide Toxicity in Surface Water 148
Factors That Affect Pesticide Movement to Surface Water 149
Distance between Application and Surface Water • Choice of Rate, Pesticide, and Application Timing • Rainfall after Application • Drift, Volatilization, and Atmospheric Transport
Conclusion 150
References 150

Sharon A. Clay
South Dakota State University

Introduction

About 1 billion pounds of pesticide were applied in the United States in 1997.[1] This extensive use of pesticides has caused concern about pollution in the environment. Indeed, pesticides have been detected in groundwater, surface water, and rain. However, the amount found normally is less than 0.1% of the amount used and occurs in seasonal cycles. Management options that limit losses from target sites will be presented.

Detection Frequency, Concentrations, and Seasonal Cycles of Pesticides Found in Surface Water

Pesticide classification may be based on the type of pest controlled. Chemicals that control weeds (herbicides) accounted for 60% of the pesticide use in 1997.[1] Chemicals that control insects (insecticides) and plant diseases (fungicides) accounted for 13% and 7%, respectively, of the chemical use.[1] Rodenticides, fumigants, and nematicides represent other pesticide classes that account for the remaining 20% of use. In a typical year, agricultural lands receive about 80% of all pesticides applied while home-owners/gardeners (8%) and the commercial market (12%) apply the rest.

The United States Geological Survey (USGS) has sampled surface water for pesticide pollution since the late 1970s. Thousands of samples have been analyzed for about 100 pesticides or breakdown products (metabolites). Herbicides are the pesticide type most commonly found in rural streams.[2] Atrazine, a broadleaf herbicide ranked number one in total pounds applied in the United States from 1987 to 1997 (about 75 million lb/yr), has been detected most often.[1] Other herbicides found in surface water include simazine and cyanazine (used for broadleaf and grass control in corn) and alachlor and metolachlor (used for grass control in corn and soybean).[2] Most of these herbicides also ranked high in U.S. agricultural use between 1987 and 1997.[1] The amount of these herbicides found in streams has typically ranged from less than 1% of the amount applied (cyanazine, metolachlor, and alachlor) up to 3% (atrazine).[2] The five insecticides frequently detected were diazinon, carbaryl, chlorpyrifos, carbofuran, and malathion.[2]

Detection of cyanazine and alachlor in the U.S. environment will decline in the future. Cyanazine is no longer labeled for use in the United States as of 2002. Alachlor is being replaced by acetochlor. However, acetochlor was detected in surface water in its first year of general use (1994) with occurrence patterns similar to other herbicides of the same family.[3] The concentration detected was lower than alachlor due to lower application rates.

Pesticides have also been detected in urban streams[4,5] In fact, the estimated contribution of insecticides to surface water contamination from urban and rural areas may be similar. Insecticides, such as malathion, carbaryl, and diazinon, are commonly used in urban settings for control of mosquitoes, turfgrass and garden insects, and termites.[4] Herbicides detected in urban streams were those commonly used for broadleaf weed control in lawns (i.e., 2,4-D, dicamba, and MCPA)[5] along with prometon and tebuthiuron, both used for total vegetation control in right-of-way areas.[4] Herbicides almost exclusively used in agricultural settings (atrazine, alachlor, metolachlor, and cyanazine) have also been detected in urban streams. Atmospheric depositions in sediment, rain, or snow or transport from agricultural watersheds upstream of urban settings are the most likely sources of these herbicides.

Pesticides in surface water have been detected throughout the year.[6] The greatest concentrations in rural streams are reported in the spring and early summer coinciding with agricultural applications[7] The mean pesticide concentrations in most months are about 1μg/L.[2] However, during peak usage, concentrations in some samples may exceed 12 μg/L. In urban streams, the mean pesticide concentration is fairly stable (<0.5 μg/L) throughout the year[2] but can differ by area of the country. In the southern United States, pesticides may be applied both earlier and later in the year due to the longer growing season compared to the northern tier of states.

Pesticide Toxicity in Surface Water

The U.S. Environmental Protection Agency[8] has established pesticide concentration criteria values for some pesticides for the protection of aquatic health. Pesticides can be toxic to aquatic invertebrates such as plankton[9] and vertebrates such as frogs and fish when above critical concentrations. The herbicides, atrazine [water quality criterion (WQC) = 1.8 μg/L] and trifluralin (WQC = 0.2 μg/L), and the insecticides, chlorpyrifos (WQC = 0.041 μg/L) and diazinon (WQC = 0.08 μg/L), were the pesticides that most often exceeded the aquatic health criteria in 37 rural streams.[2] In a survey of eight urban streams across the United States monitored from 1993 to 1995, 41 pesticides were detected. Simazine (WQC = 10 μg/L), prometon (no established criterion), atrazine, and diazinon were detected at all sites and 20 other pesticides were detected above their WQC in one or more of the streams.[4] The estimated number of days that the pesticide concentration exceeds the established standards varies by chemical.[2] Atrazine was estimated to exceed the standard in 15 rural streams from 1 day to 84 days (with an average of 36 days). In comparison, chlorpyrifos and azinphos-methyl (WQC = 0.01 μg/L) exceeded the standard in 8 streams, ranging from 1 day to 8 days (average of 3 days) and 1 day to 70 days (average of 13 days), respectively.

In addition, herbicides may be used to control plants in ponds and lakes. The reduction in vegetation may have an indirect effect on the number of weed-clinging invertebrates such as dragonflies or damselflies. However, removal of plants does not always decrease the numbers of aquatic organisms. For example, the number of aquatic organisms remain fairly constant or increase due to an increase in organisms that feed on plant debris after vegetation is controlled.[10]

Factors That Affect Pesticide Movement to Surface Water

Distance between Application and Surface Water

The distance between pesticide application and surface water can greatly affect the likelihood of surface water contamination. In most cases, buffer zones of 50–100 ft are recommended on the pesticide label. Filter strips of grass or a dense cover crop at the field edge also reduces the pesticide concentration in the runoff in two ways.[11] First, filter strips slow water movement and allow soil particles to settle out of the runoff. Second, these areas are normally high in organic matter and can sorb the pesticide out of water. Herbicide concentrations have been reduced up to 64% after flowing through a filter strip compared to concentrations upstream of the filter strip area.[11]

Choice of Rate, Pesticide, and Application Timing

Some pesticides are much more vulnerable to movement into surface water. Pesticides that degrade quickly (days to a few weeks) are less likely to contaminate surface water than those that linger in the environment (months or longer). Applications to sandy soils with low organic matter are more likely to have runoff than loamy or clayey soils with more organic matter.

The higher the rate of application the more the pesticide is available for runoff. For example, banding herbicides to only row areas and using cultivation to control weeds in interrow areas is one technique to reduce application rates. The amount of herbicide applied depends on the bandwidth. If the bandwidth is 10" on a 30" row, the amount applied is reduced by two-thirds compared to a broadcast application rate where the entire area is covered. Banding has been reported to reduce the amount of herbicide in runoff water up to 70%.[12] Applying the pesticide as a split application also reduces the amount of chemical present at one time and can reduce the potential risk of surface water contamination.

Pesticides that are applied at ounces per acre rather than pounds per acre are available. Examples of herbicides include the sulfonylureas and imidiazilinones. The low application rates of these herbicides reduce the total amount of herbicide applied to an area and therefore reduce the risk of contamination.

Rainfall after Application

The amount, intensity, and time of rainfall after application are all factors that affect the total amount and pesticide concentration in runoff. Generally, a large amount of pesticide is removed with the first rainfall after application.[13] If the soil is bare and crusted or has little or no plant canopy, the amount of runoff will be very large. Storms with high intensity rains also have more runoff than if the rain is slow and steady. The amount of water already present in the soil is another factor that affects runoff. There is little or no infiltration on saturated or very wet soils whereas dry soils will allow more water to infiltrate before runoff occurs.

Drift, Volatilization, and Atmospheric Transport

Drift is the transport of spray applications in the wind. The smaller the droplet size, the farther the droplets can travel. Very fine particles may move for miles before deposition occurs. These drops could land in lakes, streams, or other off-site areas. Wind speed is also an important factor in drift. Most applications should not take place when wind speeds are over 10mph. Pesticides should not be applied when weather conditions that cause an inversion (warm air over cold air) are present. Fine spray droplets move much farther when an atmospheric inversion exists.

There are some factors that can be used to limit drift. Applying pesticides under low pressure in high amounts of water will increase droplet size, thereby decreasing the number of droplets available for drift. Larger droplets will also lower the pesticide concentration of individual droplets. There are nozzles specifically designed to limit the amount of fine particles in a spray pattern. Antidrift chemicals are available to be mixed with pesticide application. These chemicals reduce the number of very fine droplets in the spray pattern. Lowering the spray boom limits the amount of spray subjected to the wind. Shields for the boom or individual nozzles can be attached to limit wind interception with spray patterns.

Pesticides can volatilize (change from a liquid to a gas) or sublime (change from solid to a gas) from the soil surface. The amount volatilized is a function of the vapor pressure of the chemical and the environmental conditions. For example, EPTC, a grass herbicide, is highly volatile with most of the herbicide lost to the atmosphere within a few hours if not incorporated into the soil.[14] In contrast, atrazine is not very volatile but under the right conditions, 2% of the applied chemical can be lost through this mechanism.[15] This amount is about one-half the loss due to surface runoff on a regional scale. A pesticide applied to warm moist soils on windy days will have a greater loss than if the same pesticide is applied to cool dry soils.[14]

Pesticides attached to very small soil particles are also moved into the atmosphere.[16] Soil particles and aggregates that are less than 1 mm in diameter are considered wind-erodible.[17] In the Great Plains area of the United States, the loss of soil due to wind erosion is greater than the amount lost due to water erosion.[18] The small size soil particles make up about 50% of the soil mass in the top 0.5" of soil, if the area has been chisel plowed. The amount of herbicide found on these particles can range from 50% to 200% more than the amount found on larger aggregates one day after application.[19] If there are windy conditions within days after application, the amount of pesticide lost could be very high. Shallow incorporation of pesticides limits pesticide losses into the atmosphere.

Concentrations of pesticides in the atmosphere can decrease by several methods. Dilution, removal in rain, snow, or by dry deposition, and photochemical degradation are the three ways in which pesticide concentrations can be reduced.[20] However, pesticides can move long distances from their sites of application.[21] For example, surface water and rainfall has been monitored in the remote pristine area of Isle Royale National Park, MI, located in Lake Superior[22] Atrazine was detected in rainfall from midMay to early-June, corresponding to peak application timing of atrazine in the U.S. Midwestern Corn Belt. Surface water of the lakes in Isle Royale National Park also contained atrazine at trace (part per trillion) levels. These levels are not high enough to be toxic to organisms, but their presence in this remote area point to the need to use caution in the use and application of pesticides.

Conclusion

Although other methods of pest control are important, pesticides continue to be an efficient, cost-effective method of management. The challenge is to select and apply pesticides in a manner that kills the target pest and does not harm non-target organisms or the environment. Sound and sensible management can reduce the probability of pesticides entering surface water. Precision application to only those areas needing treatment is one method to reduce the total amount of pesticide applied. Other techniques include using split applications, banding, and planting filter strips along streams and water courses.

References

1. Aspelin, A.L.; Grube, A.H. *Pesticides Industry Sales and Usage: 1996 and 1997 Market Estimates.* EPA/OPP. http://www.epa.gov/oppbeadl/pestsales/97pestsales. Updated Jan 2000. (accessed September 2001).
2. Larson, S.J.; Gilliom, R.J.; Capel, P.D. *Pesticides in Streams of the United States—Initial Results from the National Water-Quality Assessment Program.* USGS/NAWQA, WRIR98-4222, 1999, http://www.water.wr.usgs.gov/pnsp/ rep/wrir984222 (accessed September 2001).

3. Koplin, D.W.; Nations, B.K.; Goolsby, D.A.; Thurman, E.M. Acetochlor in the hydrologic system in the Midwestern United States, 1994. Environ. Sci. Technol. **1996**, *30*, 1459–1464.
4. Hoffman, R.S.; Capel, P.D.; Larson, S.J. Comparison of pesticides in eight U.S. urban streams. Environ. Toxicol. Chem. **2000**, *19*, 2249–2258.
5. Wotzka, PJ.; Lee, J.; Capel, P.D.; Lin, M. Pesticide concentrations and fluxes in an urban watershed. AWRA Technical Publication Series TPS-4; American Water Resources Association: Herndon, VA, 1994; 135–145.
6. Ferrari, M.J.; Ator, S.W.; Blomquist, J.D.; Dysart, J.E. *Pesticides in Surface Water of the Mid-Atlantic Region*. USGS-WRIR97-4280, 1997, http://www.md.water.usgs.gov/publications/wrir-97-4280 (accessed February 2001).
7. Council for Agricultural Science and Technology. Issue paper 2, April, 1994. *Pesticides in Surface and Groundwater.* Revised May, 1997. CAST. http://www.castscience.org/ pwq_ip.htm (accessed February 2001). 818 Pesticide Contamination: Surface Water.
8. U.S. Environmental Protection Agency, *National Recommended Water Quality Criteria*, USEPA 822-Z-99-001; USEPA: Washington, DC, 1999.
9. Thompson, A.R.; Edwards, C.A. Effects of pesticides on nontarget invertebrates in freshwater and soil. In *Pesticides in Soil and Water;* Guenzi, W.D., Ahlrichs, J.L., Chesters, G., Bloodworth, M.E., Nash, R.G., Eds.; Soil Sci. Soc. Amer.: Madison, WI, 1974; 341–386.
10. Harp, G.L.; Campbell, R.S. Effects of the herbicide silves on benthos of a farm pond. J. Wildl. Manag. **1964**, *28*, 308317.
11. Hall, J.K.; Hartwig, N.L.; Hoffman, L.K. Application mode and alternative cropping effects on atrazine losses from a hillside. J. Environ. Qual. **1983**, *18*, 439–445.
12. Gaynor, J.D.; van Wesenbeeck, I.J. Effects of band widths on atrazine, metribuzin, and metoachlor runoff. Weed Tech- nol. **1995**, *9*, 107–112.
13. Ma, L.; Spalding, R.F. Herbicide mobility and variation in agricultural runoff in the beaver creek watershed in Nebraska. In *Herbicide Metabolites in Surface Water and Groundwater;* ACS Symposium Series 630; Meyer, M.T., Thurman, E.M., Eds.; American Chemical Society: Washington, DC, 1996; 226–236.
14. Spencer, W.F.; Cliath, M.M. Movement of pesticides from soil to the atmosphere. In *Long Range Transport of Pesticides;* Kurtz, D.A., Ed.; Lewis Publishers: Chelsea, MI, 1990; 1–16.
15. Nash, R.G.; Hill, B.D. Modeling pesticide volatilization and soil decline under controlled conditions. In *Long Range Transport of Pesticides;* Kurtz, D.A., Ed.; Lewis Publishers: Chelsea, MI, 1990; 17–28.
16. Ciba-Geigy Corporation. *Biological Assessment of Atrazine and Metolachlor in Rainfall,* Technical Paper 1; Bern, Switzerland, 1993; 4 pp.
17. Chepil, W.S. Dynamics of wind erosion: I. Nature of movement of soil by wind. Soil Sci. **1945**, *60*, 305–320.
18. United States Department Agriculture. *Summary Report: 1997 National Resources Inventory. Revised Dec. 2000; USDA Natural Resources Conservation Ser.: Washington, DC, 2000,* http://www.nhq.nrcs.usda.gov/NRI/1997/sum- mary_report (accessed September 2001).
19. Clay, S.A.; DeSutter, T.M.; Clay, D.E. Herbicide concentration and dissipation from surface wind-erodible soil. Weed Sci. **2001**, *49*, 431–436.
20. Richards, R.P.; Kramer, J.W.; Baker, D.B.; Krieger, K.A. Pesticides in rainwater in the Northeastern United States. Nature **1987**, *327*, 129–131.
21. Kurtz, D.A. *Long Range Transport of Pesticides;* Lewis Publishers: Chelsea, MI, 1990; 462 pp.
22. Thurman, E.M.; Cromwell, A.E. Atmospheric transport, deposition, and fate of triazine herbicides and their metabolites in pristine areas of Isle Royale National Park. Environ. Sci. Technol. **2000**, *34*, 3079–3085.

23

Stormwaters: Management

Introduction 153
Stormwater Causes 153
Water Quality Characteristics of Stormwater 154
Stormwater Control 154
Low-Impact Development • Best Management Practices • Practices to Reduce Peak Flow Rate • Practices to Reduce Runoff Volume • Bioretention • Grassed Swales • Green Roofs • Infiltration Basins • Infiltration Trenches • Pervious Pavement • Rainwater Collection/Reuse • Sand Filters • Operation and Maintenance
Conclusion 163
References 163

John C. Clausen and
Michael E. Dietz
University of Connecticut

Introduction

Stormwater is excess water running off the surface of the land during and immediately after a period of rain.[1] It is also termed rainfall excess, representing water that is not infiltrated into the soil. Direct runoff and surface runoff are other synonyms. A related term is overland flow, which is water that travels over the surface of the land before it enters a water body, best management practice (BMP), or stormwater collection system.[1] Stormwater management is generally regarded as everything to do with the handling and control of stormwater. This entry emphasizes stormwater management from the perspectives of stormwater peak control and stormwater volume control. Water quality aspects of stormwater management are also explained. BMPs are briefly described and their pollutant removal effectiveness is summarized.

Stormwater Causes

Stormwater is generally caused by rainfall rates exceeding the infiltration capacity of the soil's surface. The melting of snow can also result in stormwater. Rainfall excess is commonly shown graphically by comparing infiltration rates and rainfall rates (Figure 23.1). Depression storage capacity is the volume of water required to fill surface depressions to their overflow levels.[1] If the rainfall intensity exceeds the infiltration capacity of the soil, this excess begins to fill depression storage. Infiltration is defined as the process of water entering the surface of the soil. As a rainstorm progresses, this rate typically declines over time, as shown in Figure 23.1, due to spaces in the soil being filled with water. Saturated areas in watersheds, often near streams, are more likely to be a source of runoff as compared to other parts of the watershed. These have been termed saturated source areas.

Urbanization is known to increase stormwater runoff.[2] Stormwater is commonly thought to be caused only by impervious surfaces in urbanized areas, also called total impervious area.[3] Typical impervious areas include streets, roof tops, and parking lots. Some of these areas might drain to pervious areas,

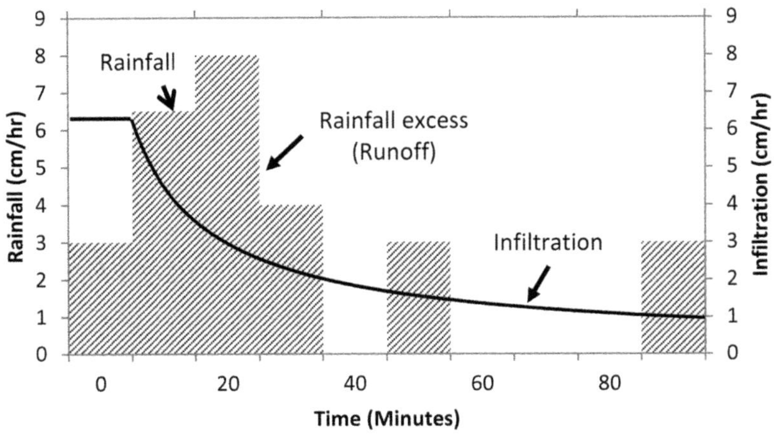

FIGURE 23.1 Hypothetical rainfall, infiltration, and runoff during a rainstorm.

and therefore are not directly connected to downstream areas. However, soil alteration from construction and compaction can also cause stormwater runoff; thus, a surface like a lawn that is commonly thought to be pervious might generate significant stormwater runoff, especially if the rain storm has a high intensity. The term "effective impervious area" refers to impervious areas that are directly connected to downstream water bodies and can include the typical impervious areas and compacted or low-permeability soils.[3]

Water Quality Characteristics of Stormwater

The water quality characteristics of stormwater are a function of its source. For example, pollutants from land surface sources that are washed into surface water resources often have components that are particulate as well as dissolved. Subsurface sources of pollutants are primarily in the dissolved form. Nutrients, such as phosphorus, and metals are found mostly in particulate forms in stormwater runoff. Nitrogen in stormwater runoff can be in several forms, and the dominant form may change within a BMP due to nitrogen cycling. Treatment of the quality of stormwater runoff must consider the pollutants involved and their form, whether particulate or dissolved. Stormwater is known to contain increased concentrations of sediment, nutrients (nitrogen, phosphorus), metals (copper, lead, zinc), bacteria, and hydrocarbons. Increasingly, chloride and other deicing chemicals are found in stormwater runoff.

Stormwater Control

The control of stormwater is conducted both at a watershed scale and through the use of individual practices. The concept of low-impact development (LID) has emerged for primarily new residential developments. The individual practices are known as BMPs. Both approaches are discussed.

Low-Impact Development

The goal of LID is the preservation of the predevelopment hydrology of a site. More specifically, predevelopment runoff peak, volume, and timing are to be maintained. To accomplish these goals, careful planning needs to begin long before construction. Design concepts include preserving functional streams and wetlands, fitting a development to the terrain of the site, and preserving as much native vegetation and soils as possible. The amount of impervious surface that is directly connected to the stormwater conveyance system is typically minimized in a LID. Common practices for impervious area reduction include bioretention, on-site infiltration, pervious pavements, vegetated swales, rainwater harvesting, and vegetated or "green" roofs. Road and driveway sizes and lengths also can be minimized, to a point consistent with safety standards.

Best Management Practices

To control the adverse effects of urbanization on stormwater, BMPs have been developed. A stormwater BMP is a practice, either structural or nonstructural, to reduce nonpoint source pollution from stormwater. A report from the National Research Council[4] recommends use of the term stormwater control measure (SCM) rather than BMP. Initially, stormwater management practices were aimed to mitigate the increased peak flow rate that occurs from development. Practices were designed to treat the 100-year, 50-year, or 25-year 24-hour rainfall amount.[5] The 100-year, 24-hour storm is the rainfall amount, based on historical records, that is expected to occur once every 100 years for a 24-hour duration. The following practices are used to reduce the peak flow rate, thereby decreasing downstream flooding and bank scouring. Stormwater prevention is perhaps the most cost efficient mechanism for stormwater control. Approaches for stormwater prevention include education, proper land use planning and site design, reduction of impervious areas, and preservation of undisturbed soils.[4] Many BMPs require pretreatment and all require regular maintenance, further described later.

Practices to Reduce Peak Flow Rate

Common practices traditionally used to control peak flow rate associated with stormwater runoff are listed in Table 23.1. Other practices are possible, but may only provide peak runoff control for more frequent, smaller storms (e.g., 2-year, 10-year). The overall approach to peak control is to first provide available storage volume, and second to provide a control structure that lets the water out more slowly at a rate that mimics predevelopment conditions. Commonly used outlet control structures include weirs, pipes, and orifices. Dry detention basins were the first practices used to reduce peak flow rates. A dry detention basin is an excavated area with an outlet control structure, and it typically does not have a permanent pool of water. Stormwater inflow is stored in the basin and released over an extended period of time, usually 48 hours or less. However, these basins do not typically reduce the volume of stormwater leaving the basin. The bottom of the basin tends to become compacted and clogged with sediment, thus reducing infiltration into the soil. Extended detention basins are intended to store the water for longer time periods, which provides more time for water quality treatment (Figure 23.2). These basins can be wet (ponded), dry, a combination, or include a wetland. Wet ponds are similar to dry ponds except that their outlet structure is elevated; therefore, the pond maintains a more permanent pool of water. Because wet ponds allow more time for particle settling, they provide additional water quality treatment.[6] Stormwater wetlands can be added to detention basin bottoms to enhance water quality treatment, at least temporarily.

Research on detention basins has shown that they have variable water quality benefits (Table 23.1). The existence of the permanent pool (Figure 23.3) of water may help to settle out more suspended solids and pollutants attached to particulates as compared to detention basins, but their ability to retain nutrients such as N and P is variable, and generally poor (Table 23.1). These systems may also increase

TABLE 23.1 Commonly Used Stormwater BMPs for Peak Control and Their Pollutant Removal Efficiencies

	Pollutant Removal (%)					
BMP	Sediment	Nitrogen	Phosphorus	Metals	Pathogens	Source
Detention basins	30–65	15–45	15–45	15–45	<30	[20]
Extended detention basins	5–90	20–60	10–55	25–65	–	[21]
Wet ponds	−30–91	5–85	10–85	10–95	–	[21]

FIGURE 23.2 **(See color insert.)** Extended detention basin.

FIGURE 23.3 **(See color insert.)** Wet pond.

the temperature of runoff waters. The addition of wetlands can improve treatment performance temporarily, but as the wetland vegetation dies and decomposes, additional N and P will be released and the wetland can become a source of N and P.

Practices to Reduce Runoff Volume

BMPs reduce runoff volume by infiltrating it into the soil, by enhancing evapotranspiration, or by water reuse (Table 23.2). Additional practices can provide some volume control on a more limited basis, such as grassed swales and buffer strips. BMPs with permanent pools generally provide little volume control because little or no infiltration occurs. Constructed wetlands and extended wet detention basins typically have permanent pools of water and volume reduction is not significant. Various studies have shown that the volume of runoff leaving a site can be reduced to predevelopment levels, at least for smaller (<25 yr) storms.[7] Pollution reduction by volume control practices vary.

Bioretention

Bioretention is a vegetated depression in the landscape, which may have an engineered media that is designed to collect and infiltrate stormwater.[8] Unlike previously used "end-of-pipe" BMPs, one major advantage of bioretention is that it can be used in residential settings to treat runoff close to the source of generation. Bioretention areas, or rain gardens as they are commonly called in a residential setting, can be functional practices that are integrated into residential landscaping (Figure 23.4).

Mechanisms for pollutant retention in bioretention include adsorption to mulch and soil media, exfiltration into native sub soils, plant uptake, and bacterial degradation. Laboratory research on bioretention has indicated the potential for excellent retention of several pollutants of concern such as N, P, copper, lead, and zinc.[9] More recent field research on bioretention has shown variable N and P removal, but high metals removal.[10,11] The variable N and P removal is due to the decomposition of dead plant material that releases nitrogen and phosphorus as well as differences in the absorptive capacity of the media. Sand, for example, does not retain P well. Chloride may not be well retained by this practice, but bacteria treatment has been successful.[11] Downstream thermal loading can be reduced simply by a reduction in volume transported. However, the effects of bioretention on stormwater temperatures have been mixed, showing little effect in one case[12] and reductions in another.[13] Bioretention use has expanded greatly throughout the Unites States since its introduction. Bioretention installations are also found in New Zealand and Australia.[14,15] By comparison, vegetated filter strips are shallow planted areas designed to spread stormwater out and slow it down. This allows more particles to settle and be filtered and more water to infiltrate into the soil.

FIGURE 23.4 Household rain garden.

FIGURE 23.5 A grassed swale.

TABLE 23.2 Commonly Used Stormwater BMPs for Volume Control and Their Pollutant Removal Efficiencies

	Pollutant Removal (%)					
BMP	Sediment	Nitrogen	Phosphorus	Metals	Pathogens	Source
Bioretention	54–99	30–99	-249–99	44–99	–	[10,11]
Green roofs	–	32	-191	-50–71	–	[17]
Grassed swales	30–88	15–45	15–45	2–67	<30	[18,20]
Infiltration basins	50–80	50–80	50–80	50–80	65–100	[20]
Infiltration trenches	50–80	50–80	15–45	50–80	65–99	[18,20]
Porous pavement	65–100	65–100	30–65	65–100	65–100	[20]
Sand filters	70	21	33	45	76	[19]

Grassed Swales

Vegetated swales can be used as an alternative to conventional curb and gutter conveyance systems (Figure 23.5). Modern swales, also termed "bioswales,"[4] are designed differently from the older ditches that were just meant to convey runoff. Modern swales are engineered to provide storage and infiltration of stormwater, but they also will convey runoff from larger events to reduce flooding potential. However, pollutant removal is generally low in swales (Table 23.2). Vegetated swales are similar to vegetated filter strips, but they are designed to convey stormwater as well. They may retain high percentages of suspended solids, N, and P, but the retention rates are variable (Table 23.2).

Green Roofs

Vegetated or "green" roofs have been used for centuries in northern Europe. Benefits of green roofs include retention of rainfall, and in some cases, energy cost savings.[16] The traditional or "intensive" green roof consists of a thick soil layer capable of supporting grasses, plants, and sometimes even trees. As would be expected, the building structure below a green roof needs to be specially engineered to support its weight safely. An "extensive" green roof consists of a shallow media (10–15 cm) of expanded clays or other light materials. The advantage of the extensive system is that the weight is much less,

FIGURE 23.6 An extensive green roof.

so the system can be applied with less structural support. However, this type and thickness of media cannot support shrubs or trees. Typically, extensive green roofs are planted with *Sedum* species, which are highly tolerant of the harsh conditions that exist on a rooftop (Figure 23.6). The amount of rainfall that is retained by a green roof depends on the type of roof, the vegetation, as well as the regional climate. Retention can be a substantial portion of annual precipitation.[17] The green roof is an ideal BMP for highly developed urban areas where little greenspace exists for traditional BMPs, but open rooftops are plentiful.

Infiltration Basins

Infiltration basins are large areas designed to infiltrate stormwater runoff. These types of basins are generally constructed in well-draining native soils, or soils that have been amended to encourage infiltration. They are designed to infiltrate most or all stormwater inflow, although they can be designed with storage and outlet control structures to provide more time for infiltration. An underdrain may also be installed to provide hydraulic relief if native soils have low infiltration capacity. Systems containing underdrains, however, typically provide little or no volume reduction. Due to the large volume of stormwater infiltrated with these systems, pollutant removal is generally high (Table 23.2).

Infiltration Trenches

Stormwater runoff can also be piped to underground storage areas, called infiltration trenches. An infiltration trench is filled either with crushed stone or housed in other types of structures (concrete or plastic). The structure or the stone act as the storage reservoir so that part or all of the water can infiltrate into surrounding soils.[18] Solids should be removed through pretreatment to avoid clogging of these types of systems. Pollutant removal is dependent on design, but generally is high due to the large storage volume and potential recharge to groundwater (Table 23.2).

Pervious Pavement

Several alternatives to traditional pavements have been developed: pervious interlocking concrete pavers (PICPs), plastic grid pavers, pervious asphalt, and pervious concrete. All are pervious and allow water to flow through the surface into the base materials. Although they differ in construction and appearance,

FIGURE 23.7 Eco-stone® pavers.

all pervious pavements have one thing in common: the preparation of the base materials below the pavement is very different than with traditional pavement. For traditional pavement, a base of processed gravel, which contains a high proportion of fine material, gets compacted to provide a stable base for the pavement. Water flow through this compacted layer is slow. Base preparation for pervious pavements involves a crushed stone reservoir, a choker course of smaller aggregate, and then the pavement. The thickness of the reservoir depends on site needs, and in some cases this reservoir can be designed to act as a below-ground retention system to contain the 100-year storm.

Many types of PICPs now exist. In concept PICPs are similar to traditional concrete paver blocks; however, they are constructed in such a way that there are spaces between the blocks when they are installed. These voids are filled with crushed stone or gravel, and allow runoff to pass by the paver block into the subsurface layers (Figure 23.7). Several types of plastic grid pavers also have been developed. They are constructed of various types of plastic, and are available in rolls or grids that lock together. The grids are highly pervious, and can be filled with topsoil and planted with turf, or filled with crushed aggregate.

Pervious asphalt has been in use for several decades (Figure 23.8), although it was not initially used as an LID tool. It was originally termed open-graded friction course (OGFC), and was used on highways to improve traction and reduce road spray. The mix is similar to traditional asphalt, except the sand component is omitted. Numerous asphaltic binders have been developed to improve adhesion, and promote stability in hot conditions. One advantage to this type of pavement is that it can be applied with the same equipment that is used for traditional asphalt. To avoid potential clogging of the pores in this material, winter sanding should be avoided. Periodic maintenance with vacuum suction or power washers may be required to remove fines that have been deposited on the surface.

Pervious concrete is a variation of traditional concrete, where the sand is omitted from the mix (Figure 23.9). The ratio of Portland cement, water, and aggregate is very specific. The mix then gets rolled or tamped to a specific density. The goal is to make sure there is sufficient water in the mix to react with the Portland cement and bond the aggregates together, without sealing the surface. The application of this type of concrete is different from traditional concrete, and should only be performed by a contractor with previous installation experience. As with pervious asphalt, winter sand application should be avoided to reduce clogging potential, and just as with salt application on traditional concrete, care should be taken to use deicing agents that will not react with the concrete.

FIGURE 23.8 Pervious asphalt.

FIGURE 23.9 A pervious concrete sidewalk.

FIGURE 23.10 Wooden rain barrel.

Rainwater Collection/Reuse

Rainwater collection has occurred for thousands of years to provide water for domestic uses. Modern rainwater harvesting is encouraged to reduce stormwater runoff volumes and provide some water for irrigation of landscaping plants. Rain barrels are the cheapest and easiest form of rainwater collection, although distribution of the water is limited to gravity draining (Figure 23.10). Larger cisterns can be installed with filters and pressurized pump systems. Much larger volumes of water can be captured and utilized with a cistern, but installation is more costly. Typically, however, rainwater harvesting does not collect a large fraction of runoff.

Sand Filters

Sand filters have been used for more than 100 years to treat water for drinking, but their use has expanded to stormwater. Sand filters are effective in removing several pollutants but provide little flow rate control.[19] Larger particles and the pollutants that are adsorbed to them can be physically strained and retained by the sand filter. Smaller particles and dissolved pollutants typically pass right through the filter without treatment. Most pollutants such as metals and P will occur in both particulate and dissolved form. Thus a sand filter can only remove a fraction of the total pollutant load. Others, such as N, and specifically nitrate nitrogen (NO_3–N), are almost all dissolved and essentially no removal will be achieved. Three designs of sand filters are in common practice: the Austin (Texas) sand filter, the Washington, DC, sand filter, and the Delaware sand filter.[19] The Austin filter is above ground or at grade while the Washington, DC, and Delaware filters are in below ground concrete tanks. The Austin sand filter is used for larger watersheds with mixed land uses up to 50 acres. The Washington, DC, and Delaware systems are used for smaller (1 acre) highly impervious watersheds. Experiments are being conducted on sand filters with peat and metal shavings for additional pollutant treatment.

Operation and Maintenance

All stormwater BMPs require maintenance, the amount dependent on the actual practice, watershed inputs, and precipitation. They also have a design life, and require rejuvenation or replacement when that time is exceeded. Additionally, BMP pretreatment may extend the life of the practice. Because stormwater BMPs are found at the low end of the landscape and water flows into them, they are often collectors of trash and debris. Such debris can interfere with some treatment processes and should be regularly removed. Basins and ponds have outlets that can also clog and require regular maintenance. Forebays to such systems may prevent them from filling up with sediment, but these should be regularly cleaned. Bioretention systems may require media replacement, mulch renewal and plant maintenance. Excess mulching could fill up these BMPs. Infiltration BMPs and sand filters are prone to clogging. Operation and maintenance of sand filters is especially important. Clogging may occur within 3–5 years after construction, and the media may need replacement. Clogging may also occur with porous pavement and prevention of clogging is an important consideration.

Operation and maintenance are complicated by ownership of the BMP. Practices on private land may not receive the attention they are given in municipalities.

Conclusion

The management of stormwater is necessitated by the effects of urbanization on infiltration rates. Stormwater prevention is the first and perhaps best tool in reducing stormwater runoff. Several BMPs have been successfully applied in reducing the peak flow rate and the volume of runoff. Some of these practices work well for large rarer rain storms, whereas others only perform well for small frequent storms. Pollution retention by BMPs varies widely and depends on the pollutant and BMP type, the watershed land uses, the original design of the BMP, pretreatment, and the performance of regular maintenance.

References

1. American Public Health Association. *Glossary: Water and Wastewater Control Engineering;* 1969; 387 pp.
2. Leopold, L.B. *Hydrology for Urban Land Planning – A Guidebook on the Hydrologic Effects of Urban Land Use.* U.S. Geological Survey Circular 554, 1968.
3. Booth, D.B.; Jackson, C.R. Urbanization of aquatic systems: degradation thresholds, stormwater detection, and the limits of mitigation. J. Am. Water Res. Assoc. **1997,** *33* (5), 1077–1090.
4. National Research Council. *Urban Stormwater Management in the United States*; The National Academies Press: Washington, DC. 2008.
5. USDA-Soil Conservation Service. *Urban Hydrology for Small Watersheds;* Technical Release 55: 1986.
6. Schueler, T.R. *Controlling Urban Runoff: A Practical Manual for Planning and Designing Urban BMPs.* Metropolitan Information Center. Metropolitan Washington Council of Governments. Washington, D.C. 1987.
7. Bedan, E.S.; Clausen, J.C. Stormwater runoff quality and quantity from traditional and low impact development watersheds. J. Amer. Water Res. Assoc. **2009,** *45,* 998–1008.
8. Clar, M.L.; Green, R. *Design Manual for Use of Bioretention in Stormwater Management;* Prince George's County, MD Department of Environmental Resources, Watershed Protection Branch, MD Department of Environmental Protection: Landover, MD. 1993.
9. Davis, A.P.; Shokouhian, M.; Sharma, H.; Minami, C. Laboratory study of biological retention for urban stormwater management. Water Environ. Res. **2001,** *73* (1), 5–14.
10. Dietz, M.E. Low impact development practices: a review of current research and recommendations for future directions. Water, Air, Soil Pollut. **2007,** *186,* 351–363.
11. Davis, A.P.; Hunt, W.F.; Traver, R.G.; Clar, M. Bioretention technology: overview of current practice and future needs. J. Environ. Eng. **2009,** *135* (3), 109–117.
12. Dietz, M.E.; Clausen, J.C. A field evaluation of rain garden flow and pollutant treatment. Water, Air, Soil Pollut. **2005,** *167,* 123–128.

13. Jones, M.P.; Hunt, W.F. Biotretention impact on runoff temperature in trout sensitive waters. J. Environ. Engr. **2009,** *135* (8), 577–585.
14. Read, J.; Wevill, T.; Fletcher, T.; Deletic, A. Variation among plant species in pollutant removal from stormwater in biofiltration systems. Water Res. **2008**, *42* (4–5), 893–902.
15. Trowsdale, S.A.; Simcock, R. Urban storm water treatment using bioretention. J. Hydro. **2011**, *397* (3–4), 167–174.
16. Hoffman, L. *Green Roofs: Ecological Design and Construction;* Schiffer Publishing Ltd.; PA. 2005.
17. Gregoire, B.G.; Clausen, J.C. Effect of a modular extensive green roof on stormwater runoff and water quality. Ecol. Engr. **2011,** *37,* 963–969.
18. ASCE. *Guide for Best Management Practice (BMP) Selection in Urban Developed Areas*; American Society of Civil Engineers: Reston, VA, 2001.
19. U.S. EPA. *Storm Water Technology Fact sheet: Sand Filters;* United States Environmental Protection Agency: Washington, D.C., EPA 832-F-99-007, 1999.
20. U.S. EPA. *Preliminary Data Summary of Urban Storm Water Best Management Practices*; United States Environmental Protection Agency: Washington, D.C., EPA-821-R-99-012, 1999.
21. U.S. EPA. *Guidance Specifying Management Measures for Sources of Nonpoint Pollution in Coastal Waters*; United States Environmental Protection Agency: EPA-840-B-92-002, 1993.

24

Stormwater: Urban Modeling

Miguel A. Medina, Jr.
Duke University

Cevza Melek Kazezyılmaz-Alhan
Istanbul University-Cerrahpaşa

Introduction 165
Rainfall Analysis and Climate Change 166
Nonstationary Time Series Analysis
Overland Flow 175
Channel Flow 176
Low Impact Development-Best Management Practices (LID-BMPS) 177
Environmental Protection Agency Stormwater Management Model (EPA SWMM) 178
LID Modeling with EPA SWMM
EPA SWMM Applications for Overland and Channel Flow 179
Duke University Campus Watersheds • Sazlidere Watershed in Istanbul, Turkey
EPA SWMM Applications for LID Implementations 184
Hydrological Model of Rainfall-Watershed-Bioretention (RWB) Experimental System • Hydrological Model of Avcılar Campus of Istanbul University-Cerrahpaşa
References 188

Introduction

Elements of the hydrologic processes that describe the transformation of precipitation into surface runoff are depicted in Figures 24.1 and 24.2. Figure 24.1 illustrates the relationship among the rainfall process, infiltration to the subsurface, generation of overland flow, and flow into the man-made drainage system. Among the many variables that describe these processes mathematically are the width, area, percent imperviousness, ground slope, roughness parameters of the land cover for both impervious and pervious fractions, and several infiltration rate parameters that depend upon methods chosen. Eventually, the runoff makes its way into inlets of the man-made drainage system.

Figure 24.2 illustrates the transport of runoff through both natural and man-made conduits, as well as routing through a detention structure. Whether natural or man made, the length, geometry, slope, and roughness characteristics must be specified for each conduit. Natural channels can be approximated with triangular, trapezoidal, and other cross-sectional shapes. For storage elements such as detention basins (there is also storage along a channel), stage discharge curves and other geometric data are required. Although the focus of this entry is to demonstrate the physical basis for the hydrologic components represented, any mathematical model is an abstraction of the actual physical system it attempts to simulate. The limitations of numerical modeling and data requirements are discussed further, as well as the need for calibration. The authors recognize that there is more to urban water management than modeling. At the same time, management without application of these powerful tools is

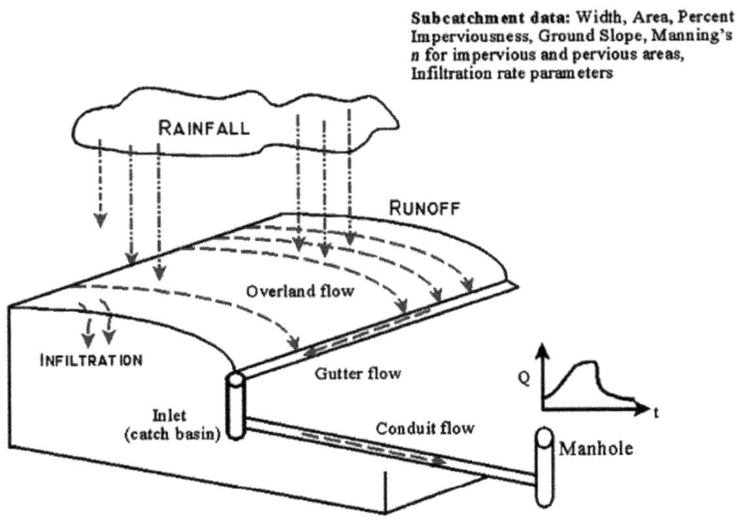

FIGURE 24.1 Rainfall, infiltration, overland flow, and flow into drainage system.

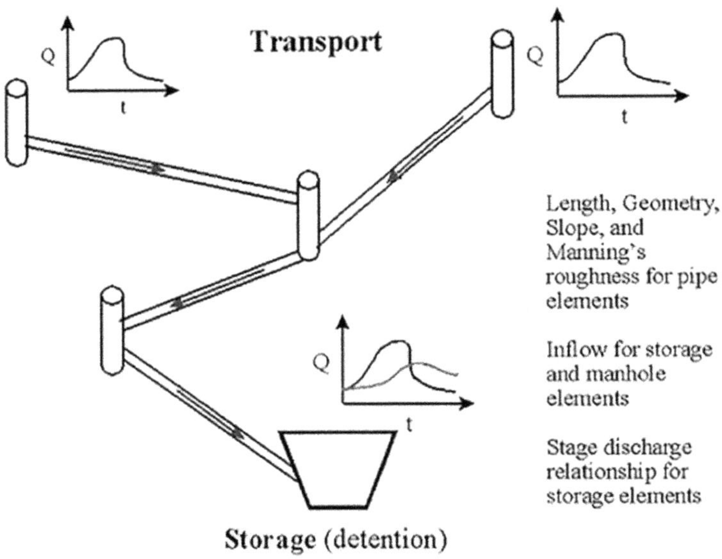

FIGURE 24.2 Transport and storage components of the drainage system.

missing a planning tool that could be used very effectively to compare alternative quantity and quality control strategies. Moreover, a complete stormwater modeling of urbanized region will include rainfall analysis and climate change of the relevant region and integration of stormwater control measures, i.e., Low Impact Developments.

Rainfall Analysis and Climate Change

A thorough rainfall analysis of historical long-term observations is an essential requirement prior to detailed urban hydrologic model simulations of the runoff generated or to be expected if climate change is taken into consideration. Until recently, statistical analyses of historical rainfall have assumed that the time series were stationary. Stationarity assumes that the statistical properties of the series (such as its mean, variance, covariance, and autocorrelation) are all constant over time. The easiest method to determine if the series is stationary or not is to examine whether a trend is present over time in the mean of the series. Some hydrologists argue that hydrologic processes have never been stationary, and will never be, since hydrologic observations typically consist of both deterministic and stochastic components [1].

A common basis for comparison among events is the average recurrence interval (ARI) in years. To determine precipitation record historical extreme events, several options are available. Event ARIs may be computed from [2] as follows:

$$T = (n+1-2\alpha)/(m-\alpha) \tag{24.1}$$

where

T = average recurrence interval (e.g., in years)
n = number of peak values (e.g., number of years)
m = relative ranking of values (largest = 1)
α = a constant ($0 < \alpha \leq 1$).

Cunnane (1978) showed [$0.375 < \alpha \leq 0.44$] where the lower bound corresponds to a normal distribution, and the upper bound is for the extreme value type I distribution. The annual exceedance probability (AEP) may be defined as:

$$p_n = 1-(1-p)^n \tag{24.2}$$

where p (=$1/T$) is the annual exceedance probability.

A reliable source of precipitation data is the NOAA NCEI (National Centers for Environmental Information, formerly the National Climatic Data Center): access to a large number of national and international rainfall stations and their data is available online. The historical record of hourly precipitation from 1948 to 2013 at Raleigh–Durham Airport (RDU), only 12 miles from Durham, NC, was analyzed. The storm event ARIs for the top 25 events, in terms of volume (depth) for statistically independent events [3] separated by at least 6 hours of no rainfall, are presented in Table 24.1. It has long been recognized that the large number of factors that affect surface runoff quantity (rainfall duration, intensity, time between events, volume, infiltration, antecedent soil moisture, etc.) and surface runoff quality (pollutant loads and buildup between storms, surface wash off, transport, kinetic interactions, etc.) limits the exclusive use of any single event for a general understanding of historical storm event characteristics. To properly analyze the time series, storm events must first be defined in terms of their statistical independence: an accepted approach is to derive a minimum intervent time (MIT), such that the intervals between storm event midpoints are nearly exponentially distributed [3]. Storm event statistics for RDU for an hourly historical record are presented in Table 24.2.

The statistics presented in Table 24.2 characterize *historical storm events* separated by a minimum intervent time (MIT), without application of a specific probability distribution. An examination of Figure 24.3, where the independent storm event (ISE) annual maxima are plotted along with a trend line, reveals a distinct upward trend over time: suggesting some nonstationary characteristics are present in the time series that warrant further analysis. A Mann–Kendall trend test is not required.

Nonstationary Time Series Analysis

If it is believed that the extremes of the data are nonstationary, then it is possible to incorporate this knowledge into the parameters. For example, we can assume that the location parameter of the extreme probability distribution selected is a linear function of time (of log-transformed values):

$$\mu_t = \mu_0 + \mu_1 t \tag{24.3}$$

TABLE 24.1 Average Recurrence Intervals for Total Storm Depth at RDU

ARI (years)	Rank	Depth (in.)	Date	Time of Day	Event Hours
70.4	1	7.45	09/04/1999	14:00	44
36.3	2	6.44	09/14/1999	23:00	35
22.2	3	5.79	10/10/2002	19:00	27
19.6	4	5.64	06/14/2006	3:00	13
17.9	5	5.53	06/06/2013	12:00	29
16.0	6	5.4	08/16/1955	18:00	34
13.3	7	5.19	09/05/2008	15:00	21
10.3	8	4.91	11/05/1963	17:00	37
9.8	9	4.86	04/25/1978	6:00	54
9.7	10	4.85	10/04/1995	8:00	27
6.5	11	4.45	11/10/2009	10:00	67
6.2	12	4.41	03/17/1998	16:00	43
6.2	13	4.4	05/11/1957	11:00	18
5.6	14	4.31	08/06/2011	4:00	7
5.3	15	4.25	07/23/1997	23:00	13
4.9	16	4.18	08/20/1986	6:00	11
4.5	17	4.1	10/14/1954	24:00	16
4.3	18	4.06	09/04/2006	20:00	7
3.9	19	4	11/20/1985	13:00	55
3.6	20	3.92	10/13/1994	13:00	32
3.4	21	3.86	03/17/1983	2:00	39
3.3	22	3.82	11/09/1962	6:00	21
2.9	23	3.73	11/12/1975	19:00	10
2.9	24	3.73	08/12/1992	16:00	10
2.8	25	3.69	08/26/2008	14:00	36

TABLE 24.2 Statistics of Storm Events, RDU, NC (1948–2013)

Station No.317069; Elevation: 416 ft.	Raleigh-Durham WSFO AP Wake County; Hourly Rainfall	Lat: N 35:52:00 Long: W 78:47:00	Record Length: 64.5 years
Storm Event Variable	**Maximum**	**Mean**	**COV**
No. of storms[1] per year[2]		99.523	0.105
Depth, in.	7.45	0.421	1.394
Intensity, in. h^{-1}	1.21	0.080	1.237
Duration, hours	67	6.852	1.095
Time between events, hours	032	87.502	1.005
Total no. of storms[1]		6.582	

[1] Number of storms is based on a minimum intervent time of **6 hours**, which yielded a **COV (Coefficient of variation) very near 1.0 for time between events** for the rainfall time series.

where μ_0 = signifies the intercept (usually the mean value of the distribution for the first year of the record) and μ_1= the slope of the linear function, while keeping the *scale* (σ) and *shape* ($\mathcal{E}$) parameters constant. The location parameter represents the most likely value of the distribution chosen. The values of μ_t must then be transformed back from log-transformed values. The use of a constant value for σ, the standard deviation of the arithmetic record peaks, will increase with the logarithmic mean distribution. This property is desirable since *nonstationarity* would likely not only alter the mean of the peaks but simultaneously also affect their associated dispersion.

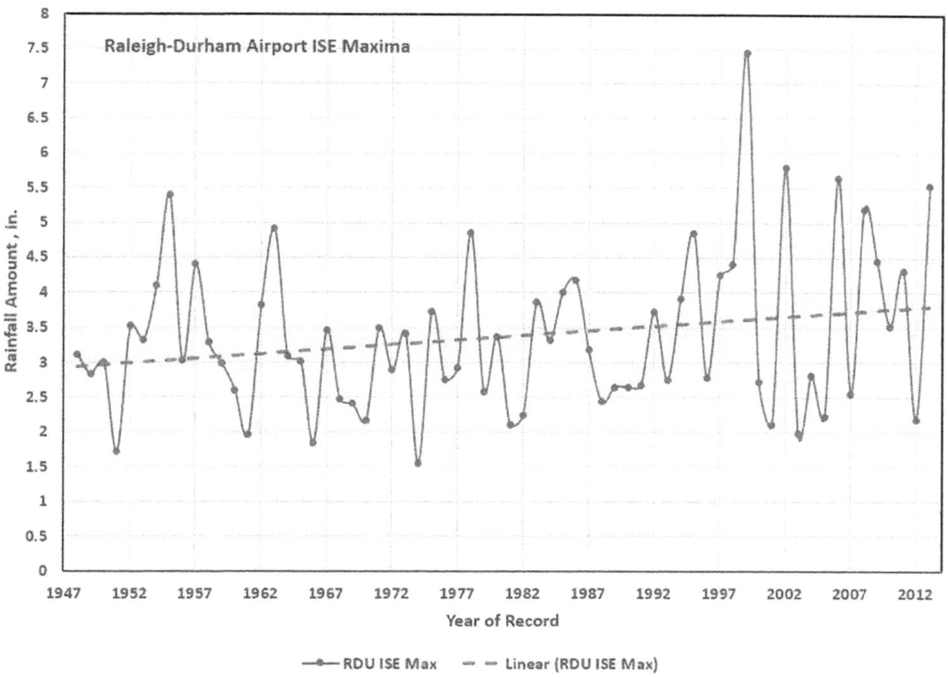

FIGURE 24.3 Computed rainfall annual maxima and trendline for RDU (1948–2013).

A record of block maxima precipitation depths, $\{P_1; \ldots ;P_n\}$, follows the **Log Pearson Type III** (LPIII) distribution if random variable $\mathbf{X} = \text{Log}_{10}(P)$ is distributed according to the PIII distribution. The PIII distribution parameters are defined by μ, σ, and, which represent the mean, standard deviation, and skewness of **X**, respectively. The probability density function, $f(x)$, and cumulative distribution function, $F(x)$, are defined as:

$$f(x) = \frac{(x-\xi)^{\varepsilon-1}\exp(-(x-\xi)/\beta)}{\beta^{\varepsilon}\Gamma(\varepsilon)} \tag{24.4}$$

$$F(x) = \frac{G(\varepsilon,(x-\xi)/\beta)}{\Gamma(\varepsilon)} \tag{24.5}$$

$$\text{where } \varepsilon = 4/\gamma^2, \beta = \sigma\gamma/2 \text{ and } \xi = \mu - 2\sigma/\gamma$$

Here, $G(\bullet)$ and $\Gamma(\bullet)$ represent the gamma and incomplete gamma functions, respectively. Further details are provided by [4]. The parameters of the LPIII distribution in the stationary and nonstationary models are estimated from the first half of the record using Bayesian inference. The second half of the record is reserved to evaluate the predictions under the models.

L-moments are linear functions of probability-weighted moments (PWMs). Unbiased sample estimators of the first four PMWs are given by:

$$\beta_0 = \frac{1}{n}\sum_{j=1}^{n} X_j = \text{mean} \qquad \beta_1 = \sum_{j=1}^{n-1} \frac{n-j}{n(n-1)} X_{(j)}$$

$$\beta_2 = \sum_{j=1}^{n-2}\left[\frac{(n-j)(n-j-2)}{n(n-1)(n-2)}\right]X_{(j)} \qquad \beta_3 = \sum_{j=1}^{n-3}\left[\frac{(n-j)(n-j-1)(n-j-2)}{n(n-1)(n-2)(n-3)}\right]X_{(j)} \tag{24.6}$$

The first four L-moments (λ_1–λ_4) are computed as follows:

$$\lambda_1 = \beta_0 \quad \lambda_2 = 2\beta_1 - \beta_0 \quad \lambda_3 = 6\beta_2 - 6\beta_1 + \beta_0$$
$$\lambda_4 = 20\beta_3 - 30\beta_2 + 12\beta_1 - \beta_0 \tag{24.7}$$

The L-moment ratios are calculated as follows:

$$\tau_2 = \frac{\lambda_2}{\lambda_1} \qquad \tau_3 = \frac{\lambda_3}{\lambda_2} \equiv L - \text{skew}$$
$$\tau_4 = \frac{\lambda_4}{\lambda_2} \equiv L - \text{kurtosis} \tag{24.8}$$

Computation of the precipitation depth X_d (or flood peak flow rate or flood peak volume) for the cumulative generalized extreme value (GEV) distribution is as follows:

$$F = \exp - \left\{ 1 - \frac{\varepsilon}{\sigma}\left[X_d(F) - \mu \right] \right\}^{1/\varepsilon} \tag{24.9}$$

where μ is the location parameter, and σ and ε are the scale and shape parameters, respectively. The parameters of the GEV distribution when estimated using L-moments [5] are:

$$Z = \frac{2}{\tau_3} - \frac{\ln(2)}{\ln(3)} \qquad \varepsilon \approx 7.8590Z + 2.9554Z^2$$
$$\sigma = \frac{\lambda_2 \varepsilon}{\left(1 - 2^{-\varepsilon}\right)\Gamma(1+\varepsilon)} \qquad \mu = \lambda_1 + \frac{\sigma}{\varepsilon}\left\{\Gamma(1+\varepsilon) - 1\right\} \tag{24.10}$$

Application of the Bayes LPIII nonstationary extreme value analysis [4] to the Raleigh–Durham Airport (1948–2013) hourly rainfall record computed ISE block maxima yields the following location parameter:

$$\mu_t = 0.45466 + 0.0015t \tag{24.11}$$

where t = year of record from 1948 to 2013. The equation above represents the most likely (mean) value, and the values are plotted in Figure 24.2, along with the computed block maxima and the predicted low (2.5%) and high (97.5%) values of the location parameter. As noted earlier, the μ_t values must be transformed back from logarithmic form: for example, $\mu_t = 0.455$ for $t = 1948$ and the most likely rainfall amount is then 100.455 = 2.85 in. The return level versus return period (ARI) curves are plotted in Figure 24.3. The nonstationary time series estimates are compared to the stationary time series estimates in Table 24.3 (Figures 24.4 and 24.5).

A similar analysis, applying the nonstationary **General Extreme Value** (GEV) distribution methodology to the RDU block maxima rainfall time series, yields a time variable location parameter equation as follows:

$$\mu_t = 0.433 + 0.007t \tag{24.12}$$

The return level versus return period curves are plotted in Figure 24.6.

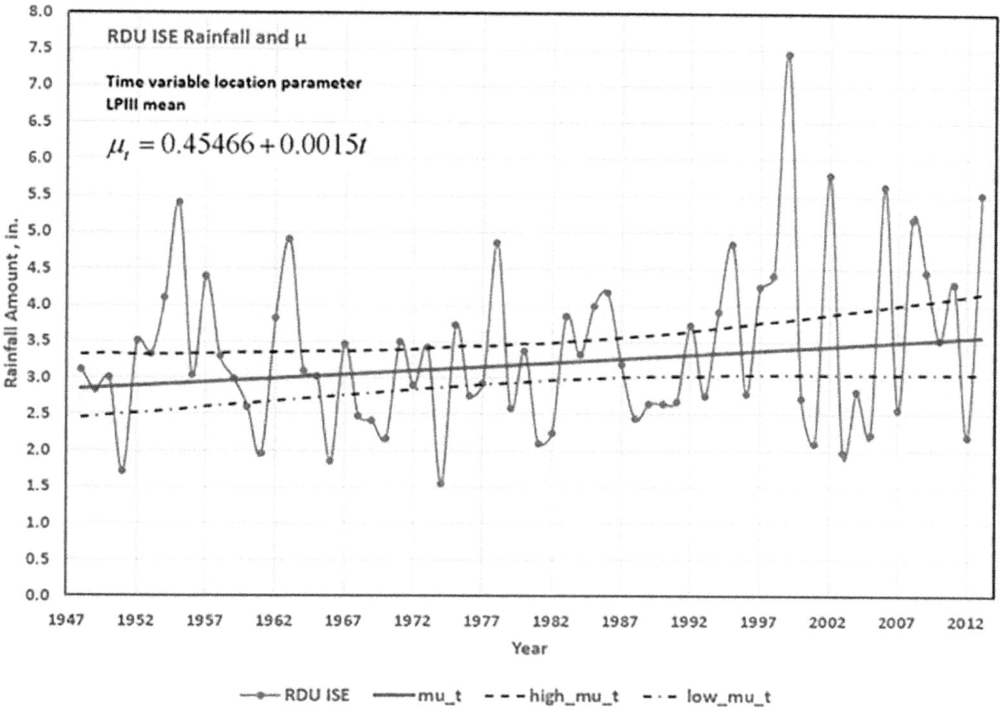

FIGURE 24.4 RDU ISE block maxima and LPIII time variable location parameter.

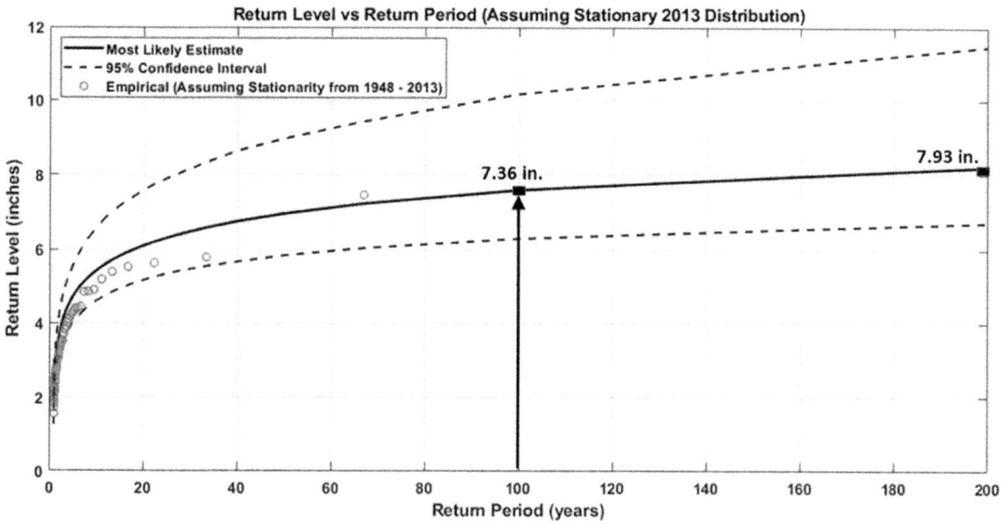

FIGURE 24.5 Return level versus return period nonstationary LPII predictions.

The results for both applications are summarized in Table 24.4. The predictions are comparable for the median (most likely values) for both distribution methodologies. The level of uncertainty increases toward the 200-year return period, such that the 95% confidence limit for the 200-year return period indicates levels as high as 12 to almost 14 in.

Extreme storm event analysis has implications for the design of urban stormwater drainage structures under a changing climate. When using the return period T (or ARI) as the design criteria, it is useful

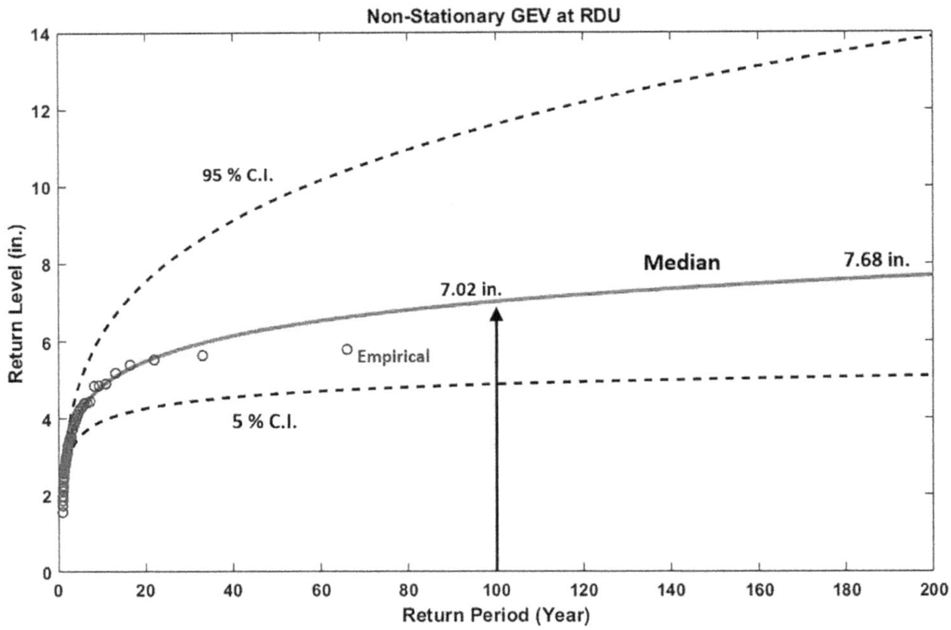

FIGURE 24.6 Return level versus return period nonstationary GEV predictions.

TABLE 24.3 Stationary versus Nonstationary Time Series Analysis Estimates (Return Levels for ISEs), Raleigh–Durham International Airport, NC (1948–2013)

NOAA Station No. 317069	Lat: N 35:52:00 Long: W 78:47:00 Elevation: 416 ft	Record Length Analyzed: 64.5 years	Stationary Log Pearson Type III
Return Period (years)	97.5% Bayes LPIII Curve (in.)	Bayes LPIII RL[1] (in.)	Sta LPIII (in.)
200	12.00	7.93	7.63
100	10.53	7.36	7.10
50	9.21	6.77	6.42
25	7.98	6.17	5.75

[1] most likely estimate for Bayes LPIII distribution location parameter return level (RL).

to remember that T is the expected waiting time (EWT), which can be extended to the nonstationary condition [1]. The derivation of EWT under nonstationarity leads to a convenient formula [1].

$$T = 1 + \sum_{x=1}^{X\max} \prod_{t=1}^{x} (1 - p_t) \tag{24.13}$$

where p_t = time-varying probabilities and an adequate value for $X_{\max} = 1{,}000$.

For a GEV distribution with the location parameter varying with time and constant scale and shape parameters, the corresponding exceedance probability is given by:

$$p_t = 1 - \exp\left\{ -\left[1 + \frac{\varepsilon}{\sigma}\left(Z_{q0} - \mu_t \right) \right]^{-1/\varepsilon} \right\} \tag{24.14}$$

where Z_{q0} = design quantile and σ (scale) and ε (shape) parameters are both time-invariant. Values of the design quantile using the GEV quantile equation are given by:

$$Z_{q0} = \mu_0 - \frac{\sigma}{\varepsilon}\left\{1 - \left[-\ln\left(1 - \frac{1}{T_0}\right)\right]^{-\varepsilon}\right\} \tag{24.15}$$

where T_0 represents the desired return period from a range of values. For each value of Z_{q0}, the time-varying probabilities (p_t) are computed from Equation (24.14). Then, using the values of p_t, the nonstationary EWT values (T) are determined from Equation (24.13). The hydrologic risk of failure R for a design life of n years is [1]:

$$R = 1 - \prod_{t=1}^{n}(1 - p_t) \tag{24.16}$$

Applying Equations (24.13) through (24.16) to the RDU block maxima, Figures 24.7 and 24.8 were derived.

Although the nonstationary versus stationary estimates presented in Table 24.4 do not appear to be dramatic, Figures 24.7 and 24.8 illustrate that there is a lower EWT for the nonstationary time series (and a higher risk of failure) for a given design return period (T_0). In Figure 24.7, for a project design life of T = 50 years, the desired level of protection under nonstationarity should be about 70 years. In Figure 24.8, for T_0 = 70 years (corresponding to T = 50), the risk R under stationarity is 51.3%. Under nonstationarity, the risk increases to 63.4%. A *climate-resilient* infrastructure approach that allows for *adaptive* design and risk management [6] is recommended.

Extreme storm event depths can also be used in intensity-duration-frequency (IDF) expressions to extract intensities. The GEV, using L-moments, was applied by [7] to derive IDF curves from either annual maxima or partial duration time series. The coefficients of the empirical IDF formula are optimized, and the derived relationship for RDU annual maxima is given below as follows:

$$i = \frac{\lambda T^K}{(d+\theta)^\eta} = \frac{36.13T^{0.218}}{(d+0.393)^{0.811}}\,\text{mm hr}^{-1} \tag{24.17}$$

The intensity can be converted to inches per hour by dividing by 25.4. For example, for T = 100 years and d = 1 hour, the intensity is 75.2 mm h^{-1} (2.96 in. h^{-1}). The IDF curves are plotted in Figure 24.9.

Butcher and Zi [8] developed an efficient method to update IDF curves to future climate projections. Their methodology replicates Atlas 14 procedures (which fit a GEV distribution to the annual

TABLE 24.4 Nonstationary Times Series Analysis Estimates (ISEs), Raleigh–Durham International Airport, FL (1948–2013)

NOAA Station No. 317069	Raleigh–Durham Annual Maxima	Lat: N 35:52:00 Long: W 78:47:00 Elevation: 416 ft	Record Length Analyzed: 64.5 years
Return Curve	Return Period (years)	Bayes LPIII RL (in.)	GEV RL (in.)
95% CI	200	12.0	13.8
	100	10.5	11.6
Median[1]	200	7.9	7.7
	100	7.4	7.0
5% C.I.	200	6.6	5.1
	100	6.2	4.9

[1] Most likely estimate for Bayes LPIII distribution location parameter return level (RL).

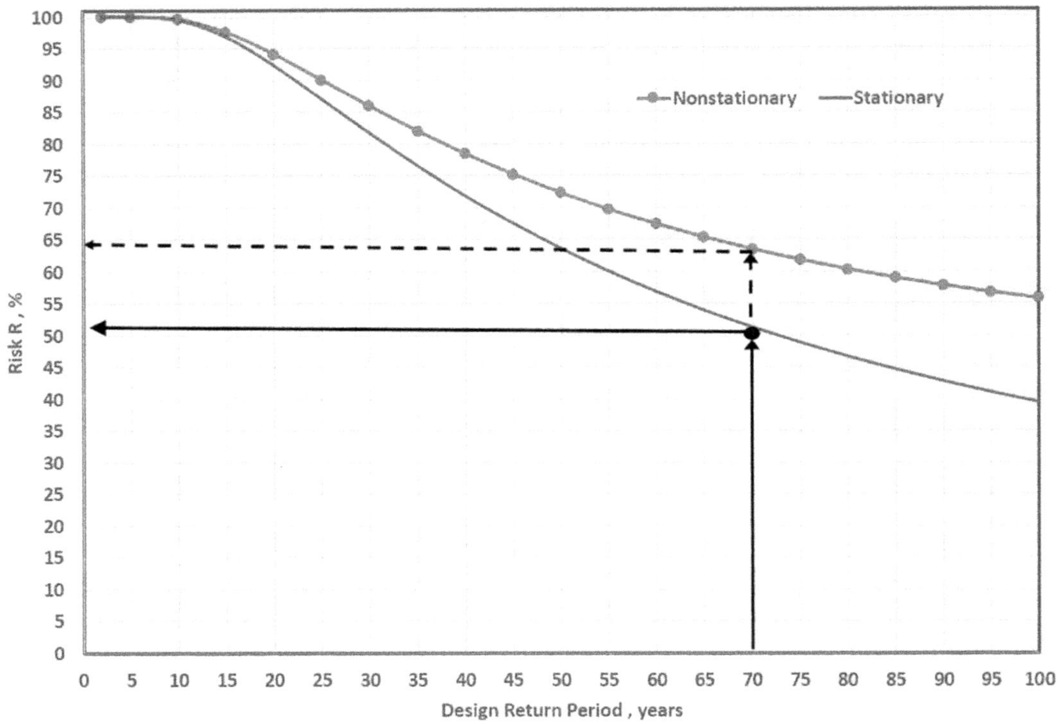

FIGURE 24.7 Nonstationary versus stationary EWTs for RDU ISE maxima.

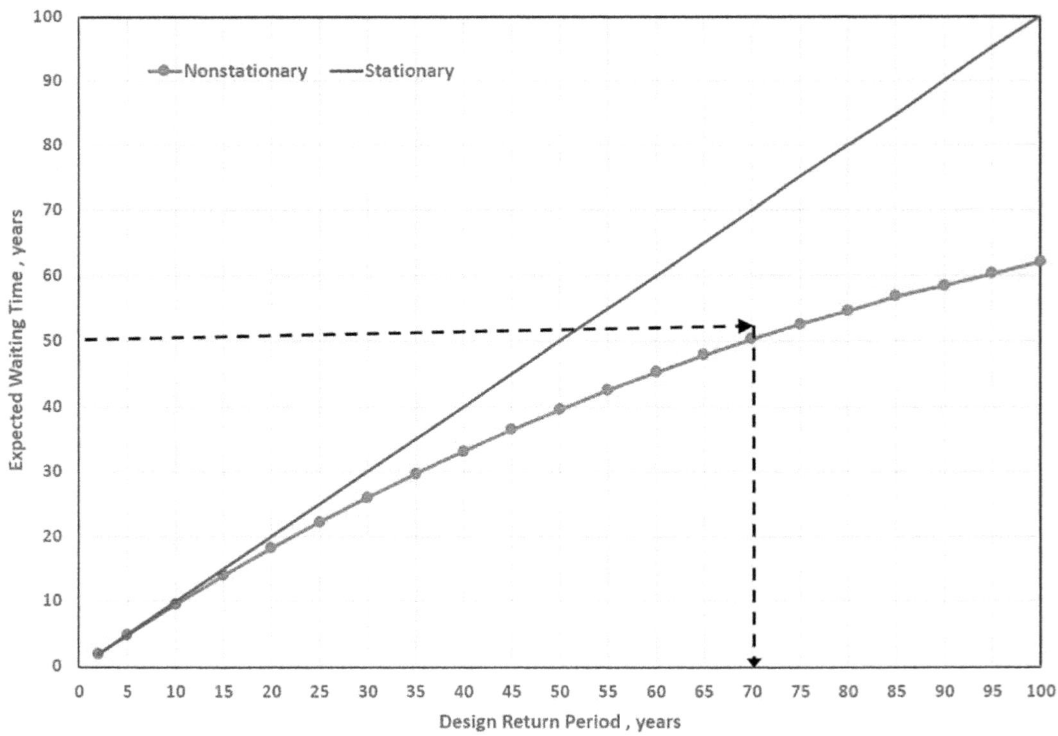

FIGURE 24.8 Nonstationary versus stationary risk percentage for RDU ISE maxima.

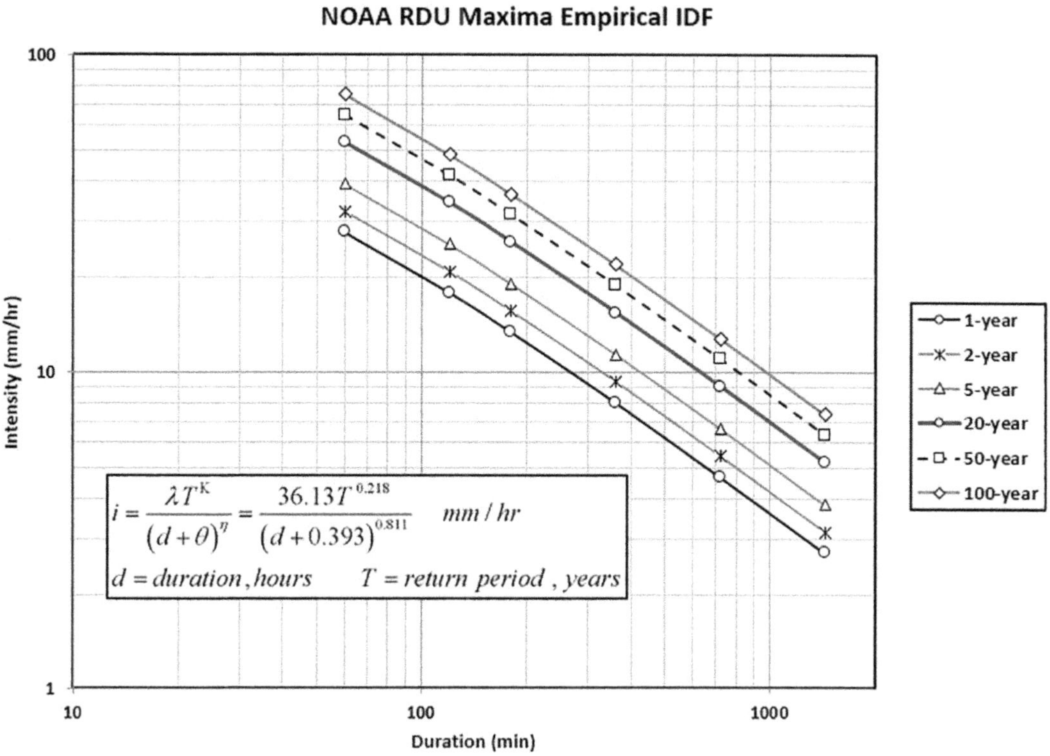

FIGURE 24.9 IDF curves for RDU annual maxima.

maximum series). Global climate model (GCM) biases are addressed through equidistant quantile mapping, mapping from a GCM's historical simulation to future predictions, then applying this mapping from the fit to data used in Atlas 14 [9].

Overland Flow

Overland flow is very thin sheet flow that develops after precipitation over a sloped land surface and infiltration to the subsurface. The depth and flow rate of this thin sheet of water depend, among others, on the rainfall intensity, ground surface characteristics (e.g., roughness, slope), and antecedent and immediate subsurface conditions (e.g., hydraulic conductivity, moisture). By neglecting the acceleration and pressure terms in the momentum equation for dynamic waves, the kinematic wave model [10] substitutes a steady uniform flow (stage discharge) relationship for the momentum equation. However, unsteady flow is preserved through the continuity equation.

The dynamic wave equations for overland flow consist of continuity and momentum equations, as follows, respectively: [2]

$$\frac{\partial y}{\partial t}+V\frac{\partial y}{\partial x}+y\frac{\partial V}{\partial x}=i-f \tag{24.18}$$

$$\frac{\partial V}{\partial t}+\frac{\partial V}{\partial x}+\frac{\partial y}{\partial x}=g\left(S_o-S_f\right)+\left[\left(i-f\right)\frac{V}{y}\right] \tag{24.19}$$

where y is depth of water (L), S_f is friction slope (L/L), S_o is bed slope (L/L), V is water velocity (L/T), i is rainfall intensity (L/T), f is infiltration rate (L/T), g is acceleration of gravity (L/T^2), t is time (T), and x is distance (L). These equations are for moderately wide overland flow and small bottom slope. An order-of-magnitude analysis of Equation (24.19) shows that $(i-f)$ has a negligible effect on the flow dynamics [11]. The flow per unit width, q (L^2/T), is equal to Vy. Thus, Equation (24.18) may be written as follows:

$$\frac{\partial q}{\partial x}+\frac{\partial y}{\partial t}=i-f \tag{24.20}$$

The simplest distributed routing method is the *kinematic wave* model, which neglects the local acceleration, convective acceleration, and pressure terms in the momentum equation (thus, friction and gravity forces essentially balance each other, ($S_f = S_o$)). The most common formulas relating water velocity, slope, and catchment roughness are the following [12]:

$$\text{Manning formula}: V=\frac{k}{n}R^{2/3}\sqrt{S_o} \tag{24.21}$$

$$\text{Chezy formula}: V=C\sqrt{R\,S_o} \tag{24.22}$$

where V is water velocity (L/T), n is the Manning friction coefficient (dimensionless), R is the hydraulic radius (L), C is the Chezy coefficient (L$^{1/2}$/T), $k = 1.486$ for English system units and 1.0 for the SI system, and the bed slope S_o replaces the friction slope S_f. For wide rectangular sections (e.g., overland flow), the hydraulic radius reduces to water depth y, and the momentum equation for kinematic waves reduces to

$$q=\alpha\, y^m \tag{24.23}$$

where for *fully turbulent* flow:

$$\text{using Manning}: \alpha=\frac{k}{n}\sqrt{S_o} \qquad m=\frac{5}{3} \tag{24.24}$$

$$\text{using Chezy}: \alpha=C\sqrt{S_o} \qquad m=\frac{3}{2} \tag{24.25}$$

Values of exponent m for laminar and mixed laminar turbulent conditions are tabulated by [12]. For overland flow problems using the kinematic wave, the continuity and momentum equations can be combined into one differential equation, as follows:

$$\left.\begin{array}{c}\dfrac{\partial y}{\partial t}+\dfrac{\partial q}{\partial x}=i-f\\ q=\alpha y^m\end{array}\right\}\Rightarrow\frac{\partial y}{\partial t}+\alpha\frac{\partial\left(y^m\right)}{\partial x}=i-f \tag{24.26}$$

Channel Flow

The distributed routing models allow us to compute flow rate and water-level variation through space and time: a major advantage over the lumped model in terms of the design criteria of any storage structure, such as a detention pond or reservoir. Similar to the overland flow problems, by eliminating some terms in the momentum equation of the Saint-Venant equations, alternative distributed flow routing models are obtained. The dynamic wave equations for channel flow consist of continuity and momentum equations, as follows, respectively [13]:

$$\frac{\partial Q}{\partial x}+\frac{\partial A}{\partial t}=q(x,t) \tag{24.27}$$

$$\frac{1}{A}\frac{\partial Q}{\partial t}+\frac{1}{A}\frac{\partial}{\partial x}\left(\frac{Q^2}{A}\right)+g\frac{\partial y}{\partial x}-g\left(S_o-S_f\right)=0 \tag{24.28}$$

where y is depth of water (L), S_f is friction slope (L/L), S_o is bed slope (L/L), Q is flow rate (L^3/T), $q\ (x,t)$ is the net lateral inflow per unit length of channel (L^2/T), g is acceleration of gravity (L/T^2), A is the channel cross-sectional area (L^2), t is time (T), and x is distance (L). In Equation (24.28), the first, second, and third terms represent local acceleration, convective acceleration, and pressure force, respectively. The bed slope and friction slope in the last term represent the gravity force and friction force.

Again, the simplest distributed routing method is also the *kinematic wave* model, which neglects the local acceleration, convective acceleration, and pressure terms in the momentum equation for dynamic waves. For open channel flows, the continuity and momentum equation and their combined form for kinematic waves are given as follows:

$$\left.\begin{array}{l}\dfrac{\partial A}{\partial t}+\dfrac{\partial Q}{\partial x}=q(x,t)\\[2ex] Q=\alpha A^m\end{array}\right\}\Rightarrow\frac{\partial A}{\partial t}+\alpha\frac{\partial(A^m)}{\partial x}=q(x,t) \tag{24.29}$$

where for *fully turbulent* flow:

$$\text{Manning:}\alpha=\frac{k}{n}\frac{\sqrt{S_o}}{P^{2/3}}\quad m=\frac{5}{3} \tag{24.30}$$

$$\text{Chezy:}\alpha=C\frac{\sqrt{S_0}}{\sqrt{P}}\quad m=\frac{3}{2} \tag{24.31}$$

Here, A is the channel cross-sectional area (L^2), Q is the flow rate (L^3/T), n is the Manning friction coefficient (dimensionless), C is the Chezy coefficient ($L^{1/2}/T$), P is the wetted perimeter (L), S_o is the bed slope (L/L), t is time (T), $q\ (x,t)$ is the net lateral inflow per unit length of channel, and x is the distance along the flow axis (L). As before, $k=1.486$ for English system units and 1.0 for the SI system.

Low Impact Development-Best Management Practices (LID-BMPS)

Rapid urbanization has a lot of adverse outcomes including environmental pollution, flooding, climate change, global warming, and heat island effects. Because of land use changes, a tremendous increase in impermeable surfaces has been observed. Therefore, specific solutions need to be explored to prevent excess runoff volume, high peak flow rate, and short concentration time. Low Impact Development-Best Management Practices (LID-BMPs) have been implemented in the last couple of decades to control the peak surface runoff and water quality, particularly in urbanized regions. LID-BMPs are nature-based and green solutions which minimize the negative effects of urbanization on nature by preserving and recreating natural landscape features, maximizing the pervious regions, and creating functional and appealing site drainage. Fundamental principles of LIDs involve generation of natural permeable layers around the impermeable areas by allowing for infiltration of the storm and snow water, reducing surface runoff, and increasing stormwater quality. Bioretention, vegetated rooftops, rain barrels, vegetative swales, infiltration trenches, permeable pavements, and stormwater wetlands are among the most important LID-BMPs. There are both numerical modeling and experimental studies and particularly experiments conducted on case study sites in the literature which are related to the hydrological and water quality improvement performance of different LID-BMPs [14–17]. Environmental Protection Agency Stormwater Management Model (EPA SWMM) is a commonly used software program in LID-BMP implementation studies [18,19].

Environmental Protection Agency Stormwater Management Model (EPA SWMM)

EPA SWMM is a dynamic rainfall-runoff simulation model that can be used for single-storm event simulation or continuous simulation of multiple storms of a specific watershed. The model calculates the flow and water quality constituents of the watershed based on hydraulic parameters set by the user and provides graphs of simulation results such as flow hydrographs and water quality concentrations. The model is widely used to plan, analyze, and control stormwater runoff; to design drainage system components; and to evaluate watershed management of both urban and nonurban areas [20,21].

The runoff component of SWMM simulates surface runoff over subcatchment areas, as generated from rainfall input. Then, SWMM routes this runoff with the transport component through a conveyance system of pipes, channels, storage/treatment devices, pumps, and regulators. SWMM calculates the quantity and quality of runoff generated within each subcatchment and the flow rate, flow depth, and quality of water in each pipe and channel during a simulation period comprised of multiple time steps. SWMM inputs include precipitation data, subcatchment delineation, pipe system characteristics, and soil properties. The watershed profile is formed by defining input data, such as slope, area, and imperviousness of subcatchments; length, cross section, height, and slope of conduits; porosity, hydraulic conductivity, and field capacity of aquifers, into the program. Outputs of the model include change of flow rate (hydrograph); change of concentration (pollute graph) through time and daily, monthly, annual, and total simulation summaries (for continuous simulation); and statistical frequency analysis. SWMM is also capable of viewing results in tabular and graphical form.

While calculating surface runoff, precipitation and flow from upstream subcatchments contribute to the inflow and infiltration and evaporation contribute to the outflow. For flow routing in conduits, steady flow, and kinematic, diffusion and dynamic wave routing options may be used. In addition, for infiltration calculation, three options exist, which are the Integrated Horton Method, the Green–Ampt Method, and the Soil Conservation Service (SCS) Curve Number Method. EPA SWMM simulates surface runoff quality by using exponential, power, and saturation functions for buildup and exponential, rating curve, and event mean concentration functions for washoff.

LID Modeling with EPA SWMM

EPA SWMM is one of the most capable simulation tools for modeling LID implementations (www.epa.gov/water-research/storm-water-management-model-swmm). EPA SWMM can explicitly model the following types of LID practices: bioretention cell, rain garden, green roofs, infiltration trenches, permeable pavement, rain barrels, rooftop disconnection, and vegetative swale [22]. EPA SWMM takes into account storage, infiltration, evaporation, and/or drainage of water through the LID depending upon its type. The definition of LID in EPA SWMM in general consists of a surface layer, soil layer, storage layer, and drain layer. Whereas bioretention cell and permeable pavement include all of these layers, an infiltration trench does not have a soil layer and a vegetative swale type of LID has only surface layer. While developing the hydrological model of a watershed, one can define different types of LID for different subcatchments to observe the effect of LID implementation. The parameters of the surface layer include berm height, vegetation volume fraction, surface roughness, and surface slope; the parameters of the soil layer include thickness, porosity, field capacity, wilting point, conductivity, conductivity slope, and suction head; the parameters of the storage layer include thickness, void ratio, seepage rate, and clogging factor; the parameters of the drain layer include flow coefficient, flow exponent, and offset height [23].

EPA SWMM Applications for Overland and Channel Flow

Duke University Campus Watersheds

The Duke University West Campus watershed is located in Durham, North Carolina, USA. Runoff from the West Campus of Duke University is simulated by using the EPA SWMM to predict surface runoff and quality and transport through the drainage system. The first formal application of EPA SWMM to the Duke Campus was completed by Mahi [24].

Calibration and Discretization

In order to be able to use EPA SWMM for surface runoff and channel flow predictions, the model needed to be calibrated against measured rainfall and runoff data. A rain gage was placed on the roof of Hudson Hall (Pratt School of Engineering) and several storm events were measured in 15-minute intervals. About 30 storms were recorded from 1994 to 1995, these were grouped into summer and winter storms for calibration purposes. The year 1997 was determined as the reference year to establish runoff *baseline conditions* (as approved by the City of Durham, NC). The last major calibration was conducted in 1996 *(and, thus, valid for 1997 baseline conditions).* In order to measure the flow rate, an aluminum compound weir was placed at the outlet of conduit 210, near node 345, a 1.83 m (6-ft.) diameter pipe (see Figure 24.10). Typically, pressure transducers were located a few feet upstream of the weirs, transmitting date, time of day, and depth of flow data to electronic data loggers. Several model parameters (e.g., infiltration, depression storage, Manning's roughness) were adjusted to match predicted versus observed flows. An example of a winter storm calibration is shown in Figure 24.11.

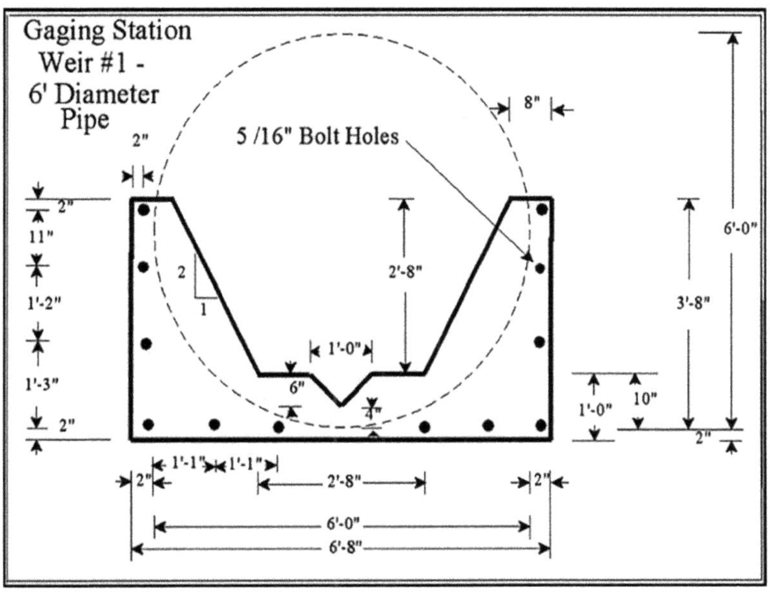

FIGURE 24.10 Compound weir at node 345.

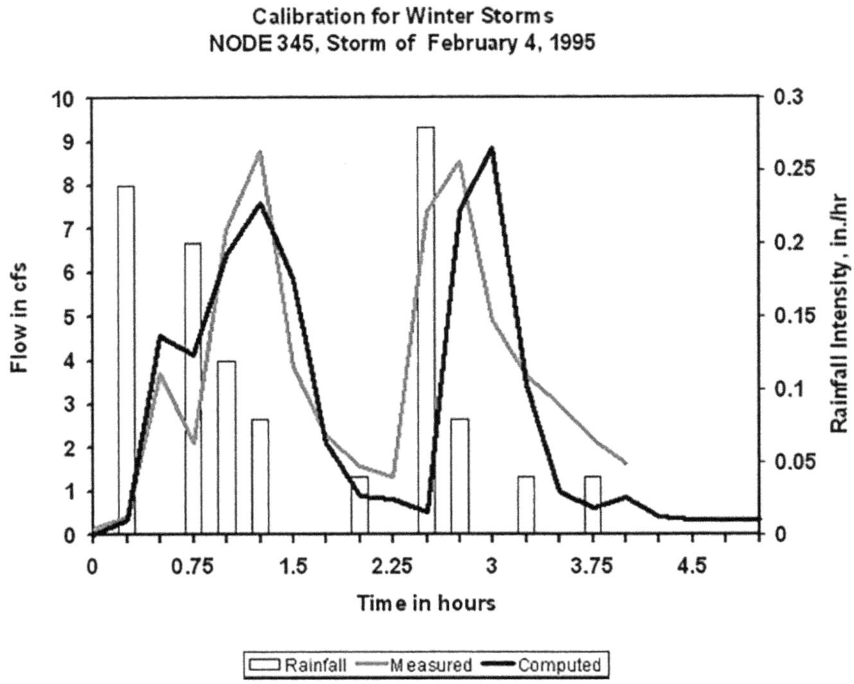

FIGURE 24.11 Observed rainfall and runoff, computed flow, February 4, 1995.

The discretization of the entire Duke University West Campus watershed using EPA SWMM is shown in Figure 24.12: it illustrates the location of the weirs used for calibration (near node 345). The rapid urbanization (medical center) upstream of this measurement station over the past 5 years resulted in the decision to consider the design and construction of a large detention pond to control stormwater flowing off-campus (at the Erwin Road Outfall) into the City of Durham.

Modeling a Best Management Practice

Detention ponds are considered one of the most effective runoff control devices; in particular, to reduce peak flows, settle particulate matter, while also reducing some pollutant concentrations in the outflow. They are not as effective in reducing dissolved fractions of contaminants. The effect of a large detention pond in capturing a relatively small storm (storm of February 15–17, 1995; for which runoff measurements were available) is presented in Figure 24.11. Conduit 203 is upstream of the simulated detention pond. Conduit 202 is downstream from the pond and shows the pond storage effects on the hydrograph. The methodology described by [25] includes a procedure for sizing the water quality control volume (WQCV). An actual pond is currently being designed by a consulting firm based on the location (between Conduit 202 and 203) and configurations simulated with SWMM5 and programs that incorporate the statistical properties of hourly rainfall recorded for the past 63 years, as recorded by National Climatic Data Center rain gauges near the campus (Figure 24.13).

Sazlidere Watershed in Istanbul, Turkey

Sazlidere Watershed is located on the European Continental side of Istanbul in Turkey and Sazlidere Lake is located downstream from the Sazlidere Watershed. Surface runoff generated over the Sazlidere Watershed (165 km^2 of drainage area) flows to Sazlidere Dam Lake which was built up in 1996. Protecting and improving the Sazlidere Watershed is of great importance as it supplies a major portion of the drinking water of Istanbul. There is a great potential of population growth and,

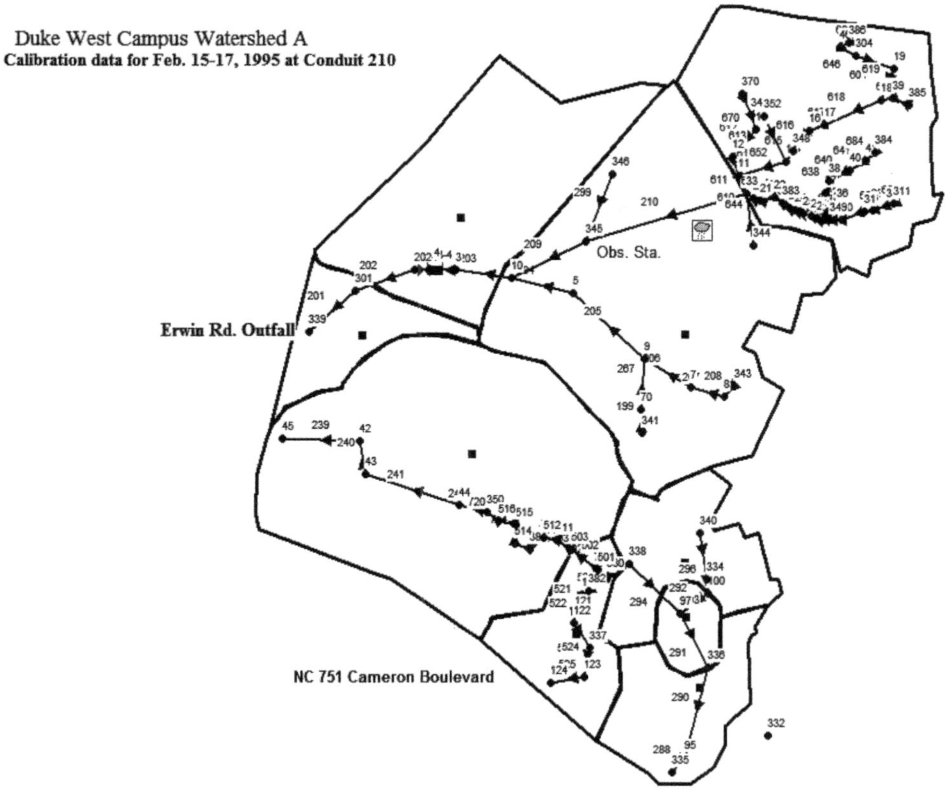

FIGURE 24.12 Watershed A discretization showing node 345 observation station.

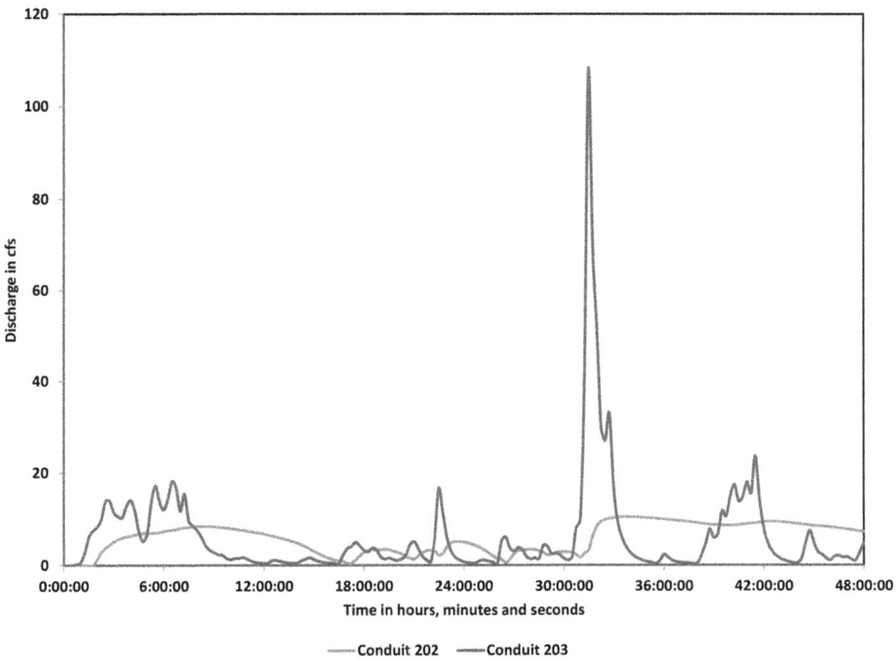

FIGURE 24.13 Hydrographs upstream and downstream of detention pond for February 15–17, 1995, storm.

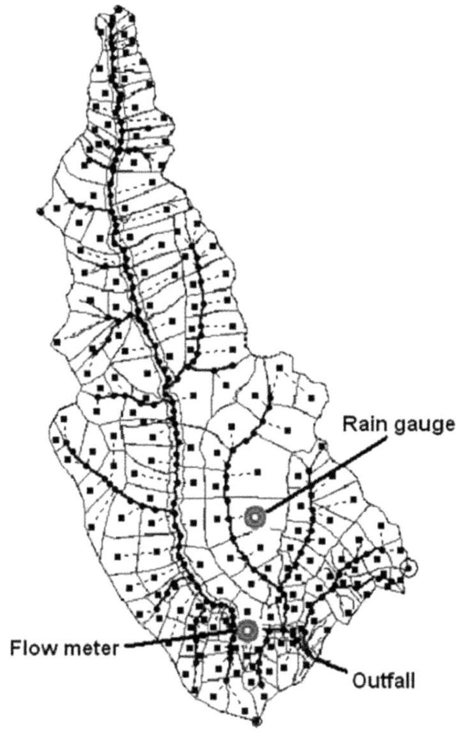

FIGURE 24.14 Hydrodynamic model of Sazlıdere Watershed in EPA SWMM.

therefore, increases in the residential area. The hydrodynamic model of the Sazlidere Watershed is depicted in Figure 24.14, using the EPA SWMM. Part of the Sazlıdere Watershed area, which feeds Türkköse Stream, is divided into 177 subcatchments by using the topographical map of the modeled area. A total number of 171 junctions and 173 conduits, which are ~44 km long in total, are defined for the hydrodynamic model. In order to calibrate and verify the hydrodynamic model, a rain gauge was set up on top of a building four stories high, located in the middle of the watershed in Haraççi, and a flow meter was set up close to the downstream segment of the Türkköse Stream (which is the main channel in the watershed). The locations of the rain gauge and flow meter are indicated by a circle in the Figure 24.14.

The flow rate was predicted during the December 26–29, 2009, rainfall event with the calibrated model at link 50, at the midpoint of the watershed where there is dense urbanization, and at link 75, which is located at a point downstream from the watershed. The rainfall data were measured at the field site. The predicted flow rates and the measured rainfall intensity are shown in Figure 24.15. The peak flow reaches ~15 $m^3 s^{-1}$ near the urbanized area. On the other hand, the channel surcharges close to the outlet of the watershed for a duration of 2 hours and with a maximum flow of 21 $m^3 s^{-1}$. In order to see the effect of extreme urbanization, the surface runoff and stream flow were also simulated during a rainfall event from December 10–13, 2009, for 80% of imperviousness on the catchment. Figure 24.16 shows the hydrographs for 10% and 80%, respectively, of urbanization for the cases with stream restoration and with no stream restoration. The calculated flow rates at the outfall for year 2010 (which corresponds to 10% urbanization) under current stream sections and under suggested stream restoration are labeled as "Q_2010_current" and "Q_2010_ suggestion 1," respectively. Under the current stream sections scenario, the channel cannot sustain the flow, and flooding occurs between the hours of 25–40 and 65–75. Therefore, a stream restoration is proposed by increasing the cross-sectional areas of the Türkköse Stream to 2.5 times its current value in order to prevent flood events on the catchment.

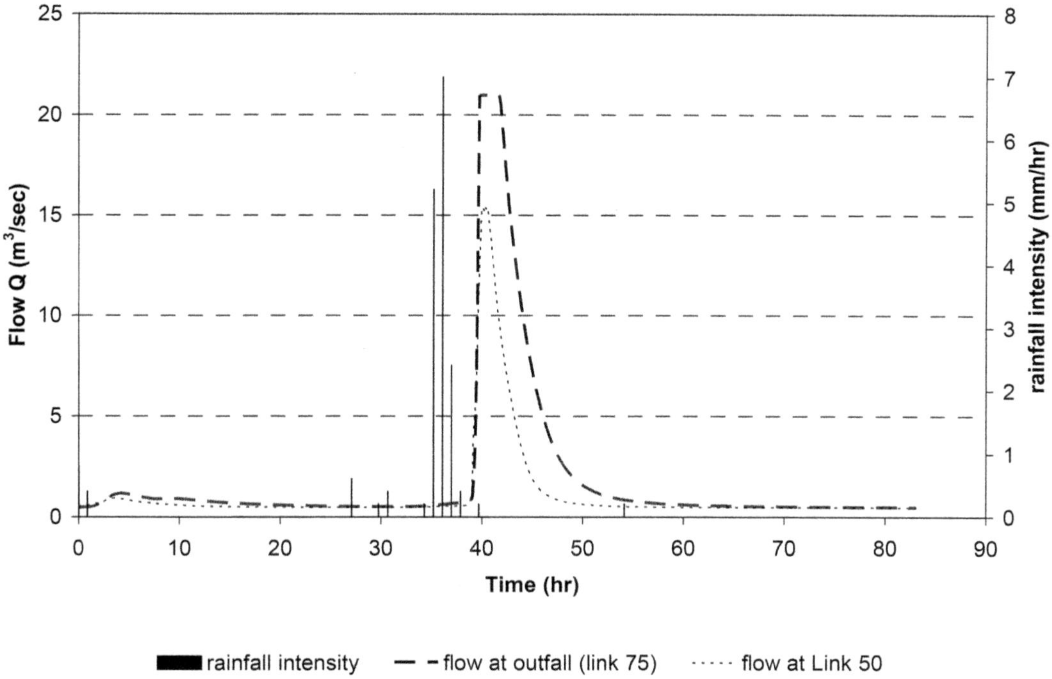

FIGURE 24.15 Rainfall intensity and predicted flow rate versus time in Sazlıdere Watershed during December 26–29 rainfall event.

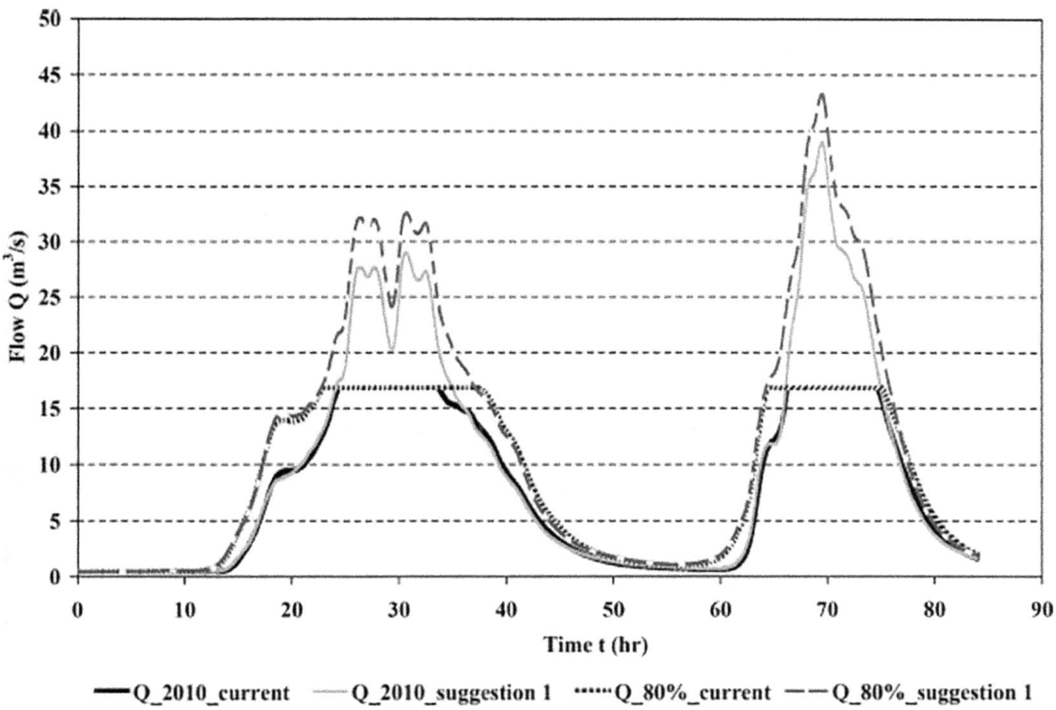

FIGURE 24.16 Comparison of the hydrographs obtained for 10% and 80%, respectively, of urbanization for the cases with stream restoration and with no stream restoration.

When the restored stream sections are considered, we see that the stream will be able to carry the expected rising flow rate and no flood will occur under the extreme urbanization with this suggested stream restoration.

EPA SWMM Applications for LID Implementations

Hydrological Model of Rainfall-Watershed-Bioretention (RWB) Experimental System

Description of RWB Experimental System

The Rainfall-Watershed-Bioretention (RWB) Experimental System was constructed at the Istanbul University-Cerrahpaşa Avcılar Campus in Istanbul, Turkey. RWB consists of 40 rainfall nozzles for rainfall simulation, $40\,m^2$ of steel construction platform for watershed simulation, and 4 bioretention columns with different soil ratios. A typical bioretention column consists of gravel, soil mixture of sand and vegetative soil, mulch, and plants. The ratio of sand and vegetative soil in the soil mixture differs in three bioretention columns and turf is added into this soil mixture to test the effect of the organic content on the hydrological performance of bioretention. In addition to these layers, local vegetation types are used [26]. RWB has also the capability of testing the hydrological performance of two other types of LIDs which are infiltration trench and vegetative swale as the three LID types are addable/detachable to/from the RWB System. A photo of the RWB System is provided in Figure 24.17.

Hydrologic Experiments with the RWB Experimental System

The hydrologic experiments with the RWB were conducted under different rainfall intensities and durations. During the experiments, both surface runoff which reaches the bioretention column as inflow and as outflow at the exit point located under the bioretention column are measured in time. In addition to flow rate, ponding depths on bioretention columns are also measured [26]. Figure 24.18 shows the surface runoff and the outflow hydrographs out of the four bioretention columns under rainfall intensities of 23 and $34\,mm\,h^{-1}$ and rainfall durations of 15 and 30 minutes. The effect of rainfall intensity and duration on peak flow reduction, arrival time, ponding depth, and shape of the hydrograph at the exit of the bioretention is observed. Long rainfalls cause an increase in ponding depth and peak outflow and the hydrograph at the exit of the bioretention expands. As rainfall intensity increases, ponding depth and

FIGURE 24.17 (See color insert.) The constructed RWB system.

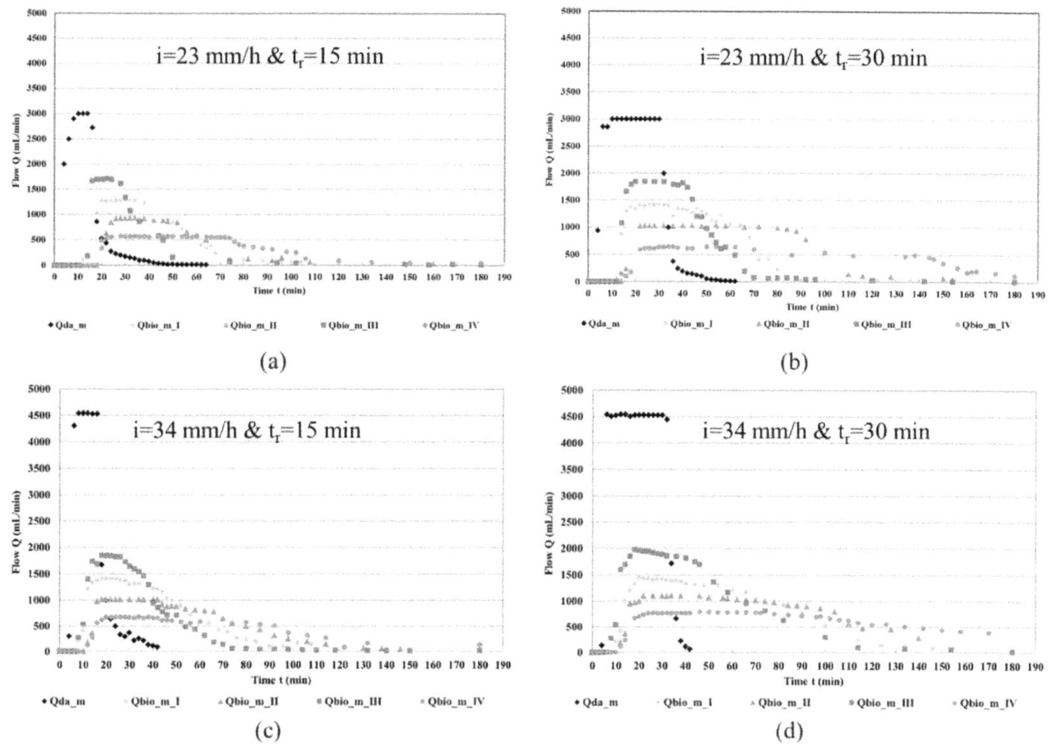

FIGURE 24.18 Measured inflow and outflow hydrographs of four bioretention columns under different rainfall intensities and durations.

peak outflow also increase while the peak outflow arrival time decreases. In addition, the hydrograph at the exit of each bioretention column has a different behavior due to the different combinations of vegetative soil, sand, and turf.

Hydrological Model of the RWB Experimental System

The EPA SWMM is used in hydrological modeling of the RWB System to demonstrate the watershed modeling capability of SWMM for LID implementations, particularly bioretention. The watershed part of the RWB is simulated by dividing the drainage area into 38 subcatchments and defining the slope, area, and width of each subcatchment in SWMM. The rainfall intensity and duration are defined using the rain gauge element. Finally, in order to model bioretention columns, parameters of berm height, soil and storage thicknesses, hydraulic conductivity, porosity, void ratio, and moisture content are defined using measured data. The rest of the parameters, i.e., vegetation volume, surface roughness, and surface slope for surface layer; field capacity, wilting point, conductivity slope, and suction head for soil layer; seepage rate and clogging factor for storage layer; and drain coefficients for drain layer are defined based on the range given by EPA SWMM Manual [27]. The modeling schematic of EPA SWMM for bioretention is illustrated in Figure 24.19.

Figure 24.20 shows the measured and calculated outflow rates at the exit of each bioretention column under rainfall intensity of 23 mm h^{-1} and rainfall duration of 15 minutes. The calculated outflow rates with the SWMM model are in general in good agreement with the measured data particularly for the rising limb of the flow curve. The calculated peak flow is slightly smaller than the measured peak flow. The calculated data of the recession limb of the flow curve is rather sharp when compared with the measured data which has a smooth decrease. Thus, EPA SWMM performs well in modeling bioretention systems, but there is room for improvement.

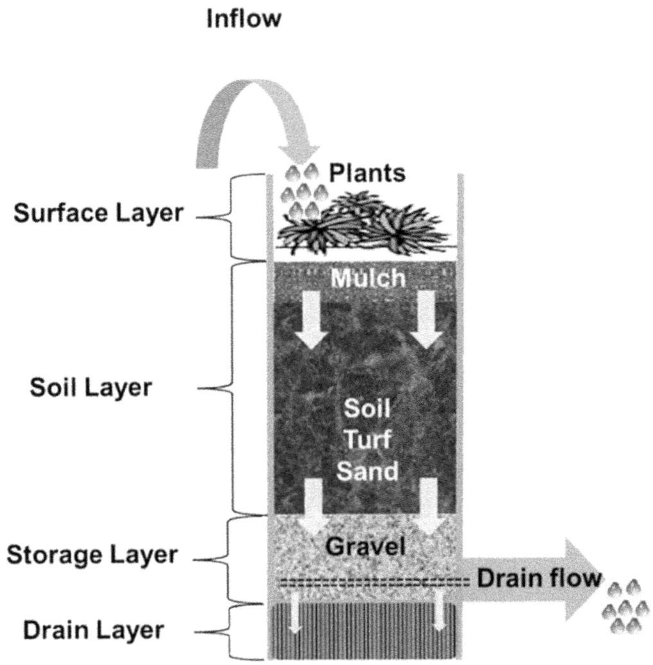

FIGURE 24.19 **(See color insert.)** Modeling of bioretention in EPA SWMM. (Adapted from Gülbaz and Kazezyilmaz-Alhan 2017.)

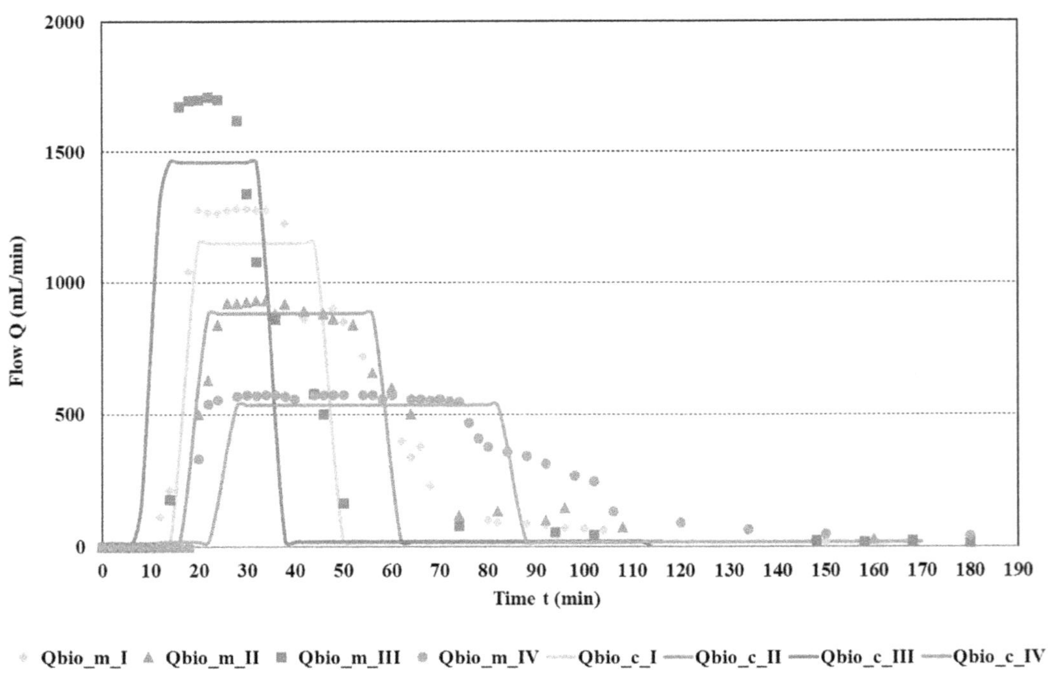

FIGURE 24.20 Measured and calculated outflow hydrographs of four bioretention columns.

Hydrological Model of Avcılar Campus of Istanbul University-Cerrahpaşa

The Avcılar Campus of Istanbul University is located on the European Continental side of Istanbul in Turkey with a campus area of 133 hectares. A major portion of the campus is forestland and has a long coast to the Küçükçekmece Lake and, thus, a major portion of the surface runoff, which develops on campus during a storm event, reaches the lake through tributary channels.

The hydrological model of the campus was generated by defining 59 subcatchments, 10 conduits, and 10 junctions in SWMM [28]. Slope, surface area, and width are defined using the topographical map of the campus; and percent imperviousness is defined using the site plan map of the campus. The amount of infiltration in forest and green land was calculated using the Horton Method. Parameters such as the Manning's roughness coefficient for pervious and impervious areas, depth of depression storage on pervious and impervious areas, Manning's roughness coefficient for conduits, and hydraulic conductivity are selected according to their typical values reported in the literature [29]. Figure 24.21 shows the hydrological model of Avcılar Campus developed by applying the EPA SWMM.

Green roof, infiltration trench, bioretention, and permeable pavement types of LID are introduced into the hydrological model of the campus where appropriate to investigate the potential effects of LID implementation on surface runoff. Green roof is defined on 11 buildings including the School of Engineering, the Veterinary School, and the School of Business Administration with a total area of 45,780 m^2.

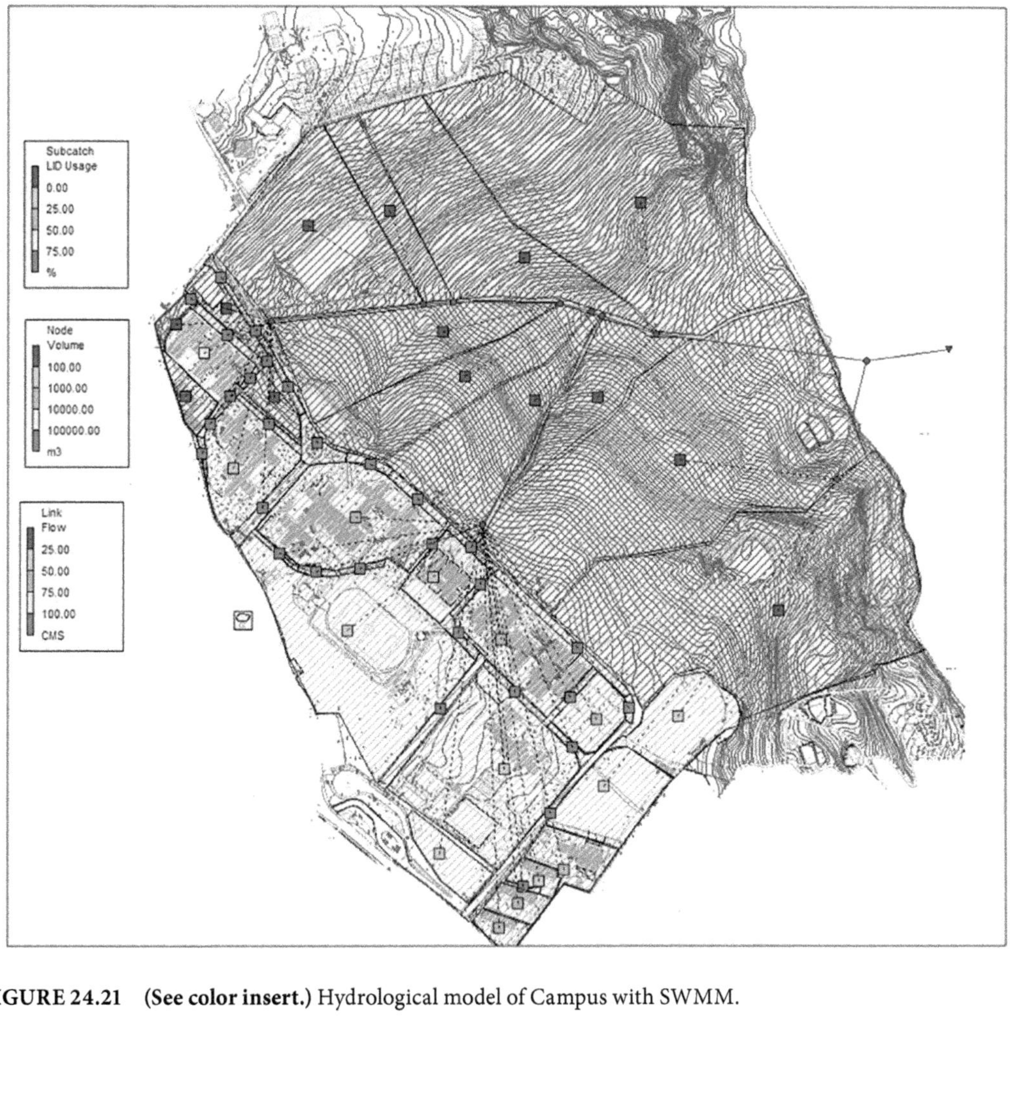

FIGURE 24.21 **(See color insert.)** Hydrological model of Campus with SWMM.

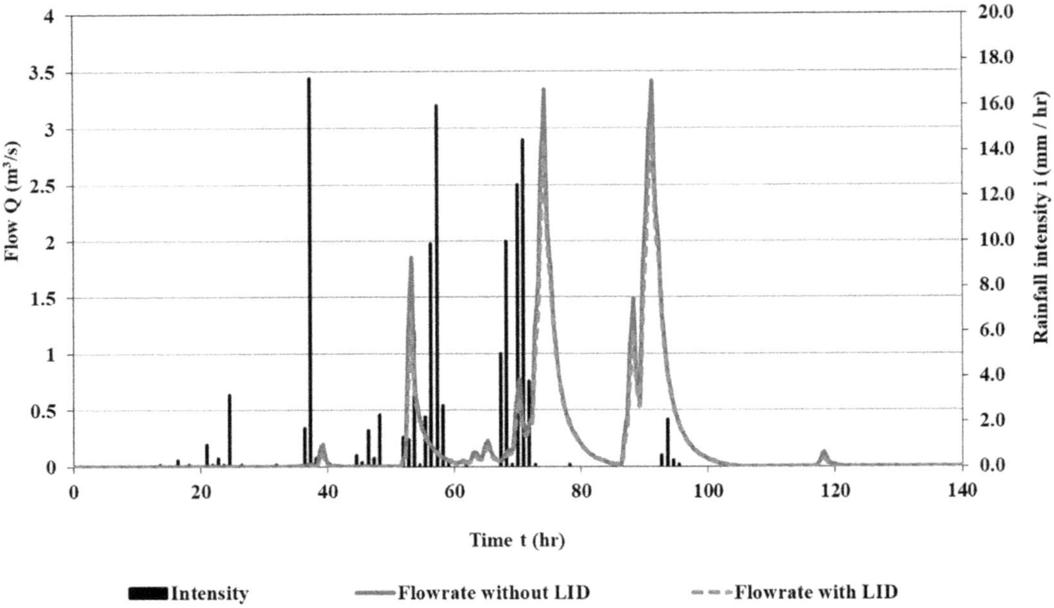

FIGURE 24.22 Predicted surface runoff at the outlet of the modeled area with LID and without LID.

An infiltration trench is defined in the region where the stadium and transportation and logistics faculty are located with a total area of 6,250 m^2. Bioretention is defined in the parking lot region with a total area of 3,760 m^2. Finally, permeable pavement is defined along the total main road of the campus.

Figure 24.22 shows the comparison of the hydrographs obtained with EPA SWMM for the cases with LID implementation and without LID implementation on the campus. The simulations of the model are done using rainfall data obtained from Florya Meteorological Station in Istanbul during the storm event occurred in September 07–09, 2009. It is observed that LID implementations on the campus result in a decrease in the peak of the hydrograph which means that LID implementation has considerable potential to control the surface runoff in the campus area during a storm event.

References

1. Obeysekera, J., Salas, J. D. 2014. Quantifying the Uncertainty of Design Floods under Nonstationary Conditions. *Journal of Hydrologic Engineering*, 19 (7), 1438–1446.
2. Cunnane, C. 1978. Unbiased Plotting Positions – A Review. *Journal of Hydrology*, 37, 205–222.
3. Restrepo-Posada, P. J., Eagleson, P. S. 1982. Identification of Independent Rainstorms. *Journal of Hydrology*, 55 (1–4), 303–319.
4. Luke, A., Vrugt, J. A., AghaKouchak, A., Matthew, R., Sanders B. F. 2017. Predicting Nonstationary Flood Frequencies: Evidence Supports an Updated Stationarity Thesis in the United States. *Water Resources Research*, 53, 5469–5494. doi:10.1002/2016WR019676.
5. Hosking, J. R. M. 1990. L-Moments: Analysis and Estimation of Distributions Using Linear Combinations of Order Statistics. *Journal of the Royal Statistical Society. Series B (Methodological)*, 52 (1), 105–124. www.jstor.org/stable/2345653.
6. Ayyub, B. M., Medina, M., Vinson, T., Walker, D., Wright, R. N., reviewed by Blue Ribbon Panel. 2018. Climate-Resilient Infrastructure: Adaptive Design and Risk Management. Committee on Adaptation to a Changing Climate (CACC), American Society of Civil Engineers (ASCE), ASCE Manuals and Reports on Engineering Practice, 140. doi:10.1061/9780784415191.

7. Chang, K. B., Lai, S. H., Faridah, O. 2013. RainIDF: Automated Derivation of Rainfall Intensity-Duration-Frequency Relationship from Annual Maxima and Partial Duration Series. *Journal of Hydroinformatics*, 15 (4), 1224–1233.
8. Butcher, J. B., Zi, T. 2019. Efficient Method for Updating IDF Curves to Future Climate Projections. *Atmospheric and Oceanic Physics*, https://arxiv.org/abs/1906.04802.
9. Bonnin, G., et al. 2006. Precipitation-Frequency Atlas of the United States. NOAA Atlas, 14, 2, Version 3, U.S. Department of Commerce, National Oceanic and Atmospheric Administration, National Weather Service, Silver Spring, Maryland.
10. Lighthill, M. J., Whitham, G. B. 1955. On Kinematic Waves. I. Flood Movement in Long Rivers. *Proceedings of the Royal Society of London. Series A. Mathematical and Physical Sciences*, 229 (1178), 281–316.
11. Eagleson, P. S. 1970. Dynamic Hydrology. McGraw-Hill, Inc.: New York.
12. Ponce, V. M. 1989. Engineering Hydrology: Principles and Practices. Prentice Hall: Englewood Cliffs, NJ.
13. Chow, V. T., Maidment, D. R., Mays, L. W. 1988. Applied Hydrology; McGraw-Hill: New York. 0-07-010810-2.
14. Gülbaz, S. 2015. An Integrated Hydrodynamic and Contaminant Transport Model Incorporating Low Impact Development Implementation, Ph.D. Thesis, Istanbul University, Institute of Science (in Turkish).
15. Gülbaz, S., Kazezyılmaz-Alhan, C. M., Copty, N. K. 2015. Evaluation of Heavy Metal Removal Capacity of Bioretention Systems. *Water Air and Soil Pollution*, 226 (11), Art. No. 376, doi:10.1007/s11270-015-2640-y.
16. Gülbaz, S., Kazezyılmaz-Alhan, C. M. 2017. Hydrological Model of LID with Rainfall-Watershed-Bioretention System. *Water Resources Management*, 31 (6), 1931–1946.
17. Davis, A. P. 2008. Field Performance of Bioretention: Hydrology Impacts. *Journal of Hydrologic Engineering*, 13 (2), 90–95.
18. Meierdiercks, K. L., Smith, J. A., Baeck, M. L., Miller, A. J. 2010. Analyses of Urban Drainage Network Structure and Its Impact on Hydrologic Response. *Journal of the American Water Resources Association (JAWRA)*, 46 (5), 932–943.
19. Chang, C. H., Wen, C. G., Lee, C. S. 2008. Use of Intercepted Runoff Depth for Stormwater Runoff Management in Industrial Parks in Taiwan. *Water Resources Management*, 22 (11), 1609–1623.
20. Huber, W. C., Dickinson, R. E. 1988. Storm Water Management Model; Version 4, User's Manual, Athens, GA. Environmental Research Laboratory, Office of Research and Development, U.S. Environmental Protection Agency (EPA).
21. Rossman, L. A. 2010. Storm Water Management Model, User's Manual; Version 5. Water Supply and Water Resources Division National Risk Management Research Laboratory, Cincinnati, Ohio, U.S. Environmental Protection Agency, EPA/600/R-05/040.
22. Rossman, L. A., Huber, W. C. 2016. Storm Water Management Model, Reference Manual, Volume III-Water Quality.
23. Rossman, L. A. 2015. Storm Water Management Model, User's Manual, Version 5.1, National Risk Management Laboratory Office of Research and Development U.S. Environmental Protection Agency 26 Martin Luther King Drive Cincinnati, OH 45268, U.S. Environmental Protection Agency, EPA/ 600/R-14/413b.
24. Mahi, H. 1994. Stormwater Management and Modeling at Duke University, M.S. Thesis. Durham, North Carolina.
25. Gironás, J., Roesner, L. A., Davis, J. 2009. Storm Water Management Model Applications Manual. Cincinnati, Ohio, EPA/600/R-09/077.
26. Gülbaz, S., Kazezyılmaz-Alhan, C. M. 2017. Experimental Investigation on Hydrologic Performance of LID with Rainfall-Watershed-Bioretention System. *Journal of Hydrologic Engineering*, 22 (1), doi:10.1061/(ASCE)HE.1943-5584.0001450.

27. Gülbaz, S., Kazezyılmaz-Alhan, C. M. 2017. An Evaluation of Hydrologic Modeling Performance of EPA SWMM for Bioretention. *Water Science and Technology*, 76, 3035–3043.
28. Gülbaz, S., Kaya, Y. E., Kazezyılmaz-Alhan, C. M. 2018. Düşük Etkili Kentleşme Uygulamalarının Yüzeysel Akışa Etkisi: İstanbul Üniversitesi Avcılar Kampüsü Örneği, *İklim Değişikliği ve Çevre*, 3, 45–50, (in Turkish).
29. Huber, W. C., Dickinson, R. E. 1988. Storm Water Management Model, Version 4, User's Manual. Athens, GA. Environmental Research Laboratory, Office of Research and Development, U.S. Environmental Protection Agency (EPA).

25

Surface Water: Nitrogen Enrichment

Tim P. Burt and
Penny E. Widdison
University of Durham

Introduction 191
Nitrate Levels in Surface Waters 191
Land Use Controls to Reduce Nitrogen Enrichment to Surface Waters 195
References 196

Introduction

Nitrogen is essential for plant growth and comprises nearly 79% of the earth's atmosphere in the form of N_2 gas. In order for nitrogen to be used for plant growth, it must be "fixed" in the form of ammonium (NH_4) or nitrate (NO_3). In the terrestrial nitrogen cycle, microbes break down organic matter to produce much of the available nitrogen in soils. Nitrate is completely soluble in water and it is not adsorbed to clay particles; therefore, it is vulnerable to being leached out of the soil by percolating rainfall or irrigation water. Generally, the movement of nitrogen can be described in three ways: 1) upward, crop uptake and gaseous loss; 2) downward, as leaching to groundwater; and 3) lateral, via surface and subsurface flow to surface waters. The nitrogen cycle under arable soils is shown in Figure 25.1.

Nitrate Levels in Surface Waters

The natural water quality of a river will be determined primarily by the catchment soil type and underlying geology to which water, falling on the catchment as rain, is exposed as it drains to the river. Climate provides an important context for nitrogen cycling by controlling the propensity for carbon and nitrogen to be stored within the catchment; thus, in the United Kingdom, upland soils tend to conserve organic matter as peat, whereas organic matter tends to decompose much more readily in lowland soils. Deviations from this baseline water quality are generally caused by the influence of human activities through point and diffuse pollution sources. Up to 40% of total nitrogen reaches the aquatic system through direct surface runoff or subsurface flow.[1] Nitrogen delivery to surface waters is further controlled by: 1) soil structure and type; 2) rainfall; 3) the amount of nitrate supplied by fertilizers; and 4) plant cover and root activity.[2]

In pristine river systems, the average level of nitrate is about 0.1 mg/L as nitrogen (mg/L N). However, in Western Europe, high atmospheric nitrogen deposition results in nitrogen levels of relatively unpolluted rivers to range from 0.1 to 0.5 mg/L.[3] In recent years, nitrate concentrations in European rivers have been rising (Figure 25.2) and "No progress has been made in reducing the concentration of nitrates in Europe's rivers."[4] High rates of nitrogen input to rivers and coastal waters are not confined to Europe. In an average year, the Mississippi River discharges 1.57 million metric tons of nitrogen into the Gulf of Mexico.[5] About 7 million metric tons of nitrogen in commercial fertilizers are applied annually in the basin leading to nitrate concentrations in agricultural drains of 20–40 mg/L or more.[5]

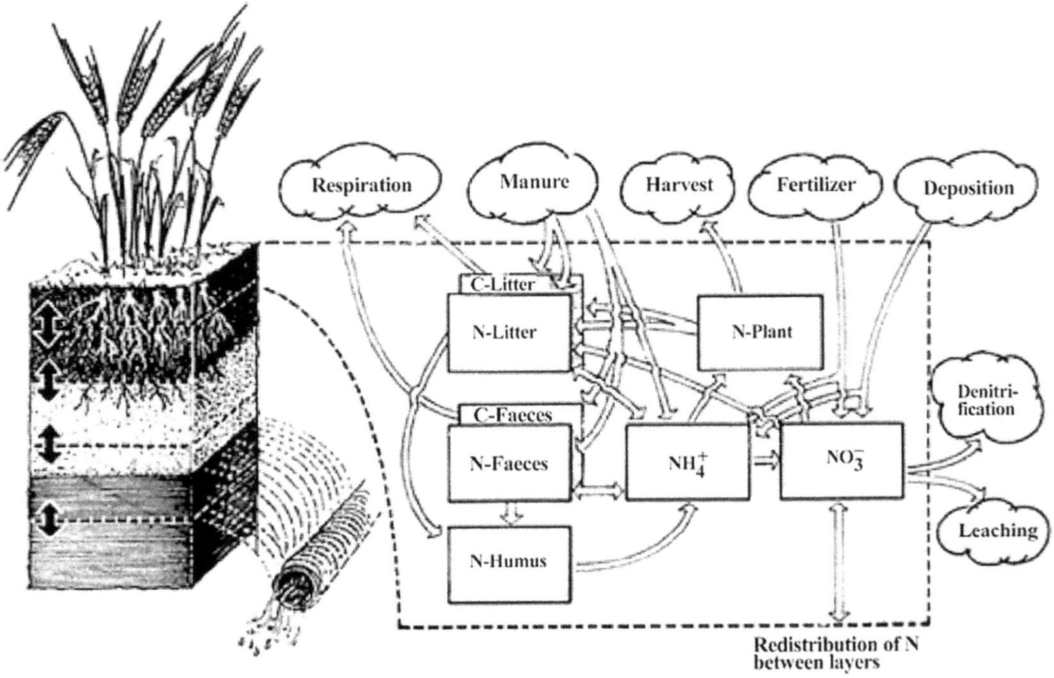

FIGURE 25.1 The nitrogen cycle under arable soils.
Source: Adapted from Burt, et al.[2]

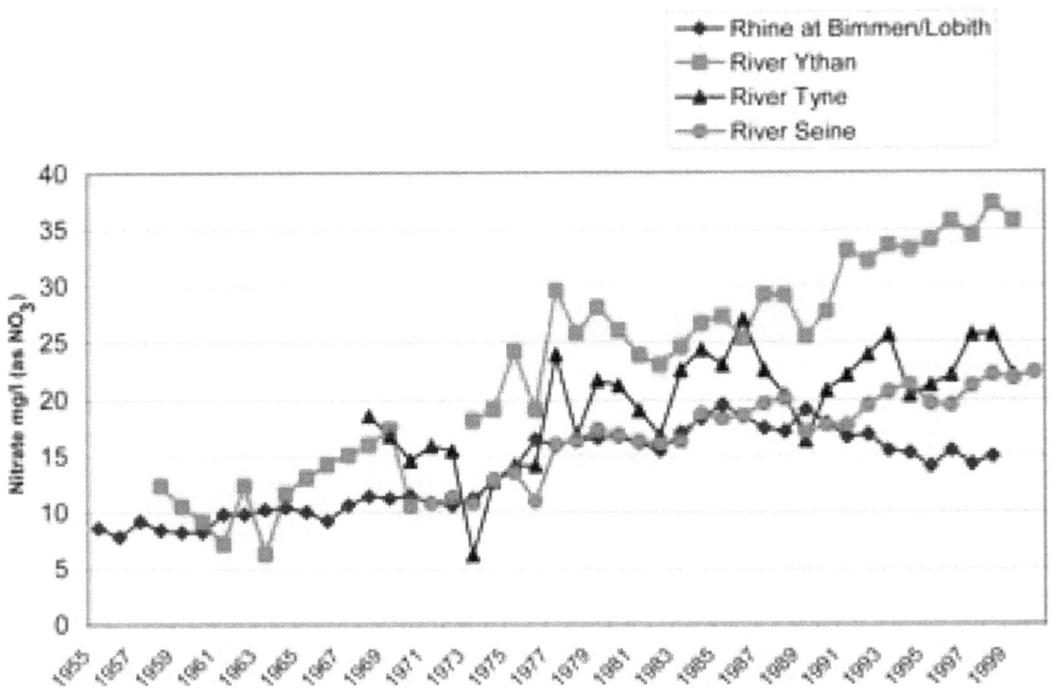

FIGURE 25.2 Nitrate concentration in selected European rivers.
Source: Adapted from European Environment Agency.[4]

TABLE 25.1 Nitrogen Inputs to Rivers and Coastal Waters

River	N Inputs to Rivers ($kg^{-2}\ yr^{-1}$)	N Exports to Coastal Waters ($kg^{-2}\ yr^{-1}$)
Mississippi	7,489	597
Amazon	3,034	692
Nile	3,601	268
Zaire	3,427	632
Zambezi	3,175	330
Rhine	13,941	2,795
Po	9,060	1,840
Ganges	9,366	1,269
Chang Jiang	11,823	2,237
Juang He	5,159	214

Source: Adapted from Norse.[11]

In the United States, in 1998, more than one-third of all river miles, lakes (excluding the Great Lakes), and estuaries did not support the uses for which they were designated under the Clean Water Act.[6] Table 25.1 illustrates N inputs to rivers and coasts in areas of America, Africa, and Asia.

It is now widely acknowledged that agriculture is the main source of N pollution in surface waters and groundwater in rural areas of Western Europe and the United States.[2,7,8] The U.K. House of Lords' report *Nitrate in Water*[9] commented on the conflicts that can arise when the use of land for farming comes into conflict with the use of land for water supply. Concern initially focused on alleged links between high nitrate concentrations in drinking water and two health problems in humans: the "blue-baby" syndrome methemoglobinemia and gastric cancer. Now, there are also concerns for environmental degradation. Nutrient enrichment in water bodies encourages the growth of aquatic plants (see Figure 25.3). Reed beds and other marginal plants may be attractive on a small scale, but when these and, particularly, underwater plant growth are excessive, this can cause a narrowing of waterways and become a nuisance to recreational users of rivers and lakes. Furthermore, eutrophication (a group of effects caused by nutrient enrichment of water bodies) can adversely affect the aquatic ecosystem. An algal bloom may cut out light to the subsurface, and when it dies back, decomposition uses the oxygen supply needed by other species. Some algae are toxic to fish, whilst others, for example, cyanobacterial species, are toxic to mammals including domestic pets.[10] Studies in Asia have demonstrated the link between increasing use of fertilizers and increasing incidence of algal blooms. Table 25.2 illustrates rates of fertilizer application for selected Asian countries. In some Chinese provinces, fertilizer application is greater than 400 kg/ha. This is usually applied as a single application, and with crop utilization efficiency as little as 30–40%, a high proportion is lost to rivers, lakes, and coastal waters.[11] The environmental impact at the regional level has led to a rise in the incidence of red tides (algal blooms). During the 1960s, less than 10 red tides per year were recorded, but in the late 1990s over 300 per year were recorded.[11]

The popular misconception that the nitrate problem is caused by farmers applying too much nitrate fertilizer is too simplistic. Nevertheless, there is now little doubt that the high concentrations of nitrate in fresh waters noted in recent years have mainly resulted from runoff from agricultural land and that the progressive intensification of agricultural practices, with increasing reliance on the use of nitrogenous fertilizer, has contributed significantly to this problem. Since 1945, agriculture in the industrialized world has become much more intensive. Fields are ploughed more frequently; more land is devoted to arable crops, most of which demand large amounts of fertilizer; grassland too receives large applications of fertilizer to ensure a high quality silage for winter feed; stocking densities in general are higher, leading to increased inputs of manure on grassland and problems of disposal of stored slurry; cattle often have direct access to water courses resulting in soil and bank erosion and direct contamination from animal waste; many low-lying fields are now underdrained, encouraging more productive use of the land and speeding the transport of leached nitrate to surface water courses. It is true that lowland

FIGURE 25.3 Choked watercourse, River Skerne, County Durham, August 2002.
Source: From P. Widdison.

TABLE 25.2 Average Fertilizer Use (kg/ha of Cropland 2000)

Country	Average Fertilizer Use kg/ha (2000)
China	255.6
Japan	301.0
Korean Rep.	407.3
Vietnam	285.3

Source: From FAO.[17]

rivers close to urban areas receive larger quantities of nitrogen from sewage effluent, but budgeting studies confirm that agriculture is the main source of nitrate in river water.[11,12]

Betton, Webb, and Walling[13] have mapped nitrate concentrations for mainland Britain. A marked northwest to southeast gradient is evident, reflecting relief, climatic conditions, and agricultural activity. Upland areas in the north and west are usually characterized by nitrate concentrations below 1 mg NO_3–N L^{-1}. This reflects the high rainfall and low temperatures of such areas: upland soils tend to conserve organic matter and mineralization rates are low. In contrast, a decreasing ratio of runoff to rainfall and an increasing intensity of agricultural land use towards the south and east of Britain result in higher mean concentrations of nitrate in river water. Many of the lowland rivers are characterized by concentrations above 5 mg NO_3–N L^{-1}; in East Anglia and parts of the Thames basin, mean nitrate concentrations in rivers are close to the E.C. limit of 11.3 mg NO_3–N L^{-1}, a level exceeded in some spring waters especially in the Jurassic limestones of the Cotswolds and Lincolnshire Wolds.[14]

The changing pattern of lowland agriculture since 1945 is reflected in long-term records of nitrate for surface and ground waters.[14] For both large and small rivers, there has been a relatively steady upward trend in nitrate concentrations, often of the order of 0.1–0.2 mg NO_3–N L^{-1}. Analyses for relatively short time series of just a few years[13] have shown that the upward trend may be interrupted, either because of climatic variability (drier years are associated with lower nitrate concentrations) or because of land use change. Nevertheless, statistical analysis of long time series shows that the main effect is a steady increase in nitrate levels over time that is independent of climate.[14] If trends continue, the mean nitrate concentration of many rivers in Europe will soon be above the E.C. limit; in many cases, this level is already exceeded during the winter when nitrate concentrations reach their maximum. In catchments where groundwater is the dominant discharge source, this long-term trend may be prolonged since it may take years for nitrate to percolate down to the saturated zone. In such basins, nitrate pollution may remain a problem for decades to come. In recent years, a number of options have been considered as a means of halting the upward trend.

Land Use Controls to Reduce Nitrogen Enrichment to Surface Waters

Trends in water management in Europe include moves toward catchment-level management, improved intersectoral co-ordination and co-operation, and frameworks facilitating stakeholder participation. This approach is developed by the European Union in its Water Framework Directive, which sets targets for good ecological status for all types of surface water bodies and good quantitative status for groundwater.[3] More localized schemes, like the U.K. Nitrate Vulnerable Zones, involve greater restrictions on farming practice, such as restricting the amount and timing of organic and inorganic fertilizer application. The EU Common Agricultural Policy is to change the way payments are made to farmers. Single-farm payments will encourage farming in a more environment friendly way. Financial payments may be available to farmers for loss of income or for changing farming practice such as improving slurry storage and fencing off watercourses to restrict livestock access.[2] Much interest currently focuses on the use of riparian land as nitrate buffer zones.[15]

The terrestrial-aquatic ecotone (boundary zone) occupies the zone between the hillslope and the river channel, usually coinciding with the floodplain. Given their position, nearstream ecotones can potentially function as natural sinks for sediment and nutrients emanating from farmland. Observed denitrification rates in floodplain sediments may be sufficient to remove all nitrate from groundwater flowing under a riparian woodland, with a floodplain width of 30 m. Saturated, anoxic soils, rich in carbon, are exposed to nitrate-rich groundwater. Rates of denitrification are high within this zone since the nutrients required by denitrifying bacteria are abundant. Wetlands and wet meadows (defined as areas where the water table is at or above land surface for long enough each year to promote the formation of hydric soils and support the growth of aquatic vegetation) also have potential as nitrogen sinks.[16] High production rates by wetland vegetation result in an abundance of carbon providing an organic substrate for bacterial processes. Wetland plants transport oxygen into anaerobic sediments, which can enhance denitrification leading to losses of nitrogen as N_2O or N_2 from wetland sediments.

The type of vegetation found on the floodplain controlling the efficiency of nitrate absorption is the subject of much debate (see, for example,[15]). Several studies have argued the presence of trees is crucial; yet others state the role of surface vegetation is secondary to the presence of saturated conditions together with a carbon-rich sediment. Denitrifying bacteria operate best at the junction of anaerobic/aerobic zones where both carbon and nitrate are abundant. It is clear that nitrate losses may be reduced by creating a nutrient-retention zone between the farmland and the river. Given that many floodplains around the world are part of an intensive agricultural system, creating permanently vegetated buffer strips between field and water courses is an idea that should be actively promoted. However, buffer strips will only be successful nutrient sinks if they are managed in an appropriate way. Underlying artificial drainage should be broken or blocked up to prevent a direct route to the watercourse for solutes, and grassland strips need maintenance to prevent them becoming choked with sediment and losing their sediment retention potential.

Solving the problem of nutrient enrichment of surface waters cannot be seen in the short term. Long-term land use change is needed. Taking farm land immediately adjacent to water courses out of production is one option that could go some way to allow modern agriculture and water supply to coexist in the same basin.

References

1. Heathwaite, A.L. Nitrogen cycling in surface waters and lakes. In *Nitrate: Processes, Patterns and Management*; Burt, T.P., Heathwaite, A.L., Trudgill, S.T., Eds.; Wiley: 1993.
2. Burt, T.P.; Heathwaite, A.L.; Trudgill, S.T.; Eds. *Nitrate: Processes, Patterns and Management.* Wiley, 1993.
3. WHO. *Water and Health in Europe: A Joint Report from the European Environment Agency and the WHO Regional Office for Europe*; World Health Organisation: 2002.
4. European Environment Agency. *Nitrogen Concentrations in Rivers*; European Environment Agency: 2001.
5. United States Geological Service. Nitrogen in the Mississippi basin-estimating sources and predicting flux to the Gulf of Mexico. In *USGS Fact Sheet No. 135–00*; USGS: U.S.A., 2000.
6. Ribaudo, M. Non-point source pollution control policy in the USA. In *Environmental Policies for Agricultural Pollution Control*; Shortle, J.S., Abler, D.G., Eds.; CAB International: 2001.
7. Power, J.F.; Wiese, R.; Flowerday, D. Managing farming systems for nitrate control: a research review from management systems evaluation areas. J. Environ. Qual. **2001**, *30*, 1886–1880.
8. Royal Society. The *Nitrogen Cycle of the United Kingdom*; Royal Society: London, 1983.
9. House of Lords. *Nitrate in Water: Sixteenth Report from the European Communities;* HMSO: London, 1989.
10. Addiscott, T.M. Fertilizers and nitrate leaching. In *Agricultural Chemicals and the Environment*; Hester, R.E., Harrison, R.M., Eds.; Royal Society of Chemists, 1996.
11. Norse, D. Fertilisers and world food demand. Implications for environmental stress. IFA-FAO Agriculture Conference. Rome, 2003.
12. Burt, T.P.; Johnes, PJ. Managing water quality in agricultural catchments. Trans. Inst. Br. Geogr. **1997**, 22(1).
13. Betton, C.; Webb, B.W.; Walling, D.E. *Recent Trends in NO3-N Concentration and Load in British rivers*; IAHS Publication, 1991; 203, 169–180.
14. Johnes, PJ.; Burt, T.P. Nitrate in surface waters. In *Nitrate: Processes, Patterns and Management*; Burt, T.P., Heathwaite, A.L., Trudgill, S.T., Eds.; Wiley: 1993; 269–317.
15. Haycock, N.E. et al., Eds. Buffer zones: their processes and potential in water protection. Quest Environmental: Harpenden.
16. Howard-Williams, C.; Downes, M.T. Nitrogen cycling in wetlands. In *Nitrate: Patterns, Processes and Management*; Burt, T.P., Heathwaite, A.L., Trudgill, S.T., Eds.; Wiley: 1993; 141–167.
17. Food and Agriculture Organisation. Earth Trends Data Tables: Agricultural Inputs 2003; http://earthtrends. wri.org.

26

Surface Water: Nitrogen Fertilizer Pollution

Introduction 197
Problems Caused by Nitrogen Pollution of Surface Waters 197
Sources of N Pollution 198
Fertilizers 198
What Happens When Fertilizer Is Applied to Soil? 198
Biochemical Processes
Hydrologic Processes 198
Spatial Variability 200
Watershed Scale Analyses 200
Regional Nitrogen Input–Output Analyses
Hydrologic Process Models 201
The Role of Fertilizer Use Efficiency 201
Additional Needs and Approaches for Reducing Nitrogen Transport ... 202
Conclusions 202
References 203

Gregory McIsaac
University of Illinois

Introduction

The use of industrially manufactured nitrogen (N) fertilizers increased rapidly in developed countries between 1960 and 1980. This facilitated a large increase in the production of feed and food grains (maize, wheat, and rice) per unit of cultivated land, but in some regions it also contributed to enrichment of surface and groundwater with various forms of nitrogen. Fertilizer, however, is not the only source of nitrogen that can cause contamination of surface waters. Biological nitrogen fixation, mineralization of soil organic nitrogen, and animal wastes can also contribute to nitrogen enrichment of water bodies. Additionally, under some conditions, nitrogen applied to the soil may be converted to gaseous or immobile forms of nitrogen that do not contribute to surface water contamination. Because of these various sources and transformations of nitrogen, the severity of surface water contamination by nitrogen fertilizer has been difficult to precisely quantify. Existing research indicates that the amount of contamination from fertilizer varies depending on the amount of fertilizer applied, and characteristics of the soils, crops, climate, and the receiving water bodies.

Problems Caused by Nitrogen Pollution of Surface Waters

There are three water quality concerns associated with different forms of nitrogen. First, the combined concentrations of nitrate (NO_3^-) plus nitrite (NO_2^-) in excess of 10 mg N L^- can contribute to methemoglobinanemia ("blue baby syndrome") in infants if ingested.[1] To guard against this, the U.S. Public Health

Service limits nitrate plus nitrite concentration in public drinking water supplies to 10 mg N L^-. Secondly, unionized ammonia (NH_3) may be toxic to fish at concentrations as low as 0.02 mg N L^-. Finally, elevated total nitrogen concentrations (including nitrate, ammonia, and organic forms) in rivers can promote the process of cultural eutrophication in coastal waters, whereby increased production and decomposition of algae, leads to reduced oxygen concentrations. This, in turn, may reduce the abundance and diversity of marine life and may promote the outbreak of nuisance algae.[2]

Sources of N Pollution

Nitrogen contamination may come from a variety of sources: municipal sewage, animal manure, atmospheric deposition, biological N fixation, soil organic N, and/or nitrogen fertilizers. The consequences of contamination in a specific water body will depend upon the amount of contamination from all sources and characteristics of the receiving waters. Shallow rivers, wetlands, lakes, and reservoirs, have some capacity to remove nitrogen by microbial denitrification. The susceptibility of estuaries and coastal waters to eutrophication depends on temperature, availability of phosphorus and silica for algae production, and the rate of water exchange with the open ocean.

Fertilizers

The contribution of inorganic fertilizer to surface water N contamination increased after 1960 as the widespread and intensive use of inorganic N fertilizers rapidly expanded.[3] The use of N fertilizer has allowed greater production of feed and food crops per unit area cultivated. In the United States, 75% of N fertilizer is applied to maize, while in other countries, N fertilizer is primarily used on wheat and rice. Prior to 1960, nitrogen for crop production was obtained primarily by using crop rotations that included legumes such as clover and alfalfa, which can establish a symbiotic relationship with soil bacteria that can convert atmospheric N_2 gas to biologically available forms of N.

Commercial nitrogen fertilizer is primarily manufactured as gaseous ammonia (NH_3), using the Ha–Bosch process in which gaseous nitrogen is reacted with gaseous hydrogen under pressure. The gaseous ammonia may be injected into the soil, which is a common fertilizer application practice in the United States. Additionally, a wide variety of granular and aqueous fertilizer products containing nitrogen are manufactured from manufactured ammonia.

What Happens When Fertilizer Is Applied to Soil?

Biochemical Processes

In the soil, ammonia reacts with water and is largely converted to ammonium (NH_4^+), which tends to be strongly adsorbed on soil particles. This adsorption inhibits the movement of ammonium through the soil. Ammonium is an energy rich substance and certain soil bacteria can utilize this energy by decomposing the ammonium to nitrate (NO_3^-). Unlike ammonium, nitrate is not adsorbed to soil particles and, therefore, moves readily with water in the soil. Nitrate that is not taken up by plant roots or soil microorganisms can be transported to groundwater and surface water by a variety of mechanisms.

Hydrologic Processes

Rainfall, snow melt, or irrigation water input to the soil periodically exceeds the water holding capacity of the soil in the root zone. Depending on the characteristics of the soil, this may lead to one or more of the following: 1) saturation of the root zone with water; 2) surface runoff; and 3) drainage of water through the soil profile to groundwater and/or surface water bodies. Each of these has different consequences for transport of nitrate to surface waters.

If the soil becomes saturated, oxygen may become scarce and in anoxic conditions, denitrifying bacteria may convert the nitrate to nitrogen gases (NO, N_2O, and N_2). Nitrogen converted to these gases becomes unavailable for plant uptake or for surface water contamination. Additionally, saturated soil during the growing season is harmful to many crops like maize that cannot tolerate low oxygen concentrations in the root zone for more than a few days.

Surface runoff has the capacity to transport soil, vegetation, and surface applied granular fertilizers from agricultural fields to surface water bodies. If a granular form of nitrogen fertilizer had been applied immediately prior to the event that caused the surface runoff to occur, nitrate and ammonia concentrations in runoff can be very high.[4] This does not appear to be a common phenomenon, however. Small rainfall events are much more common than large events that typically produce surface runoff. After granular fertilizer is applied, it is likely that a series of small rainfall events will dissolve the granules and move the nitrogen into the soil profile, where it is less likely to contaminate surface runoff. Surface runoff usually has a low nitrate concentration but it can be high in organic and particulate N derived from soil and vegetation.

Drainage of water through the soil profile to groundwater and surface water appears to be the hydrologic pathway that most frequently leads to problematic nitrate contamination of surface waters in agricultural watersheds. This can occur in two ways: by natural drainage where ground water contributes to stream flow and river flow, and by artificial subsurface drainage, where perforated pipes (sometimes called tile drains) have been buried in the soil for the purpose of removing water to reduce damage caused by saturated conditions and thereby enhance crop production (Figure 26.1).

Artificial subsurface drainage improves aeration of the soil root zone and increases the length of time that machinery can be used on the soil.[6] It is a common practice in the north central United States, and in northwestern Europe, where flat and swampy land has been converted to cropland. The water removed from the soil by artificial drainage is usually directed to surface ditches, streams, and rivers. This water can have high concentrations of nitrogen, principally in the nitrate form, especially where

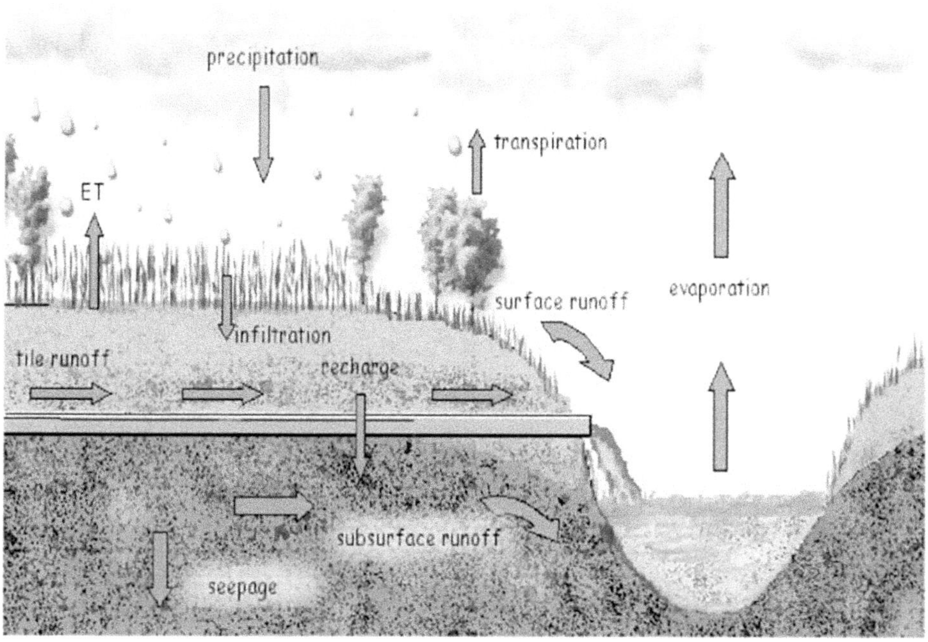

FIGURE 26.1 Illustration of the hydrologic cycle with artificial subsurface drainage (tile runoff) contributing to surface channel flow.
Source: Adapted from Steinheimer, et al.[5]

nitrogen fertilizers are applied in excess to the amount necessary for crop production.[7,8] This nitrate can also be derived from microbial conversion of soil organic matter to inorganic N in the process of mineralization. Mineralized soil organic N may come from crop residues (unharvested leaves, stalks, and roots). Of course, some of the N in crop residues may have originated from fertilizer applied in previous years, but it can also derive from biological N fixation or animal manures applied on a field.

Spatial Variability

In most agricultural settings, commercial fertilizer provides only one source of N used for crop production. Animal manure, biological N fixation, mineralization from soil organic N, and deposition of N from the atmosphere can also contribute to soil fertility and surface water contamination. Because there are multiple sources and sinks of N in the soil, the relationship between N fertilizer application rate and nitrogen loss in drainage water is not always consistent across locations and across studies. If denitrification and plant and microbial uptake of N are large, nitrate concentrations in subsurface drainage may be low in spite of high fertilizer N inputs. If mineralization of soil organic matter is large, nitrogen in drainage water may be large without N fertilizer input. High rates of mineralization of soil organic N occur after the initial cultivation of virgin land, and after a leguminous forage crop such as alfalfa or clover is cultivated into the soil. Appropriate use of N fertilizer should take all of these N sources into account, as should studies examining the relationship between N fertilizer use and water quality.

Watershed Scale Analyses

Regional Nitrogen Input–Output Analyses

Howarth et al.[10] developed an approach for estimating the net nitrogen inputs to a region N that is highly correlated with average nitrogen transport in the rivers draining temperate regions (Figure 26.2). Net N input to a region was defined as sum of N in fertilizer used, biological N fixation of agricultural

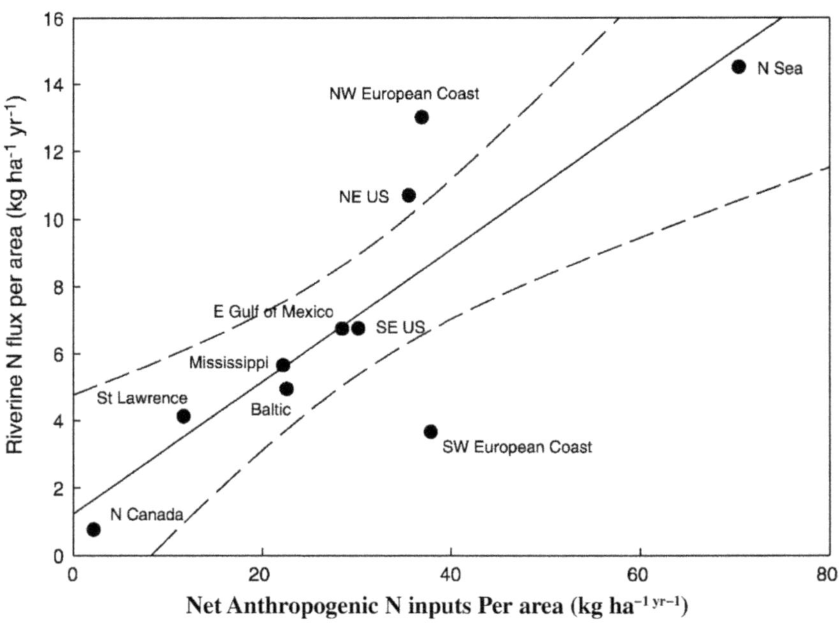

FIGURE 26.2 Average annual riverine total N flux as a function of net anthropogenic N inputs to temperate regions draining to the North Atlantic Ocean.
Source: Adapted from Howarth, et al.[10]

crops, oxidized N in atmospheric deposition in the region, and the N in food and feed imported to the region minus the N in food and feed exported from the region. This approach assumes that there is no net gain or loss of N from soil organic matter. This assumption appears to be reasonable in regions where most soils have been under continuous cultivation for 60 years or more, at which time, annual mineralization of soil organic N is roughly replaced by organic N returned to the soil in crop residues and microbial biomass.[11]

In temperate regions, riverine N transport is, on average 25% of the net N input to the region. The fate of the other 75% of the net N is unknown, but much of it is probably converted to gaseous forms of N by microbial denitrification. The high net N input in countries draining to the North Sea, most notably the Netherlands, is in part due to high density of domestic animals as well as use of N fertilizers. In tropical regions, riverine N flux is much greater than 25% of net N inputs, even in regions where little N fertilizer was used. The reasons for this are not precisely known but it is believed to be due, in part, to greater rates of biological N fixation in both cultivated and non-cultivated land in the tropics. This may also be due to the recent conversion of forest, wetlands, and grasslands to crop production, which leads to high rates of mineralization of soil organic N to nitrate which is highly mobile.

Hydrologic Process Models

The quantity of nitrate transported in rivers is also related to the quantity of water flowing in the rivers per unit of land area, which is also known as water yield. Caraco and Cole[12] demonstrated that riverine nitrate N transport in major rivers in the world was a function of water yield, fertilizer use, population density, and atmospheric deposition of oxides of N. Building on these results, McIsaac et al.[13] developed the following model of annual nitrate discharge in the Lower Mississippi River 1960–1998:

$$NF_m = 0.66WY^{0.93}e^{\left(0.13NNI^{2-5} + 0.06NNI^{6-9}\right)} \tag{26.1}$$

where NF_m = annual nitrate N flux in Lower Mississippi River (kg N ha^{-1} yr^{-1}), NNI^{2-5} = average annual net N input during the previous 2–5 years (kg N ha^{-1} yr^{-1}), NNI^{6-9} = average annual net N input during the previous 6–9 years (kg N ha^{-1} yr^{-1}), and WY = annual water yield (m yr^{-1}).

This equation accounted for 95% of the annual variation in nitrate flux in the Mississippi River from 1960 to 1998 and suggested that riverine nitrate in a given year was correlated with net N input averaged over the previous 2–9 years. Furthermore, calculations with the model suggest that if the N fertilizer use in the basin had been 12% lower than actual during this period, nitrate flux to the Gulf of Mexico would have been 33% less than observed (Figure 26.3), assuming crop yields were not limited by N shortages.

The Role of Fertilizer Use Efficiency

The efficiency of fertilizer used for maize production in the major maize producing states (Illinois, Iowa, Indiana, Minnesota, and Nebraska) in the Mississippi River Basin increased between 1986 and 2000. Maize yields have increased about 20% from 1986 to 2000, while N fertilizer use has remained roughly constant. Between 1976 and 1986, an average of 0.023 kg N of fertilizer was applied for each kg of maize harvested. Between 1996 and 2000, an average of 0.018 kg N of fertilizer was applied per kg of maize harvested. If this improvement in fertilizer use efficiency had not occurred, nitrate flux in the Mississippi River in 1996–1998 would have been about 50% greater than the measured flux, according to the model of McIsaac et al.[13] (Figure 26.3).

Farmers face two major uncertainties when making fertilizer application decisions: they do not know what their yields will be nor how weather conditions might influence the availability of N fertilizer to the crop. The cost of nitrogen fertilizer has been relatively low in relation to the value of the increased yields, and consequently many farmers have believed that applying more N fertilizer than necessary

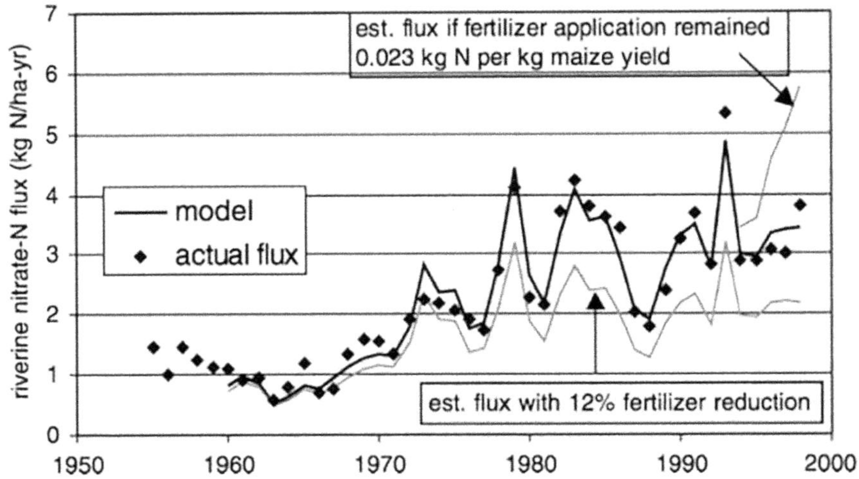

FIGURE 26.3 Annual riverine nitrate flux in the Lower Mississippi River at St. Francisville, Louisiana, as determined from measurements (diamonds) as estimated from Equation (26.1) (thick line), as estimated from Equation (26.1) assuming a 12% reduction in N fertilizer input (thin lower line), and assuming fertilizer applications remained 0.023 kg N per bushel of harvested maize rather than declining to 0.018 kg N per bushel of harvested maize (thin upper line).
Source: Adapted from McIsaac.[13]

provides "cheap insurance" against the uncertainties. In some instances, farmers did not consider N available from animal manure or from previously harvested legume crops like soybeans.

A number of factors are likely responsible for the increased fertilizer use efficiency. Research and outreach efforts have provided farmers with better information for making N fertilizer decisions. Water quality concerns have focused attention on the need for improved nutrient management. Weather during the 1990s was generally more favorable for corn production than the 1980s, when three major droughts occurred in the corn growing region of the Mississippi River Basin.

Additional Needs and Approaches for Reducing Nitrogen Transport

A recent improvement in the efficiency of N fertilizer use has also been observed in wheat production in the United Kingdom and rice production in Japan.[3] However, improved fertilizer use efficiency alone may not be sufficient to address water quality problems in some settings. Jaynes et al.[8] reported that even with N fertilizer rates at recommended levels, nitrate concentration in tile drainage water sometimes exceeded the drinking water standard of 10 mg N L^{-} in Iowa.

Zucker and Brown[9] recommended several additional practices that can reduce nitrogen contributions in tile drainage: water table management, treatment of drainage water in wetlands, and use of crop rotations that reduce N losses. Additional monitoring and documenting the changes in water quality associated with changing fertilizer management practices are needed to improve our understanding of the connections between N fertilizer use and water quality in different geographic settings.

Conclusions

In many settings nitrogen enrichment of surface water bodies has increased following the increased use of N fertilizers. The precise contribution of nitrogen fertilizers to surface water nitrogen has been difficult to quantify because there are multiple sources of nitrogen contributing to most water bodies, and, depending on environmental conditions, a certain portion of soil nitrogen may be converted to gaseous or immobile forms. In general, however, agricultural regions with extensive artificial subsurface drainage systems or with sandy soils tend to have the most nitrogen enriched surface waters.

The efficiency of nitrogen fertilizer used for crop production increased in many areas in the 1990s and this has very likely limited or reduced the subsequent contamination of surface waters. Continued improvements in fertilizer use efficiency, and the use of wetlands for removing nitrogen from surface waters will help alleviate problems caused by nitrogen enrichment. Additional monitoring and research are needed to more precisely quantify how nitrogen management practices influence surface water nitrogen concentrations in different settings. A more precise understanding of the causal relationships between nitrogen inputs to the land and the contamination of surface waters could provide more effective guidance for management, policies, and programs intended to protect aquatic resources while maintaining optimal use of land resources.

References

1. Skipton, S.; Hay, D. *Drinking Water: Nitrate and Methemoglobinemia ("Blue Baby" Syndrome);* Publication G98-1369; University of Nebraska Cooperative Extension: Lincoln, NE, 1998, http://www.ianr.unl.edu/pubs/water/g1369.htm (accessed February 2002).
2. National Research Council. *Clean Coastal Waters: Understanding and Reducing the Effects of Nutrient Pollution;* National Academy Press: Washington, DC, USA, 2000.
3. Smil, V. *Enriching the Earth: Fritz Haber, Carl Bosch and the Transformation of World Food Production;* MIT Press: Cambridge, MA, USA, 2001.
4. Romkens, M.J.M.; Nelson, D.W.; Mannering, J.V. Nitrogen and phosphorus composition of surface runoff as affected by tillage method. J. Environ. Qual. **1973,** *2* (2), 292–295.
5. Steinheimer, T.R.; Scoggin, D.K.; Kramer, L.A. Agricultural chemical movement through a field-size water-shed in Iowa: subsurface hydrology and distribution of nitrate in groundwater. Environ. Sci. Technol. **1998,** *32,* 1039–1047.
6. Badiger, S. Integrated Numerical Modeling of Spatial and Time-Variant Hydrologic Response in Subsurface Drained Watersheds. Ph.D. thesis. Department of Agricultural Engineering, University of Illinois: Urbana, 2001.
7. Stevenson, F.J.; Ed. *Agricultural Drainage;* American Society of Agronomy (Monograph # 38): Madison, WI, USA, 1999.
8. Jaynes, D.B.; Colvin, T.S.; Karlen, D.L.; Cambardella, C.A.; Meek, D.W. Nitrate loss in subsurface drainage as affected by nitrogen fertilizer rate. J. Environ. Qual. 2001, 30, 1305–1314.
9. Zucker, L.A.; Brown, L.C. *Agricultural Drainage: Water Quality Impacts and Subsurface Drainage Studies in the Midwest;* Ohio State University Extension Bulletin: 1998; 871, http://ohioline.osu.edu/b871/ (accessed February 2002).
10. Howarth, R.W.; Billen, G.; Swaney, D.; Townsend, A.; Jaworski, N.; Lajtha, K.; Downing, J.A.; Elmgren, R.; Caraco, N.; Jordan, T.; Berendse, F.; Freney, J.; Kudeyarov, V.; Murdoch, P.; Zhao-Liang, Z. Regional nitrogen budgets and riverine N&P fluxes for the drainages to the North Atlantic Ocean: natural and human influences. Biogeochemistry **1996**, 35, 75–139.
11. Stevenson, F.J.; Ed. Origin and distribution of nitrogen in the soil. *Nitrogen in Agricultural Soils*; Agronomy Monograph 22; American Society of Agronomy: Madison, WI, 1982; 1–42.
12. Caraco, N.F.; Cole, J.J. Human impact on nitrate export: an analysis using major world rivers. Ambio **1999**, 28, 167–170.
13. McIsaac, G.F.; David, M.B.; Gertner, G.Z.; Goolsby, D.A. Nitrate flux in the Mississippi River. Nature **2001**, *414,* 166–167.

27

Transpiration

Thomas R. Sinclair
University of Florida

Background 205
Stomata 205
Leaf Transpiration Rate 207
Canopy Transpiration Rate 208
References 209

Background

Carbon dioxide (CO_2) assimilation in photosynthesis was greatly facilitated, and thereby allowing rapid plant growth, when plants invaded the Earth's land masses. No longer was it necessary for CO_2 to diffuse at very slow rates through the water surrounding aquatic plant life, but rather CO_2 could be absorbed directly from the atmosphere into individual cells. While photosynthesis rates and plant growth were enhanced substantially by allowing direct exposure of photosynthetic cells to the atmosphere, a potentially fatal consequence was that water could be evaporated at very high rates from exposed cell surfaces.

The problem of rapid evaporation from cell surfaces can be solved by having either an especially effective plant structure to rapidly transport large quantities of water in order to replenish each cell with water, or mechanisms to substantially inhibit water loss (and also CO_2 assimilation rates) from cell surfaces. While evolution "experimented" with each of these approaches, an innovative third, anatomical solution dominates. The photosynthetic cells are packaged inside thin broad tissues, i.e., leaves, that can throttle water vapor diffusion between the cells inside leaves and the atmosphere outside the leaves (Figure 27.1).

The exterior of leaf epidermal cells is coated with a cuticle containing waxy materials that effectively block water loss directly from epidermal cells to the atmosphere. Consequently, virtually all water lost by leaves must move through stomatal pores that are scattered in the leaf epidermis (Figure 27.2). Stomata regulate gas diffusion through the epidermis by adjustments in the dimension of the stomatal pore. The apertures of stomatal pores adjust to maintain a fairly stable CO_2 concentration inside leaves. When leaf photosynthetic rates are high, stomata adjust to increase the size of the pore aperture for rapid diffusion of CO_2 into leaves. On the other hand, when photosynthetic rates are low, aperture size decreases. Since water vapor diffuses through the same stomatal aperture, adjustments to accommodate photosynthesis cause changes in transpiration rate.

Stomata

Stomata are embedded in the leaf epidermis and are formed by a pair of cells called guard cells. The guard cells in monocots plants, as shown in Figure 27.2, tend to resemble barbells with bulbous structures at each end. The guard cells of dicot plants shown in Figure 27.3, have a kidney shape. In both cases, the pair of guard cells is attached to each other at both ends. Swelling of the bulbous end of guard cells in monocots causes the cylindrical midsections to move apart increasing the aperture of the pore.

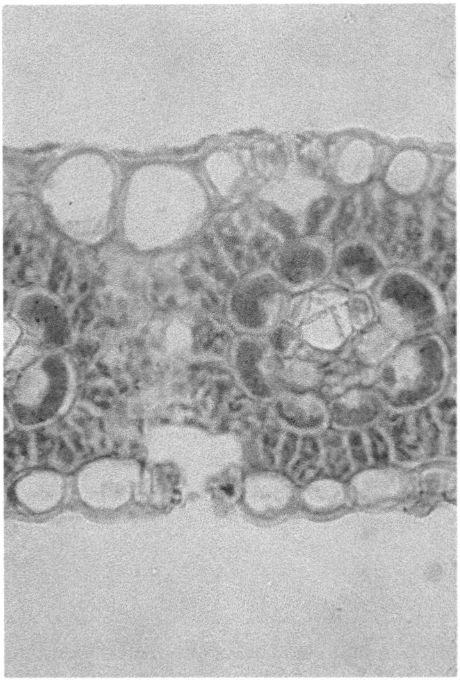

FIGURE 27.1 Cross-section of monocot leaf showing stomata in epidermis of both sides of the leaf.

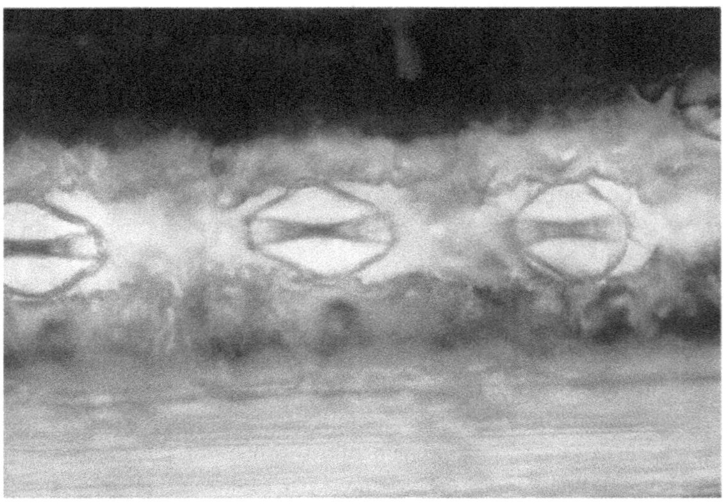

FIGURE 27.2 Photograph showing stomata in the leaf surface of sorghum.

The entire kidney-shaped cells of dicots swell and, as a result of a specialized cell wall structure bordering the pore, the aperture of the pore increases. Conversely, a decrease in the size of guard cells in both cases results in a decrease in pore aperture.

The shrinking and swelling of guard cells are under active control by plants as a result of changes in the concentrations of specific compounds in guard cells and neighboring cells. An increase in solute concentration in guard cells causes water to flow into the cells so that guard cells swell. There appears to be two mechanisms in guard cells that result in changes in solute concentration.[1] One mechanism is based

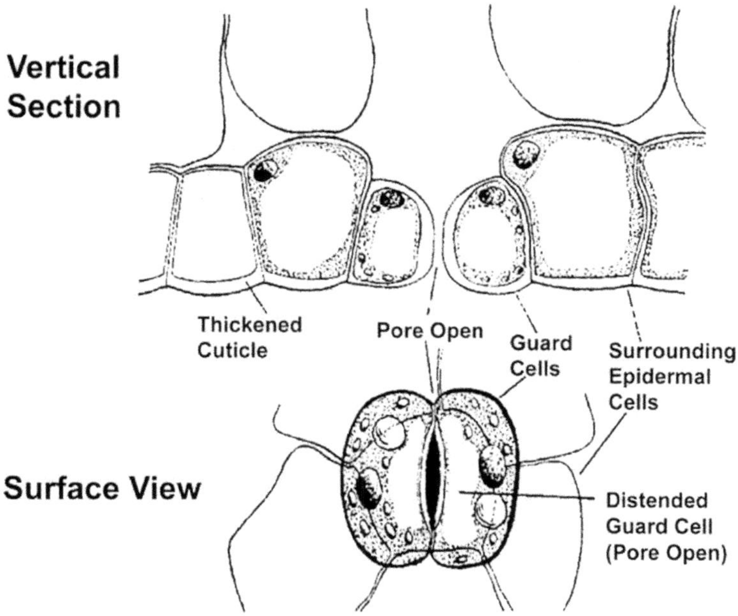

FIGURE 27.3 Drawing of cross-section and surface view of a dicot stomata.
Source: Adapted from Sinclair & Gardner.[10]

on potassium-malate transport and appears to be particularly involved with stomata opening in response to light at sunrise. The second mechanism involves sucrose accumulation in guard cells and is associated with maintenance of stomata aperture during the day. Both mechanisms are potentially sensitive to changes in the CO_2 environment of the leaf interior.[1] The sucrose mechanism seems especially sensitive to leaf photosynthetic rates so that stomata aperture can be fine tuned throughout the day as the photosynthetic rate of the leaf responds to changing environmental conditions. Of course, changes in aperture to match CO_2 diffusion to leaf photosynthetic capacity also results in changes in transpiration rate.

Leaf Transpiration Rate

Leaf transpiration involves the diffusion of water vapor through stomatal pores, so transpiration rate is dependent on conductance of vapor through and above the pores, and on the gradient of water vapor across the pores. Not surprisingly, conductance of an individual stomatal pore (g_p, $cm^3\ sec^{-1}$) is directly dependent on the dimensions of the pore aperture and can be expressed quantitatively[2] as

$$g_p \approx \pi abD/d \tag{27.1}$$

where a is the semilength of the major axis (cm), b, the semilength of the minor axis (cm), d, the depth of the pore (cm), and D, the molecular diffusion coefficient of water vapor ($0.24\,cm^2\ sec^{-1}$ at 20°C).

The semilength and depth of the pore usually remain fairly stable, so the main variable influencing g_p is the semilength of the minor axis of the pore.

The overall stomata conductance of the leaf (g_s, $cm\ sec^{-1}$) is calculated by incorporating into Equation (27.1) stomatal density (n, stomata cm^{-2}) and the influence of "end effects" exterior to the leaf.[2] Consequently,

$$g_s \approx \pi abD/\left(d + b/2\ln(4a/b)\right) \tag{27.2}$$

An estimate of maximum g_s can be calculated for fully open stomata ($b \approx 3 \times 10^{-4}$cm) by assuming that $a = 8 \times 10^{-4}$cm, $d = 10 \times 10^{-4}$cm, and $n = 10 \times 10^3$ stomata cm^{-2}. Equation 27.2 gives 1.3 cm sec^{-1} for a leaf with stomata on only one side, and 2.5 cm sec^{-1} for a leaf with stomata on two sides. Of course, when the stomatal pore is closed (b = 0), Equation (27.2) gives a conductance of zero.

In addition to the restriction on water vapor diffusion resulting from stomatal conductance, there is also a limitation on water loss resulting from the aerodynamic boundary layer around leaves (g, cm sec^{-1}). The value of g_{bl} is dependent on wind speed and leaf dimensions. In the case of a 100 cm sec^{-1} wind speed and 8 cm wide leaf, the value of g_{bl} is approximately 2.5 cm sec^{-1}.[3] Consequently, in this case, the value of g_{bl} is equal to that of g_s.

As g_s and g_{bl} are in series, the inverse of the two conductances are added together to calculate leaf transpiration rate, Tr_L (g cm^{-2} sec^{-1}). The combined conductance is multiplied by the vapor pressure difference between the interior of the leaf, which is calculated as the saturated vapor pressure at leaf temperature and atmospheric vapor pressure (P_a).[3]

$$T_{rL} = \varepsilon\left(g_s g_{bl}/\left(g_s + g_{bl}\right)\right)\left(P_L^* - P_a\right)/H_v \tag{27.3}$$

where ε is the molecular weight of water (18 g mol^{-1}) and H_v the heat of vaporization (44 kJ mol^{-1} at 25°C).

Canopy Transpiration Rate

In principle, the transpiration of a leaf canopy (Tr_C) can be calculated by summing Tr_L for all the individual leaves in the canopy. This is a formidable task, however, because values of g_s, g_{bl}, and leaf temperature are required for each individual leaf. Consequently, "summary" expressions have been developed in an effort to express the transpiration rates of the entire canopy.

One of the more popular approaches relies on predictions of the energy balance of the leaf canopy. The Penman-Monteith equation[4] gives an explicit solution for canopy transpiration rate based on the energy balance of the entire canopy. The difficulty with this approach is that it requires an estimate of the canopy boundary layer conductance and the "canopy conductance" for the entire canopy, which must be appropriately weighted to represent vapor transfer through all the stomata distributed in the canopy. There is really no independent method for measuring canopy conductance, and estimates are obtained only by back-solving the energy balance equation. In practice, the Penman-Monteith is often applied as an empirical equation where boundary layer and canopy conductances are estimated from empirical functions.3

A recent innovation has been developed to calculate Tr_C based on the water use efficiency of leaf canopies. Water use efficiency has been a topic of research since at least 1699[5] giving it a longer history of study than virtually any other plant trait. Roughly 100 years ago, there was a particularly intensive period of study in both Europe and the United States on plant water use efficiency, culminating in the classic investigations by Briggs and Shantz.[6,7] In an analysis of much of the data from this period, deWit[8] found a highly stable relationship within each species between accumulated plant mass and cumulative transpiration over a wide range of conditions when normalized by "atmospheric demand" (Figure 27.4).

Tanner and Sinclair[9] extended the analysis of deWit[8] by deriving a mechanistic expression for transpirational water use efficiency of canopies. Their derivation defined a specific transpirational water use efficiency coefficient (k, Pa) for each crop species dependent on the photosynthetic pathway and the biochemical composition of the plant products. Their estimates of k for maize, wheat, and soybean were 12 Pa, 5 Pa, and 4 Pa, respectively. Then, for each species, water use efficiency based on accumulated plant mass (M) was stable when expressed in the following equation:

$$M/\mathrm{Tr_C} = k/\left(P_a^* - P_a\right) \tag{27.4}$$

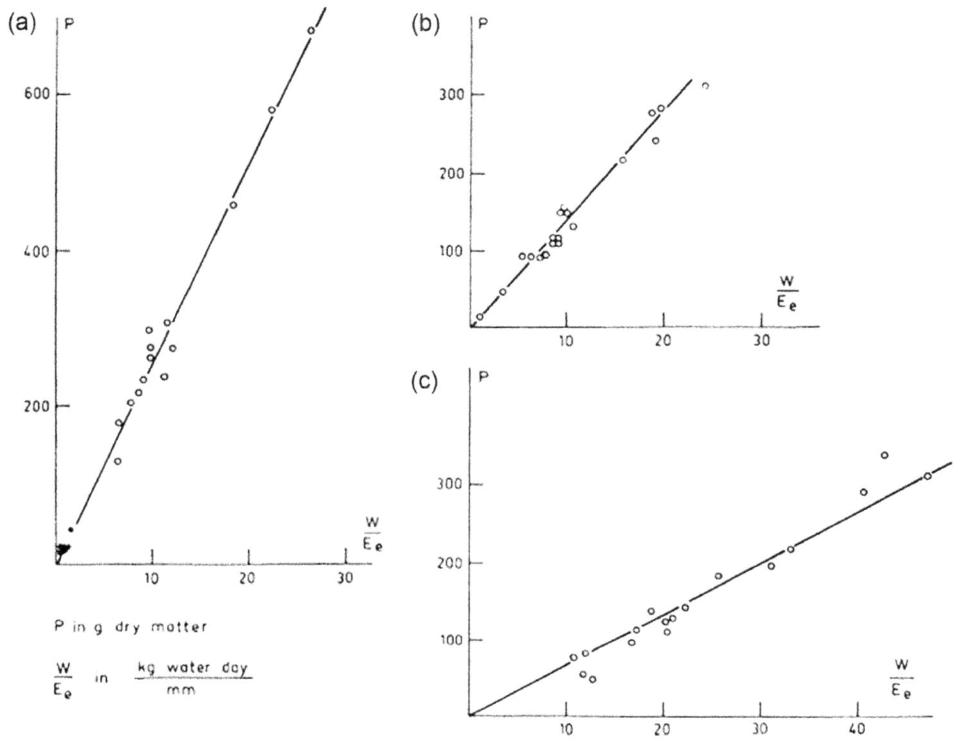

FIGURE 27.4 Plot of crop growth (*P*) against transpiration water loss (*W*) divided by pan evaporation for (**a**) sorghum, (**b**) wheat, and (**c**) alfalfa.
Source: Adapted from deWit.[8]

Tanner and Sinclair[9] approximated for daily transpiration by assuming this value was 75% of the maximum atmospheric vapor pressure deficit calculated at the daily maximum temperature.

A solution for TrC is obtained directly by the rearrangement of Equation (27.4):

$$\mathrm{Tr_C} = M\left(P_a^* - P_a\right)/k \tag{27.5}$$

This equation can be readily used based on estimates of *M*, which in turn can be calculated based on the interception of solar radiation and the radiation use efficiency of the canopy. Consequently, Equation (27.5) is consistent with the energy balance approach in that it is sensitive to the amount of solar radiation intercepted by the canopy.

Not only is Equation (27.5) easier in principle to implement that an energy balance approach, but it is conceptually much more compatible with the understanding of stomata regulation of transpiration. That is, Equation (27.5) is calculated from a direct dependence of $\mathrm{Tr_C}$ on CO_2 assimilation, which is consistent with the fact that stomata are regulated for CO_2 assimilation and the rate of water loss from leaves is simply a consequence of this process. Equation 27.5 has been used very effectively to calculate transpiration rates and evaluate limitations of water availability on crop yields.

References

1. Talbott, L.D.; Zeiger, E. The role of sucrose in guard cell osmoregulation. J. Exp. Bot. **1998**, *49*, 329–337.
2. Parlange, J.; Waggoner, P.E. Stomatal dimensions and resistance to diffusion. Plant Physiol. **1970**, *46*, 337–342.

3. Sinclair, T.R. Theoretical considerations in the description of evaporation and transpiration. In *Irrigation of Agricultural Crops*; Stewart, B.A., Nielsen, D.R., Eds.; American Society of Agronomy: Madison, WI, 1990; 343–361.
4. Monteith, J.L. Evaporation and environment. In *The State and Movement of Water in Living Organisms, XIX Symp. Soc. Exp. Biol.* Academic Press: NY, 1965; 205–234.
5. Woodward, J. Some thoughts and experiments concerning vegetation. Philos. Trans. R. Soc. London **1699**, *21*, 193–227.
6. Briggs, L.J.; Shantz, H.L. The water requirement of plants. II. A review of the literature. U.S. Dep. Agric. Bur. Plant Ind. Bull. **1913**, *285*, 1–96.
7. Briggs, L.J.; Shantz, H.L. The water requirement of plants as influenced by environment. Proc. Second Pan-American Science Congress, Washington, DC, 1917; 95–107.
8. deWit, C.B. *Transpiration and Crop Yields;* Institute of Biological and Chemical Research on Field Crops and Herbage: Wageningen, the Netherlands, 1958; Vol. 64.6, 1–88.
9. Tanner, C.B.; Sinclair, T.R. Efficient water use in crop production: research or re-search? In *Limitations to Efficient Water Use in Crop Production;* Taylor, H.M., Jordan, W.R., Sinclair, T.R., Eds.; American Society of Agronomy: Madison, WI, 1983; 1–27.
10. Sinclair, T.R.; Gardner, F.R. *Principles of Ecology in Plant Production;* CAB International: Wallingford, UK, 1998; 1 pp.

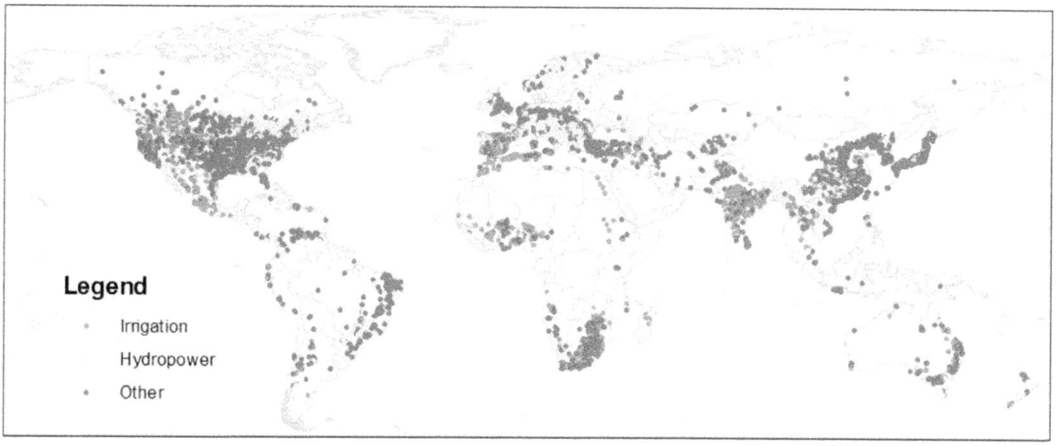

FIGURE 3.1 The global distribution of large dams and their main purpose as archived in the GRanD database.

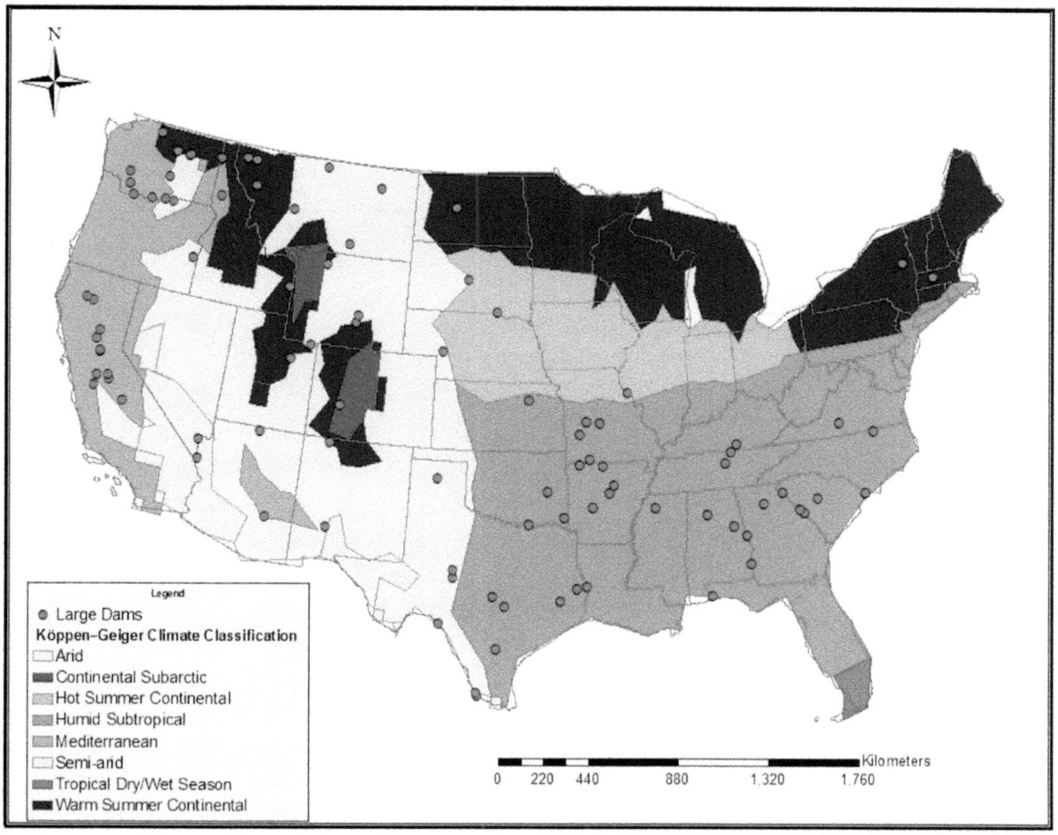

FIGURE 3.3 Location of the 92 large dams according to the Köppen climate map.
Source: Reprinted with kind permission from American Society of Civil Engineers.

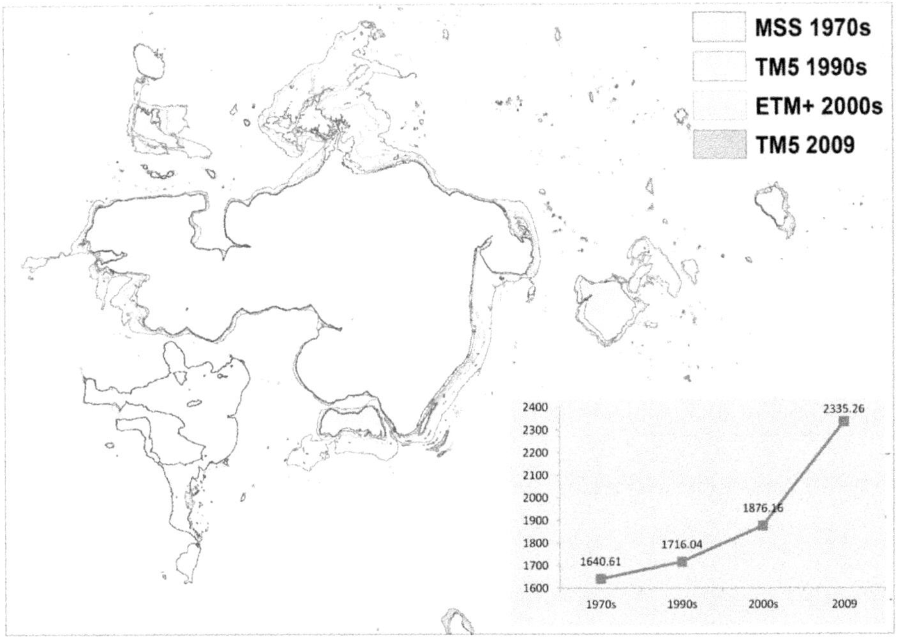

FIGURE 5.3 Decadal area changes of Siling Co as monitored by remote sensing. Since 1976, Siling Co has expanded steadily (1990 and 2000) and became the largest lake in 2009 with an area of 2,335 km^2, a 42% expansion from the 1970s.

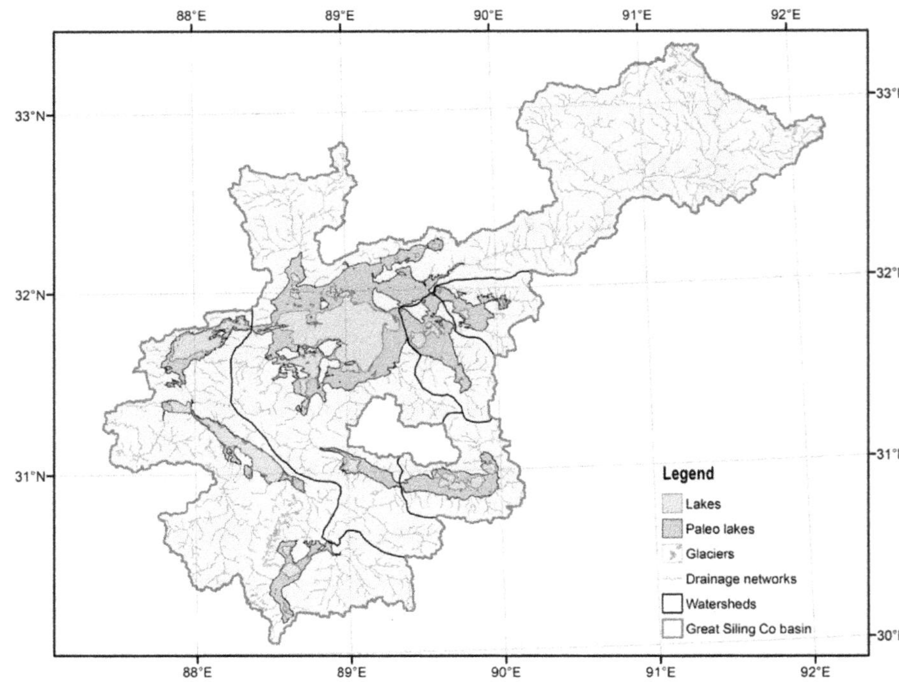

FIGURE 5.6 Paleolake recovery in Great Siling Co basin. The extents of four paleolakes (in dark gray) were recovered in the basin using current Landsat images and digital elevation data. They had shrunk in Holocene up to ~79% in area with an up-to-56 m level drop in the basin.

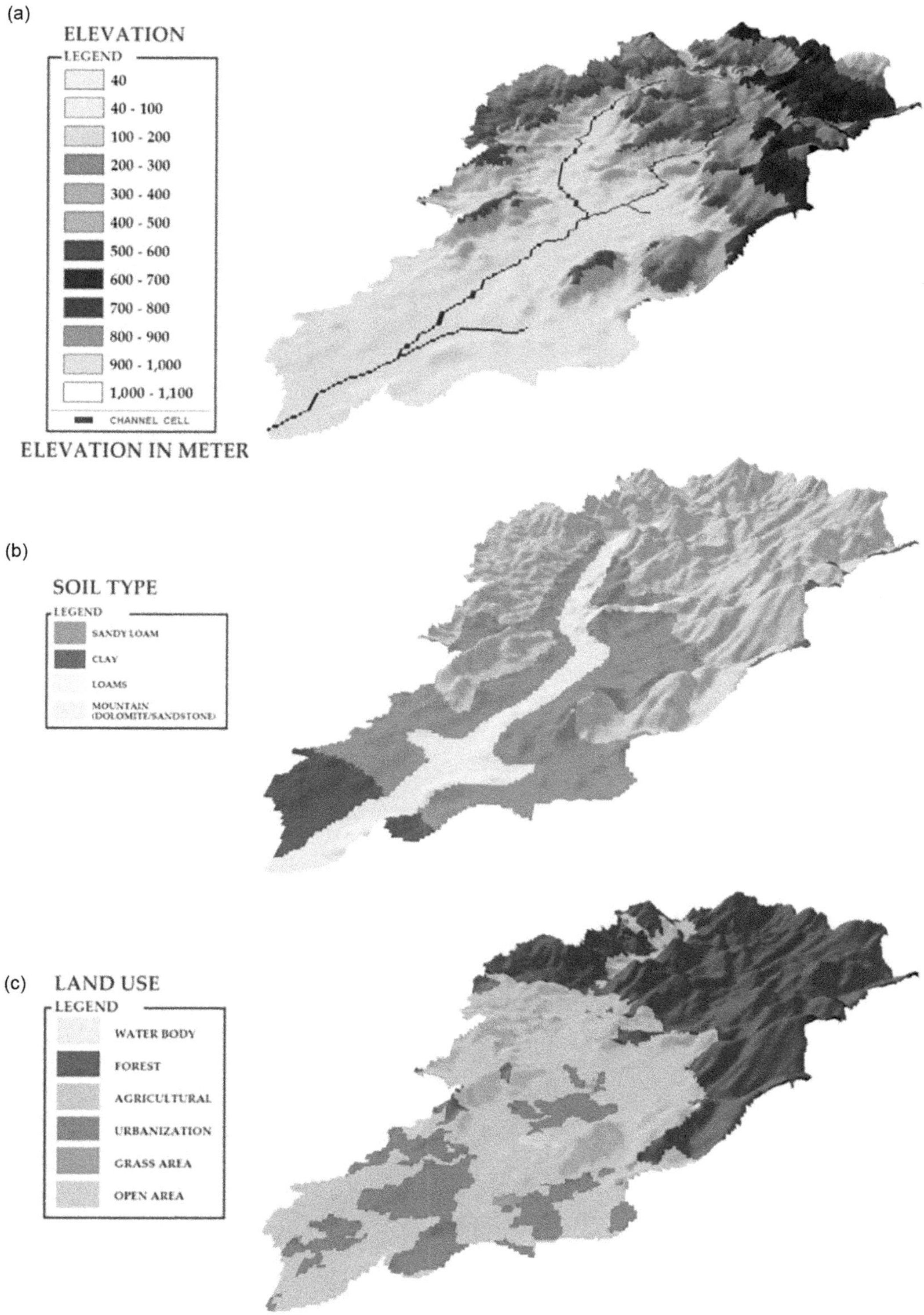

FIGURE 11.2 Sungai Semenyih 90 m TREX GIS data layers: (**a**) elevation grid in meter and channels cells; (**b**) soil-type classes; (**c**) land use classes.

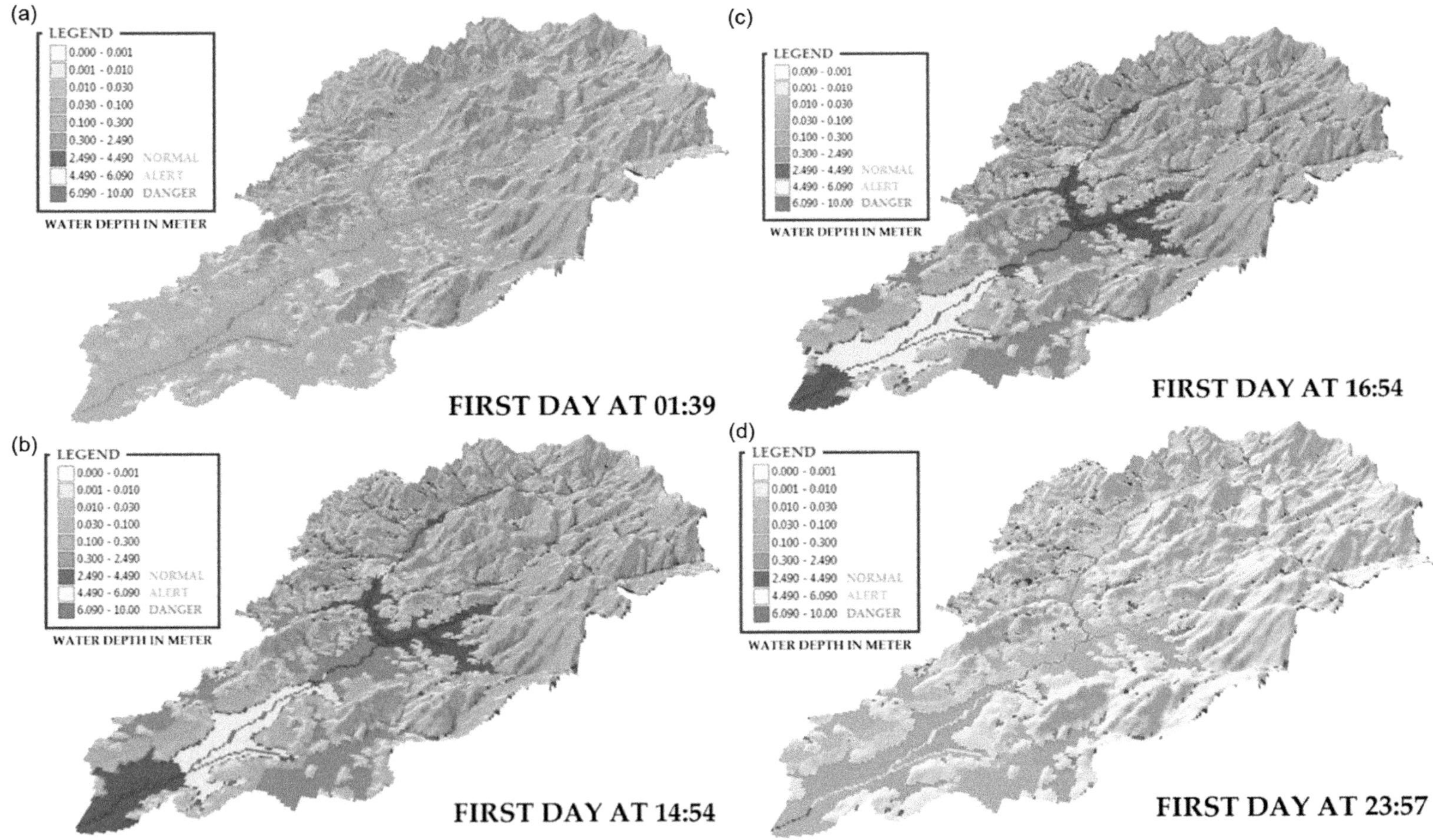

FIGURE 11.6 Spatial distribution of water depth (m) with world's greatest precipitation event (68.24 mm/hr at 16 hr duration).

FIGURE 14.2 Impervious surfaces derived through on-screen digitizing using high-resolution aerial imagery.

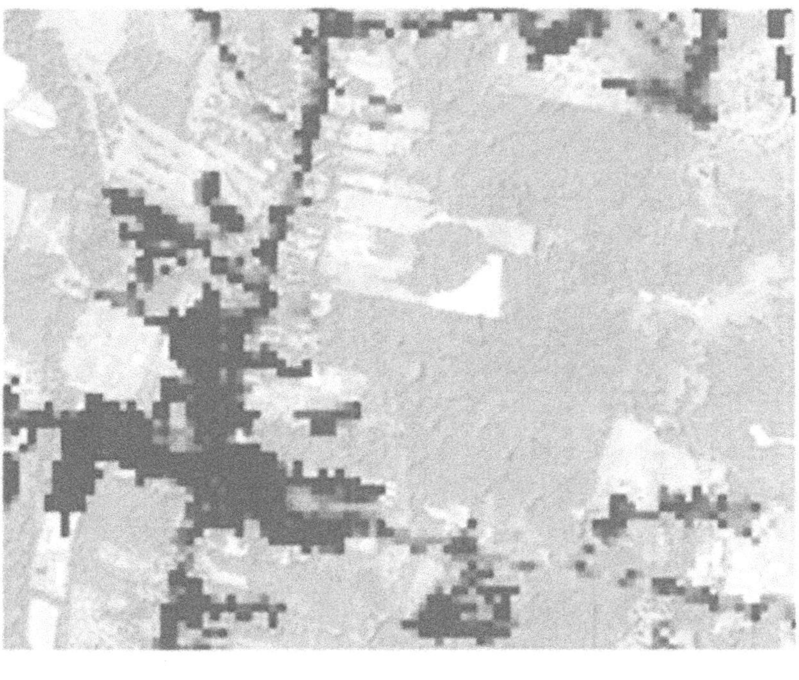

FIGURE 14.3 Percent impervious cover derived from processing of 30 m resolution Landsat satellite remote sensing data.
Source: Adapted from the United States Geological Survey National Land Cover Dataset, USGS NLCD.

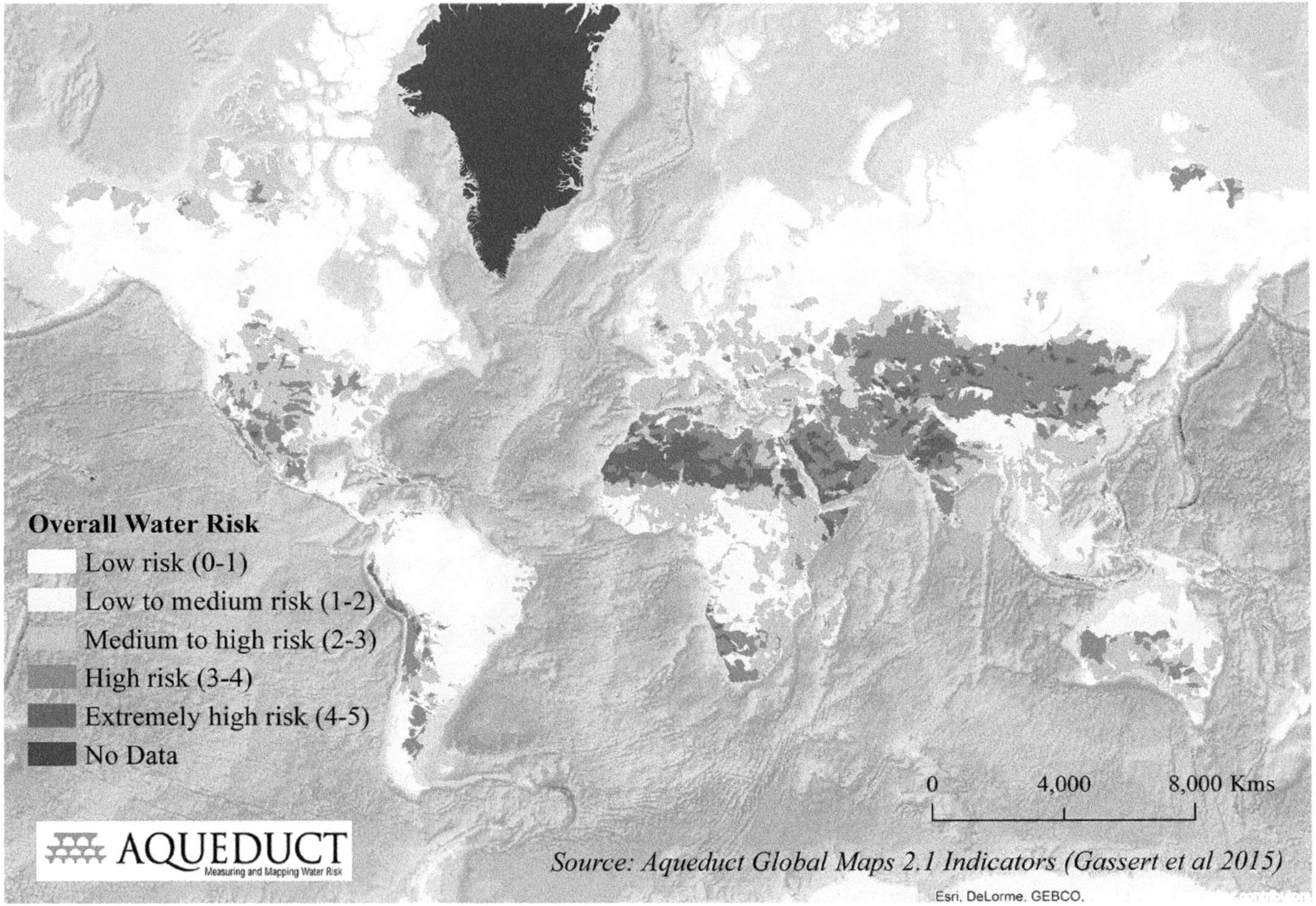

FIGURE 20.3 Overall water risk map.
Source: WRI Aqueduct 2014 (Gassert et al. 2014); www.wri.org/applications/maps/aqueduct-atlas (accessed July 22, 2019).

FIGURE 23.2 Extended detention basin.

FIGURE 23.3 Wet pond.

FIGURE 24.17 The constructed RWB system.

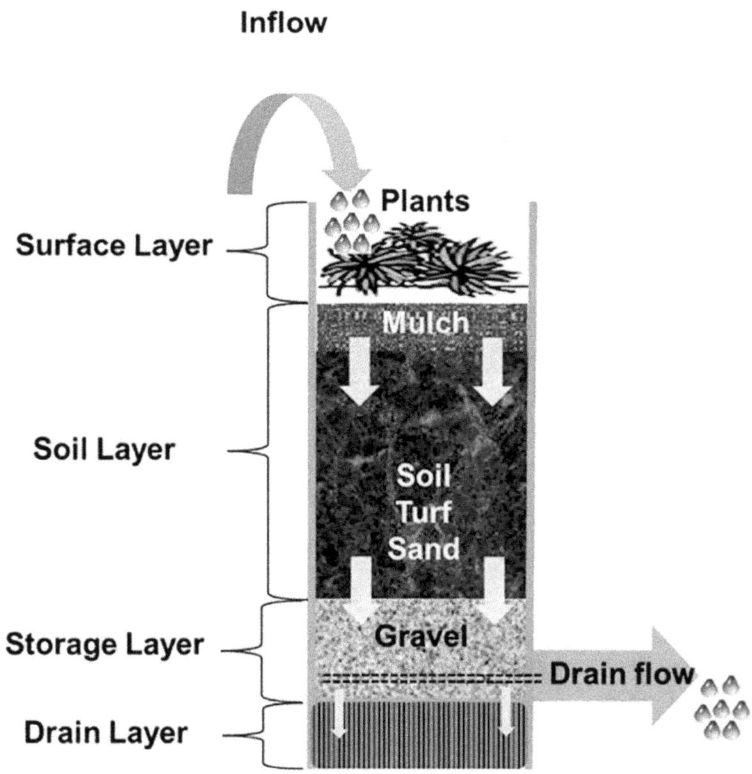

FIGURE 24.19 Modeling of bioretention in EPA SWMM. (Adapted from Gülbaz and Kazezyilmaz-Alhan 2017.)

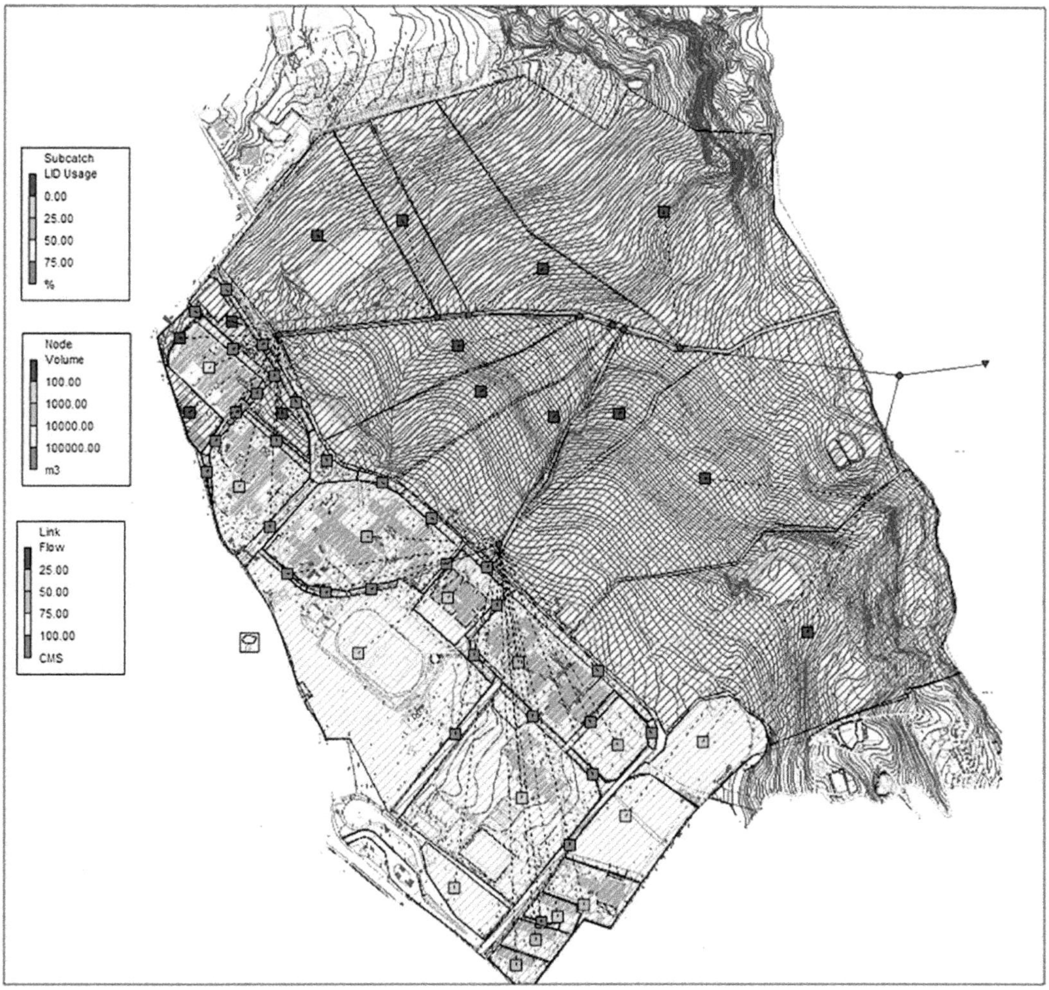

FIGURE 24.21 Hydrological model of Campus with SWMM.

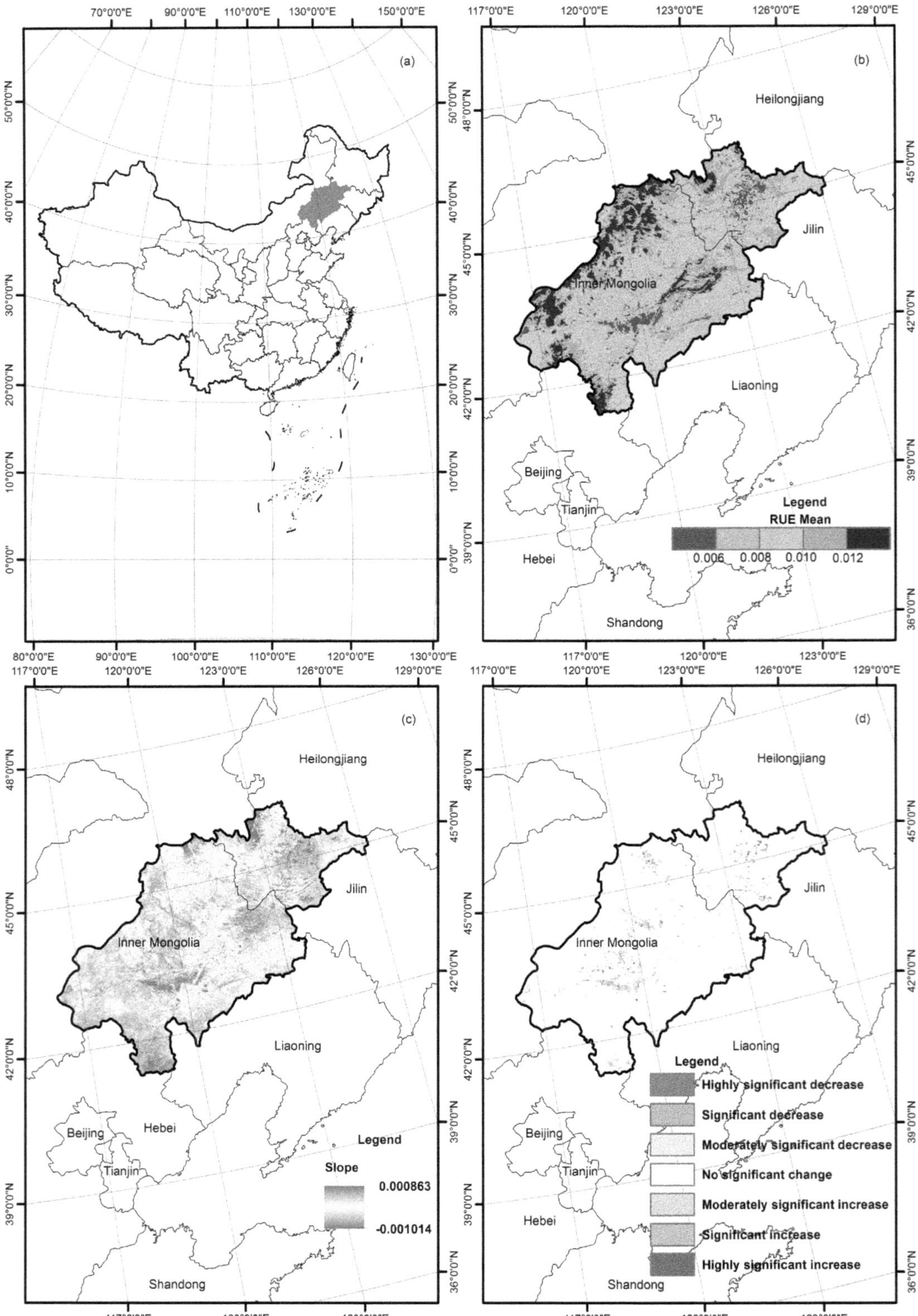

FIGURE 31.1 Satellite-based RUE and the changes in the west of Songliao Plain, China.

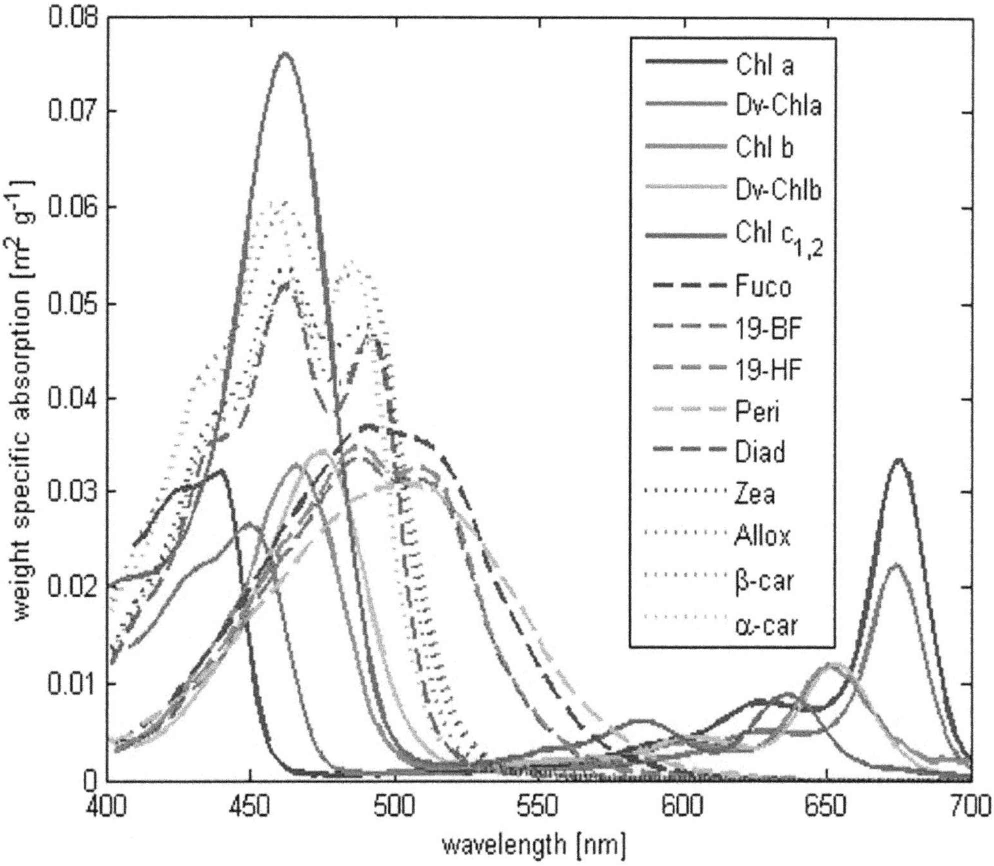

FIGURE 34.2 Absorption spectra showing the primary and secondary absorption peaks for key phytoplankton pigment groups including Chl *a*, divinyl chlorophyll (Dv-Chl *a*), chlorophyll *b*, divinyl chlorophyll *b* (Dv-Chl *b*), chlorophyll c_1 and c_2 (chl $c_{1,2}$), 19'-butanoyloxyfucoxanthin (19-BF), 19'-hexanoxyfucoxanthin (19-HF), fucoxanthin (Fuco), peridinin (Peri), diadino (Diad), zeaxanthin (Zea), alloxanthin (Allo), beta-carotene (β-car), and alpha-carotene (α-car). (Reproduced with permission from www.oceanopticsbook.info based on Bidigare et al., 1990.).

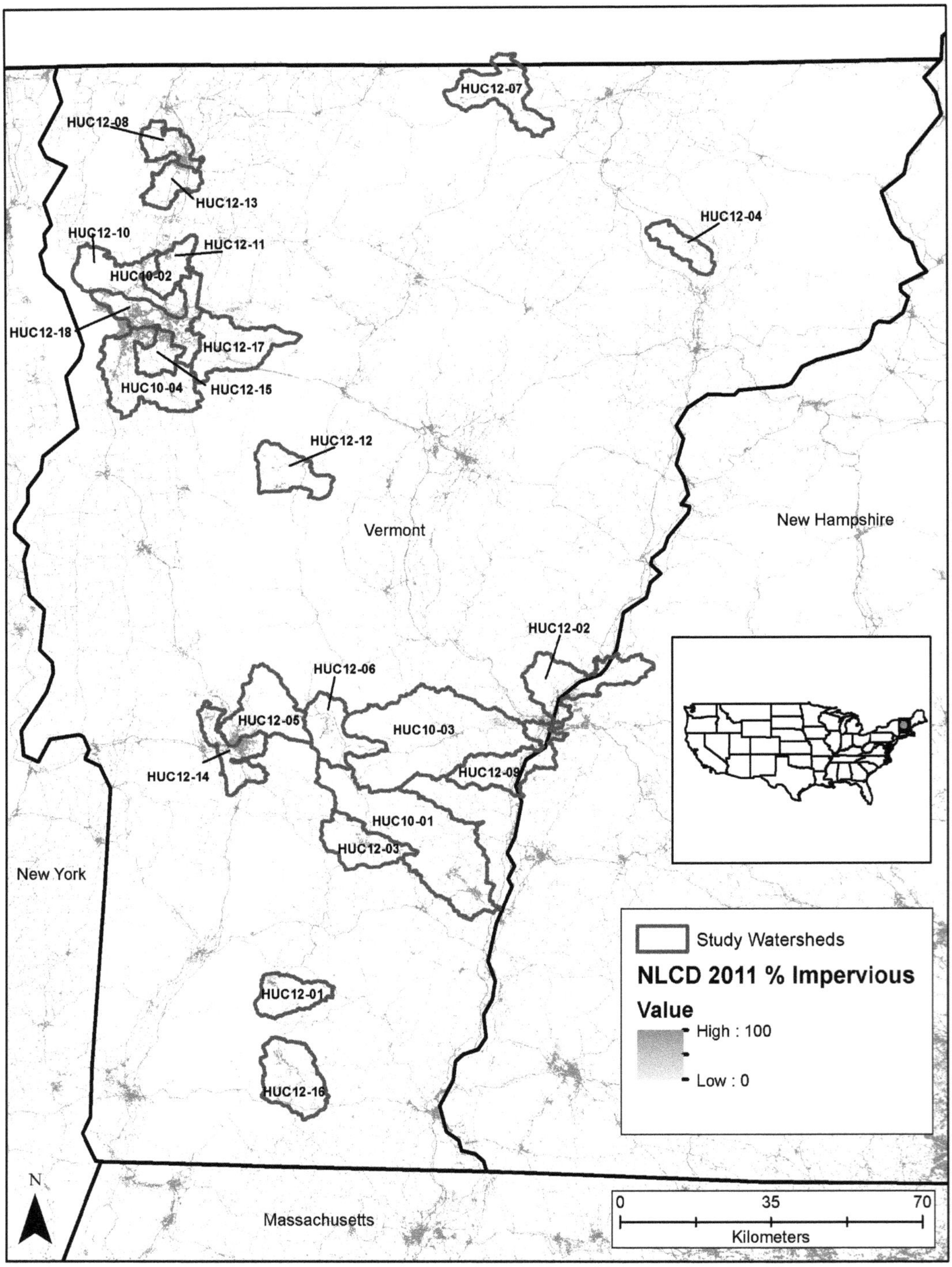

FIGURE 35.1 Map of study area—watershed labels correspond to catchment and catchment number in Table 35.1.

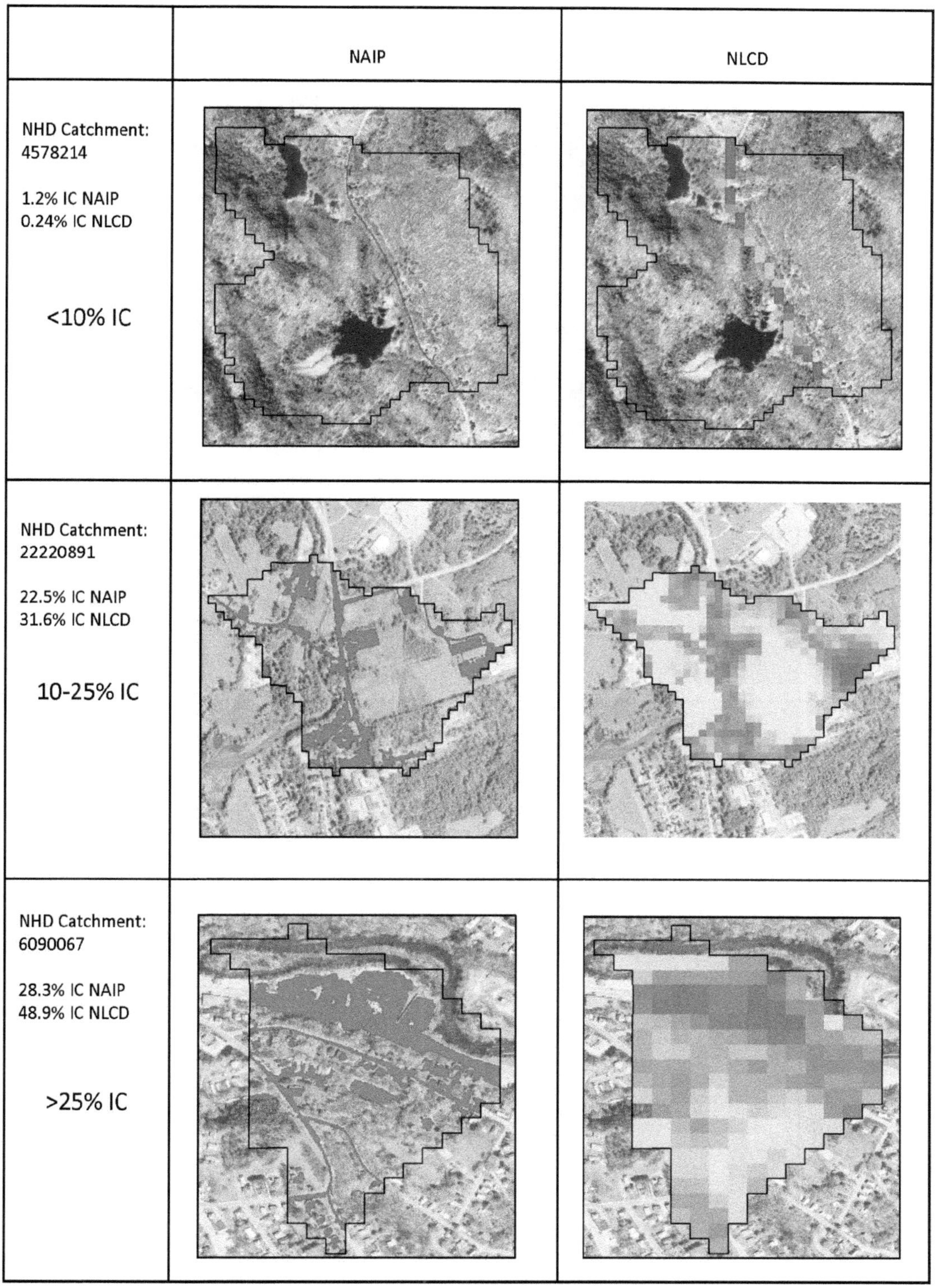

FIGURE 35.2 Comparison of NAIP and NLCD IC classifications in select NHDPlus catchments.

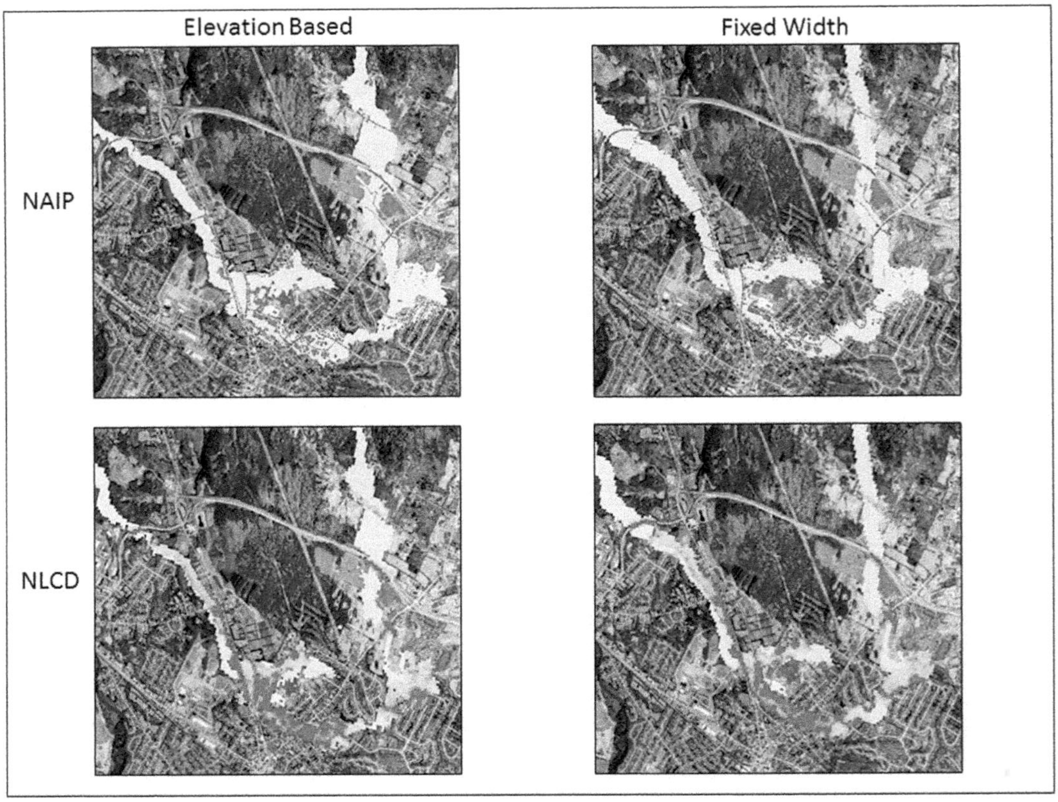

FIGURE 36.2 Comparison of delineated riparian buffer zone by data (NAIP, NLCD) and buffer type (elevation based, fixed width).

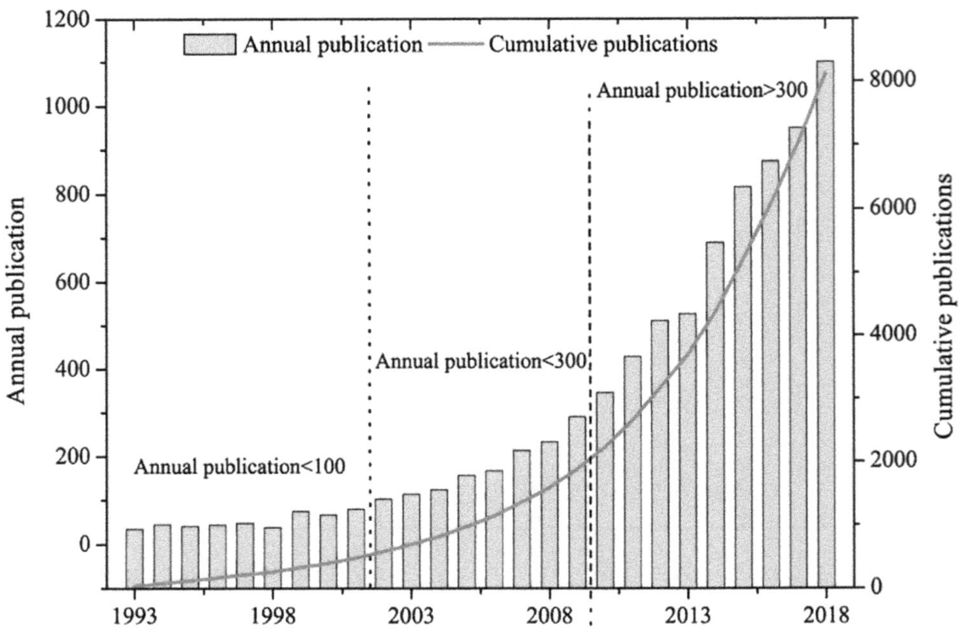

FIGURE 38.1 Annual publication and cumulative number in water footprint studies, 1993–2018.

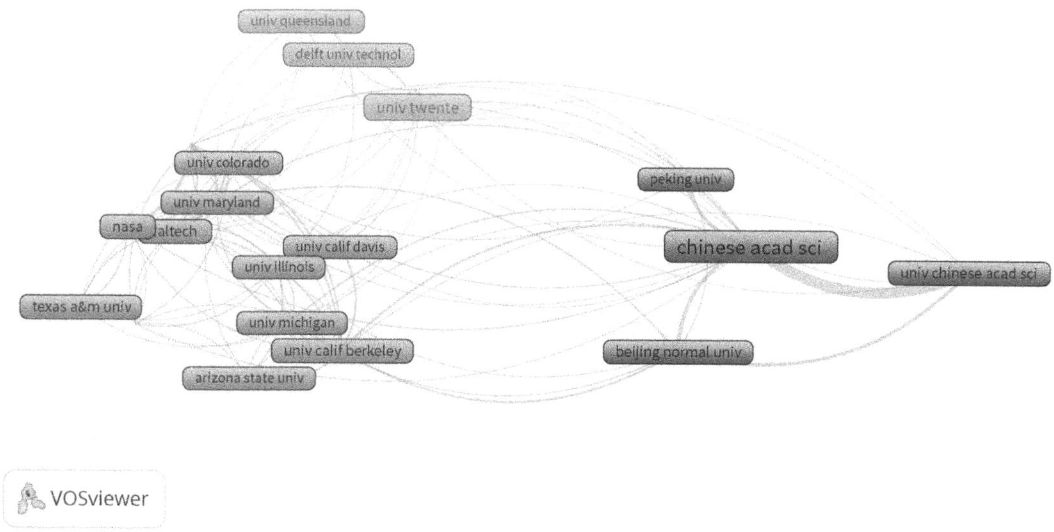

FIGURE 38.2 Co-authorship network among productive organizations.

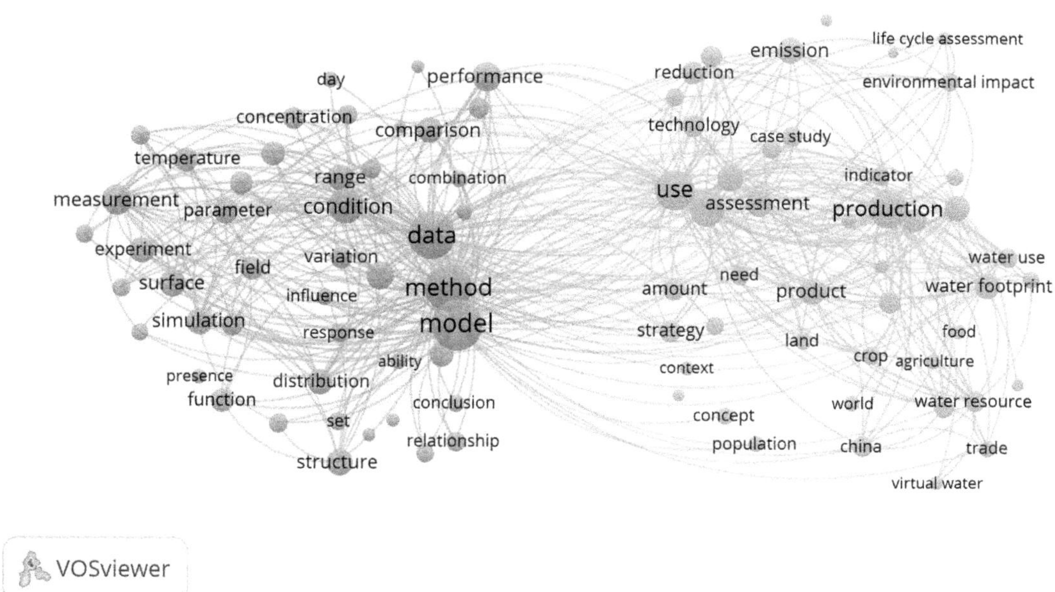

FIGURE 38.4 The terms co-occurrence network map of water footprint.

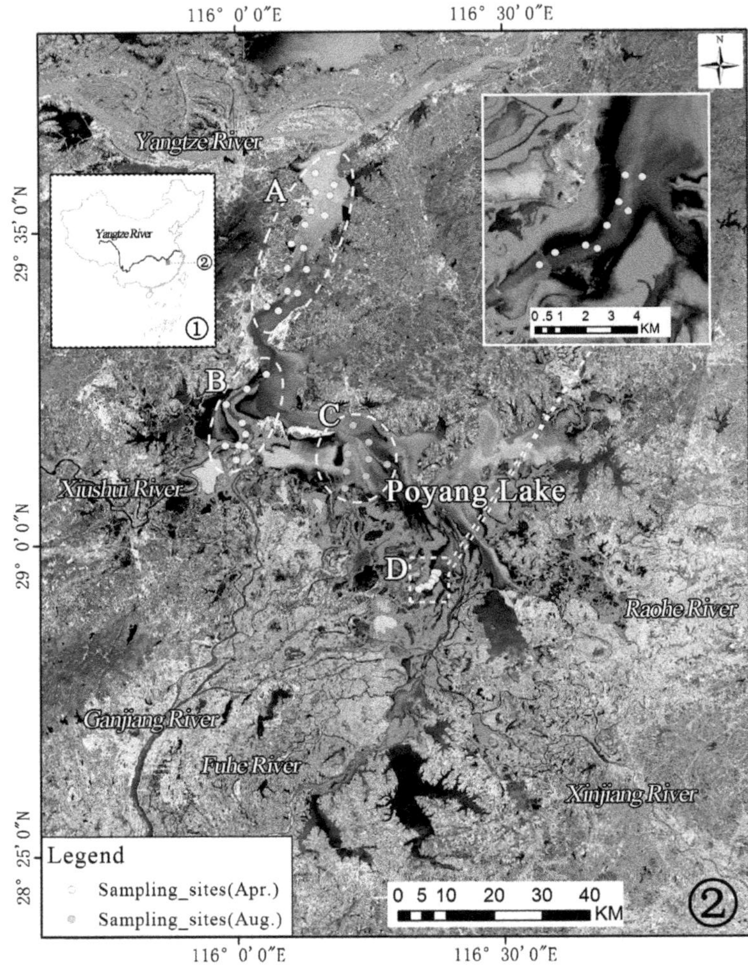

FIGURE 40.1 Locations of representative sections and sampling sites. The background display is the false color image (red: 770–890 nm, green: 630–690 nm, blue: 520–590 nm) of Chinese high spatial resolution satellite data acquired on April 14, 2015.

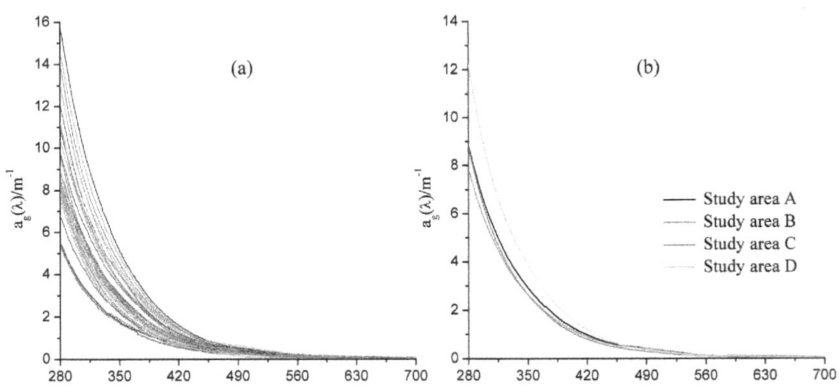

FIGURE 40.2 Absorption spectra of CDOM: (a) CDOM absorption spectra of all sampled sites; (b) averaged CDOM absorption spectra of each section.

28

Transpiration: Water Use Efficiency

Development of the Concept of Water Use Efficiency 211
Water Use Efficiency on an Evapotranspiration Basis 211
Water Use Efficiency on a Transpiration Basis 212
Transpiration Efficiency Units 212
Transpiration Efficiency and Comparison Across Seasons 212
Instantaneous Transpiration Efficiency Measured at an Individual Leaf Level 212
Physiology of Transpiration Efficiency 212
Environmental Influences on Transpiration Efficiency 213
Water Deficit Effect • Rising CO_2
Transpiration Efficiency in Drought Research 213
References 213

Miranda Y. Mortlock
University of Queensland

Development of the Concept of Water Use Efficiency

Early last century, Briggs and Shantz studied the water requirements (WRs) of crops by weighing containers and working out WR per plant; however, they did not use an area basis.[1] Water requirement is the inverse of water use efficiency. Viets confined his definition of water use efficiency to the ratio of plant production to ET measured on the same area.[2] Tanner and Sinclair summarized some early studies and defined water use efficiency as the bio mass of water accumulated per unit of water transpired and evaporated per unit crop area.[1] Biomass is expressed as total yield or economic yield. To compare species with different chemical composition (protein vs. carbohydrate products), grams of glucose *equivalents* are used. In irrigation studies, water use efficiencies may be referring to a broader definition of water, with efficiencies comparing situations where soil water drainage, surface run-off, or soil evaporation are considered. These types of water use efficiency are not included in this discussion.

Water Use Efficiency on an Evapotranspiration Basis

Water use can be defined per unit of evapotranspiration (ET), and under field conditions, this is more practical than the narrower definition using just transpiration. Measures of ET integrate both soil and crop factors for the season, which confound the respective efficiencies of the plant and soil evaporation. Timing and frequency of irrigations and rain, the soil type and plant or mulch cover can affect soil evaporation. The transpiration part of ET use is a measure of crop performance.

Water Use Efficiency on a Transpiration Basis

Another definition of water use efficiency is in terms of transpired water only. Measuring the transpiration component is hard to do in practice, as it is difficult to prevent soil evaporation. Deep soil drainage also needs to be measured or prevented. It is only possible to measure transpiration on an experimental basis with the use of weighing lysimeters. Typically, the lysimeter is a large pot in the greenhouse and may weigh up to 80 kg. The lysimeter is weighed frequently over the crop season, and known quantities of water are added which is, both costly and limits the practical size of the trial. When extended to field studies (lysimeters within a growing crop), a limited number of comparisons are made due to the setup cost and expense of rainout facilities at sites.

Transpiration Efficiency Units

Water use efficiency can be expressed as $g\ kg^{-1}$ of water transpired. Typical values may be 1.6–2.4 $g\ kg^{-1}$ for sun flower or around 9 $g\ kg^{-1}$ for sorghum. It can also be expressed by using a molar scale. Transpiration may be typically 1–5 $mmol\ m^{-2}\ sec^{-1}$ and photosynthetic rates for C_3 20–25 $pmol\ m^{-2}\ sec^{-1}$ or 40 $pmol\ m^{-2}\ sec^{-1}$ for C_4 plants. Hence, C_4 plants have higher transpiration efficiencies than C_3 plants.

Transpiration Efficiency and Comparison Across Seasons

As the evaporative demand of the atmosphere will vary from place to place and from season to season, transpiration efficiency (TE) from particular trials are adjusted for the vapor pressure deficit (VPD) of the atmosphere:

$$Y/T = k/\left(e^{*} - e_{i}\right)$$

where Y is the yield, T, the transpiration, k, the transpiration coefficient, e^*, the saturated vapor pressure, and e_i, the vapor pressure of the atmosphere.[1] The coefficient (k) is estimated as 9 kPa for sorghum.

Typically, a mean value of VPD for hours of daylight over the stress period season can be used. The mean daily value of the 9.00 hr VPD and VPD at the time of maximum temperature can been used to compute a seasonal VPD over the stress period.[3]

Instantaneous Transpiration Efficiency Measured at an Individual Leaf Level

Transpiration and photosynthesis measured by using a canopy gas exchange system are used to compute an instantaneous measurement of TE at the leaf level. This rarely correlates with TE computed for a season as there are many processes integrated within the plant alone and over the cropping season.

Water use efficiency is the molar ratio of CO_2 uptake (A) to transpiration (E) and can be written as

$$A/E = \left(c_a - c_i\right)/1.6\Delta_w$$

where c_a is external and c_i is the internal partial pressure of CO_2, respectively. Δ_w is the leaf-air VPD.[4]

This shows that the internal partial pressure of CO_2 is linked to water use efficiency.

Physiology of Transpiration Efficiency

C_3 and C_4 plants vary in TE as the carboxylation pathways give rise to different efficiencies.[5–7] C_4 plants, such as maize and sorghum, have higher transpiration efficiencies than C_3 plants. Legumes have a lower TE than cereal crops due to the metabolic cost of symbiotic N fixation. Variations in TE for wheat correlate with carbon isotope dis crimination.[8] Carbon isotope discrimination (Δ) as determined on dried leaf tissue of C_3 plants has shown to be correlated to TE. This is useful, but is not necessarily practical as carbon isotope discrimination is a costly measurement and therefore not ideal as a selection criterion. Progress has been made in C_3 crops, towards obtaining a selection index for high TE lines. In peanut, the carbon isotope discrimination was linearly related to the specific leaf area of the leaves. Thus, a surrogate for Δ is available and has been used to breed high TE lines.[9]

Environmental Influences on Transpiration Efficiency

Water Deficit Effect

Under moderate water deficits, TE increased in grain sorghum.[3,10] The leaf area index is irreversibly reduced under stress, cell density is maintained but cell enlargement is irreversibly affected.[11] Some plants may be able to adjust osmotically which may contribute to resistance to water deficits.

Rising CO_2

As the CO_2 concentration doubles, transpiration will decline and photosynthesis will increase.[12]

Transpiration Efficiency in Drought Research

If this trait can be identified as significant and suitable selection criteria can be developed, more efficient crops will result through the incorporation of high TE lines in breeding programs. Benefits of high TE lines, anticipated from crop simulation modeling of grain sorghum, suggest a 10% increase in yield from moderate to good environments.[13] Genetic differences in TE appear to be detectable in a range of crops. The combination of C_4 productivity in marginal environments along with improved drought efficiency may have useful economic benefits in the future.

References

1. Tanner, C.B.; Sinclair, T.R. Efficient water use in crop production: research or re-search? In *Limitations to Efficient Water Use in Crop Production;* ASA-CSSASSSA: Madison, WI, 1983.
2. Viets, F.G., Jr. Fertiliser and the efficient use of water. Adv. Agron. **1962**, *14*, 223–264.
3. Mortlock, M.Y.; Hammer, G.L. Genotype and water limitation effects on transpiration efficiency in sorghum. In *Water Use in Crop Production;* Kirkham, M.B., Ed.; Food Products Press, The Haworth Press: Binghamton, New York, 1999; 265–286.
4. Griffiths, H. Carbon isotope discrimination. In *Photosynthe sis and a Changing Environment: A Field and Laboratory Manual;* Hall, D.O., Sturlock, O., Bolhàr-Nordenkampf, R., Leegood, R.C., Long, S.P., Eds.; Chapman & Hall: London, 1993.
5. Farquhar, G.D.; O'Leary, M.H.; Berry, J.A. On the relationship between carbon isotope discrimination and the intercellular carbon dioxide concentration in leaves. Aust. J. Plant Physiol. **1982**, *9*, 121–137.
6. Farquhar, G.D. On the nature of carbon isotope discrimination in C4 species. Aust. J. Plant Physiol. **1983**, *10*, 205–226.
7. Hubick, K.T.; Hammer, G.L.; Farquhar, G.D.; Wade, L.; von Caemmerer, S.; Henderson, S. Carbon isotope discrimination varies genetically in C_4 species. Plant Physiol. **1990**, *91*, 534–537.
8. Farquhar, G.D.; Richards, R.A. Isotopic composition of plant carbon correlates with water use efficiency in wheat genotypes. Aust. J. Plant Physiol. **1984**, *11*, 539–552.
9. Wright, G.C.; Hubick, K.T.; Farquhar, G.D. Discrimination in carbon isotopes of leaves correlates with water-use efficiency of field grown peanut cultivars. Aust. J. Plant Physiol. **1988**, *15*, 815–825.

10. Donatelli, M.; Hammer, G.L.; Vanderlip, R.L. Genotype and water limitation effects on phenology, growth, and transpiration efficiency in grain sorghum. Crop Sci. **1992**, *32* (3), 781–786.
11. Hsiao, T.C.; Bradford, K.J. Physiological consequences of cellular water deficits. In *Limitations to Efficient Water Use in Crop Production*; ASA-CSSA-SSSA: Madison, WI, 1983.
12. Nobel, P.S. Leaves and fluxes. In *Physicochemical and Environmental Plant Physiology;* Academic Press: New York, 1991.
13. Hammer, G.L.; Butler, D.; Muchow, R.C.; Meinke, H. Integrating physiological understanding and plant breeding via crop modelling and optimisation. In *Plant Adaptation and Crop Improvement*; Cooper, M., Hammer, G.L., Eds.; CAB International: Wallingford, U.K., ICRI-AT, IRRI, 1996.

29

Water Deficits: Development

Kadambot H. M. Siddique and Helen Bramley
University of Western Australia

Introduction 215
The Soil, Plant, and Atmosphere Continuum 215
Water Status of Crop Plants 217
Mechanisms for Control of Water Transport 217
Measurement of Plant–Water Relations 218
Effects of Water Deficits 218
Conclusion 219
References 219

Introduction

Despite water being the Earth's most abundant compound, water deficit is the single most important factor limiting crop yields worldwide. Of the Earth's total water supply less than 1% exists in freshwater aquifers, rivers, and lakes. Agriculture consumes more than two-thirds of this available fresh water worldwide. The increasing global demands on both food and water resources require more efficient agricultural water use to maintain, or even improve production. To enable more efficient and sustainable use of our water resources, a greater understanding of plant water use is needed, especially the way in which water deficits develop and affect plant growth and productivity.

The Soil, Plant, and Atmosphere Continuum

In plants, water deficits develop because of water loss from leaves and insufficient resupply.[1] Water evaporates through open stomata during the uptake of CO_2 from the atmosphere for photosynthesis. This transpired water is replaced by water taken up by the roots and transported to the shoots and leaves. Water absorbed from the soil by roots has to travel a tortuous pathway, crossing cells via the symplasm and cell walls (apoplast), and flowing longitudinally through xylem vessels to reach the sites of evaporation within the leaves.[2] This pathway forms a hydraulic continuum between the soil, plant, and atmosphere. The rate of water flow through this continuum depends on the resistance of the pathway to water flow and the size of the force driving the water flow.

Water moves passively down a chemical potential gradient, which is the free energy associated with water and is more generally called water potential by plant scientists. Hence, gradients in water potential can drive water flow through the soil–plant–atmosphere continuum. The unsaturated atmosphere has an extremely negative water potential, whereas a soil saturated with fresh water has a water potential close to zero. As the soil dries out or if the water is salty, the water potential decreases. Accordingly, a plant must lower its water potential in order to absorb water from the soil.

There are two mechanisms that generate the water potential gradients within plants to induce soil water uptake and drive water upward to the leaves. In the first mechanism, solutes are released into the root xylem, thereby lowering the xylem water potential and inducing root water uptake.[3] As more water is drawn into the xylem a positive pressure develops that pushes water upward. Woody species tend to have low pressures but some herbaceous crop species can generate root pressures up to 300 kPa,[4] which would be sufficient to push water to the top of the plant. However, the high volumes of water that flow through a plant when it is transpiring dilute the xylem sap so that root pressure is negligible during the day.

The second and dominant mechanism generating water potential gradients within plants is due to transpiration. As water evaporates from the interstitial tissues of leaves, surface tension is generated at the air-water interface. Surface tension places the mass of water behind the air-water interface under negative pressure, which is transmitted throughout the liquid continuum due to the cohesion of water molecules,[5] drawing water into the roots and upward to the leaves. As more water evaporates, the greater the surface tension, the more negative the hydrostatic pressure and the greater the pulling force. Hydrogen bonds between water molecules give water its unique tensile strength that can withstand the high tensions generated in transpiring plants. Tensions within the xylem are typically –1 to –2 MPa, but can be as extreme as –10 MPa.[6]

Although water columns in the xylem do not break under the tensions generated by transpiration, the water column can be disrupted by cavitation.[6] Cavitation is generally thought to occur by air-seeding through intervessel pit membranes.[6] This is particularly problematical in dry and saline soils, and hot dry atmospheric conditions, when tensions in the plant are at their largest. Once cavitated, the vessel quickly embolizes and water flow is obstructed. Embolisms can occur in herbaceous species at fairly high water potentials in comparison to woody species, but the reason behind this is not known. However, cavitation may not be as detrimental to crop plants compared with woody species because embolized vessels of herbaceous species are possibly refilled overnight by root pressure, but many trees rely on new growth in the spring to produce water- filled vessels.

Large-water-potential gradients are needed to drive water uptake from the soil if the soil is dry or saline, or if the soil–root interface and the plant tissue through which water flows form significant resistances to water flow. The difference in water potential between the plant and the soil and progressive changes in soil– and plant-water potential as the soil dries out are presented schematically (Figure 29.1).

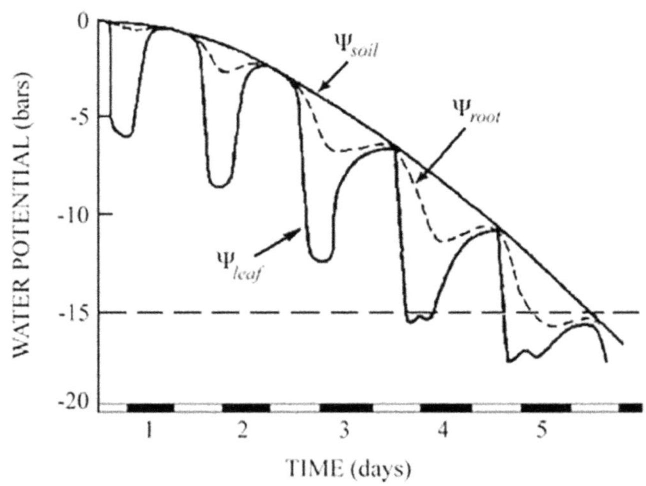

FIGURE 29.1 Schematic representation of changes in leaf-water potential (ψ_{leaf}), root-water potential (ψ_{root}), and soil-water potential (ψ_{soil}) as transpiration proceeds during a drying cycle. The same evaporative conditions are considered to prevail each day; the horizontal dashed line indicates the value of ψ_{leaf} at which wilting occurs.
Source: From Slatyer, R.O. 1967. *Plant–Water Relations*, Academic Press, New York.

Water Status of Crop Plants

The water status of a crop plant is usually defined in terms of its relative water content or water potential. Relative water content is the amount of water in a leaf relative to that at full hydration and standardized by unit dry weight. The net water potential of a plant comprises of two principal components: osmotic pressure and hydrostatic pressure. A gravitational component may also be included, but this is only relevant for tall trees. Osmotic pressure is defined by the concentration of solutes in solution, whereas hydrostatic pressure results from the pressure exerted by water on its surroundings. Within plant cells, water is drawn into the cell due to the concentration of solutes in the cytoplasm and tonoplast. As more water accumulates, the cell volume resists expansion because of the rigidity of cell walls, and a positive hydrostatic pressure develops within the cell called turgor pressure.

Maintenance of leaf hydration and turgor is paramount to maintaining growth and stomatal opening. Plants with higher relative water contents at lower leaf water potentials are generally more resistant to drought, but typically have slow growth rates. Many crop plants are known to accumulate solutes in leaves and roots to maintain turgor in a process called osmotic adjustment. The elasticity of plant cells also determines their capacity for water retention and relative changes in turgor with changes in water content.[7] Plant cells with high volumetric elastic modulus (low elasticity) hold less water at full turgor and for any given change in water content the changes in turgor are larger.

Leaf expansion is one of the first processes that is affected by water deficit and is inhibited at higher leaf water potentials than photosynthesis or metabolism, causing a change in the sink for photoassimilates.[8] Maintenance of plant water status is also important for reproductive processes and fruit or grain growth and development. The grains of legumes are isolated from the water relations of the maternal plant, such that seed-coat turgor is maintained despite changes in the water potential of leaves, stems, and pod walls.[9] The water relations of cereal grains have been shown to decouple from the maternal plant during the grain-filling stage, about two weeks after flowering, which is when the water content of the grain is at its highest.[10] Whether hydraulic isolation occurs because of xylem discontinuity into the grain, formation of an apoplastic barrier, or some other process, the mechanism may ensure continued seed filling and assimilate redistribution, even when the water potential of the environment has reduced the current availability of assimilates from the leaves.

Mechanisms for Control of Water Transport

The relative magnitude of water supply and water loss, as well as the water-holding capacity determine the degree to which plants are water deficient. However, because plants are living organisms, it is important to realize they do not solely provide a pathway for the movement of water from the soil to the atmosphere. In addition to mechanisms for osmotic regulation, plants can also effectively control the flow of water according to their water status through stomatal regulation, changes in anatomy and aquaporin activity.

Water loss from plant communities is usually termed evapotranspiration, because of the combined soil evaporation and plant transpiration. Evapotranspiration from nonstressed crops is largely governed by meteorological conditions. The plant, in contrast to the soil, exercises control over water loss through its stomata and cuticle, and also through its ability to reduce the radiation load through active and/or passive changes in leaf orientation. Because these controls are located at the leaf surface, the internal water status of the plant is closely coupled to that of the soil and less closely coupled to the atmosphere.[1]

A high cuticular resistance is associated with either thicker cuticles or with cuticles enriched with hydrophobic materials. Although the plant cannot flexibly control this, cuticular resistance has been found to decrease with increasing temperature and relative humidity of the atmosphere. Thicker or waxier cuticles may develop in response to drought, which in addition to reducing cuticular transpiration can also increase light reflectance and reduce radiation load.

The density of stomata on the leaf surface as well as stomata apertures determine stomatal conductance to water vapor, CO_2, and other gases. Stomatal density is determined during leaf development, but stomatal aperture is controlled by the turgor pressure of two flanking guard cells, which enables the plant to reversibly modify gas exchange with the atmosphere, with a time scale of only a few minutes. This short-term regulation may both optimize the dry matter production and offer some protection from xylem embolism or other detrimental effects of very low leaf water potential.[11] There is wide variation between species in the critical threshold level in leaf-water potential or relative water content when stomata close. Stomata of crops such as lupin close before any detectable changes in leaf water potential occur in response to depletion of soil water, whereas the stomata of other crops such as wheat gradually close with decreasing soil moisture content and leaf water potential.[12] These differences are, in part, related to differing sensitivity to chemical signals originating in dehydrating roots and conveyed toward the stomatal complexes that can control stomatal apertures.

Crop canopy photosynthesis, transpiration, or leaf extension growth may show midday depressions caused by high evaporative demand if the soil water supply is depleted below a threshold of ca. 50% of total extractable soil water. The exact point of the threshold water content is difficult to distinguish and is usually extrapolated from measurements taken when evapotranspiration is clearly reduced by drought.[13] Because stomata act as regulators for CO_2 exchange, water deficits sufficient to close stomata must also depress photosynthesis. Photosynthesis declines initially as a result of stomatal closure, but prolonged and severe water stress can lead to depression of chloroplast and enzyme activity and to nonstomatal effects on photo-synthesis.[1] In shoot tissues, when water deficits limit the supply of CO_2 for photosynthesis, a number of reversible mechanisms downregulate the efficiency of excitation energy transfer to photosystem II.

The rate of water transport through the plant in the liquid phase is constrained by the hydraulic resistance of the plant tissue. Tracheids and xylem vessels have relatively low resistances in comparison with living tissues. Anatomical changes, such as suberin deposition in cell walls, may occur in roots and leaves in response to soil water deficit, which help conserve water. However, these changes are irreversible, slow, and growth dependent, and increase the resistance to water flow. It is now generally accepted that changes in aquaporin activity can regulate rapid and reversible changes in root or leaf hydraulic resistance.[2] Aquaporins are membrane-embedded proteins that form water-conducting channels. By increasing the abundance or opening the water channels in roots, the plant can decrease the resistance to water flow to match transpiration demands, minimize the drop in water potential needed to drive water flow and aid in the maintenance of leaf water status. Conversely, a decrease in aquaporin activity may help conserve water in adverse conditions.

Measurement of Plant–Water Relations

Any study of plant–water relations requires accurate methods for measuring them. Indirect methods that have been used to characterize or infer leaf water status include visible wilting, color change, leaf rolling, leaf temperature, and leaf thickness. Pressure chambers and psychrometers directly, but destructively, measure water potential, although new hygrometers can be clamped directly to stems and leaves. Infrared gas analyzers are traditionally used to measure gas exchange. Acoustic emission sensors detect cavitation events and pressure probes can directly measure turgor of plant cells, although cell pressure probes are impractical for field measurements. Infrared thermography and remote sensing are rapidly being adopted to monitor canopy temperature as a surrogate for stomatal conductance in crop phenotyping programs. New sensors such as magnetic probes, which detect relative changes in leaf hydration, and modified time-domain reflectance probes, which detect stem water content, are being used for irrigation scheduling and increasing sustainable water use.

Effects of Water Deficits

Long-term severe water deficits can have a negative effect on crop growth and yield, and may even result in senescence of the plant. It has been suggested that drought resistance of crops may be improved via breeding for specific phenological, morphological, physiological, or biochemical characteristics that improve yields in water-limited environments. These characteristics include early vigor, transpiration efficiency, stomatal control, abscisic acid accumulation, proline accumulation, osmotic adjustment, carbon remobilization, rapid grain growth, changes in hydraulic conductance, membrane stability, and lethal water potential.[14] It is, however, important to note that water deficits do not always lead to detrimental effects on yield. Mild or early water deficits may in fact increase yields or at least have no effect on yield, especially in indeterminate species.

Conclusion

Water deficit is an inevitable consequence of life on dry land for plants. A variety of mechanisms have evolved to control plant water status, particularly the regulation of water loss, turgor maintenance, and rate of water transport. Many of these mechanisms are being selected by breeding programs to increase crop productivity and drought resistance under water-limited conditions. With increasing global atmospheric CO_2, regional changes in climate are expected. These include warmer and drier conditions, changes to length of growing season, precipitation, and evaporation; all of which will affect plant–water relations.[15] Changes in land use and regional climate brought about by anthropogenic incursions mean that crop productivity in more marginal rainfall areas will be of increasing importance in world agriculture. Expansion to these areas requires the ability to identify and breed for characteristics that better enable crop plants to cope with limited water availability.

References

1. Begg, J.E.; Turner, N.C. Crop water deficits. Adv. Agron. **1976**, *28*, 161–217.
2. Bramley, H.; Turner, D.W.; Tyerman, S.D.; Turner, N.C. Water flow in the roots of crop species: the influence of root structure, aquaporin activity and waterlogging. Adv. Agron. **2007**, *96*, 133–196.
3. Zholkevich, V.N. Root pressure. In *Plant Roots: The Hidden Half;* Waisel, Y.; Eshel, A.; Kafkafi, U., Eds.; Marcel Dekker Inc.: New York, 1991; pp. 589–603.
4. Steudle, E.; Jeschke, W.D. Water transport in barley roots. Planta. **1983**, *158*, 237–248.
5. *Dixon, H.H. Transpiration and the Ascent of Sap in Plants. 1914, Macmillan: London.*
6. Tyree, M.T.; Sperry, J.S. Vulnerability of xylem to cavitation and embolism. Ann. Rev. Plant Physiol. Plant. Mol. Biol. **1989**, *40*, 19–36.
7. Cosgrove, D.J. In defence of the cell volumetric elastic modulus. Plant Cell. Environ. **1988**, *11*, 67–69.
8. Boyer, J.S. Leaf enlargement and metabolic rates in corn, soybean and sunflower at various leaf water potentials. Plant Physiol. **1970**, *43*, 1056–1062.
9. Shackel, K.A.; Turner, N.C. Seed coat cell turgor in chickpea is independent of changes in plant and pod water potential. J. Exp. Bot. **2000**, *51* (346), 895–900.
10. Barlow, E.W.R.; Lee, J.W.; Munns, R.; Smart, M.G. Water relations of the developing wheat grain. Aust. J. Plant Physiol. **1980**, *7*, 519–525.
11. Smith, J.A.C.; Griffiths, H. Integrating Plant Water Deficits from Cell to Community. In *Water Deficits: Plant Responses From Cell to Community;* Smith, J.A.C.; Griffiths, H., Eds.; BIOS Scientific Publishers Ltd.: Oxford, UK, 1993; 1–4.
12. Henson, I.E.; Jensen, C.R.; Turner, N.C. Leaf gas exchange and water relations of lupins and wheat. I. Shoot responses to soil water deficits. Aust. J. Plant Physiol. **1989**, *16*, 401–403.
13. Ritchie, J.T. Water dynamics in the soil-plant-atmosphere system. Plant Soil **1981**, *58*, 81–96.
14. Turner, N.C. Further progress in crop water relations. Adv. Agron. **1997**, *58*, 293–338.
15. Jarvis, P.G. Global Change and Plant Water Relations. In *Water Transport in Plants Under Climatic Stress*; Borghetti, M.; Grace, J.; Raschi, A., Eds.; Cambridge University Press: Cambridge, UK, 1993, 1–13.

30

Watershed Hydrology and Land-Use and Land-Cover Change (LULCC)

Yuyu Zhou
Joint Global Change Research Institute

Introduction 221
LULCC Impact 221
Hydrologic Impact of LULCC 222
Hydrologic Modeling 223
Conclusions 223
References 223

Introduction

Carbon dioxide concentration increase and land-use and land-cover change (LULCC) are two major factors of human-induced change in our global environment.[1] LULCC, mainly urbanization, agriculture practice, and deforestation, has a profound impact on natural resources and the environment.[2,3] Change in watershed hydrology is one of the major impacts from LULCC, and it causes a series of environmental problems such as soil erosion and nonpoint source pollution. The hydrologic impact of LULCC varies with climate, location, scale, and other related factors. This entry reviews human-induced LULCC, its hydrologic impact, and methods used to study the hydrologic impact of LULCC.

LULCC Impact

Land cover refers to the physical material on the Earth's surface, such as water and vegetation, whereas land use is a more complex term that describes how people use the land.[4] LULCC is a process of human modification of earth's terrestrial surface.[5] LULCC is driven by factors such as population growth and migration, economic development, culture, regional and local policies, environmental conditions, their interactions, and people's responses to economic opportunities.[6] Remote sensing was identified as a major source of consistent and continuous data and has been used to derive information about LULCC across a variety of temporal and spatial scales.[7–9]

LULCC plays an important role in many processes such as biodiversity loss, climate change, and water and carbon cycle. LULCC is one of the most important factors contributing to biodiversity loss via processes such as forest fragmentation induced by human activities. For example, it was found that richness of plant species was significantly negatively related to nitrogen input, an indicator of land-use intensity.[10] LULCC impacts climate significantly across a variety of spatial and temporal scales in mainly two ways. First, LULCC-induced greenhouse gas emissions increase the concentration of carbon dioxide, an important driver of climate change, in the atmosphere. Second, LULCC-related albedo change can alter the surface energy balance. Pielke et al.[11] found that land-use change influenced regional and global climate through surface energy budget, as well as carbon cycle. Change in land-use and land-cover is also an important factor impacting watershed hydrology and associated environmental problems, such as soil erosion and water pollution. The impact of LULCC on watershed hydrology, the focus of this entry, is discussed in the following sections.

Hydrologic Impact of LULCC

Human-induced LULCC affects hydrologic processes such as evapotranspiration, interception, infiltration, surface flow, ground flow and stream flow, response time, and watershed annual and seasonal water yield. Urban and suburban development, agricultural practice, and deforestation and reforestation are typical forms of human-induced LULCC, and they influence watershed hydrology in different ways owing to their special characteristics. Due to the variability of agricultural practice, the results about its hydrologic impact are very different across numerous studies.[12] In this section, the focus is on the LULCC of urbanization and deforestation for which some consensus has been reached.

Impervious surface area (ISA), a key indicator of urban and suburban development, is often used to study the environmental impact of urbanization.[13,14] Increasing ISA influences a series of hydrologic processes (Figure 30.1), and it reduces infiltration, raises runoff volumes and velocities, increases peak flow and duration, decreases baseflow, and increases sediment loading. As a result, these processes bring environmental issues such as flooding, erosion, sedimentation, increased temperatures, habitat changes, and loss of fish populations. Although the magnitude of hydrologic change due to urbanization varies significantly in numerous studies, a consensus has been reached that increasing peak flow is one of the major hydrologic impacts of urbanization.

A number of studies have reported that reforestation and deforestation impact various aspects of the hydrologic cycle. Reforestation and deforestation change the hydrologic processes such as interception of precipitation, evapotranspiration, and soil moisture, and the impact varies with spatial and temporal scales and local conditions. Inconsistent conclusions have been drawn about how reforestation and deforestation impact the watershed hydrology in previous studies. According to Siriwardena et al.,[15] deforestation increases the runoff, and thus the annual water yield. A study by Lin and Wei[16] indicated that forest harvesting increased the mean and peak flows in annual and spring periods but the mean and peak flows in summer and winter periods were not significantly affected.

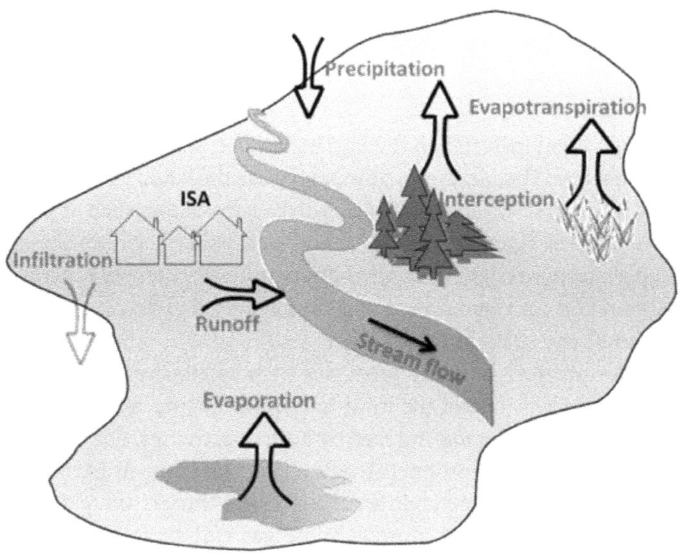

FIGURE 30.1 Typical hydrologic processes in the urbanizing environment.

Hydrologic Modeling

Although there is no doubt that human-induced LULCC influences watershed hydrology, the magnitude and mechanism of the impact are not clear for most LULCC types. Numerous methods have been developed to understand the hydrologic consequences of LULCC. Hydrologic modeling is the most widely used method for this purpose.

Numerous hydrologic models have been used effectively to investigate the hydrologic impact of LULCC with great flexibility. For example, the Variable Infiltration Capacity (VIC) model has been employed to study the hydrologic effects of long-term land and water management in North America and Asia over the past 300 years.[17] The hydrologic effects of LULCC in four catchments within the Columbia River basin have been studied with the topographically explicit distributed hydrology–soil–vegetation model (DHSVM) and a comparison between the results using this method and those from the experiment using the VIC model have also been conducted.[18] In particular, because distributed hydrologic models can take spatial heterogeneities into account and provide detailed descriptions of the hydrologic processes in a watershed, the impact of LULCC and its spatial pattern on hydrology can be investigated through these models in a more spatially explicit way. For example, a distributed rainfall–runoff simulation model was developed to evaluate the impact of urbanization on runoff using high spatial resolution ISA data.[19] With the capacity of hydrologic models, the hydrologic impact from LULCC can be studied with the consideration of other exogenous variables such as climate in the evaluation, and therefore understanding of the hydrologic impact of LULCC can be improved. It is worth noting that the calibration of hydrologic model is important and it may introduce uncertainties in the evaluation.

Conclusions

Due to people's different responses to economic opportunities, a variety of LULCC types such as urbanization, agriculture practice, and deforestation and reforestation occur across spatial and temporal scales on the Earth's surface. LULCC with spatial and temporal characteristics has a profound impact on the Earth's environment and can be effectively studied with the help of remote sensing data and technology.

LULCC affects biodiversity, climate change, watershed hydrology, and other processes on the Earth. The change in watershed hydrology induced by LULCC not only has direct influence on the water system, but also brings about a series of associated environmental issues such as water pollution. Watershed hydrology is complex and impacted by a number of factors, such as climate and soil condition, as well as LULCC. Hydrologic models provide the capability to comprehend the complex hydrologic processes and have been widely and effectively used to understand the hydrologic impact of LULCC.

Although much progress has been made in the study of hydrologic impact of LULCC, there are still numerous uncertainties and even controversies. There are uncertainties from data sources, LULCC quantification, and hydrologic modeling in the evaluation. The hydrologic impact of LULCC may vary with study area, climate condition, geography, and spatial scale, and it is also dependent on the temporal scale analyzed. All of these factors make the evaluation of the hydrologic impact of LULCC more challenging.

Although a great deal of effort has focused on the improvement of LULCC quantification using remote sensing across spatial and temporal scales and distributed and process-based hydrologic modeling, interdisciplinary and multidisciplinary collaborations are still necessary to comprehend the response of watersheds with different size, topography, soil moisture, and local condition to LULCC, thus contributing to an improved understanding of the hydrologic impact of LULCC.

References

1. Bronstert, A. Rainfall-runoff modelling for assessing impacts of climate and land-use change. Hydrol. Process. **2004,** *18* (3), 567–570, doi:10.1002/hyp.5500.

2. Foley, J.A.; DeFries, R.; Asner, G.P.; Barford, C.; Bonan, G.; Carpenter, S.R.; Chapin, F.S.; Coe, M.T.; Daily, G. C.; Gibbs, H.K.; Helkowski, J.H.; Holloway, T.; Howard, E.A.; Kucharik, C.J.; Monfreda, C.; Patz, J.A.; Prentice, I.C.; Ramankutty, N.; Snyder, P.K. Global consequences of land use. Science **2005**, *309* (5734), 570–574, doi:10.1126/ science.1111772.
3. DeFries, R.; Eshleman, K.N. Land-use change and hydrologic processes: A major focus for the future. Hydrol. Pro- ces. **2004**, *18*, 2183–2186, doi: 10.1002/hyp.5584.
4. Fisher, P.; Comber, A.J.; Wadsworth, R. Land use and land cover: Contradiction or complement. In *Re-Presenting GIS.;* Fisher, P.; Unwin, D., Eds.; Wiley: London, UK, 2005; 85–98.
5. Ellis, E. Land-use and land-cover change. In *Encyclopedia of Earth. Environmental Information Coalition,* Cleveland, C.J., Ed.; National Council for Science and the Environment: Washington, D.C., 2012.
6. Lambin, E.F.; Turner, B.L.; Geist, H.J.; Agbola, S.B.; Angelsen, A.; Bruce, J.W.; Coomes, O.T.; Dirzo, R.; Fischer, G.; Folke, C.; George, P.S.; Homewood, K.; Imbernon, J.; Leemans, R.; Li, X.; Moran, E.F.; Mortimore, M.; Ramakrishnan, P.S.; Richards, J.F.; Skånes, H.; Steffen, W.; Stone, G.D.; Svedin, U.; Veldkamp, T.A.; Vogel, C.; Xu, J. The causes of land-use and landcover change: moving beyond the myths. Global Env. Change **2011**, *11* (4), 261–269, doi:10.1016/s0959-3780(01)00007-3.
7. Singh, A. Digital change detection techniques using remotely-sensed data. Int. J. Remote Sensing **1989**, *10* (6), 989–1003.
8. Brink, A.B.; Eva, H.D. Monitoring 25 years of land cover change dynamics in Africa: A sample based remote sensing approach. Appl. Geogr. **2009**, *29* (4), 501–512. doi:10.1016/j.apgeog.2008.10.004.
9. Xiao, J.; Shen, Y.; Ge, J.; Tateishi, R.; Tang, C.; Liang, Y.; Huang, Z. Evaluating urban expansion and land use change in Shijiazhuang, China, by using GIS and remote sensing. Landscape Urban Plan. **2006**, 75 (1–2), 69–80. doi:10.1016/j. landurbplan.2004.12.005.
10. Kleijn, D.; Kohler, F.; Báldi, A.; Batáry, P.; Concepción, E.D.; Clough, Y.; Díaz, M.; Gabriel, D.; Holzschuh, A.; Knop, E.; Kovács, A.; Marshall, E.J.P.; Tscharntke, T.; Ver- hulst, J. On the relationship between farmland biodiversity and land-use intensity in Europe. Pro. R. Soc. B: Biol. Sci. **2009**, *276* (1658), 903–909, doi:10.1098/rspb.2008.1509.
11. Pielke, R.A.; Marland, G.; Betts, R.A.; Chase, T.N.; Eastman, J.L.; Niles, J.O.; Niyogi, DdS; Running, S.W. The influence of land-use change and landscape dynamics on the climate system: relevance to climate-change policy beyond the radiative effect of greenhouse gases. Philos. Transact. A Math. Phys. Eng. Sci. **2002**, *360* (1797), 1705–1719, doi:10.1098/rsta.2002.1027.
12. Li, L.; Jiang, D.; Li, J.; Liang, L.; Zhang, L. Advances in hydrological response to land use/land cover change. J. Nat. Resour. **2007**, 22 (2), 211–224.
13. Shuster,W.; Bonta, J.; Thurston, H.; Warnemuende, E.; Smith, D. Impacts of impervious surface on watershed hydrology: A review. Urban Water J. **2005**, *2* (4), 263–275, doi:10.1080/15730620500386529.
14. Zhou, Y.; Wang, Y.Q. Extraction of impervious surface areas from high spatial resolution imagery by multiple agent segmentation and classification. Photogramm. Eng. Rem. S. **2008**, *74* (7), 857–868.
15. Siriwardena, L.; Finlayson, B.L.; McMahon, T.A. The impact of land use change on catchment hydrology in large catchments: The Comet River, Central Queensland, Australia. J. Hydrol. **2006**, *326* (1–4), 199–214.
16. Lin, Y.; Wei, X. The impact of large-scale forest harvesting on hydrology in the Willow watershed of central British Columbia. J. Hydrol. **2008**, *359* (1–2),141–149, doi:10.1016/j.jhydrol.2008.06.023.
17. Haddeland, I.; Skaugen, T.; Lettenmaier, D.P. Hydrologic effects of land and water management in North America and Asia: 1700–1992. Hydrol. Earth Syst. Sci. **2007**, *11* (2), 1035–1045, doi:10.5194/ hess-11-1035-2007.
18. VanShaar, J.R.; Haddeland, I.; Lettenmaier, D.P. Effects of land-cover changes on the hydrological response of interior Columbia River basin forested catchments. Hydrol. Process. **2002**, *16* (13), 2499–2520, doi:10.1002/ hyp.1017.
19. Zhou, Y.; Wang, Y.; Gold A.; August, P. Modeling watershed rainfall-runoff relations using impervious surface-area data with high spatial resolution. Hydro- geol. J. **2010**, *18* (6), 1413–1423, doi:10.1007/s10040-010-0618-9.

31

Rain-Use Efficiency: Remote Sensing

Fang Huang
Northeast Normal University

Introduction 225
Calculation of Rain-Use Efficiency by Satellite Observations 226
Conclusions 228
Acknowledgments 228
References 229

Introduction

Understanding how vegetation productivity responds to variations in precipitation is critical for evaluating the impacts of global change on terrestrial ecosystems. Rain-use efficiency (RUE), also known as precipitation-use efficiency (PUE), is defined as the ratio of net primary production (NPP) or aboveground NPP (ANPP) to precipitation on the basis of field measurements [1]. It calculates the number of kilograms aerial dry matter phytomass produced over 1 ha in a year per 1 mm of rain fallen, usually expressed in kg $ha^{-1}year^{-1}mm^{-1}$. RUE is related to another index, the water-use efficiency (WUE) [2], generally estimated by the ratio of gross primary productivity (GPP) or NPP to evapotranspiration (ET). WUE is widely used in plant ecophysiology and irrigation management to measure the dry matter increment of vegetation per unit of transpired water for a stand of vegetation, a single plant, or a leaf [3]. Like WUE, RUE can describe the relationship between water use and biomass production, especially for dryland ecosystems [4], thus help improve the knowledge of the potential trade-offs between carbon and water cycles of an ecosystem [5].

RUE involves transpiration, evaporation of bare soil, water loss from canopy interception and direct evaporation, infiltration and drainage from the rooting zone, runoff, storage in the rooting zone, plus any deeper water connected to the root zone by continuous capillary water, and variations in the water content of the vegetation [3]. RUE may be influenced by biotic and abiotic factors, including vegetation composition, climatic and soil conditions, terrain topography, and biogeochemical constraints [1,6–10]. It is reported that communities with higher diversity of plant species and functional groups may have higher RUE because of the increased ANPP [11]. Soil properties such as texture, water-holding capacity, depth, and fertility have effects on RUE [1,12–13]. Additionally, the interpretation of RUE may also require information on human activities such as population, management regime, and economic development [10,14]. In general, RUE patterns and the associate driving factors may vary among different regions.

Since the 1980s, the increasing availability of earth observation satellites allows estimating vegetation RUE over large-scale areas and throughout long time periods [15]. RUE derived from remote sensing data has significant advantages when there is a lack of reliable field-based vegetation measurement and precipitation data. For a given ecosystem preserved from any kind of functional change through time, the RUE values should be stable over time [1]. Under the same precipitation condition, degraded ecosystems may have less biomass and produce more runoff than undegraded ecosystems, leading to smaller RUE [16]. RUE and its change trends have been suggested to diagnose nonprecipitation-related land degradation and even desertification in arid or semiarid regions [3,15–22]. More recently, RUE and its derived index (e.g., maximum RUE) are used to evaluate ecological multifunctioning and the loss of ecosystem function [23,24].

Calculation of Rain-Use Efficiency by Satellite Observations

RUE was developed for use with field measurements of rainfall and crop or natural vegetation production [3]. It can also be calculated from remotely sensed primary productivity and rainfall data [3,7,16,24–30]. Estimating ANPP over large regions needs upscaling of ground data using remote sensing methods [15]. The temporal sums of satellite measurements of vegetation indices could be used to estimate vegetation productivity. Normalized Difference Vegetation Index (NDVI) has been widely used to detect changes in vegetation greenness, biomass, productivity, and density [31–34]. There may exist a near-linear relationship between primary productivity and NDVI for relatively homogeneous biomes [25,31–32,34–37]. The seasonal/annual summed NDVI is often assumed to be an appropriate proxy for ANPP [3]. Most of the studies have been carried out by satellite-based NDVI data, collected from different sensors such as the NOAA AVHRR (National Oceanic and Atmospheric Administration Advanced Very High Resolution Radiometer), GIMMS (Global Inventory Modeling and Mapping Studies), Terra/Aqua MODIS (Moderate Resolution Imaging Spectroradiometer), and SPOT VGT (Système Pour l'Observation de la Terre Vegetation) [3,15–16,20,25,38–40]. In addition, satellite-driven NPP/GPP products based on empirical and physical models derived from above Earth observation data may also be used to estimate RUE [41–44].

Precipitation data can be acquired from ground meteorological stations. The site-level data can be converted into gridded precipitation data with the same spatial resolution as NDVI dataset through appropriate spatial interpolation method. In recent decades, the advancement of satellite remote sensing has enabled the generation of many precipitation estimate products. Some large-scale precipitation datasets, such as blended gauge-satellite GPCP product (Global Precipitation Climatology Project, with a spatial resolution of 2.5°) [45], CMAP data (Climate Prediction Center or CPC Merged Analysis of Precipitation) produced by the CPC of the National Weather Service [46,47], RFE data (Rainfall Estimate, with a spatial resolution of 0.1°) provided by NOAA-CPC [47], and the global gridded precipitation dataset with a spatial resolution of 0.5° from Climate Research Unit (CRU) [48], have been utilized to estimate RUE in reported studies [20,21,24].

Therefore, the RUE can generally be determined by the following equation, respectively:

$$RUE = ANPP/PPT \tag{31.1}$$

where RUE ($g \cdot m^{-2} \cdot mm^{-1}$) represents rain-use efficiency, ANPP ($g \cdot m^{-2}$) is aboveground net primary productivity, and PPT (mm^{-1}) is mean or summed precipitation during a certain time period (i.e., the entire year or the growing season). The accumulated NDVI (ΣNDVI) within the above time period can be used as a substitute for ANPP.

In this chapter, we describe the characteristics of remote sensing RUE in the west of Songliao Plain, China, calculated using SPOT-VGT NDVI time series and precipitation data from 1999 to 2011 [14]. This region is a typical farming-pastoral transitional zone and sensitive to global change (Figure 31.1a). Figure 31.1b shows the spatial pattern of average RUE. The average RUE of saline-alkali land and open shrub land was the lowest (<0.006), while relative high RUE values (>0.010) were observed in some grassland and cropland. Annual mean RUE increased slightly with an undulating trend, indicating the ecosystem function tended to improve in this region.

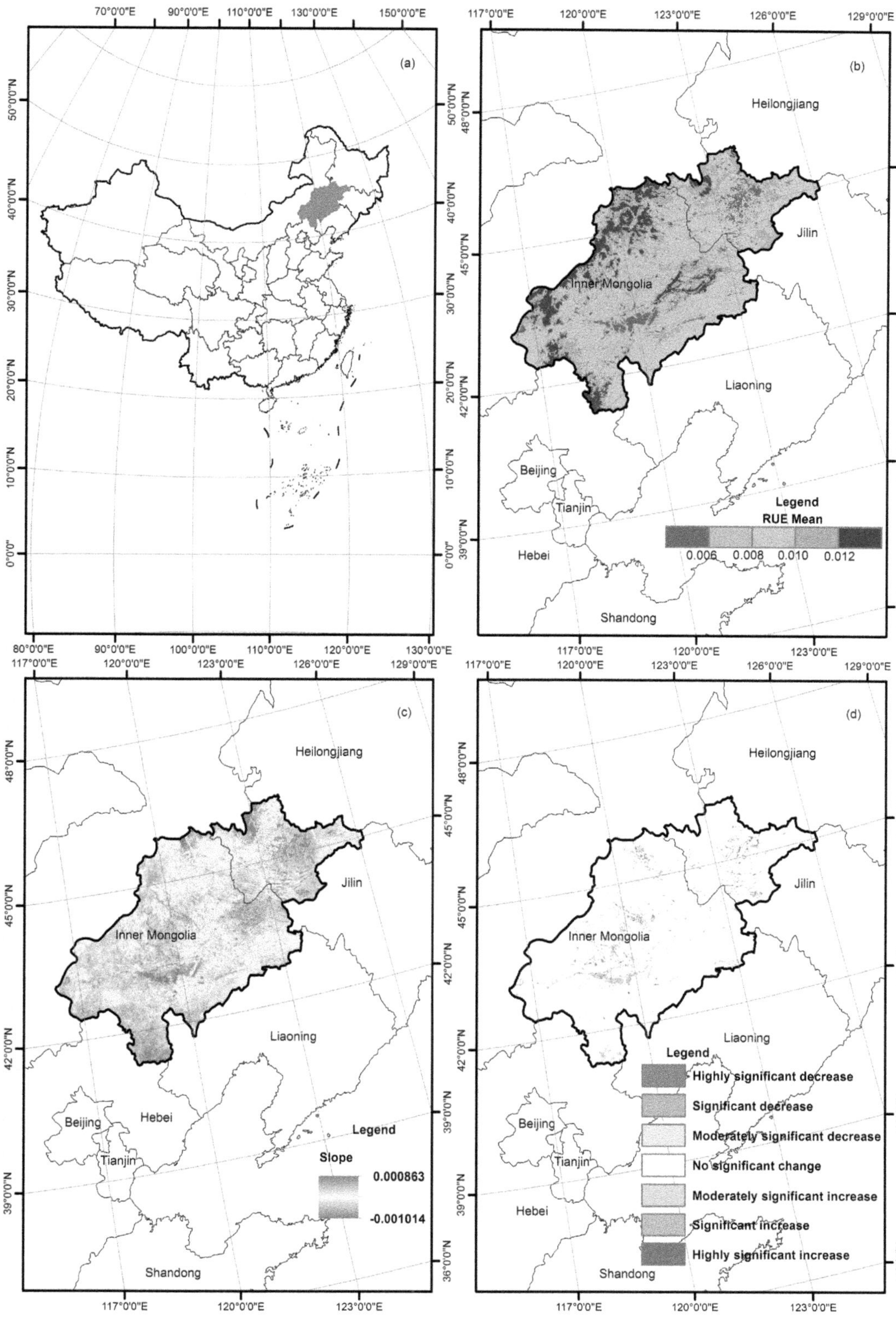

FIGURE 31.1 **(See color insert.)** Satellite-based RUE and the changes in the west of Songliao Plain, China.

TABLE 31.1 Rain-Use Efficiency (RUE) Change Trend in the West Songliao Plain, China (1999–2011)

RUE Change Trend	Significant Level (p)	Area (%)
Highly significant decrease	Slope < 0, $p \leq 0.01$	0.07
Significant decrease	Slope < 0, $0.01 < p \leq 0.05$	0.41
Moderately significant decrease	Slope < 0, $0.05 < p \leq 0.1$	0.81
No significant change	$p > 0.1$	94.47
Moderately significant increase	Slope > 0, $0.05 < p \leq 0.1$	2.25
Significant increase	Slope > 0, $0.01 < p \leq 0.05$	1.66
Highly significant increase	Slope > 0, $p \leq 0.01$	0.32

Source: Data adapted from [14].

Spatial change trends of RUE for each pixel were estimated using a linear regression model ($y = a + bx$), in which year, and RUE as the independent variable (x) and dependent variable (y), respectively. The regression slope value of each pixel represents its change trend. Figure 31.1c shows the spatial pattern of the change slope of RUE. Of the total area, 59.53% experienced an RUE increase, while the area with a negative SLOPE occupied the rest of the area. The common interpretation is that statistically significant negative slopes indicate land degradation. Conversely, positive slopes may imply improved ecosystem function.

Table 31.1 and Figure 31.1d show the change trends of RUE and are classified into seven types by the significance test of the regression model. Most of the region (94.47%) had no significant change of RUE. About 4.23% of the west Songliao plain showed a significant RUE increase, while 1.29% experienced a significant decrease. From Figure 31.1d, the area where RUE significantly increase was mainly distributed in saline-alkali land in the counties of Jilin province and the shrub land and grassland in Ongniud Bannar, Balinyou Banner, and Ar Horqin Banner. The negative trends were mainly found in the north central area of Horqin Youyizhong Banner and the south of Horqin Zuoyihou Banner. The findings may reflect the positive effects owing to the implementation of ecological conservation and restoration projects at national and local levels, such as the "Grain for Green Project," the "Natural Forest Protection Project," and the "Beijing and Tianjin sandstorm source controlling project" [49–53].

Conclusions

Vegetation affects several processes, including water balance, absorption and reemission of solar radiation, latent and sensible heat fluxes, the carbon cycle, etc. The demand for latest information on the vegetation cover with regard to climate and ecosystems change has increased recently. RUE is not only an important variable for linking the carbon and water cycles of terrestrial ecosystems but also a way of regulating ecological populations and systems responding to climate change. Although there might be some debates about the use of remote sensing data for ANPP estimation, its ecological interpretation and the spatiotemporal pattern of RUE, satellite remote sensing provides an effective approach of estimating long-term RUE over large-scale regions. Satellite-based RUE is widely used to assess land degradation, especially in arid or semiarid ecosystems. The case of the west of Songliao Plain, it has been characterized by a slight increase in RUE over the period 1999–2011. The average RUE tended to fluctuate from year to year, implying the instability of the water use status. The areas experienced a significant increase in RUE, probably suggesting improved ecosystem function. We infer that increased vegetation activity may be attributed to the effectiveness of some ecological and environmental restoration projects implemented during the past decade.

Acknowledgments

This research was supported by the National Natural Science Foundation of China (Grant no. 41571405, 41630749 and 41571115).

References

1. Le Houèrou H.N. Rain use efficiency: A unifying concept in arid-land ecology. *Journal of Arid Environments* 1984, 7, 213-247.
2. Viets F.G. Fertilizers and the efficient use of water. *Advances in Agronomy* 1962, 14, 223–264.
3. Prince S. D., Colstoun, D., Brown E., Kravitz L. Evidence from rain-use efficiencies does not indicate extensive Sahelian desertification. *Global Change Biology* 1998, 4, 359–374.
4. Loik M.E., Breshears D.D., Lauenroth W.K., Belnap J. A multi-scale perspective of water pulses in dryland ecosystems: Climatology and ecohydrology of the western USA. *Oecologia* 2004, 141, 269–281.
5. Li H.X., Liu G.H., Fu B.J. Spatial variations of rain-use efficiency along a climate gradient on the Tibetan Plateau: A satellite-based analysis. *International Journal of Remote Sensing* 2013, 34, 7487–7503.
6. Lauenroth W.K., Sala, O.E. Long-term for age production of North American shortgrass steppe. *Ecological Applications* 1992, 2, 397–403.
7. Paruelo J.M., Lauenroth W.K., Burke I.C., Sala O.E. Grassland precipitation-use efficiency varies across a resource Gradient. *Ecosystems* 1999, 2, 64–68.
8. Huxman T.E., Smith M.D., Fay P.A., Knapp A.K., Shaw M.R., Loik M.E., Smith S.D., Tissue D. T., Zak J.C., Weltzin J.F., Pockman W.T., Sala O.E., Haddad B.M., Harte J., Koch G.W., Schwinning S., Small E.E., Williams D.G. Convergence across biomes to a common rain-use efficiency. *Nature* 2004, 429, 651–654.
9. Bai Y.F., Wu, J.G., Xing Q., Pan Q.M., Huang J.H., Yang D.L., Han, X.G. Primary production and rain use efficiency across a precipitation gradient on the Mongolia Plateau. *Ecology* 2008, 89, 2140–2153.
10. Hein L., Ridder D.E. Desertification in the Sahel: A reinterpretation. *Global Change Biology* 2006, 12, 751–758.
11. Hooper D.U., Vitousek P.M. The effects of plant composition and diversity on ecosystem processes. *Science* 1997, 277, 1302–1305.
12. Noy-Meir, I. Desert ecosystems: environment and producers. *Annuals Reviews in Ecology Systems* 1973, 4, 25–51.
13. Sala O.E., Parton W.J., Joyce L.A., Lauenroth W.K. Primary production of the central grassland region of the United States. *Ecology* 1988, 69, 40–45.
14. Huang F., Xu S.L. Spatio-Temporal variations of rain-use efficiency in the west of Songliao Plain, China. *Sustainability* 2016, 8, 308.
15. Dardel C., Kergoat L., Hiernaux P., Grippa M., Mougin E., Ciais P., Nguyen C. Rain-use-efficiency: What it tells us about the conflicting Sahel greening and Sahelian paradox. *Remote Sensing* 2014, 6, 3446–3474.
16. Wessels K.J., Prince S.D., Malherbe J., Small J., Frost P.E., VanZyl, D. Can human-induced land degradation be distinguished from the effects of rainfall variability? A case study in South Africa. *Journal of Arid Environments* 2007, 68, 271–297.
17. Justice C.O., Dugdale G., Townshend J.R.G., Narracott A.S., Kumar, M. Synergism between NOAA-AVHRR and Meteosat data for studying vegetation development in semi-arid West Africa. *International Journal of Remote Sensing* 1991, 12, 1349–1368.
18. Verón S.R., Paruelo J.M., Oesterheld M. Assessing desertification. *Journal of Arid Environments* 2006, 66, 751–763.
19. Hu Z.M., Yu G.R., Fan J.W., Zhong H.P., Wang S.Q., Li S.G. Precipitation-use efficiency along a 4500-km grassland transect. *Global Ecology and Biogeography* 2010, 19, 842–851.
20. Fensholt R., Rasmussen K. Analysis of trends in the Sahelian 'rain-use efficiency' using GIMMS NDVI, RFE and GPCP rainfall data. *Remote Sensing of Environment* 2011, 115, 438–451.

21. Fensholt R., Rasmussen K., Kaspersen P., Swinnen E., Huber S., Horion S., Swinnen E. Assessing land degradation/recovery in the African Sahel from long-term earth observation based primary productivity and precipitation relationships. *Remote Sensing* 2013, 5, 664–686.
22. Kundu A., Patel N.R., Saha S.K., Dutta D. Desertification in western Rajasthan (India): An assessment using remote sensing derived rain-use efficiency and residual trend methods. *Natural Hazards* 2017, 86, 297–313.
23. Zhao Y.C., Wang X.Y., Vázquez-Jiménez R. Evaluating the performance of remote sensed rain-use efficiency as an indicator of ecosystem functioning in semi-arid ecosystems. *International Journal of Remote Sensing* 2018, 39, 3344–3362.
24. Du L., Mikle N., Zou Z.H., Huang Y.Y., Shi Z., Jiang L.F., McCarthy H.R., Liang J.Y., Luo Y.Q. Global patterns of extreme drought-induced loss in land primary production: Identifying ecological extremes from rain-use efficiency. *Science of the Total Environment* 2018, 628, 611–620.
25. Nicholson S.E., Tucker C.J., Ba M.B. Desertification, drought, and surface vegetation: An example from the West African Sahel. *Bulletin of the American Meteorological Society* 1998, 79, 815–829.
26. Tucker C.J., Justice C.O., Prince S.D. Monitoring the grasslands of the Sahel 1984–1985. *International Journal of Remote Sensing* 1986, 7, 1571–1582.
27. Justice C.O., Eck T., Tanre D., Holben, B. The effect of water vapor on the normalized difference vegetation index derived for the Sahelian Region from NOAA AVHRR data. *International Journal of Remote Sensing* 1991, 12, 1165–1188.
28. Pickup, G. Estimating the effects of land degradation and rainfall variation on productivity in rangelands: An approach using remote sensing and models of grazing and herbage dynamics. *Journal of Applied Ecology* 1996, 33, 819–832.
29. Holm A.M., Cridland S.W., Roderick M.L. The use of time-integrated NOAA NDVI data and rainfall to assess landscape degradation in the arid shrubland of Western Australia. *Remote Sensing of Environment* 2003, 85, 145–158.
30. Chang J., Tian J.X., Zhang Z.X., Chen X., Chen Y.Z., Chen S., Duan Z. Changes of grassland rain use efficiency and NDVI in Northwestern China from 1982 to 2013 and its response to climate change. *Water* 2018, 10, 1689.
31. Prince S.D. Satellite remote sensing of primary production: Comparison of results for Sahelian grasslands 1981–1988. *International Journal of Remote Sensing* 1991, 12, 1301–1311.
32. Prince S.D. A model of regional primary production for use with coarse resolution satellite data. *International Journal of Remote Sensing* 1991, 12, 1313–1330.
33. Schmidt H., Karnielli A. Remote sensing of the seasonal variability of vegetation in a semi-arid environment. *Journal of Arid Environments* 2000, 45, 43–59.
34. Wang J., Rich P.M., Price K.P., Kettle W.D. Relations between NDVI and tree productivity in the central Great Plains. *International Journal of Remote Sensing* 2004, 25, 3127–3138.
35. du Plessis W.P. Linear regression relationships between NDVI, vegetation and rainfall in Etosha National Park, Namibia. *Journal of Arid Environments* 1999, 42, 235–260.
36. Wessels K.J., Prince S.D., Frost P.E., van Zyl D. Assessing the effects of human-induced land degradation in the former homelands of northern South Africa with a 1km AVHRR NDVI time-series. *Remote Sensing of Environment* 2004, 91, 47–67.
37. Li H.X., Liu S.Z. Research on Grassland Degradation Assessment Model based on ETM+ Image: A case study in Naqu Country of Tibet. *Journal of Desert Research* 2007, 3, 412–418.
38. Roukos C., Papanikolaou K., Kandrelis S., Mygdalia A., Chatzitheodoridis F. A GIS-based assessment of rain-use efficiency factor and grazing capacity in Preveza prefecture, Greece. *Journal of Agricultural Research* 2011, 49, 97–107.
39. Hein L., de Ridder N., Hiernaux P., Leemans R., de Wit A., Schaepman M. Desertification in the Sahel: Towards better accounting for ecosystem dynamics in the interpretation of remote sensing images. *Journal of Arid Environments* 2011, 75, 1164–1172.

40. Herrmann S.M., Anyamba A., Tucker C.J. Recent trends in vegetation dynamics in the African Sahel and their relationship to climate. *Global Environmental Change* 2005, 15, 394–404.
41. Williams C.A., Hanan N.P., Baker I., Collatz G.J., Berry, J., Denning A.S. Interannual variability of photosynthesis across Africa and its attribution. *Journal of Geophysical Research-Biogeosciences* 2008, 113.
42. Sabbe H., Veroustrate F. Demonstration of a standard Net Primary Productivity product for the SPOT 4-VEGETATION instrument. *VITO* 2000.
43. Running S.W., Nemani R.R., Heinsch F.A., Zhao M.S., Reeves M., Hashimoto H. A continuous satellite-derived measure of global terrestrial primary production. *Bioscience* 2004, 54, 547–560.
44. Zhang X., Moran M.S., Zhao X., Liu S.H., Zhou T., Ponce-Campos G. E., Liu F. Impact of prolonged drought on rainfall use efficiency using MODIS data across China in the early 21st century. *Remote Sensing of Environment* 2014, 150, 188–197.
45. Adler R.F., Huffman G.J., Chang A., Ferraro R., Xie P.P., Janowiak J.E., Rudolf B., Schneider U., Curtis S., Bolvin D., Gruber A., Susskind J. The version-2 global precipitation climatology project (GPCP) monthly precipitation analysis (1979–present). *Journal of Hydrometeorology* 2003, 4, 1147–1167.
46. Xie P., Arkin P.A. Analysis of global monthly precipitation using gauge observations, satellite estimates, and numerical model prediction. *Journal of Climate* 1996, 9, 840–858.
47. Xie P.P., Arkin P.A. Global precipitation: A 17-year monthly analysis based on gauge observations, satellite estimates, and numerical model outputs. *Bulletin of the American Meteorological Society* 1997, 78, 2539–2558.
48. Harris I., Jones P.D., Osborn T.J., Lister D.H. Updated high-resolution grids of monthly climatic observations-the CRU TS3.10 dataset. *International Journal of Climatology* 2014, 34, 623–642.
49. Zhang G.L., Dong J.W., Xiao X.M., Hua Z.M., Sheldon S. Effectiveness of ecological restoration projects in Horqin Sandy Land, China based on SPOT-VGT NDVI data. *Ecological Engineering* 2012, 38, 20–29.
50. Peng F., Wang T., Xue, X. Study of human impact on vegetation in desertified regions based on rainfall use efficiency-A case study of Horqin region, Inner Mongolia. *Journal of Desert Research* 2010, 30, 896–902.
51. Yin H., Li Z.G., Wang Y.L., Cai, F. Assessment of desertification using time series analysis of hyper-temporal vegetation indicator in Inner Mongolia. *Acta Geographica Sinica* 2011, 66, 653–661.
52. Stokes A., Sotir R., Chen W., Chestem M. Soil bio- and eco-engineering in China: Past experience and future priorities preface. *Ecological Engineering* 2010, 36, 247–257.
53. Huang L., Xiao T., Zhao Z.P., Sun C.Y., Liu J.Y., Shao Q.Q., Fan J.W., Wang J.B. Effects of grassland restoration programs on ecosystems in arid and semiarid China. *Journal of Environment Management* 2013, 117, 268–275.

32

Water Management: South Asia

Soni M. Pradhanang and Marzia Tamanna
University of Rhode Island

Introduction 233
Integrated Water Resources Management (IWRM) 235
Conclusions 251
References 251

Introduction

The Ganges, Brahmaputra, and Meghna (GBM) river system originating in the Himalayas and draining into the Bay of Bengal, runs through five countries of South Asia—Bangladesh, Bhutan, China, India, and Nepal—categorized by huge population, inadequate land resources, and recurrent floods and natural hazards (Figure 32.1 and Table 32.1). Although the GBM region is rich in water resources, this is one of the poorest areas in the world. In this region, from economy to human and environmental health, all depend on water, and water is, therefore, integral to sustainable development, economic progress, and poverty depletion [32].

There are several factors that complicate water management in South Asia [38]. The few important factors are presented below.

1. Water management intersects many sectors such as, agriculture, energy, environment, health, industry, and land use. Due to its inter-sectoral importance, it is also impacted largely by policies and actions taken outside the water sector. Addressing this requires a multi-sectoral approach, beyond the confines of the water sector, to equitably allocate water and control and regulate use within sustainable limits.
2. Water resource management encompasses working collaboratively across numerous political and administrative jurisdictions—villages, cities, districts, states, provinces, and nations.
3. There is a need of integrated approach to water resource management to ensure adequate management and protection of this valuable resource.
4. Water resource management issues are diverse, and responses, are, therefore full of challenges. Diverse climate and topographic regimes spanning across South Asia leads to the difficulty in managing water resources.
5. The systematic planning, management, and development are essential to address water-related natural disasters—droughts and floods.

This chapter succinctly describes two country-specific cases that have highlighted coordination and management of water resources through global partnership. The analysis of two very contrasting results of managing transboundary rivers in South Asia, a most successful one in Bhutan and India, and a missed opportunity in Nepal, India, and Bangladesh is portrayed in [7]. Bhutan and India have shown that, water can be successfully used as an engine for economic growth when there is goodwill and trust between the shared countries. This can bring substantial benefits to the people of both the countries.

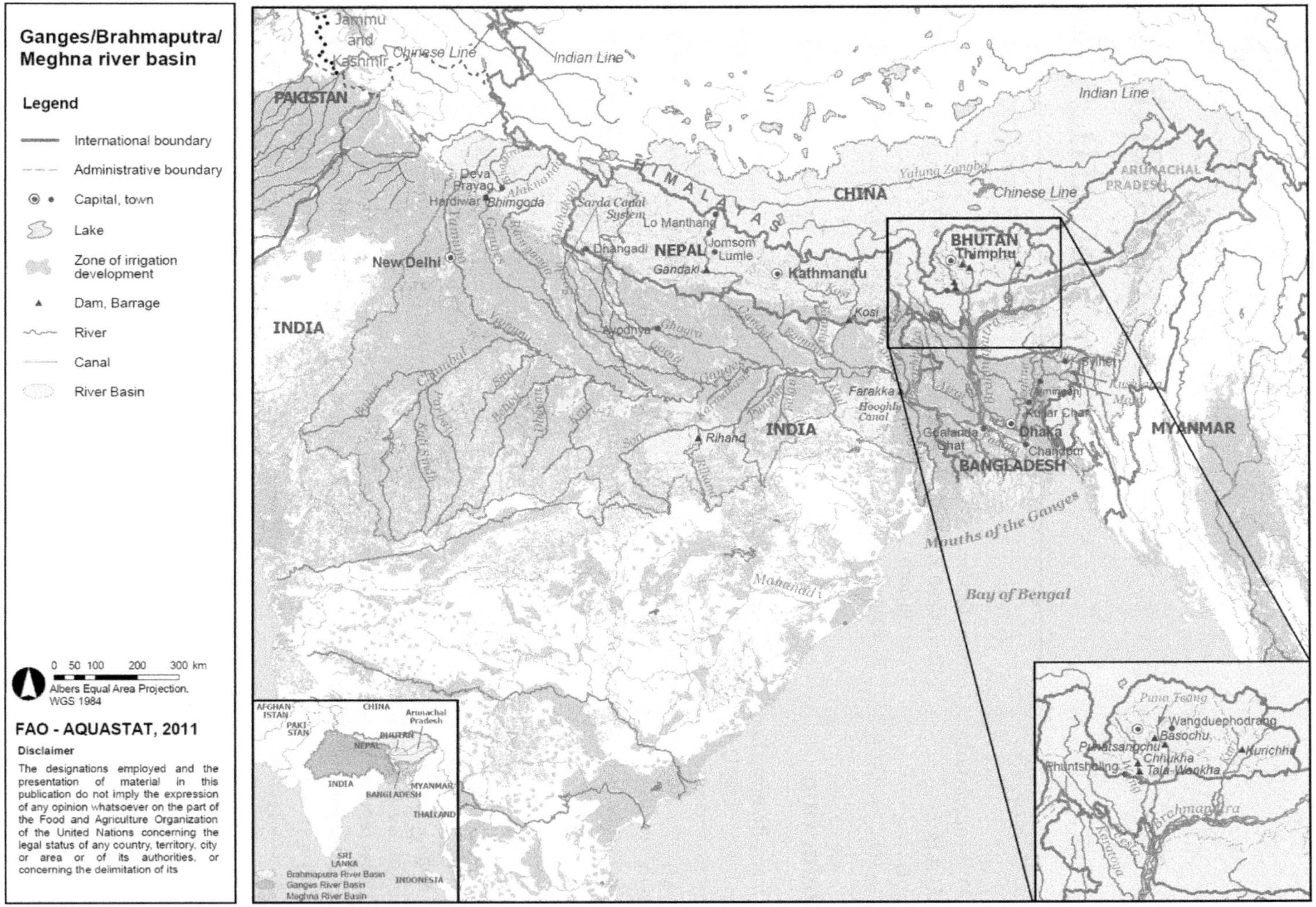

FIGURE 32.1 Map of Bangladesh and international river basins [18].

TABLE 32.1 Key Indicators Related to Agriculture and Water in South Asia

Indicators	2007[a]	2050 Projection
Population (millions)	1,520	2,242
Population density (per km²)	352	-
Annual population growth rate (%)	1.5	0.53
Population below USD 1.25 a day (million)	596 (2005)	14.1
Poverty ratio (below USD 1.25 a day)	40.3 (2005)	-
Per capita GDP growth (%)	3.6 (1995–1997)	-
Undernourished people (millions)	331	93
Undernourished population (%)	21.8	4.2
Population without access to safe water (millions)	269	-
Total land area (million km²)	4.47	-
Cultivable land area (million km²)	2.19	-
Cultivated area (million km²)	2.04	-
Arable land (million ha)	204	213
Irrigated area (million ha)	104.3 (2000)	135.2
Total rain-fed area	98 (2000)	110
Annual growth rate in irrigated area (%)	1.6	0.1
Cultivated area (% of total area)	93	-
Cultivated land (ha per person)	0.12	0.08
Cultivated land irrigated (%)	39	-
Contribution of irrigated agriculture in total food production (%)	60–80	-
Agriculture growth rate (%)	2.4	1.3
Total water withdrawal (km³)	1,023.40	-
Annual Water Withdrawal by Sector (%)		
Agriculture	91	-
Municipalities	7	-
Industry	2	-
Total water consumption in agriculture sector (km³)	1,479 (2000)	1,922
Total water withdrawal for irrigation (km³)	1,095 (2000)	1,817
Per capita water withdrawal (m³)	631	-
Agricultural use (% of total withdrawal)	82.30	-
Total irrigated area (million ha)	88.60	-
Total irrigated area (%)	47.48	-
Irrigation		
Surface water contribution	36.7 (2009)	-
Groundwater contribution	54.7 (2009)	-

Sources: Data from Rasul [31], FAO [19], de Fraiture and Wichelns [11], Lal [26].
[a] All data are for 2007 unless otherwise specified.

In contrast, lack of trust between Nepal, India, and Bangladesh has compounded the deprivation of the region through underdevelopment.

Integrated Water Resources Management (IWRM)

Integrated Water Resources Management (IWRM) is defined by the Global Water Partnership (GWP) as such "a process which promotes the coordinated development and management of water, land, and related resources, with the aim of maximization of the resultant economic and social welfare in an equitable manner without compromising the sustainability of vital ecosystems [24]."

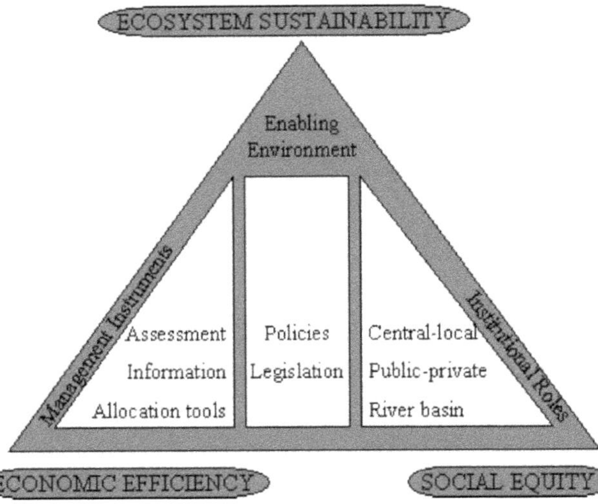

FIGURE 32.2 Complementary elements for IWRM [24].

Therefore, IWRM is a process that addresses the following three prevailing elements:

- Economic efficiency in water use: water needs to be used with maximum likely economic efficiency because of the increasing shortage of water and financial resources, as a resource, the finite and vulnerable nature of water, and the growing demands upon it.
- Social justice and equity concerns: access to water of adequate quantity and quality is the basic right of all people for the sustenance of human well-being.
- Environmental and ecological sustainability: the existing use of the resource should be managed in a way that sustains the vital life support systems, thus making sure of having adequate resource available for use by future generations [10,25].

Das Gupta et al. [10] mentioned about a framework for moving toward IWRM (Figure 32.2) proposed by the GWP-Technical Advisory Committee (GWP-TAC). Simultaneous development and strengthening of the following three elements are required:

"Enabling environment" is made of national, provincial, and local policies and legislation. These establish the "rules of the game," which qualify all stakeholders to play their respective roles.

"Institutional roles" state distinct boundary of responsibilities between actors, the parting of regulation from service provision functions, suitable coordination mechanisms, filling jurisdictional gaps, and eradicating overlaps and matching responsibilities both to authority and to capacities for action.

"Management instruments" encompass water resources assessment, demand management, social change instruments, conflict management, regulatory instruments, economic instruments, and information and communication instruments.

Case Study 1: Implementation of IWRM—Bangladesh

Bangladesh (56,980 square miles) is the eighth-most populous country of world with a population of 164.7 million [38]. It is one of the South Asian countries. The country is called as a land of rivers with 5,000 miles of inland waterways. In reference to Atapattu and Kodituwakku [2], annual renewable water resources in Bangladesh is 1,210.6 $km^3 year^{-1}$, out of which 76.35 km^3 of water withdrawal is for agricultural sector (Table 32.2). Thus, water management in Bangladesh is a critical issue due to the growing demand and increasing conflict between alternative uses. Chowdhury [9] identified various geographic, socioeconomic, and environmental factors that shape the water management issues of Bangladesh. The rapid growing demand for water is originating from agricultural sector particularly from irrigation for cereal production, the urban and industrial development, fishery, inland navigation, and salinity control. The huge quantity of water is being withdrawn for the irrigation purposes (Figure 32.3). The supply of fresh and unpolluted water has fallen far short of demand as a result of insufficient flows in

TABLE 32.2 Irrigation Water Use per Country in South Asia

	Annual Renewable Water Resource (km^3 year $^{-1}$)[a]	Irrigation Water Requirements (km^3)	Water Requirement Ratio in Percentages	Water Withdrawal for Agriculture (km^3)	Water Withdrawal as Percentage of Renewable Water Resources	Agricultural Use (m^3 p^{-1} $year^{-1}$)[b]		Per Capita Withdrawal (%)[b]
Afghanistan	65.0[c]	8.78	38%	22.84	35%	765	779	98
Bangladesh	1,210.6[d]	19.09	25%	76.35	6%	538	560	96
Bhutan	95.0[e]	-	-	-	-	187	199	94
India	1,907.8[d]	303.24	54%	558.39	29%	506	585	86
Nepal	210.2[d]	2.45	25%	9.82	5%	0	9	0
Maldives	0.03[d]	-	-	-	-	362		96
Pakistan	233.8	72.14	44%	162.65	73%	254	375	90
Sri Lanka	50.0[d]	2.92	24%	12.00	24%	579	1,072	95

Source: Data from Atapattu and Kodituwakku [2].

Note these data are typically comprised of both renewable surface water and groundwater supplies, including surface inflows from neighboring countries. The FAO refers to this as total natural renewable water resources. Flows to other countries are not subtracted from these numbers.

[a] FAO [17] (data refers to 2000).

[b] Gleick [23]. The World's Water 2006–2007: online at www.worldwater.org/data.html-2007 Pacific Institute. Population data. Population Division of the Department of Economic and Social Affairs of the United Nations Secretariat. 2005. UNDP [36].

[c] Estimates from FAO [15]. Water Resources of the Near East Region: A Review. Food and Agriculture Organization, United Nations, Rome, Italy.

[d] FAO [16].

[e] Estimates from Belyaev, Institute of Geography, Union of Soviet Socialist Republics (USSR, 1987).

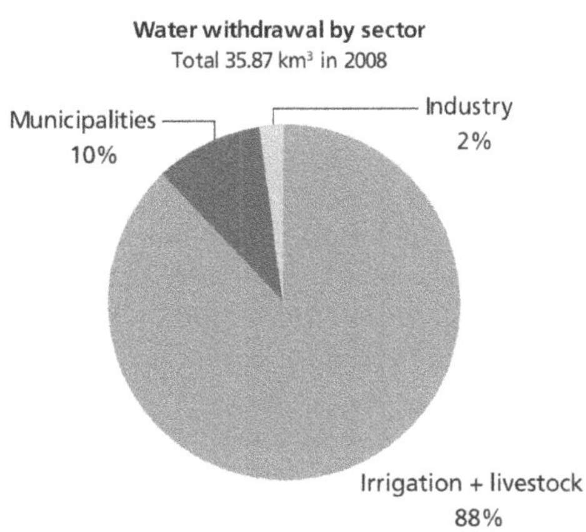

FIGURE 32.3 Water withdrawal by sector in Bangladesh, 2008 [20].

the Ganges, pollution caused by the disposal of effluents and chemicals, and arsenic contamination and salinity intrusion in the coastal area. Additional, the availability of fresh water is highly seasonal depending on the presence and duration of the monsoon. River morphology is deeply affected by the incidence of both flood and drought in a yearly cycle. Das Gupta et al. [10] also identified the main problems in water management arising from the flat topography where no potential storage of water is possible and because of the marked difference between the wet and dry seasons.

The Government of Bangladesh has taken several initiatives such as flood control, drainage, and irrigation (FCDI) projects at various scales, structural flood proofing, disaster management, flood and cyclone forecasting/warning, etc. to deal with the water problems mentioned above. Major FCDI projects in Bangladesh as compiled by the Water Resources Planning Organization (WARPO) are illustrated in Figure 32.4 [10].

Even though the main IWRM challenge is to find a right mix of management tools appropriate to a specific case, at present, the IWRM concept is accepted and practiced at the national level of the country. Yet, the process has not been taken into account in the planning and management of water resources [10]. Table 32.3 presents the insights of major policies for water management in Bangladesh.

People's participation is one of the main features highlighted in the IWRM concept. But, in Bangladesh, people's participation is weak and must be encouraged to take account of the needs and priorities of the local population. The seasonal pattern of rainfall, geographic position, and topography brings about enormous constraints on the economic development of Bangladesh. The priorities for water resources development and management in an integrated manner should be assessed according to the economic and social values of water with special consideration to the poor and disadvantageous groups of society. Three main key elements are, therefore, urgently needed for the IWRM process in Bangladesh: stakeholder participation with elected representatives; the pricing of water to put an economic value on it; and a knowledge base to support decision-making at different levels [10].

Das Gupta et al. [10] summarize the Bangladesh point of view on IWRM concept. To do so, Das Gupta et al. [10] mentioned about Albert [1] who carried out a personal discussion and interviews in order to evaluate the viewpoint of Bangladesh on IWRM. In this process, total 17 organizations associated with the water sector in Bangladesh were consulted to assess the status of the IWRM implementation and find their respected views on water management in Bangladesh.

The survey findings show that regional cooperation among shared countries for transboundary rivers management is not well developed. But, the Ganges Water Sharing Treaty of 1996 is considered as the milestone agreement on the way to hydrosolidarity.

As per the survey respondents, there are three major problems faced by the water sector in Bangladesh—these are floods, river erosion, and arsenic in groundwater.

Based on the survey results, the majority of people admit that though the principles and concepts of IWRM are well understood, they are not so far practical. The National Water Policy (NWPo) is considered as a first step toward the initiation of the IWRM process. Yet, the policy itself is not sufficient; it needs to be executed, supervised continuously, and updated with varying natural and socioeconomic condition.

It is specified by the survey participants that overall the water resources management division is fragmented and people's participation has only recently taken place in some of the programs. The primary data collection is quite insufficient for planning and decision-making. In addition to that, the interview results reveal that the WARPO and the Planning Commission act as coordinating bodies. Land acquisition is seen as complex, time consuming, and cumbersome in the water resources development process. Sectoral planning with inter-sectoral priorities is designated as "fair," and operation and management are defined as "weak."

Although the questionnaires were given to individuals with different educational and professional backgrounds from different institutions and with different involvement in the water sector, responses to IWRM-related queries were similar. People strongly believe and hope that the IWRM process could help find the solution to the water problems in Bangladesh.

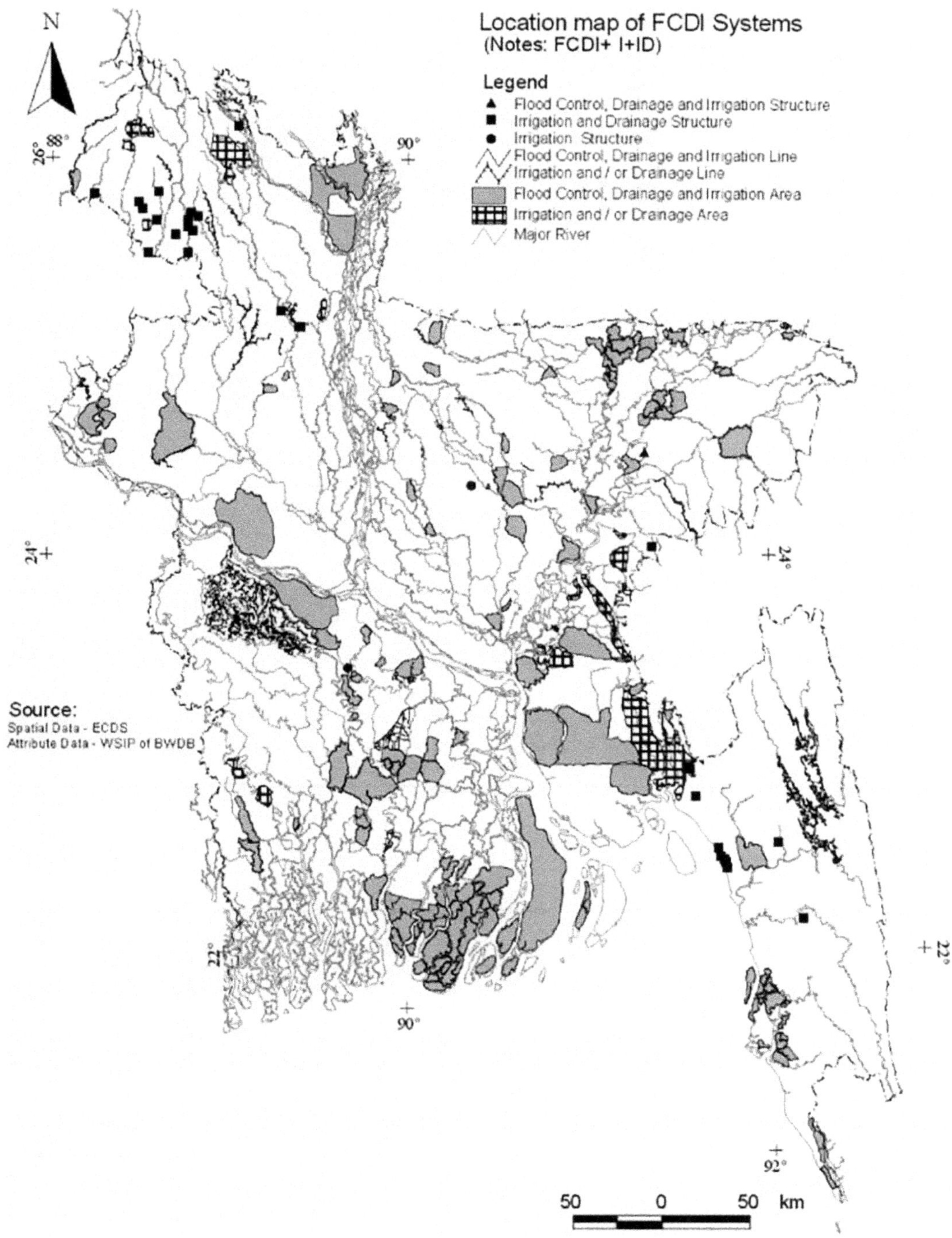

FIGURE 32.4 Major FCDI projects in Bangladesh [37].

TABLE 32.3 Insights of Major Policies for Water Management in Bangladesh

Key Policies	Year	Insights of Key Policies
IECO Master Plan	1964	Shifts from fragmented management to top-down centralized approach; high engineering approach for FCDI; major focus is on surface water
IBRD Report	1972	Besides engineering approach, introduction of small-scale water resources projects; the conjunctive use of both surface and groundwater
National Water Plan (NWP) phase I	1983–1986	Projected the future demand for water by different sectors
Flood Action Plan (FAP) reports	1989–1995	Introduced "stakeholder involvement" in water management
National Water Plan (NWP) phase II	1991	Introduced catchment scale planning and carried out comprehensive assessment for water resources
National Water Policy	1999	Accelerating the development of sustainable public and private water delivery systems, with appropriate legal and financial measures and incentives including formulation of water rights and water pricing
National Water Management Plan	2004	Multiuse approach of water resources; provides guidelines for implementation of water management programs
Coastal Zone Policy	2005	Developed based on Integrated Coastal Zone Management (ICZM) approach for management of coastal regions of the country
National Adaptation Programme of Action (NAPA)	2005	Developed adaptation strategies following sustainable development goals and objectives through participation of stakeholders
Bangladesh Climate Change Strategy and Action Plan (BCCSAP)	2009	Integrate climate change issues to support economic growth and poverty reduction
Vision 2021	2010	Development scenario for achieving a higher standard of living through better education, social justice, protection of environment
Sixth Five-Year Plan	2011	A key focus on strategies, policies, and institutions to help guide the private sector in helping Bangladesh achieve the goals set in Vision 2021
Bangladesh Water Act	2013	Enacted to integrate the management, development, utilization, and protection of water resources

Source: Data from Gain et al. [22].

Case Study 2: Implementation of IWRM—Nepal

Nepal (56,827 square miles) is a landlocked country in South Asia with a population of 29.3 million [38]. It is located largely in the Himalayas but also comprises parts of the Indo-Gangetic Plain. Water is the principal renewable resource which not only provides sustenance to millions of people living in the high mountain regions of Nepal and other countries of the Hindu Kush Himalayas (HKH) but also transforms the economics of these countries with its enormous potential for harnessing it for clean energy [8].

Nepal has more than 6,000 rivers, which provide a dense network with steep topographic conditions. All rivers in Nepal drain into the Ganges River.

In Nepal, traditional water resources management is focused on the supply side; in contrast, technical solutions were considered only in order to meet the rising demand for water. Isolated projects for irrigation, hydropower, drinking water supply and sanitation, flood control, and other uses were settled. Evaluation was primarily based on economic criteria, while the environmental and social impacts were barely considered. Authorities from independent sectors mostly controlled these projects on the basis of command and control. Brief of Projects on Irrigation, Drainage and Flood Control in Nepal is listed in Table 32.4. The results, so far, have not been reasonable, resulting in the generation of inter-sectoral,

TABLE 32.4 Brief of Projects on Irrigation, Drainage, and Flood Control in Nepal

Name of Project	Client	Funding Agency	Period	Approx. Value of Services
Program Management and Implementation Consultant (PMISC), Small Irrigation Program (SIP) Phase-I	Department of Local Infrastructure & Agriculture Roads/SDC	SDC/Government of Switzerland	March, 2015 to March, 2019	CHR 936,440
Detailed Survey, Design and Estimate of Dodhara-Chandani Area of Mahakali Irrigation Stage III Area	Department of Irrigation	GoN	April, 2012 to January, 2014	NRs 3,778,000
Detailed Engineering Planning and Design of Irrigation System of Siddaniya Branch Canal of Sikta Irrigation Project	Department of Irrigation	GoN	February, 2012 to July, 2013	NRs 7,018,430
Preparation of Subproject Preparation Report including IEE, Agricultural Development Plan, and Livelihood Plan of 42 Irrigation Subprojects of Batch II; Community Managed Irrigated Agricultural Sector Project	Department of Irrigation	ADB/OPEC	July, 2009 to December, 2010	NRs 19,000,000
Canal Alignment Survey and Detail Engineering Design of Surkhet Valley Irrigation Development Project (First Phase)	Department of Irrigation	GoN	May, 2007 to January, 2008	NRs 758,569
Detailed Study for the Preparation of River Training Master Plan of West Rapti River, Banke	Department of Water Induced Disaster Prevention	GoN	June, 2007 to May, 2009	NRs. 2,034,000
Irrigation Development Project, Mid-Western Development Region	European Commission	EU	January, 1998 to July, 2002	NRs. 58,280,000
Praganna Irrigation Project [KFAED Loan No. 552]	Department of Irrigation	KFAED	April, 2000 to December, 2000	NRs. 10,805,765
Community Groundwater Irrigation Sector Project	CIDA	CIDA	August, 1998 to April, 2006	NRs. 10,000,000
Capacity Building Training for Water Users' Association in East Rapti Irrigation Project [TA No 1969-NEP]	Asian Development Bank	ADB	August, 1998 to November, 1998	NRs. 1,600,000
Study on Socioeconomic and Irrigated Agricultural Aspects in the West Gandak Irrigation System	Department of Irrigation	GoN	February, 1996 to June, 1996	NRs. 181,000,000
Master Plan Preparation of Bagmati–Bishnumati River Improvement Works	Department of Irrigation	GoN	April, 1996 to July, 1996	NRs. 823,100,000

(Continued)

TABLE 32.4 (*Continued*) Brief of Projects on Irrigation, Drainage, and Flood Control in Nepal

Name of Project	Client	Funding Agency	Period	Approx. Value of Services
Rajapur Irrigation Rehabilitation Project	Department of Irrigation	ADB	March, 1994 to December, 2000	NRs. 11,500,000
Feasibility Study on the Rehabilitation for Government-Developed Irrigation Scheme in the Kathmandu Valley	JICA	JICA	April, 1994 to May, 1994	NRs. 176,700
Preparation Mission on Intensification of Small Farming and Ethnic Group Development of the Mid and Far Western Terai	IFAD/FAO	FAO	May, 1994 to June, 1994	NRs. 220,000
Technical Assistance to Irrigation Line of Credit Pilot Project	Department of Irrigation	WB	June, 1994 to July, 1997	NRs. 12,120,000
Mahakali Irrigation Project—Stage II (Construction Phase)	Department of Irrigation	WB	December, 1994 to December, 1996	NRs. 650,000
Feasibility Study on Rajkudwa Irrigation Project	JICA	JICA	January, 1993 to February, 1993	NRs. 200,000
Feasibility Study, Detail Design and Project Preparation of Brahamadev Irrigation Project	Department of Irrigation:	GoN	April, 1993 to May, 1994	NRs. 3,600,000
Master Plan Study on the Rehabilitation for Government-Developed Irrigation Scheme in the Kathmandu Valley	JICA	JICA	April, 1993 to June, 1993	NRs. 610,000
Groundwater Study for Small Farm and Ethnic Group Development in Mid and Far West Terai	IFAD	FAO	December, 1993 to January, 1994	NRs. 117,150
Strengthening PPME in Narayani Zone Irrigation Project (NZIDP)	Department of Irrigation	SDC	February, 1992 to June, 1993	NRs. 2,100,000
Reappraisal of Khutiya River Irrigation Project, Phase II	UNCDF/UNDP	UNCDF	February, 1992 to March, 1992	NRs. 74,550
Irrigation Management Transfer Project (ADB TA 1435-NEP)	ADB	ADB	November, 1992 to June, 1993	NRs. 2,250,000
Irrigation Management Project	Department of Irrigation	USAID	January, 1991 to June, 1995	NRs. 20,850,000

(*Continued*)

TABLE 32.4 (*Continued*) Brief of Projects on Irrigation, Drainage, and Flood Control in Nepal

Name of Project	Client	Funding Agency	Period	Approx. Value of Services
Case Studies for a Manual on Appropriate Design of Hill Irrigation Schemes	ILO	ILO	September, 1991 to December, 1991	NRs. 225,000
Budget and Accounting Review of the Department of Irrigation	Department of Irrigation	WB	October, 1990 to December, 1990	NRs. 175,000
Feasibility Study of Naya Belhani Irrigation Project	Department of Irrigation	GoN	January, 1989 to July, 1989, 1989	NRs. 400,000
Feasibility Study of Dhoba Khola Irrigation Project	Department of Irrigation	GoN	February, 1989 to July, 1989	NRs. 250,000
Medium-Scale Irrigation: Arjun Khola	European Economic Community	UNCDF/UNDP	February, 1989 to July, 1989	NRs. 360,000
Feasibility Study of Fawa Khola Irrigation Project	Department of Irrigation	GoN	March, 1989 to July, 1989	NRs. 250,000
East Rapti Irrigation Project	Nippon Koei Co. Ltd.	ADB	March, 1989 to June, 1989	NRs. 1,150,000
Mahakali Irrigation Project—Stage II (Construction Phase)	Department of Irrigation	WB	September, 1989 to December, 1993	NRs. 650,000
Irrigation Sector Project–Subproject Assessment Study	Department of Irrigation	ADB	December, 1989 to March, 1990	NRs. 521,000
Detail Engineering of Khutiya River Irrigation Project, Phase II (NEP/89/001)	UNCDF/UNDP	UNCDF/UNDP	December, 1989 to February, 1990	NRs. 248,650
Mahakali Irrigation Project Stage II (Design Phase)	Sir William Halcrow & Partners	WB	May, 1987 to August, 1987	NRs. 225,000
Rehabilitation Study of Rampurtar Irrigation Project	Department of Irrigation	GoN	June, 1987 to July, 1987	NRs. 195,000

Source: Data from https://geoceconsultants.com/briefs-of-projects-on-irrigation-drainage-and-flood-control/.

ADB, Asian Development Bank; CHR, Swiss Frank; FAO, Food and Agriculture Organization; GoN, Government of Nepal; IFAD, International Fund for Agricultural Development; ILO, International Labour Organization; JICA, Japan International Cooperation Agency; NRs, "Nepalese Rupee" [1 NRs = 0.0088 United States Dollar]; SDC, The Swiss Agency for Development and Cooperation; UNCDF, United Nations Capital Development Fund; UNDP, United Nations Development Program; USAID, United States Agency for International Development; WB, World Bank.

inter-regional, and riparian conflicts. The invention of the Water Resources Strategy (WRS) is built on certain recognized policy principles including IWRM. Two of the policy principles relevant to River Basin Management (RBM) are: (i) Development and management of water resources shall be commenced in all-inclusive and systematic manner, depending on IWRM; (ii) water utilization shall be sustainable to safeguard conservation of resources and protection of the environment. Table 32.5 presents the insights of major policies for water management in Nepal.

TABLE 32.5 Insights of Major Policies for Water Management in Nepal

Key Policies	Year	Insights of Key Policies
Essential Commodity Protection Act	1955	• Deems drinking water an essential commodity and strictly protects drinking water. • Prohibits any unauthorized use or misuse, stealing, damaging, etc. of drinking water.
Muluki Ain	1963	• Sets out the order of priority of use of water for irrigation. • Regulates traditional farmer-managed irrigation systems.
Soil and Watershed Conservation Act	1982	• Defines "Soil and Watershed Conservation" as acts to prevent or save any area from being destroyed from natural calamities such as flood, landslide, and soil erosion and keep the volume and flow of water in a normal condition or keep on maintaining cleanliness by preventing the flow of water from being muddy. • The Act permits to carry out by a conservation officer to maintain the soil fertility and the cleanliness of water and environment in a balanced manner and carry out such other soil and watershed conservation related acts as prescribed by Government of Nepal. • Classification of lands within conserved watershed area • Farming to be done by following land use system • Prohibit acts of throwing away solid wastes or similar other detritus contaminating the environment or build a site for collecting or keeping such solid wastes or detritus. • Powers to shift industry, business, and settlement of habitation within a conserved watershed area or to acquire the land where such industry, business, and settlement of action are situated.
Solid Waste (Management and Resource Mobilization) Center Act	1987	• Establishes the Solid Waste Management and Resource Mobilization Center as the responsible authority for the management of solid waste. • Deals with the pollution of water by solid waste.
Solid Waste (Management and Resource Mobilization) Regulation	1989	• Deals with the collection, transportation, and disposal of solid waste. • Deals with the provision of public toilets and bath houses.
Nepal Water Supply Corporation Act	1989	• Establishes the Nepal Water Supply Corporation as the perpetual, autonomous government-controlled corporation responsible for the supply of drinking water. • Prohibits certain acts and provides penalties/punishment for violation.
The Constitution of the Kingdom of Nepal	1990	• Guarantees the right to life and property. • Provides for the acquisition of property under certain circumstances and for compensation.
Water Resource Act	1992	• The Umbrella Act governing water resource management. • Declares the order of priority of water use. • Vests ownership of water in the state. • Provides for the formation of Water User Associations and establishes a system of licensing. • Prohibits water pollution.

(*Continued*)

TABLE 32.5 (*Continued*) Insights of Major Policies for Water Management in Nepal

Key Policies	Year	Insights of Key Policies
Electricity Act	1992	• Governs the use of water for hydropower production. • Establishes a system of licensing.
Industrial Enterprises Act	1992	• Requires permission for the extension and diversification of environmentally sensitive industries. • Provides financial incentives for industrial enterprises that minimize harmful effects on the environment.
Forest Act	1993	• The trees cannot be felled in such a way that it may cause significant adverse effect in the environment or damage or loss publicly or soil erosion in the watershed area. • Government of Nepal may impose restriction on collection, cutting, consumption, transportation sale, distribution, or foreign export of the prescribed forest products for the protection of biodiversity and environment upon publishing a notice in the Nepal Gazette.
Water Resource Regulation	1993	• The Umbrella Regulation governing water resource management. • Sets out the procedure to register a Water User Association and to obtain a license. • Sets out the rights and obligations of Water User Associations and license holders. • Establishes the District Water Resource Committee.
Environment Protection Act	1996	• Defines "Pollution" as the activities that significantly degrade, damage the environment, or harm on the beneficial or useful purpose of the environment, by changing the environment, directly or indirectly. • To carry out Initial Environmental Examination or Environmental Impact Assessment. • Prohibition on implementation of proposal without approval. • To submit proposal for approval. • Prevention and control of pollution (nobody shall create pollution in such a manner as to cause significant adverse impacts on the environment or likely to be hazardous to public life and people's health). • Designate Environmental Inspectors to inspect as to whether or not the acts of mitigation, avoidance, or control of pollution have been carried out in accordance with this act or the rules framed under this act. • Protection of national heritage. • Declare Environment Protection Area. • Establishment of a laboratory to help in the activities related to environment protection and pollution control. • To allow collection of samples to study, examine, or analyze the pollution, sound, heat, and wastes to be or likely to be generated from any industry, factory, machine, vehicle, etc. • Establishment and operation of environment protection fund. • Power to constitute environment protection council. • Punishments (e.g., cash penalty up to one hundred thousand Nepalese Rupees).
Environment Protection Regulation	1997	• Lists the water-related projects required to conduct an Environmental Impact Assessment (EIA) or Initial Environmental Examination (IEE). • Deals with the control of water pollution and pollution control certificate.
Drinking Water Regulation	1998	• Regulates the use of drinking water. • Provides for the formation of Drinking Water User Associations and sets out the procedure for registration. • Deals with licensing of use drinking water. • Deals with the control of water pollution and maintenance of quality standards for drinking water. • Sets out the conditions of service utilization by consumers.

(*Continued*)

TABLE 32.5 (*Continued*) Insights of Major Policies for Water Management in Nepal

Key Policies	Year	Insights of Key Policies
Local Self Governance Act	1999	• Establishes a decentralized governance structure. • Sets out the powers, functions, and duties of the Village Development Committee (VDC), municipality, and District Development Committee (DDC) in relation to water and sanitation.
Local Self Governance Regulation	1999	• Sets out the powers, functions and duties of VDC, municipality, and DDC in relation to water and sanitation. • Establishes the procedure for the formulation of water-related plan and project implementation.
Irrigation Regulation	2000	• Deals with Irrigation Water User Associations and the transfer of projects to Irrigation Water User Associations. • The environmental effect evaluation or preliminary environment examination shall be conducted as necessary under the prevailing law through public hearing in connection with the negative impact on the environment from the irrigation project. • In the course of using water for irrigation from any river/rivulet, only the remaining water shall be utilized for irrigation purpose upon leaving minimum water in the same river/rivulet after determining so that it will not have negative effects on the local biodiversity. • Monitoring, study, and research work in the water quality shall be conducted and water in irrigation shall be conserved effectively without negative impact in the environment. • Available groundwater resources shall be developed and utilized as like the surface water reservoirs, and arrangements shall be made for conservation, promotion, and control in quality.
Water Resources Strategy Nepal	2002	• Goal: Living conditions of Nepali people are significantly improved in a sustainable manner. • Drinking water is the basic minimum need of all human beings. • Entact and enforce standards and regulatory mechanisms for water quality and effluent discharge. • Appropriate sanitation services in rural and urban areas will be provided through community awareness programs • Ensure compliance with enviromental regulations • Implement and enforce equitable mechanisms for cost sharing. • Increase the access of the population to drinking water through rainwater harvesting programs, community-based water supply and sanitation sector projects, rural water supply projects/water resource management programs, a small town water supply and sanitation program, a water quality improvement program, etc.
National Wetland Policy	2003	• Defines wetlands as perennial water bodies that originate from underground sources of water or rains. It means swampy areas with flowing or stagnant fresh or salt water that are natural or man made, or permanent or temporary. Wetlands also mean marshy lands, riverine floodplains, lakes, ponds, water storage areas, and agricultural lands. • Goals are (i) to conserve and manage wetlands resources wisely and in a sustainable way with local people's participation and (ii) to put the conservation and management aspects of wetlands conservation within the framework of broader environmental management. • Regulate the extraction of underground water and prevent water pollution. • Make provisions for environmental impact assessment according to prevalent laws prior to approving development programs and activities around wetlands. • Ensure environmental protection by regularly monitoring development programs and activities around wetlands

(*Continued*)

TABLE 32.5 (*Continued*) Insights of Major Policies for Water Management in Nepal

Key Policies	Year	Insights of Key Policies
Rural Water Supply and Sanitation National Policy	2004	• Recognizes that all people have a right to access to basic water supply and sanitation services and that these services are necessary for socioeconomic development and to combat waterborne diseases.
National Water Plan—Nepal	2005	• The National Water Plan (NWP) has been prepared to operationalize the WRS of Nepal, approved by the HMGN in January 2002. • The major doctrines of the NWP are integration, coordination, decentralization, popular participation, and implementation of water-related programs within the framework of good governance, equitable distribution, and sustainable development. • The NWP includes subsector-wise action programs in water-induced disasters, environmental action plan on management of watershed and aquatic ecosystem, water supply, sanitation and hygiene, irrigation for agriculture, hydropower development, industries, tourism, fisheries, and navigational uses, water-related information systems (Decision Support System for River Basin Planning and Management), legal frameworks, and institutional mechanisms. • Environment Management Plan, a strategic document for the implementation of environmental protection measures (including downstream water pollution and groundwater quality, erosion/landslide and sedimentation, water pollution and sanitation, effect on aquatic life and wetland ecosystem), monitoring (baseline, impacts, and compliance), environmental auditing and institutional and procedural arrangements.
Water Supply Management Board Act	2006	• Defines "Water supply service" as the provision of water to the users upon processing and purifying the natural water. • Defines "Sanitation service" as an act of destroying, discharging, processing, or purifying dirty water discharged from human beings or from domestic, trade, or industrial use and all kinds of dirty substances mixed with such water. • The Board prevents the misuse of potable water and prevents pollution of potable water. • The Board carry out, or cause to be carried out, study, research, and survey on the source, distribution of potable water, and sanitation. • The Board may punish a person causing adverse effect to the public health by contaminating the potable water, a fine of up to Rs 25,000 in view of the degree of the offense.
The Constitution of the Kingdom of Nepal	2015	• Rights regarding clean environment. • Right to healthcare. • Right to food. • Policies regarding the conservation, management, and use of natural resources. • Policies regarding the basic needs of citizens.

Source: Data from www.wepa-db.net/policies/law/nepal/list.htm.

The huge quantity of water is being withdrawn for the irrigation purposes (Figure 32.5). The 20-year Agricultural Perspective Plan (APP) adopted in 1995 focuses on ways to reduce food deficits by increasing food production. The APP gave top priority to groundwater development policy mainly by installing shallow tube wells in the Terai. However, the farmers' requests for shallow tubewells unexpectedly declined because of the removal of the subsidy and the weakening trend of government investment in repair and maintenance of irrigation infrastructure [20].

Suhardiman et al. [35] looked at how the IWRM concept has been applied in Nepal by highlighting the normative approach in IWRM policy formulation (Table 32.6). The authors identified the existing institutional barriers behind the execution of IWRM and how these resulted in the framing of IWRM "implementation" as just a compilation of donor-funded projects. Recent discourse on IWRM highlights the need to shift the emphasis from national policy formulation to local adaptive, pragmatic approaches to IWRM.

A list of most relevant peer-reviewed literature related to IWRM implementation in Bangladesh and Nepal is presented in Table 32.7.

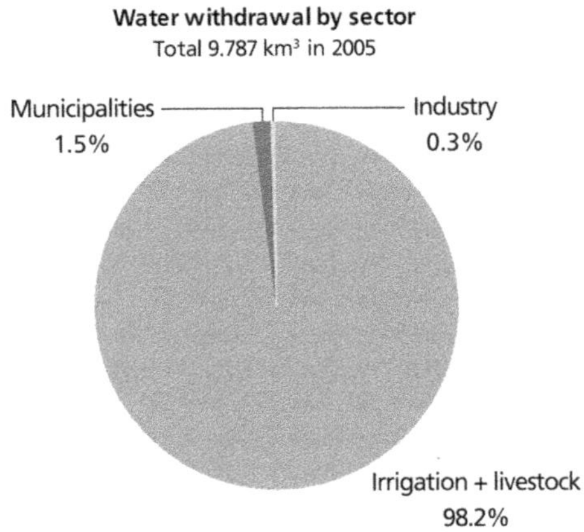

FIGURE 32.5 Water withdrawal by sector in Nepal, 2005 [20].

TABLE 32.6 Views on IWRM of Respondents from Various Government Agencies in Nepal

Organization	Is IWRM Needed in Nepal?	What Aspects of IWRM Are Needed or Not Needed?	Why Is IWRM Needed or Not Needed?
Water and Energy Commission Secretariat	Yes, in theory	Coordination across all users	Maximizing the use of water without compromising the environment.
Ministry of Energy	Yes, generally, but not for hydropower development, except in spring-fed rivers	Coordination across uses (hydropower, domestic use, and irrigation)	Lack of conflicts in water management and especially water allocation among water uses—rooted in the geographical separation of large hydropower development from settlements.
Department of Electricity Development	Yes	Inter-sectoral coordination	To assess the impacts of other projects on hydropower; waste of money.
Department of Irrigation	Not currently	Coordination across uses (hydropower and irrigation)	Ministries can do their work separately without causing any problem or conflict. There are no conflicting demands for water for agriculture or energy during the dry season. Distinct geographical areas of development for irrigation and hydropower.
		Better water management	Efficiency is required only when resources are limited and scarce. Nepal needs infrastructure not better management.
	Yes, in the future	Coordination across uses (irrigation, hydropower and domestic water)	To be cost-effective; potential impacts of hydropower development and operation on irrigation supply.
		Better water management	Increases in population and emerging problems such as groundwater exploitation and pollution.
Department of Water Induced Disaster Prevention	Yes	Coordination across all uses	We need IWRM for future generations.
Department of Local Infrastructure Development and Agricultural Roads	Yes	Coordination across sectors in the water, sanitation, and hygiene sector	To minimize negative impacts on the environment and avoid replication.

Source: Data from Suhardiman et al. [35].

TABLE 32.7 A List of Most Relevant Peer-Reviewed Literature Related to IWRM Implementation in Bangladesh and Nepal

Author(s)	Objectives	Methods	Major Findings
Rouillard et al. [34]	Evaluation of how IWRM approaches help enhance adaptive capacity to climate change impacts on water resources in Bangladesh.	Derivation of an evaluative framework from key IWRM principles and their roles in modulating adaptive capacity.	Policy principles and implementation in favor of IWRM can be a source of success but also of failure for adaptive capacity.
Evans and Varma [14]	Review the initial findings of a participatory action planning process for managing wastewater for agricultural use in Bangladesh.	Study of various existing relevant literature.	Stakeholders were willing to cooperate but that considerable external support was required.
Nowreen et al. [28]	To identify and mitigate the problems and issues related to the current approach of Small-Scale Water Resources Development Sector Projects (SSWRDSP) of Local Government Engineering Department (LGED) in Bangladesh.	The contrasting scenarios of the two neighboring projects have been deeply and critically studied by collecting both primary and secondary data.	Project objectives (to enhance agriculture and alleviate poverty) have not been fully achieved mostly due to dysfunctional Water Management Cooperative Association (WMCA).
Gain et al. [22]	Evaluation of the current status of IWRM in Bangladesh.	Analysis of policy shifts, institutional transitions, and project transformations using seven key dimensions of IWRM.	As a whole, the IWRM dimensions are reflected in recent policies, institutional reforms, and project formulation in Bangladesh. Still, to solve the complex water problems, basin scale management through transboundary cooperation and equity and social issues needs to be implemented at institutional and project levels.
Emmanuel and Huq [13]	Establishment of a relation among three basic components, i.e., IWRM, poverty, and human development.	Analysis based on secondary data available in published and unpublished materials.	The roots of the crisis in water can be traced to poverty, inequality, and unequal power relationships, along with flawed water management policies that exacerbate scarcity.
Gain et al. [21]	Assessment of the (water–energy–food) WEF nexus approaches to investigate the extent of "integration" achieved in the national policy of Bangladesh.	Review of the emerging literature on the WEF nexus and then analyze the nexus in the context of Bangladesh.	The WEF nexus is not yet recognized in the policy documents of Bangladesh, while conflicts over these resources are growing.
Biggs et al. [6]	Exploring why Nepal, a nation with adequate water availability, is struggling to ensure water security.	Study of various existing relevant literature.	In spite of adequate water supplies, a nation may still experience severe water insecurity, particularly if effective governance and equitable access are not prioritized.
Duncan et al. [12]	A comprehensive spatiotemporal analysis of precipitation time series data for Nepal and discuss the contribution of precipitation change to water resources management for this land-locked Himalayan nation.	Trend analysis and field significance testing for a range of precipitation variables across Nepal.	Water resources management needs to focus on population and environmental pressures, rather than specifically mitigating for precipitation change.

(Continued)

TABLE 32.7 (*Continued*) A List of Most Relevant Peer-Reviewed Literature Related to IWRM Implementation in Bangladesh and Nepal

Author(s)	Objectives	Methods	Major Findings
Rees et al. [33]	To assist water resource practitioners in the Himalayan region to meet the competition for water for irrigation, domestic supply, industrial uses, and hydropower.	By developing a software system	Standard display and output formats provide a consistent set of reporting tools.
Bhattarai et al. [5]	Analysis of the constraints and opportunities of managing the water resources in Nepal, on the principle of IWRM.	Case studies conducted in the Indrawati river basin by IWMI/Nepal and Water and Energy Commission Secretariat/Nepal (WECS/Nepal).	There is a lack of coordination among various local institutional stakeholders, for managing the water resources in the basin.
Rahaman and Varis [30]	Analyzes the current status of Brahmaputra water resources and identifies the perspectives of riparian countries (China, India, Bhutan, and Bangladesh) regarding the development of the Brahmaputra basin. Also identifies the opportunities for cooperation and regional development through integrated water development and management of the basin.	Analysis based on both primary data and secondary data.	It is essential to develop an IWRM approach involving all riparians to foster regional development and overcome the prospect of severe water conflict along the Brahmaputra basin.
Rahaman [29]	To identify the constraints and prospects of the IWRM concept for the development of the Ganges basin (shared by Nepal, India, and Bangladesh).	Analysis based on both primary data and secondary data.	Water is strongly interlinked with the overall development framework of the basin. Achieving IWRM in the basin requires cooperation between the riparian countries, and an integrated water management approach would offer prosperity for the whole basin.
Bandaragoda [4]	A five-country (China, Indonesia, Nepal, Philippines, and Sri Lanka) river basin study in Asia to learn about contextual processes, as well as for stakeholder consultation to develop action plans.	Used a participatory method for diagnostic investigations.	The method was found to be exceptionally effective and had many advantages over the conventional methods of field research and action planning where the stakeholders are treated as objects of research and passive recipients of development messages.
Merrey et al. [27]	Examines the weaknesses in the current understanding of IWRM from the perspective of livelihoods.	Study of various existing relevant literature.	IWRM needs to be placed in the broader context of both modern Integrated Natural Resource Management (INRM) and the livelihoods approach, which together take a holistic and people-centered approach.
Babel et al. [3]	To assess the vulnerability of water resources in the Bagmati River Basin in Nepal.	Adopts an indicator-based approach wherein vulnerability is expressed as a function of water stress and adaptive capacity.	While the northern parts need attention to reduce pollution loading and conserve vegetation cover, the southern parts need improvements in physical capacity, i.e., water infrastructures.

Conclusions

South Asia's abundant human and physical geography are tightly bound to the rivers that radiate out and down from the high Himalaya and the extensive Ganga, Indus, Brahmaputra, and Meghna basin aquifers. These large transboundary basins have primarily contributed to the development of human civilizations and provided livelihood security for several millennia. Water is a critical element of many South Asian culture and economy, but its many large headwater rivers frequently flood and create problems for local communities and downstream population. Simultaneously, a shortage of fresh water occurs in the dry and winter seasons. Therefore, valuing water—environmentally, economically, socially, and culturally—is essential. These water resources are now rapidly changing, and this change brings heightened risk and uncertainty. IWRM is, therefore, an absolute must for managing these large transboundary water resources.

References

1. Albert, X. Integrated water resources management in a water abundance-scarcity cycle regime: A case study of Bangladesh. *Doctoral dissertation*, Thesis No. WM-00-11, Asian Institute of Technology, Thailand. 2001.
2. Atapattu, S.S. and Kodituwakku, D.C. Agriculture in South Asia and its implications on downstream health and sustainability: A review. *Agricultural Water Management* 2009, 96(3), 361–373.
3. Babel, M.S., Pandey, V.P., Rivas, A.A. and Wahid, S.M. Indicator-based approach for assessing the vulnerability of freshwater resources in the Bagmati River basin, Nepal. *Environmental Management* 2011, 48(5), 1044.
4. Bandaragoda, D.J. Stakeholder participation in developing institutions for integrated water resources management: Lessons from Asia. IWMI Working Paper No. 618-2016-41234. 2005.
5. Bhattarai, M., Pant, D., Mishra, V.S., Devkota, H., Pun, S., Kayastha, R.N. and Molden, D. Integrated development and management of water resources for productive and equitable use in the Indrawati River Basin, Nepal. IWMI Working Papers No. H030393, Vol. 41. IWMI 2002.
6. Biggs, E.M., Duncan, J.M., Atkinson, P.M. and Dash, J. Plenty of water, not enough strategy: How inadequate accessibility, poor governance and a volatile government can tip the balance against ensuring water security: The case of Nepal. *Environmental Science & Policy* 2013, 33, 388–394.
7. Biswas, A.K. Cooperation or conflict in transboundary water management: Case study of South Asia. *Hydrological Sciences Journal* 2011, 56(4), 662–670.
8. Chalise, S.R., Kansakar, S.R., Rees, G., Croker, K. and Zaidman, M. Management of water resources and low flow estimation for the Himalayan basins of Nepal. *Journal of Hydrology* 2003, 282 (1–4), 25–35.
9. Chowdhury, N.T. Water management in Bangladesh: An analytical review. *Water Policy* 2010, 12(1), 32–51.
10. Das Gupta, A., Singh Babel, M., Albert, X. and Mark, O. Water sector of Bangladesh in the context of integrated water resources management: A review. *International Journal of Water Resources Development* 2005, 21(2), 385–398.
11. de Fraiture, C. and Wichelns, D. Looking ahead to 2050: scenarios of alternative investment approaches. In: Molden, D. (Ed.), Water for Food, Water for Life: A Comprehensive Assessment of Water Management in Agriculture. 2007. Earthscan and IWMI, London and Colombo, 279–310.
12. Duncan, J.M., Biggs, E.M., Dash, J. and Atkinson, P.M. Spatio-temporal trends in precipitation and their implications for water resources management in climate-sensitive Nepal. *Applied Geography* 2013, 43, 138–146.

13. Emmanuel, M.N. and Huq, N. Integrated water resources management and poverty eradication-policy analysis of Bangladesh and Cameroon. *Journal of Water Resource and Protection* 2010, 2(3), 191–198.
14. Evans, A. and Varma, S. Practicalities of participation in urban IWRM: Perspectives of wastewater management in two cities in Sri Lanka and Bangladesh. *Natural Resources Forum* 2009, 33(1), 19–28.
15. FAO. Water Resources of the Near East Region: A Review. Food and Agriculture Organization of the United Nations, Rome, Italy, 1997.
16. FAO. Irrigation in Asia in Figures. Food and Agriculture Organization, United Nations, Rome, Italy, Water Report, 1999.
17. FAO. Statistical data base accessed in July 2006. (AQUASTAT). Online http://www.fao.org/nr/water/aquastat/data/query/index.html. 2006.
18. FAO. AQUASTAT Transboundary River Basins – Ganges-BrahmaputraMeghna River Basin. Food and Agriculture Organization of the United Nations (FAO), Rome, Italy, 2011.
19. FAO. Irrigation in Southern and Eastern Asia in Figures AQUASTAT Survey – 2011. FAO Water Reports 37. Food and Agriculture Organization of the United Nations, Rome, 2012.
20. Frenken, K. Irrigation in Southern and Eastern Asia in figures: AQUASTAT Survey-2011. *Water Reports* 2012, (37).
21. Gain, A.K., Giupponi, C. and Benson, D. The water–energy–food (WEF) security nexus: the policy perspective of Bangladesh. *Water International* 2015, 40(5–6), 895–910.
22. Gain, A., Mondal, M. and Rahman, R. From flood control to water management: A journey of Bangladesh towards integrated water resources management. *Water* 2017, 9(1), 55.
23. Gleick, P.H. The World's Water 2006-2007. The Biennial Report on Freshwater Resources. Island Press, ISBN: 9781597261067, Online http://www.worldwater.org/data.html. 2006.
24. GWP-TAC. Integrated Water Resources Management. Background Paper No. 4 (Stockholm: Global Water Partnership/Technical Advisory Committee) 2000.
25. Hossain, N. and Bahauddin, K.M. Integrated water resource management for mega city: A case study of Dhaka city, Bangladesh. *Journal of Water and Land Development* 2013, (19), 39–45.
26. Lal, R. Soil degradation and environment quality in south Asia. *International Journal of Ecology and Environmental Sciences* 2007, 33(2–3), 91–103.
27. Merrey, D.J., Drechsel, P., de Vries, F.P. and Sally, H. Integrating "livelihoods" into integrated water resources management: Taking the integration paradigm to its logical next step for developing countries. *Regional Environmental Change* 2005, 5(4), 197–204.
28. Nowreen, S.M., Khan, S.A. and Huq, H. Development of an operational IWRM framework in a selected small scale water resources development sector project in Bangladesh. *South Asian Water Studies* 2011, 2(2), 6–26.
29. Rahaman, M.M. Integrated water resources management in the Ganges basin: Constraints and opportunities. Department of Civil and Environmental Engineering, Laboratory of Water Resources 2005.
30. Rahaman, M.M. and Varis, O. Integrated water management of the Brahmaputra basin: Perspectives and hope for regional development. *Natural Resources Forum* 2009, 33(1), 60–75.
31. Rasul, G. Food, water, and energy security in South Asia: A nexus perspective from the Hindu Kush Himalayan region. *Environmental Science & Policy* 2014, 39, 35–48.
32. Rasul, G. Water for growth and development in the Ganges, Brahmaputra, and Meghna basins: An economic perspective. *International Journal of River Basin Management* 2015, 13(3), 387–400.
33. Rees, H.G., Holmes, M.G., Fry, M.J., Young, A.R., Pitson, D.G. and Kansakar, S.R. An integrated water resource management tool for the Himalayan region. *Environmental Modelling & Software* 2006, 21(7), 1001–1012.

34. Rouillard, J.J., Benson, D. and Gain, A.K. Evaluating IWRM implementation success: are water policies in Bangladesh enhancing adaptive capacity to climate change impacts? *International Journal of Water Resources Development* 2014, 30(3), 515–527.
35. Suhardiman, D., Clement, F. and Bharati, L. Integrated water resources management in Nepal: Key stakeholders' perceptions and lessons learned. *International Journal of Water Resources Development* 2015, 31(2), 284–300.
36. UNDP. Human Development Report 2004. United Nations Development Programme, New York, 2004.
37. WARPO. National Water Management Plan, Draft Development Strategy (DDS), Annex M: Technical Options. Water Resources Planning Organization, Dhaka, 2000.
38. The World Bank. South Asia Water Initiative, Annual Report, July 2016–June 2017. The World Bank Group, Washington, DC, 2017.

III

Water and Watershed Monitoring

33

Water Quality and Monitoring

Introduction 257
Monitoring Network Design 258
Water Quality Problem 258
Objectives 259
Statistical Designs 259
Scale of Study 260
Variable Selection 260
Sample Type 260
Sampling Location 260
Sampling Frequency and Duration 261
Sample Collection and Analysis 261
Sample Collection • Sample Preservation and Transport • Quality Assurance
Land Use and Management Monitoring 262
Conclusion 262
References 262
Bibliography 263

John C. Clausen
University of Connecticut

Introduction

Water quality monitoring is the collection of information on the physical, chemical, biological, sometimes radiological characteristics of water.[1] Water quality monitoring is conducted for several reasons. Monitoring is used to determine baseline values or examine trends in the data. Best-known examples of this approach includes individual state monitoring in response to Section 305(b) of the Clean Water Act (PL92–500), and monitoring conducted by the U.S. Geological Survey at its National Stream Quality Accounting Network stations.[2] Monitoring is used to determine the fate and transport of pollutants. For example, how far has a contaminant moved in a ground water aquifer? Monitoring has been used to define critical areas needing treatments, especially in agricultural watersheds. The results of such monitoring can then be used to target practices. Compliance with water quality plans and permit provisions is a common objective of monitoring. For example, wastewater treatment plants often monitor discharge as a requirement of their NPDES permit. Measuring effectiveness of best management practices, either on an individual basis or on a programmatic watershed basis is often a goal of monitoring. Wasteload allocations from point sources and load allocations from nonpoint sources are requirements of total maximum daily loads (TMDLs). Determining these loads and the maximum loading capacity of the receiving water body require monitoring. When the source of a problem is unknown, water quality monitoring is used to determine the cause of impairment. Monitoring is used to validate and calibrate models as well as to provide input data for models used to predict transport rates and concentrations regionally. One such model in common use is Spatially Referenced Regressions on Watershed attributes (SPARROW).[2] Researchers also conduct monitoring to address specific research questions and test hypotheses. Monitoring water quality is expensive and time consuming because water is constantly flowing and changing. Care is needed in interpreting the results of water quality monitoring because the samples collected may not represent the full range of values that may occur. Weather conditions have an important influence on water quality. The important components to consider in water quality monitoring are outlined in this entry.[3]

Monitoring Network Design

Much has been written regarding how to design and conduct water quality monitoring networks.[4] The Intergovernmental Task Force on Monitoring Water Quality[5] recommended a strategy for nationwide water quality monitoring that focused primarily on understanding the condition of the Nation's waters to support sound water quality decision making. The National Water Quality Monitoring Council was formed subsequent to the intergovernmental task force to continue guidance and coordination of water quality monitoring. Examples of monitoring networks include major programs by the U.S. EPA and the U.S. Geological Survey. EPA conducts National Aquatic Resource Surveys of lakes and streams that are probability-based on a multiyear revolving approach. Lakes are selected by ecoregion using stratified random sampling based on the density of lakes in the region. Streams are stratified by Strahler order. A proportion of sites are resampled during each rotation. The U.S. Geological Survey monitors nearly 400 National Stream Quality Accounting Network stations throughout the United States.[2] Several approaches have been used to design monitoring networks other than the fixed station and randomized methods. Optimization approaches have been developed to design networks based on economic and logistic considerations, hydrodynamics, and proximity to sources.[6] A critical sampling points methodology[7] has been developed to locate sampling sites in watersheds that is based on GIS, fuzzy logic, and a model that considers loading from the land and upstream areas. The remainder of this entry describes components to consider in water quality monitoring for all purposes and designs.

Water Quality Problem

When formulating a water quality problem statement, it is important to distinguish between a problem and a symptom. A water quality problem is a water quality issue requiring a solution, often stated in the form of a question. A symptom is a characteristic or condition of a water body indicating a problem. For example, algal blooms might be symptomatic of excessive nutrient loadings. For every water quality problem, there are typically several symptoms. Table 33.1 summarizes some typical symptoms and problems. A complete water quality problem statement should include information about the specific water body, the cause of the problem, and how the problem manifests itself, and the source of the causal agents. A good example of a definition of a water quality problem is: *The problem is eutrophication of St. Albans Bay due to excessive phosphorus loading from agricultural sources.*

TABLE 33.1 Water Quality Symptoms and Problems[3]

Symptom	Problem
Color	Algae, sediment, organic acids
Excess algae	Nutrients
Excess macrophytes	Nutrients, light
Hypoxia	Nutrients
Low biotic diversity	Toxics, nutrients
Taste	Salinity, algae, metals
Turbidity	Algae, sediment

Objectives

The formulation of monitoring objectives is a "decisionmaking" process, beginning with a problem definition and ending with the action to begin water quality monitoring. It is important that the objectives address the identified water quality problem. Formulating objectives can be viewed as a series of three steps that involve 1) identifying the objectives, 2) developing an objective hierarchy, and 3) specifying attributes to measure the level of achievement of these objectives.

An objective describes the answer to the following question: "What must be done?" By definition, an objective includes an object as part of the statement, an infinitive, and sometimes a constraint to limit all opportunities. For example, a complete objective, with a '+' to indicated parts of the statement, could be: *To determine + the effect of implementing conservation practices + on fecal coli-form levels in Long Lake.*

Statistical Designs

Major water quality study designs include the plot, single watershed, above and below, two watersheds, paired watershed, multiple watershed, and trend station (Table 33.2). Plots are generally small areas (fractions of an acre) that are replicated on the landscape. All plots are treated alike except for the factor(s) under study. A major advantage of the plot design is that it has a control. A control is a plot that is monitored like all others but does not receive the treatment. The primary statistical approach used is the analysis of variance of a randomized complete block design. A single watershed has sometimes been used to evaluate the water quality changes both before and after a treatment is applied. This design should not be used because the effect due to the treatment cannot be separated from other confounding effects such as year-to-year differences in climate. Two watersheds, one with a treatment and one without, have been incorrectly used to evaluate the effects of a practice on water quality. This design should always be avoided because differences might be inherent in the two watersheds. The above-and-below watershed approach is sometimes viewed as a single watershed with monitoring above and below a practice. In actuality, there are two watersheds being monitored, one nested within the other. If the above- and-below approach is applied both before and after the practice is installed, this approach can be analyzed as a paired watershed design as described later. The above-and- below design is analyzed as a *t*-test of the differences between paired observations at the above and below stations.

The paired watershed approach requires a minimum of two watersheds—control and treatment—and two periods of study—calibration and treatment.[8] The control watershed serves as a check over year-to-year climate variations, and receives no changes during the study. During the calibration period, the two watersheds are treated identically and paired water quality data are collected. During the treatment period, one randomly selected watershed is treated, while the control watershed remains in the original management. The primary statistical approach is analysis of covariance.[9]

The multiple watershed approach involves several watersheds[10,11] that are selected with the treatments already in place from across the geographic region of interest. The greatest advantage of the

TABLE 33.2 Statistical Designs for Water Quality Monitoring

Design	Characteristic	Advantage	Disadvantage	Statistics
Plot	small replicated	uses control	not transferable	ANOVA
Single	before and after	simple	temporal confounding	t-test
Above–below	nested watersheds	no climate effects	may not be independent	t-test
Two watersheds				
Paired watersheds	two treatment periods	variation controlled	no replication	ANCOVA
Multiple watersheds	> two watersheds	results transferable	large area	unpaired t-test
Trend stations	single watershed	large area	climate confounding	many

multiple watershed approach is that the results are transferable to the region included in the monitoring. The basic statistical approach is the comparison of the means of two populations using the *t*-test.

A trend is a persistent change in the water quality variable(s) of interest over time. Trend stations are single independent watersheds where a group of treatments might be implemented gradually over time or where the response might take a long time. A control watershed for trend detection is needed to distinguish climate trends. It is generally recommended to perform several different techniques before reaching a conclusion,[12] such as 1) a time plot, 2) a least-square fit regression, 3) comparison of annual means, 4) comparing cumulative distribution curves, 5) comparing quartiles (Q–Q plot), 6) double mass analysis, 7) paired regressions, 8) time series analysis, and 9) the seasonal Kendall test.[13]

Scale of Study

Three scale categories are commonly recognized: plot, field, and watershed; although, all three scales are in reality watersheds. Plots are appropriate monitoring units when the objective is to replicate several treatments. Limnocorrals and artificial stream channels are plots of a mesocosm scale[14] for lakes and streams. Monitoring on a field scale implies a larger area than an individual plot. Watershed scale monitoring is desired if the water resource system of concern is groundwater, a stream, or a lake/estuary.

Variable Selection

Several methods for selecting variables include variable matrices, variable cross-correlations, and the probability of exceeding standards. Certain water quality characteristics are highly related, but the analysis cost of one is much cheaper than the other. Correlations between these variables and the probability of exceeding a standard can be used to reduce the variable list.[15]

Sample Type

There are four types of water quality samples that can be collected: grab, composite, integrated, or continuous. The variable to sample influences the sample type. For example, bacteria samples must be taken as grab samples with sterilized bottles. Other variables change dramatically during storage and therefore are inappropriate for compositing. These include all dissolved gases, chlorine, pH, temperature, and sulfide.[16]

A grab sample is a discrete sample that is taken at a specific point and time, usually by hand.[16,17] A series of grab samples, usually collected at different times or at flow intervals and lumped together, are considered composite samples, usually taken with an automatic sampler. A depth-integrated sample accounts for velocity- or stratification-induced changes with depth, especially for sediment sampling or stratified lakes.[18] Near continuous monitoring can be used for any water quality variable that is measured using electrometric methods, such as conductivity or dissolved oxygen.

Sampling Location

Sampling locations should be stable and representative of the system being monitored. Stream sampling locations must account for vertical, horizontal, and longitudinal differences in water quality due in part to velocity differences. Lake systems also are heterogeneous due to vertical stratification, longitudinal gradients, and currents caused by winds and density differences. In order to describe lake water quality at a particular point, samples will be needed from each stratified layer in the system. Samples may have to be located within each bay or section of the lake. Groundwater monitoring must also consider vertical, horizontal, and longitudinal water quality differences.

Site selection criteria for a stream include accessibility, power available, ownership, streambed stability, permeability and gradient, uniform and straight section, control at all stages, and within a confined channel. Groundwater sampling sites consider the location of the water table divide, barrier locations (stream, strata), direction of flow, stratified concentrations, and depth to any confining layer. Lake sites are also concerned with stratified layers, longitudinal gradients, bays and beaches, and water circulation patterns.

Sampling Frequency and Duration

One of the most frequently asked questions is "How many samples and for how long?" Unfortunately, the correct response is as follows: "It depends." The sample size can be calculated from the relationship:

$$n = \frac{t^2 s^2}{d^2} \tag{33.1}$$

where n is the calculated sample size, t is Student's 't' at $n-1$ degrees of freedom and a selected confidence level, s is the estimate of the population standard deviation, and d is the allowable difference from the mean.[19]

Sample Collection and Analysis

Sample Collection

A sample should be representative of the water it came from and not contaminated before analysis. Bottles are often rinsed with the sample water two or three times before filling.[16] If samples are collected from pipes under pressure, the system should be flushed for a sufficient period to guarantee that new water is being sampled. Bacteria samples are collected in sterilized bottles. Stream and lake grab samples should be collected just below the surface of the water to avoid the surface film.

Collection of samples from wells might involve purging or sampling at the rate of well recovery. Sampling of volatile substances requires specialized sampling equipment in wells.

Samples should be collected so that the bottle is completely full to reduce volatilization losses. An exception to this would be if the sample was to be frozen, in which case room for expansion upon freezing should be left in the container. Sampling of toxic substances requires extra precautions, including personal protection.

Sample Preservation and Transport

Collected samples require preservation to slow rates of changes in composition through chemical, physical, and biological processes; however, these changes may not be preventable and rapid analysis is recommended.[20] Examples of physical changes include settling of solids, adsorption of certain cations on container walls, and loss of dissolved gases. Chemical changes could include precipitation, dissolution from sediments, complexation with other ions, and changes in valence state. Biological reactions could result in both the uptake and/or release of certain constituents. Predation within the sample may occur or the specimens may degrade. Microbial activity may change the species of nitrogen present.[16] Primary preservation methods are acidification, refrigeration, filtration, and preventing light from reaching the sample.[16,20] Glass containers may leach sodium and silica; plastic containers may sorb organics.[16] Transportation to the laboratory should be direct and use some method of preservation.

Quality Assurance

Quality assurance (QA) is "the total integrated program for assuring the reliability of monitoring and measurement data."[21] QA is composed of quality control and quality assessment. Quality control (QC) refers to activities conducted to provide high-quality data.[22] Quality assessment refers to techniques used to evaluate the effectiveness of the program.[23] Table 33.3 summarizes the major components of quality control and quality assessment programs. Indicators of data quality include: a) precision,

TABLE 33.3 Components of a Quality Control and Quality Assessment Program[3]

Quality Control	Quality Assessment
Good laboratory practices (GLPs)	Internal
Standard operating procedures (SOPs)	Duplicate samples
	Standard additions (spikes)
Education/training	Tests of sampling frequency
Sample custody procedures	Tests of reason with
Calibration and maintenance	comparable data
Supervision	Missing analysis records
	Standard curves Internal audit
	External
	Exchange sample with other lab
	External known materials
	External audit

b) ccuracy, c) representativeness, d) comparability, and e) completeness. Transfer of sample custody takes place upon delivery of samples to the laboratory by signing over custody to a laboratory person receiving the samples. Performance audits use EPA-approved unknown ampules. Field equipment also requires calibration and is subject to sample blanks.

Land Use and Management Monitoring

An essential element of water quality monitoring is the tracking of land use and management activities in the watershed being monitored. Land use and management data are needed to explain any water quality changes that may occur. It is important to be able to attribute the water quality changes to the management practice and not other confounding influences such as climate or a point source. There are four basic approaches for monitoring land treatment data: personal observations, field logs, personal interviews, and remote sensing.

Conclusion

Water quality monitoring is an important element of local, state, and federal programs as well as for research and education. It is expensive and therefore should be conducted in an appropriate manner to maximize utility.

References

1. Sanders, T.G.; Ward, R.C.; Loftis, J.C.; Steele, T.D.; Adrian, D.D.; Yevjevich, V. *Design of Networks for Monitoring Water Quality;* Water Resources Publications: Littleton, CO, 1983; 336 p.
2. Smith, R.A.; Schwarz, G.E.; Alexander, R.B. Regional interpretation of water-quality monitoring data. Water. Res. Res. **1997**, *33* (12), 2781–2798.
3. USDA, Natural Resources Conservation Service. Water quality monitoring system design, Part 614, *National Water Quality Handbook;* U.S. Department of Agriculture; Washington, DC, 2003.
4. Strobl, R.O.; Robillard, P.D. Network design for water quality monitoring of surface freshwaters: A review. J. Environ. Mgmt. **2008**, *87*, 639–648.
5. Intergovernmental Task Force on Monitoring Water Quality. *The Strategy for Improving Water-Quality Monitoring in the United States. Final Report of the Intergovernmental Task Force on Monitoring Water Quality*; U.S. Geological Survey, Office of Water Data Coordination: Reston, VA. 1995.

6. Telci, I.T.; Nam, K.; Guan, J.; Aral, M.M. Optimal water quality monitoring network design for river systems. J. Environ. Mgmt. **2009**, *90*, 2987–2998.
7. Strobl, R.O.; Robillard, P.D.; Debels, P. Critical sampling point methodology: case studies of geographically diverse watersheds. Environ. Monit. Assess. **2007**, *129*, 115–131.
8. Wilm, H.G. How long should experimental watersheds be calibrated? Amer. Geophys. Union Trans. Part II. **1949**, *30* (2), 618–622.
9. Reinhart, K.G. Watershed calibration methods. In *Proceeding Internship Symposium on Forest Hydrology*; Sopper, W.E., Lull, H.W., Eds.; Pergamon Press: Oxford, 1967; 715–723.
10. Striffler, W.D. *The Selection of Experimental Watersheds and Methods in Disturbed Forest Areas.* Publisher No. 66. *I.*A.S.H. Symposium of Budapest: 1965; 464–473.
11. Wicht, C.L. The validity of conclusions from South African multiple watershed experiments. In *Proceeding International Symposium on Forest Hydrology;* Sopper, W.E., Lull, H.W., Eds.; Pergamon Press: Oxford, 1967; 749–760.
12. World Health Organization. *Water Quality Surveys-A Guide for the Collection and Interpretation of Water Quality Data*; IHD-WHO Working Group on Quality of Water. UNESCO/WHO: Geneva, Switzerland, 1978.
13. Hirsch, R.M.; Slack, J.R.; Smith, R.A. Techniques of trend analysis for monthly water quality data. Water Res. Res. **1982,** *18* (1), 107–121.
14. Odum, E.P. The mesocosm. BioScience **1984**, *34* (9), 558–562.
15. Moser, J. H.; Huibregtse, K.R. *Handbook for Sampling and Sample Preservation of Water and Wastewater;* U.S. Environment Protection Agency: 600/4–76-049, 1976.
16. Clesceri, L.S.; Greenberg, A.E.; Eaton, A.D. *Standard Methods for the Examination of Water and Wastewater* 17th Ed.; American Public Health Association: Washington, D.C., 1989.
17. Ponce, S.L. *Water Quality Monitoring Programs;* WSDG Technical Paper WSDG-TP-0002; Watershed Systems Development Group, USDA Forest Service: Fort Collins, CO, 1980.
18. U.S. Geological Survey. Chemical and Physical Quality of Water and Sediment. Chap. 5 In *National Handbook of Recommended Methods for Water-Data Acquisition*; Office of Water Data Coordination, Geological Survey: Reston, VA, 1977.
19. Snedecor, G.W.; Cochran, W.G. *Statistical Methods,* 7th Ed.; The Iowa State University Press: Ames, IO, 1980.
20. U.S. Environmental Protection Agency. *Methods for Chemical Analysis of Water and Wastes*; EPA-600/4–79-020. Office of Research and Development: Cincinnati, OH, 1983.
21. U.S. Environmental Protection Agency. *Guide for the Preparation of Quality Assurance Project Plans for the National Estuary Program*; EPA 556/2–88-001. Office of Marine and Estuarine Protection: Washington, D.C., 1988.
22. Lawrence J.; Chau, A.S.Y. *Quality Assurance for Environmental Monitoring*; Symposium on monitoring, modeling, and mediating water quality. American Water Resources Association: 1987; 165–175.
23. Taylor. J.E. Essential features of a laboratory quality assurance program. In *Statistics in the Environmental Sciences*; Gertz, S.M.; London, M.D., Eds.; American Society for Testing and Materials: 1984; 66–73.

Bibliography

USDI, Geological Survey. National field manual for the collection of water-quality data. *Techniques of Water-Resources Investigations*; Book 9. Handbooks for Water-Resources Investigations. http://pubs.water.usgs.gov/twri9A/.

Ward, R.C.; Loftis, J.C.; McBride, G.B. *Design of Water Quality Monitoring Systems*; Van Nostrand Reinhold: New York, NY. 1990.

34

Algal Bloom Monitoring: Remote Sensing

Introduction 265
Monitoring Phytoplankton Abundance Using the Spectral Signature of Chlorophyll 266
Forecasting Chlorophyll Concentrations Using Ocean Color Algorithms 268
Reflectance Classification Algorithms 268
Reflectance Band Ratio Algorithms 268
Spectral Band Difference Algorithms 268
Bio-Optical Models 269
Estimating Bloom Initiation and Duration 269
Methods to Estimate Bloom Timing
Monitoring Algal Blooms in Inland Lakes and Drinking Water Reservoirs 269
Satellite Monitoring of Harmful Algal Blooms in Freshwater Water Bodies 270
Conclusions 271
Acknowledgments 271
References 272

Darryl J. Keith
U.S. Environmental Protection Agency

Introduction

Algal blooms are characterized by excessive phytoplankton biomass and/or growth rate. Phytoplankton bloom abundance is commonly expressed in terms of community biomass with chlorophyll concentration as an index of abundance against which rate processes (e.g., productivity) are often normalized (Smayda, 1997). In a lengthy definition, Blondeau-Patissier et al. (2014) defined a phytoplankton bloom "as a biological event composed of micro-algal species that is sustained both over time and space and that results in noticeable changes in satellite radiances at wavelengths used for algal bloom proxies due to an increase in biomass (in comparison to surrounding algal bloom-free waters." In the coastal and open ocean as well as in shallow, eutrophic, or hypereutrophic near-coastal and inlands waters, increases in phytoplankton cell numbers result from favorable environmental conditions which include water column stratification and increases in light availability (Gohin et al., 2003), water temperature (Thomas et al., 2003), and nutrient levels (Santoleri et al., 2003).

Ocean color remote sensing, the passive satellite-based measurement of visible light reflected from the ocean surface has provided more than two decades of near real-time synoptic and recurrent measurements of global phytoplankton abundance (Blondeau-Patissier et al., 2014). Remote sensing offers substantial advantages over traditional monitoring methods mainly because the synoptic coverage provided by satellites offers frequent views of inland and coastal water, facilitating quantitative assessments of algal biomass and valuable insights into bloom conditions in turbid waters (Binding et al., 2018b). The simplest technique used to identify bloom presence from satellite imagery is visual observation, especially when blooms occur in locations where they are regularly expected (Srokosz and Quartly, 2013) or in areas where they are not expected such as in oligotrophic waters (Wilson and Qiu, 2008). This chapter will discuss (1) the importance of the spectral character of chlorophyll in estimating phytoplankton abundance based on the optical properties of ocean, coastal, and inland waters; (2) the algorithm types used for the detection, mapping, and analysis of algal blooms from passive multispectral ocean color imagery; (3) the methods to forecast phytoplankton bloom initiation and duration; and (4) monitoring cyanobacteria blooms in inland waters and drinking water reservoirs.

Monitoring Phytoplankton Abundance Using the Spectral Signature of Chlorophyll

Phytoplankton produce chlorophyll, a pigment used to harvest sunlight and produce energy via photosynthesis. As seen in Figure 34.1, chlorophylls cause two dominant peaks in the visible absorption spectra of phytoplankton, a primary peak in the blue region (at 440 nm) and a secondary absorption peak in red region of the spectra (at 675 nm).

As seen in Figure 34.2, other pigments that are present (depending on species and taxa) will cause the broadening of blue peak and the appearance of additional absorption maxima. Spectra of phytoplankton

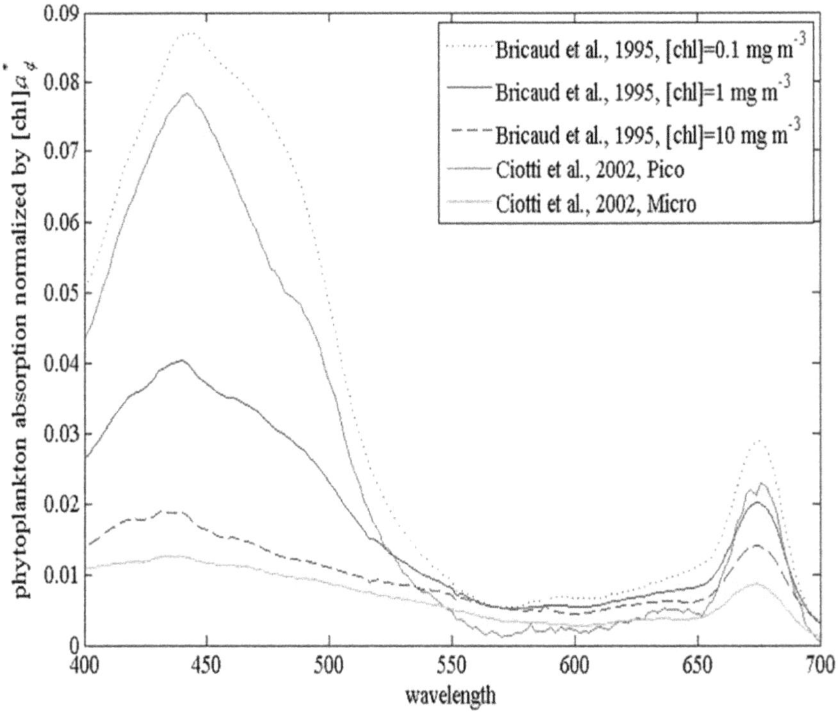

FIGURE 34.1 Phytoplankton absorption spectra normalized by chlorophyll which illustrates two dominant peaks: a primary peak in the blue region (at 440 nm) and a secondary absorption peak in red region of the spectra (at 675 nm). (Reproduced with permission from www.oceanopticsbook.info based on Bricaud et al., 1995 and Ciotti et al., 2002.)

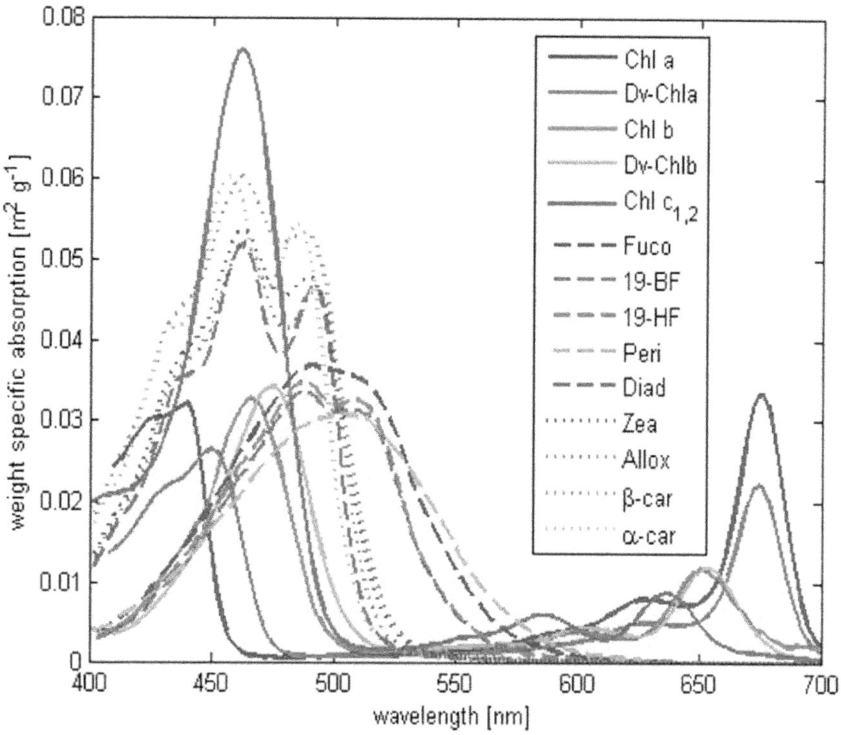

FIGURE 34.2 (See color insert.) Absorption spectra showing the primary and secondary absorption peaks for key phytoplankton pigment groups including Chl *a*, divinyl chlorophyll (Dv-Chl *a*), chlorophyll *b*, divinyl chlorophyll *b* (Dv-Chl *b*), chlorophyll c_1 and c_2 (chl $c_{1,2}$), 19'-butanoyloxyfucoxanthin (19-BF), 19'-hexanoxyfucoxanthin (19-HF), fucoxanthin (Fuco), peridinin (Peri), diadino (Diad), zeaxanthin (Zea), alloxanthin (Allo), beta-carotene (β-car), and alpha-carotene (α-car). (Reproduced with permission from www.oceanopticsbook.info based on Bidigare et al., 1990.).

absorption vary in magnitude and shape according to factors such as different cellular pigment compositions and pigment packaging (Bricaud and Stramski, 1990; Hoepffner and Sathyendranath, 1993; Ciotti et al., 2002). Specific pigment–protein complexes present in the cell will cause changes in absorption spectra and magnitudes. Furthermore, the increase in cellular pigment concentration and cell size (i.e., packaging effect) will flatten the specific absorption spectra (Kirk, 1976).

In the open ocean (Case 1 waters, Morel and Prieur, 1977), the water-leaving reflectance has been observed to "green" in a predictable fashion that allows for the creation of empirical relationships between phytoplankton concentrations and the ratio of reflectance in the blue-to-green wavelengths of light (Schofield et al., 2004). The blue–green end of the spectrum is, however, significantly affected by absorption by optically active constituents (OACs) including colored dissolved organic matter (CDOM), suspended matter, and phytoplankton pigments that are common in the biologically productive, turbid waters of fresh water, estuarine, and coastal environments (Case 2 waters, Morel and Prieur, 1977). For these environments, algorithms have been developed based on the properties of the phytoplankton reflectance peaks in the red and near-infrared (NIR) portions of the visible spectrum near 700 nm (Gitelson, 1992; Odermatt et al., 2012) often taking the ratio of remotely sensed reflectances at 705 and 670 nm (e.g., Gitelson, 1992). Due to the influence of OACs on light absorption, algorithms for chlorophyll (Chl *a*) developed for satellite imaging of the open ocean regularly overestimate chlorophyll in the coastal ocean, estuaries, and lakes.

Forecasting Chlorophyll Concentrations Using Ocean Color Algorithms

Interest in retrieving accurate retrievals of Chl *a* concentrations in open, coastal ocean and inland waters from satellite imagery has driven extensive research, yielding algorithms using different features of the reflectance spectrum to detect phytoplankton (e.g., Moses et al., 2009a). There are a wide variety of operational algorithms for ocean color satellite data to assist in the detection and monitoring phytoplankton blooms (Schalles et al., 1998; Keith, 2010; Keith et al., 2018; Klemas, 2012). Empirical algorithms have long been used to derive phytoplankton pigment concentrations in open ocean waters (O'Reilly et al., 1998; Campbell and Feng, 2005) because of their use in determining trophic status of waters and in the derivation of products such as phytoplankton primary production. These types of algorithms include reflectance-based classification algorithms, spectral band ratios, spectral band difference algorithms, and bio-optical models (Blondreau-Patissier et al., 2014). An overview of each class of algorithms follows. A more complete review of the applications, techniques, and ocean color sensor used to detect phytoplankton bloom events is found in Blondreau-Patissier et al., (2014). See Binding et al. (2018a) and Odermatt et al. (2012) for an overview of algorithms which forecast Chl *a* concentrations from satellite imagery.

Reflectance Classification Algorithms

Studies have shown that information to detect algal blooms can be retrieved from spectral reflectances in the blue, green, yellow, red, and NIR portion of the visible spectrum (Steemann Nielsen, 1963). These algorithms rely on the detection of specific spectral features and are often well suited for algal blooms in principle, but in practice, the reliability of measured reflectance can be hampered by atmospheric correction errors, bottom reflectance, and suspended particles (Blondeau-Patissier et al., 2014).

Reflectance Band Ratio Algorithms

Band ratio algorithms make use of chlorophyll primary and secondary absorption maxima at 442 nm and around 665 nm (Bricaud et al., 1995) as well as a reflectance peak around 700 nm and its fluorescence emission band at 681 nm (Gower et al., 1999) to estimate phytoplankton biomass. The first use of reflectance band ratios was in studies of the near-surface distribution of phytoplankton blooms in the open ocean and their relationship to oceanographic conditions (e.g., Gordon et al., 1983, 1980; Banse and English, 2000; Feldman et al., 1984). In open ocean waters, the blue and green regions of the visible spectrum are used to estimate Chl *a* concentrations. In coastal ocean waters, rivers, lakes, and estuaries, spectral bands near 670 nm have been widely used for the retrieval of Chl concentrations because reflectance increases in the NIR beyond 700 nm due to increased scattering from algal biomass (Blondreau-Patissier et al., 2014). The strength of the correlation of Chl *a* with the 700/670 nm band ratio is based on the interaction between backscattering from particulate matter (phytoplankton) and the strong absorption of water, which both increase toward the infrared. Absorption by water causes an offset to backscattering and a sharp peak near 700 nm in highly turbid or productive waters (Matthews, 2011). The height and position of the peak is known to be well correlated with Chl with the peak shifting to greater wavelengths as Chl *a* increases (Gitelson, 1992). See the following sources for more details: Werdell et al. (2018), Blondreau-Patissier et al. (2014), and Odermatt et al. (2012).

Spectral Band Difference Algorithms

Spectral band difference algorithms use band triplets from the red NIR or blue–green spectral regions to compare changes in the reflectance spectrum of waters with algal blooms with that of bloom-free waters. In theory, two adjacent reflectance spectral bands may have similar backscatter properties but will differ in absorption characteristics. The spectral difference quantifies the absorption feature. One of the most used ocean color spectral band difference algorithms is the Fluorescence Line Height (FLH) algorithm which estimates Chl *a* concentration from the height of the fluorescence maximum peak near 685 nm along a linear baseline drawn between two points on either side of the peak (Xing et al., 2007; Gower, 1980; Gitelson et al., 1994). Matthews (2011) indicated that the FLH algorithm has limited suitability in high biomass waters because the backscattering peak at 700 nm overwhelms fluorescence which makes it difficult to differentiate the signal for particulate backscattering from solar-induced fluorescence. This characteristic confines its use to low biomass waters with Chl *a* concentrations less than 30 mg m^{-3} (Matthews, 2011).

Bio-Optical Models

Bio-optical algorithms provide accurate assessments of Chl *a* concentrations from satellite data in optically complex waters where the presence of suspended sediments and CDOM interfere with the phytoplankton signal and confound band ratio algorithms (Shanmugam, 2011). Bio-optical models relate the optical characteristics of in-water constituents (e.g., Chl *a*, CDOM absorption) to the inherent optical properties (IOPs) of absorption and backscattering and to apparent optical properties (AOPs) such as water-leaving radiances (L_w) and remote sensing reflectance (R_{rs}). These relationships can be derived using either forward modeling or inverse modeling approaches which use empirical and semi-analytic algorithms to obtain Chl *a*, CDOM, and suspended matter concentrations from ocean color data or radiance/reflectance data. In a forward modeling approach, empirical algorithms link measured radiances or reflectances to Chl *a* concentrations (Sathyendranath et al., 2000; O'Reilly et al., 1998). Semi-analytical algorithms use an inverse modeling approach to retrieve in-water constituents (e.g., Chl *a*) from L_w or R_{rs} using empirical relationships (D'Sa et al., 2006; Neumann et al., 2000).

Estimating Bloom Initiation and Duration

Methods to Estimate Bloom Timing

In the open ocean environment, biomass-based threshold (TH) methods have been most widely used in conjunction with ocean color data to identify bloom initiation at basin and global scales (Brody et al., 2013). Brody et al. (2013) compared patterns of bloom initiation dates based on (i) a rate of change (ROC) method for finding bloom start dates (BSDs) (HANTS-FFT, Harmonic ANalysis of Time Series-Fast Fourier Transform (Roerink et al., 2000), (ii) the TH bloom initiation method which finds the yearly median of a chlorophyll time series, then identifies the BSD as the first point at which chlorophyll levels rise a certain percentage above the median (Siegel et al., 2002), and (iii) the cumulative sum (CS) method based on a cumulative biomass curve of chlorophyll time series which uses the time the curve rises above a preset percentage of the total biomass as the BSD. In contrast, Kahru et al. (2007) used satellite data from the Baltic Sea over a period of 15 summers, the period when cyanobacteria blooms were known to occur, to map their variability in time and space in relation to environmental conditions such as nitrogen and phosphorous ratios, sea surface temperature, water column stratification, and wind speed.

Monitoring Algal Blooms in Inland Lakes and Drinking Water Reservoirs

Remote sensing has long been recognized as having the potential to complement conventional approaches to monitoring algal blooms in inland lakes and drinking water reservoirs (Bukata, 2013). Satellite observations have mainly been used operationally to measure ocean color, while such monitoring of inland waters has historically been limited by spatial resolution of available data (Dekker et al., 1991). Higher spatial resolution data from satellite sensors which were designed primarily for land applications (e.g., Vincent et al., 2004) has become available more recently, though with a trade-off in spectral coverage. Moreover, the radiometric sensitivity of high-resolution sensors is not optimal for many applications over inland waters (Palmer et al., 2015).

Additional challenges to inland water remote sensing of algal blooms include the optical complexity of inland waters, atmospheric correction issues, adjacency effects (mixed land and water pixels at shoreline interfaces), and the high variability of in-water constituents between and within water bodies. The optical complexity of inland waters arises from the fact that they are typically dominated by huge concentrations of phytoplankton ranging from 1 to 100 mg m^{-3} (Gitelson et al., 1993) and in some locations with "algal scum" at concentrations that can reach up to 350 mg m^{-3} (Quibell, 1992). Mineral particles, detritus, and CDOM concentrations co-occur and typically vary over time (Palmer et al., 2015). These issues compound to complicate the development of algorithms for inland waters and typically limit their applicability across different water bodies (Palmer et al., 2015). While major challenges exist, several recent studies have demonstrated that water column constituents can be successfully retrieved from optically complex waters using algorithms developed from a variety of sensors (e.g., Moderate Resolution Imaging Spectrometer (MODIS); Medium Resolution Imaging Spectrometer (MERIS), Hyperspectral Instrument for the Coastal Ocean (HICO); Landsat 8 (Kutser, 2009; Moses et al., 2009b; Matthews, 2011; Odermatt, et al., 2012; Keith et al., 2018)). A more detailed discussion of spectral approaches for estimating bloom concentrations can be found in Matthews (2011).

Satellite Monitoring of Harmful Algal Blooms in Freshwater Water Bodies

Harmful algal blooms (HABs) are environmental events in lakes and drinking water sources that happen when algal populations occur in sufficiently high densities to cause environmental or health consequences (Smayda, 1997). HABs occur worldwide and have been documented in lakes and drinking water sources across the United States and on every continent (Loftin et al., 2016; Clark et al., 2017; Kahru et al., 2007; Verschuren et al., 2002; Stumpf et al., 2012; Dorra et al., 2010; Hitzfeld et al., 2000). HABs are composed of both phytoplankton and cyanobacteria (aka blue–green algae) of which some are toxin-producing genera (*Microcystis, Anabena*, and *Planktothrix*) (Becker et al., 2009; Rinta-Kanto et al., 2005). The conditions which favor the formation of HABs are not completely understood although it is believed that several factors (such as sunlight amount, increased nutrient concentrations due to alterations in land use practices, precipitation events, warming water temperatures, zooplankton grazing pressure, and water depth) collectively promote and enhance their development (Wynn et al., 2010; Lunetta et al., 2015; Becker et al., 2009; Havens, 2008; Paerl, 2008).

Many states in the United States find it difficult to develop HAB monitoring programs because of the need to cover large geographic areas and numerous isolated water bodies with limited financial and personnel resources (Schaeffer et al., 2018). Satellite remote sensing presents an opportunity to assess and monitor HAB abundance for spatially resolvable lakes and public water supplies using algal pigments as surrogates for HAB abundance to detect changes.

Cyanobacteria contain Chl *a* and the accessory pigment phycocyanin (PC; Fay, 1983) which strongly absorbs in the 620–630 nm wavelength region (Sathyendranath et al., 1987) and gives cyanobacteria their blue–green color (Becker et al., 2009). This accessory pigment can be used to identify potentially toxic cyanobacterial blooms in the presence of planktonic blooms (Becker et al., 2009). There have been numerous attempts to develop empirical algorithms specifically for detecting PC based on a distinct absorption minimum in water-leaving reflectance at 620 nm (Dekker, 1993; Jupp et al., 1994). A more detailed discussion of spectral approaches for estimating PC can be found in Matthews (2011). For monitoring phytoplankton and cyanobacteria in inland waters, Chl *a* is still the preferred pigment, since most existing satellite-borne sensors are better suited to capture the Chl *a* reflectance peak in the NIR region of the visible spectrum than the narrow peak associated with PC (Keith et al., 2018; Reinart and Kutser, 2006; Kutser, 2004).

Attempts to use satellite data to remotely map and monitor the distribution of cyanobacteria in inland waters according to their spectral signature have been primarily limited by the course spatial resolutions (300 m^2–1 km^2 pixel size) of ocean color sensor systems (Sea-viewing Wide Field-of-view Sensor (SeaWiFS), Moderate Imaging Spectrometer (MODIS), and MERIS). This limitation has resulted in the use of data from satellite sensors with higher spatial resolutions (10–30 m^2) that were designed for land applications (e.g., the Landsat series). While these sensors have adequate spatial resolutions to resolve smaller lakes and drinking water reservoirs, their spectral coverage and radiometric sensitivity are not optimal for many applications over inland waters (Palmer et al., 2015). More discussions of research using satellite remote sensing for cyanobacteria monitoring are found in Clark et al. (2017) and Becker et al. (2009).

Several ecological software tools exist which hindcast HAB events, eutrophication status, and phytoplankton biomass using archived satellite data and readily available Chl *a* data (e.g., see Coad et al., 2014; Sadeghian et al., 2018; Kim et al., 2014; Malve et al., 2007; Norton et al., 2012; Zhang et al., 2013). However, these tools are not equipped for repeated, intuitive, and rapid assessment of inland waters for HAB monitoring. These software packages are usually desktop computer-based, require some scientific knowledge in the field of ocean color remote sensing, and need sufficient computer hardware to handle satellite images and processing capabilities. Further, computer coding experience is required to batch process large numbers of satellite image files. The products of these packages are water quality products such as HAB abundance.

Unlike the previous software packages, the US Environmental Protection Agency (USEPA) has partnered with the National Aeronautics and Space Administration (NASA), National Oceanic and Atmospheric Administration (NOAA), and US Geological Survey (USGS) to develop a mobile device application (CyAN mobile app) that reduces barriers of entry such as the need for scientific expertise in ocean color interpretation, hardware requirements associated with the use of satellite data, and the computational requirements necessary to produce cyanobacteria abundance data. The CyAN app represents the first attempt to make satellite-derived cyanobacteria concentrations directly available for monitoring inland waters using an Android mobile device. The CyAN app also is the first of its kind to provide a cost-effective delivery system for satellite-derived HAB data products in a simple display format that is intuitive and applicable across multiple water bodies. More complete discussions of the CyAN app, its development, and application as a monitoring tool are found in Schaeffer et al. (2018) and Mattas-Curry et al. (2015).

Conclusions

Satellite imagery provides frequent, synoptic information for mapping phytoplankton blooms in ocean, coastal, and inland environments. A wide variety of algorithms and approaches for process ocean color imagery are used to retrieve Chl *a* concentrations. Relationships with phytoplankton biomass data have been used for the derivation of ecological baselines for change detection to assess bloom conditions in relation to biological and physical triggering events and to produce maps which allow managers to create rankings of HAB occurrence. Outputs can be used to quantify and prioritize locations that may require additional monitoring and adaptive management solutions (Clark et al., 2017). With the advent of next-generation satellite sensors specifically designed to solve the issues of high spatial and spectral resolution, forecasts of phytoplankton and cyanobacteria concentrations in coastal and small inland water bodies should become more reliable and readily available.

Acknowledgments

I would like to thank Nate Merrill, Laura Erban, Steven Rego, Timothy Gleason, and Wayne Munns for their review and comments of the manuscript. This is contribution number ORD-032215 of the Atlantic Ecology Division, National Health and Environmental Effects Research Laboratory, Office of Research and Development, US Environmental Protection Agency. Any mention of trade names or commercial products does not constitute endorsement or recommendation for use.

References

Banse, K. and D.C. English. 2000. Geographical differences in seasonality of CZCS-derived phytoplankton pigment of the Arabian Sea for 1978–1986. *Deep Sea Research*, 47, 1623–1677.

Becker, R.H., M.I. Sultan, G.L. Boyer, M.R. Twiss, and E. Konopko. 2009. Mapping cyanobacterial looms in the Great Lakes using MODIS. *Journal of Great Lakes Research*, 35, 447–453.

Bidigare, R.R., M.E. Ondrusek, J.H. Morrow, and D. Kiefer. 1990. In vivo absorption properties of algal pigments. *Ocean Optics X Proceedings of SPIE*, 1302, 290–302.

Binding, C.E., T.A. Greenberg, G. McCullough, S.B. Watson, and E. Page. 2018a. An analysis of satellite-derived chlorophyll and algal bloom indices on Lake Winnipeg. *Journal of Great Lakes Research*, 44, 436–446.

Binding, C., R.P. Stumpf, B.A. Schaeffer, A. Tyler, and P. Hunter. 2018b. Chapter 2: Introduction to Deriving Water Quality Measures from Satellites. In *Reports and Monographs of the International Ocean Colour Coordinating Group (IOCCG)*, S. Greb, A. Dekker, and C. Binding eds., IOCCG Report Number 17, pp. 15–28.

Blondeau-Patissier, D., J.F.R. Gower, A.G. Dekker, S.R. Phinn, and V. Brando. 2014. A review of ocean color remote sensing methods and statistical techniques for the detection, mapping, and analysis of phytoplankton blooms in coastal and open oceans. *Progress in Oceanography*, 123, 123–144.

Bricaud, A., M. Babin, A. Morel, and H. Clustre. 1995. Variability in the chlorophyll-specific absorption coefficients of natural phytoplankton: Analysis and parametrization. *Journal of Geophysical Research: Oceans*, 100, 13321–13332.

Bricaud, A. and D. Stramski. 1990. Spectral absorption coefficients of living phytoplankton and non-algal biogenous matter: A comparison between the Peru upwelling area and the Sargasso Sea. *Limnology and Oceanography*, 35, 562–582.

Brody, S.R., M.S. Lozier, and J.P. Dunne. 2013. A comparison of methods to determine phytoplankton bloom initiation. *Journal of Geophysical Research: Oceans*, 118(5), 2345–2357.

Bukata, R.P. 2013. Retrospection and introspection on remote sensing of inland water quality: "Like déjà vu all over again". *Journal of Great Lakes Research*, 39, 2–5.

Campbell, J. W. and H. Feng. 2005. The empirical chlorophyll algorithm for MODIS: Testing the OC3M algorithm using NOMAD data. In ocean color bio-optical algorithm mini-workshop, 27–29 September 2005, New England Center, University of New Hampshire, Durham, NH, 1–9.

Ciotti, A.M., M.R. Lewis, and J.J. Cullen. 2002. Assessment of the relationships between dominant cell size in natural phytoplankton communities and the spectral shape of the absorption coefficient. *Limnology and Oceanography*, 47, 404–417.

Clark, J.M., B.A. Schaeffer, J.A. Darling, E.A. Urquhart, J.M. Johnston, A.R. Ignatuis, M.H. Mayer, K.A. Loftin, P.J. Werdell, and R.A. Stumpf. 2017. Satellite monitoring of cyanobacterial harmful algal bloom frequency in recreational waters and drinking water sources. *Ecological Indicators*, 80, 84–95.

Coad, P., B. Cathers, J.E. Ball, and R. Kadluczka. 2014. Proactive management of estuarine algal blooms using an automated monitoring buoy coupled with an artificial neural network. *Environmental Modelling & Software*, 61, 393–409.

D'Sa, E., R.L. Miller, and C.D. Castillo. 2006. Bio-optical properties and ocean color algorithms for coastal waters influenced by the Mississippi River during a cold front. *Applied Optics*, 45 (28), 7410–7428.

Dekker, A.G. 1993. Detection of optical water quality parameters for eutrophic waters by high resulton remote sensing. PhD thesis, Free University, Amsterdam.

Dekker, A.G., T. Malthus, and E. Seyhan. 1991. Quantitative modeling of inland water quality for high-resolution MSS systems. *IEEE Transactions on Geoscience and Remote Sensing*, 29, 89–95.

Dorra, F.A., E. Pintoa, R.M. Soaresh, and S.M. Feliciano de Oliveira e Azevedo. 2010. Microcystins in South American aquatic ecosystems: Occurrence, toxicity and toxicological assays. *Toxicology*, 56, 1247–1256.

Fay, P. 1983. The Blue-greens. Vol 160. *Studies in Biology*. Hodder, London, ISBN 10:071312878X/ISBN 13: 9780713128789.

Feldman, G., D. Clark, and D. Halpern, 1984. Satellite color observations of the phytoplankton distribution in the eastern equatorial Pacific during the 1982–1983 El Nino. *Science*, 226, 1069–1071.

Gitelson, A. 1992. The peak near 700 nm on radiance spectra of algae and water: Relationships of its magnitude and position with chlorophyll concentrations. *International Journal of Remote Sensing*, 13, 3367–3373.

Gitelson, A., G. Garbuzov, F. Szilgyi, K-H. Mittenzwey, A. Karnieli, and A. Kaiser. 1993. Quantitative remote sensing methods for real-time monitoring of inland waters quality. *International Journal of Remote Sensing*, 14, 1269–1295.

Gitelson, A., M. Mayo, Y.Z. Jacobi, A. Parparov, and T. Berman. 1994. The use of high-spectral-resolution radiometer data for detection of low chlorophyll concentrations in Lake Kinneret. *Journal of Plankton Research*, 16, 993–1002.

Gohin, F., L. Lampert, J.F. Guillaud, A. Herbland, and E. Nézan. 2003. Satellite and in situ observations of a late winter phytoplankton bloom in the Northern Bay of Biscay. *Continental Shelf Research*, 23, 1117–1141.

Gordon, H.R., D.K. Clark, J.W. Brown, O.B. Brown, R.H. Evans, and W.W. Broenkow. 1983. Phytoplankton pigment concentrations in the Middle Atlantic Bight: Comparison of ship determinations and coastal zone color scanner measurements. *Applied Optics*, 22, 4175–4180.

Gordon, H.R., D.K. Clark, J.L. Mueller, and W.A. Hovis. 1980. Phytoplankton pigments from the Nimbus-7 coastal zone color scanner: Comparisons with surface measurements. *Science*, 210, 23–66.

Gower, J. 1980. Observations of *in situ* fluorescence of chlorophyll-*a* in Saanich Inlet. *Boundary-Layer Meteorology*, 18, 235–245.

Gower, J.F.R., R. Doerffer, and G.A. Borstad. 1999. Interpretation of the 685 nm peak in water-leaving radiance spectra in terms of fluorescence, absorption, and scattering and its observation by MERIS. *International Journal of Remote Sensing*, 20(9), 1771–1786.

Havens, K.E. 2008. Cyanobacteria blooms: Effects on aquatic ecosystems. In: Hudnell, H.K. (Ed.) *Cyanobacterial Harmful Blooms: State of the Science and Research Needs*. Springer, New York, NY.

Hitzfeld, B.C., C.S. Lampert, N. Spaeth, D. Mountfort, H. Kaspar, and D.R. Dietrich. 2000. Toxin production in cyanobacterial maps from ponds on the McMurdo Ice Shelf, Antartica. *Toxicon*, 38, 1731–1748.

Hoepffner, N. and S. Sathyendranath, 1993. Determination of the major groups of phytoplankton pigments from the absorption spectra of total particulate matter. *Journal of Geophysical Research Oceans*, 98, 22789–22803.

Jupp, D.L.B., J.T.O. Kirk, and G.P. Harris. 1994. Detection, identification and mapping of cyanobacteria – using remote sensing to measure optical quality of turbid inland waters. *Australian Journal of Marine and Freshwater Research*, 45, 511–522.

Kahru, M., O.P. Savchuk, and R. Elmgren. 2007. Satellite measurements of cyanobacterial bloom frequency in the Baltic Sea: Interannual and spatial variability. *Marine Ecological Progress Series*, 343, 15–23.

Keith, D. 2010. Chapter 7. Estimating Chlorophyll Conditions in Southern New England Coastal Waters from Hyperspectral Remote Sensing. In: Y. Wang (Ed.) *Remote Sensing of Coastal Environments*, CRC Press, Boca Raton, FL.

Keith, D, J. Rover, J. Green, B. Zalewsky, M. Charpentier, G. Thursby, and J. Bishop. 2018. Monitoring algal blooms in drinking water reservoirs using the Landsat-8 Operational Land Imager. *International Journal of Remote Sensing*, 39 (8), 2818–2846.

Kim, Y., H.S. Shin, and J.D. Plummer. 2014. A wavelet-based autogressive fuzzy model for forecasting algal blooms. *Environmental Modeling Software*, 62, 1–10.

Kirk, J.T.O. 1976. *Light and Photosynthesis in Aquatic Ecosystems*. Cambridge University Press, Cambridge.

Klemas, V. 2012. Remote sensing of algal blooms: An overview with case studies. *Journal of Coastal Research*, 28, 34–43.

Kutser, T. 2004. Quantitative detection of chlorophyll in cyanobacterial blooms by satellite remote sensing. *Limnology and Oceanography*, 49, 2179–2189.

Kutser, T. 2009. Passive optical remote sensing of cyanobacteria and other intense phytoplankton blooms in coastal and inland waters. *International Journal of Remote Sensing*, 30, 4401–4425.

Loftin, K.A., J.L. Graham, E.D. Hilborn, S.C. Lehmann, M.T. Meyer, J.E. Dietz, and C.B. Griffith. 2016. Cyanotoxins in inland lakes of the United States: Occurrence and potential recreational health risks in the EPA national lakes assessment. *Harmful Algae*, 56, 77–90.

Lunetta, R.S., B.A. Schaeffer, R.P. Stumpf, D. Keith, S.A. Jacobs, and M.S. Murphy. 2015. Evaluation of cyanobacteria cell count detection derived from MERIS Imagery across the eastern USA. *Remote Sensing of Environment*, 157, 24–34.

Malve, O., M. Laine, H. Haario, T. Kirkkala, and J. Sarvala. 2007. Bayesian modelling of algal mass occurrences during adaptive MCMC methods with a lake water quality model. *Environmental Modeling Software*, 22, 966–977.

Mattas-Curry, L., B. A. Schaeffer, R. N. Conmy, and D.J. Keith. 2015. A space satellite perspective to monitor water quality using your mobile phone. North American Lake Management Society. *Lakelines*, Summer, 28–31.

Matthews, M.W. 2011. A current review of empirical procedures of remote sensing in inland and near-coastal waters transitional waters. *International Journal of Remote Sensing*, 32 (21), 6855–6899.

Morel, A. and L. Prieur, 1977. Analysis of variations in ocean color. *Limnology and Oceanography*, 22 (4), 709–722.

Moses, W.J., A.A. Gitelson, S. Berdnikov, and V. Povazhnyy. 2009a. Satellite estimation of chlorophyll-*a* concentration using the red and NIR bands of MERIS – the Azov Sea cast study. *IEEE Geoscience and Remote Sensing Letters*, 6, 845–849.

Moses, W.J., A.A. Gitelson, S. Berdnikov, and V. Povazhnyy. 2009b. Estimation of chlorophyll-*a* concentration in case II waters using MODIS and MERIS data – successes and challenges. *Environmental Research Letters*, 4, 045005.

Neumann, A., R. Doerffer, H. Krawcyzk, M.D. Dowell, R. Arnone, C.O. Davis, M. Kishino, A. Tanaka, C. Hu, R.P. Bukata, H.R. Gordon, J. Campbell, and S. Sathyendranath. 2000. Algortihms for Case 2 Waters. In *Remote Sensing of Ocean Color in Coastal, and Other Optically-Complex Waters.* Sathyendranath, S. (Ed.) reports of the International Ocean-Colour Coordinating Group, No. 3, IOCCG, Dartmouth, Canada.

Norton, I., J.A. Elliott, S.C. Maberly, and L. May. 2012. Using models to bridge the gap between land use and algal blooms: An example from the Lowes water catchment, UK. *Environmental Modeling Software*, 36, 64–75.

Odermatt, D., A. Gitelson, V.E. Brando, and M. Schaepman. 2012. Review of constituent retrieval in optically deep and complex waters from satellite imagery. *Remote Sensing of Environment*, 118, 116–126.

O'Reilly, J.E., S. Maritorena, B.G. Mitchell, D.A. Siegel, K.L. Carder, S.A. Graver, M. Kahru, and C. McClain. 1998. Ocean color algorithms for SeaWiFS. *Journal of Geophysical Research*, 103, 24, 937–953.

Paerl, H.W. 2008. Nutrient and other environmental controls of harmful cyanobacterial blooms along the freshwater-marine continuum. In: Hudnell, K.H. (Ed.) *Cyanobacterial Harmful Algal Blooms: State of the Science and Research Needs. Advances in Experimental Medicine and Biology*, 619, 217–237. Springer Science+Business Media, New York.

Palmer, S.C.J., T. Kutser, and P.D. Hunter. 2015. Remote sensing of inland waters: Challenges, progress and future directions. *Remote Sensing of Environment*, 157, 1–8.

Quibell, G. 1992. Estimating chlorophyll concentrations using upwelling radiance from different freshwater algal genera. *International Journal of Remote Sensing*, 13, 2611–2621.

Reinart, A. and T. Kutser. 2006. Comparison of different satellite sensors in detecting cyanobacterial blooms events in the Baltic Sea. *Remote Sensing of Environment*, 102, 74–85.

Rinta-Kanto, J.M., A.J.A. Oullette, M.R. Twiss, G.L. Boyer, T.B. Bridgeman, and S.W. Wilhelm. 2005. Quantification of toxic *Microcystis* spp. During the 2004 and 2004 blooms in Western Lake Erie Using Quantitative Real-Time PCR. *Environmental Science and Technology*, 39 (11), 4198–4205.

Roerink, G., M. Menenti, and W. Verhoef. 2000. Reconstructing cloud-free NDVI composites using Fourier analysis of time series. *International Journal of Remote Sensing*, 21, 1911–1917.

Sadeghian, A., S.C. Chapra, J. Hudson, H. Wheater, and K.E. Lindenschmitt. 2018. Improving in-lake water quality modeling using variable chlorophyll *a*/algal biomass ratios. *Environmental Modeling and Software*, 101, 73–85.

Sathyendranath, S., R.P. Bukata, R. Arnone, M.D. Dowell, C.O. Davis, M. Babin, J.-F. Berthon, O.V. Kopelevich, and J.W. Campbell. 2000. Chapter 2. Colour of Case 2 Waters. In *Remote Sensing of Ocean Color in Coastal, and Other Optically-Complex Waters*. Sathyendranath, S. (Ed.) Reports of the International Ocean-Colour Coordinating Group, No. 3, IOCCG, Dartmouth, Canada.

Sathyendranath, S., L. Lazzara, and L. Prieur. 1987. Variations in the spectral values of specific absorption of phytoplankton. *Limnology and Oceanography*, 32 (2), 403–415.

Santoleri, R., V. Banzon, S. Marullo, E. Napolitano, F. D'Ortenzio, and R. Evans. 2003. Year-to-year variability of the phytoplankton bloom in the Southern Adriatic Sea (1998–2000): Sea-viewing wide field-of-view sensor observations and modeling study. *Journal of Geophysical Research*, 108, 8122.

Schaeffer, B.A., S.W. Bailey, R.N. Conmy, M. Galvin, A.R. Ignatius, J.M. Johnston, D.J. Keith, R.S. Lunetta, R. Parmar, R.P. Stumpf, E.A. Urquhart, P.J. Werdell, and K. Wolfe. 2018. Mobile device application for monitoring cyanobacteria harmful algal blooms using Sentinel-3 satellite Ocean and Land Color Instruments. *Environmental Modeling and Software*, 109, 93–103.

Schalles, J.F., A.A. Gitelson, Y.Z. Jacobi, and E. Kroenke. 1998. Estimation of chlorophyll-*a* from time series measurements of high spectral resolution reflectance in a eutrophic lake. *Journal of Phycology*, 34 (2), 383–390.

Schofield, O., T. Bergman, W.P. Bissett, M.A. Moline, and C. Orrico. 2004. Inverting inherent optical signatures in the near shore coastal waters at the Long Term Ecosystem Observatory. *Journal of Geophysical Research*, 109, C12S04.

Shanmugam, P. 2011. A new bio-optical algorithm for the remote sensing of algal blooms in complex ocean waters. *Journal of Geophysical Research*, 116, C04016.

Siegel, D., S. Doney, and J. Yoder. 2002. The North Atlantic spring phytoplankton bloom and Sverdrup's critical depth hypothesis. *Science*, 296 (5568), 730–733.

Smayda, T.J. 1997. What is a bloom? A commentary. *Limnology and Oceanography*, 42, 1132–1136.

Srokosz, M.A. and G.D. Quartly. 2013. The Madagascar Bloom – a serendipitous study. *Journal of Geophysical Research*, 118(1), 14–25.

Steemann Nielsen, E. 1963. Productivity, definition and measurement. In: Hill, M.N. (Ed.) *The Sea*. Wiley, London.

Stumpf, R.P., T.T. Wynne, D.B. Baker, and G.L. Fahnenstiel. 2012. Interannual variability of cyanobacteria blooms in Lake Erie. *PLoS One*, 7, e42444.

Thomas, A.C., D.W. Townsend, and R. Weatherbee. 2003. Satellite-measured phytoplankton variability in the Gulf of Maine. *Continental Shelf Research*, 23, 671–989.

Verschuren, D., T.C. Johnson, H.J. Kling, D.N. Edington, P.R. Leavitt, E.T. Brown, M.R. Talbot, and R.E. Hecky. 2002. History and timing of human impact on Lake Victoria, East Africa. *Proceedings of the Royal Society of London B: Biological Science*, 269, 289–294.

Vincent, R.K., X.M. Qin, R.M. McKay, J. Miner, K. Czajkowski, J. Savino, and T. Bridgeman. 2004. Phycocyanin detection from Landsat TM data for mapping cyanobacterial blooms in Lake Erie. *Remote Sensing of Environment*, 89 (3), 381–392.

Werdell, P.J., L.I.W. McKinna, E. Boss, S.G. Ackleson, S.E. Craig, W.W. Gregg, Z. P. Lee, S. Maritorena, C.S. Roesler, C.S. Rousseaux, D. Stramski, J.M. Sullivan, M.S. Twardowski, M. Tzortziou, and X. Zhang. 2018. An overview of approaches and challenges for retrieving marine inherent optical properties from ocean color remote sensing. *Progress in Oceanography*, 160, 186–212.

Wilson, C. and X. Qiu. 2008. Global distribution of summer chlorophyll blooms in the oligotrophic gyres. *Progress in Oceanography*, 78 (2), 107–134.

Wynn, T.T., R.P. Stumpf, M.C. Tomlinson, and J. Dyble. 2010. Characterizing a cyanobacteria bloom in western Lake Erie using satellite imagery and meteorological data. *Limnology and Oceanography*, 55, 2025–2036.

Xing, X-G., D.Z. Zhao, Y.G. Liu, J.H. Yang, P. Xiu, and L. Wang. 2007. An overview of remote sensing of chlorophyll *a* fluorescence. *Ocean Science Journal*, 42, 49–59.

Zhang, H., W. Hu, K. Gu, Q. Li, D. Zheng, and S. Zhai. 2013. An improved ecological model and software for short term algal bloom forecasting. *Environmental Modelling Software*, 48, 152–162.

35

Mapping Impervious Cover in Catchments Using High Spatial Resolution Aerial Imagery

Jessica Morgan
United States Coast Guard Academy

Yeqiao Wang
University of Rhode Island

Naomi Detenbeck
U.S. Environmental Protection Agency

Introduction 277
Methods 279
Study Area and Site Selection • Data Preparation • Impervious Cover Classifications • Post-Classification Accuracy Assessment • Statistical Analysis
Results 284
Post-Classification Accuracy Assessment • Impervious Cover Classification Comparison
Discussion 286
Acknowledgments 288
References 288

Introduction

Nonpoint source pollution is one of the leading causes of water pollution in the United States (US EPA 2002). The relationship between land use/land cover, water pollution, and stream ecosystem health is well documented (Harding et al. 1998; Allan 2004; Walsh et al. 2005; Johnson and Host 2010; Zhang et al. 2013). The extent of impervious cover (IC) in a catchment is a known stressor on stream biota and habitat and has been used as a metric for estimating the effects of urbanization on stream health (Schueler 1994, 2003; Arnold and Gibbons 1996; Brabec 2002; Morse et al. 2003; Roy and Shuster 2009; Schueler et al. 2009; Arnold et al. 2010). Urban development drastically alters the landscape and affects stream ecosystem health by changing the magnitude and timing of stormwater runoff, leading to higher peak stream discharges, less infiltration of stormwater, changes in channel morphology, increases in nutrient and pollutant inputs, decreases in biotic richness and subsequent increases in pollutant-tolerant species, and altered thermal regimes (Wang and Kanehl 2003; Walsh et al. 2005; Nelson and Palmer 2007).

The effects of combined anthropogenic stressors from urbanization on freshwater streams can be difficult to quantify. IC as an indicator metric can help to quantify these effects on stream biota and habitat (Schueler 1994; Arnold and Gibbons 1996; Schueler and Claytor 1997; Brabec 2002; Schueler et al. 2009). The extent of IC in a catchment has been used to guide stormwater management practices, to develop stormwater management utility rates and credits, and to develop an IC Total Maximum Daily Load (TMDL) for mitigating water quality impacts as mandated by the US Clean Water Act (Parikh et al. 2005; Schueler et al. 2009; Arnold et al. 2010; Wickham et al. 2016).

Towns and municipalities, already restricted by tight budgets, are increasingly being tasked with addressing stormwater management and water quality issues. The US Municipal Separate Storm Sewer System (MS4) stormwater rule requires publicly owned agencies, such as cities, towns, counties, or states, to implement six minimum control measures within a stormwater management plan. These measures include public education and outreach, public participation or involvement (community cleanups, citizen water quality monitoring, etc.), detection and elimination of illicit discharges, controls for stormwater runoff both pre- and post-construction, and pollution prevention measures for municipal operations (US EPA 2016).

Increasingly, stormwater management utility districts are being established to fund stormwater projects. These utilities charge a user fee for stormwater management, similar to a sewer service charge for wastewater. The fees are often based on the extent of IC on each property in the district (van der Tak et al. 2014). The shared fees provide a dedicated funding source for stormwater improvement projects including flood mitigation, water quality impairment remediation, and replacement of aging stormwater infrastructure (NH DES 2014). Higher spatial resolution estimates of IC are required for accurate fee structures and more effective decisions in the context of integrated water resources management.

Fine-scale mapping of IC at broad spatial extents presents unique challenges. Site-specific estimates of IC are often derived from digitized aerial imagery and the use of Light Detection and Ranging (LIDAR) technology. Object-oriented classification methodologies, using programs such as eCognition and the ArcGIS Feature Analyst extension, have yielded impressive results for classifying IC and other land use/land cover classes (O'Neil-Dunne 2013; Stueve et al. 2015). Multiple agent segmentation and classification (MASC) using submodels of segmentation, shadow effect, Multivariate Analysis of Variance (MANOVA)-based classification, and post-classification have been successfully utilized to classify IC over broad spatial extents (Zhou and Wang 2008). For broad-scale studies, estimates of IC are commonly obtained from the National Land Cover Database (NLCD). This 30 m spatial resolution dataset, produced by the Multi-Resolution Land Characteristics Consortium, classifies land surface data into thematic classes, (e.g., high intensity developed, open space developed, agriculture), percent IC, and percent tree canopy cover from Landsat imagery (Homer et al. 2015).

A 10%–12% IC threshold has been well supported in the literature for quantifiable impacts to water quality indicators at the catchment scale (Schueler 2003, Schueler et al. 2009). High spatial resolution IC data have also been shown to improve hydrologic simulation models (Zhou et al. 2010; Zhou et al. 2014). However, recent analyses suggest that macroinvertebrate communities are being impacted at much lower levels of IC (Baker and King 2010; Detenbeck et al. 2013; Smucker et al. 2013). It has been suggested that the 30 m resolution NLCD data are underestimating IC, particularly in suburban areas where impervious surfaces may be masked by vegetation and trees (Roy et al. 2003; Jones and Jarnagin 2009; Claggett et al. 2013). Stueve et al. (2015) found that the NLCD underestimated tree cover and overestimated water classes in a 25 km^2 Minnesota catchment. An assessment of the 2001 NLCD demonstrated that the NLCD underestimated both canopy cover and IC when compared to photographic interpreted imagery (Nowak and Greenfield 2010). Recent studies have found that the NLCD underestimates IC in less developed watersheds and overestimates IC in more developed watersheds (Smucker et al. 2016). A timely, cost-effective method for accurately mapping IC from high spatial resolution data over broad scales is needed to accurately depict the impact of urbanization on stream response variables.

Using readily available 1 m spatial resolution National Agriculture Imagery Program (NAIP) data (USDA 2013), we developed a method for characterizing IC over broad spatial extents. IC was extracted and quantified at the Hydrologic Unit Code (HUC) 10 and HUC 12 levels (USGS and USDA 2013) as well as at the National Hydrography Dataset (NHDPlus) catchment level (USGS and US EPA 2010) for select catchments in Vermont, using supervised image classification. Post-classification accuracy assessments were conducted on the 1 m NAIP classifications to quantify the accuracy of the dataset. The 1 m NAIP estimates were then compared to the 30 m NLCD estimates and validated against established high spatial resolution IC estimates from the University of Vermont (UVM). The data and methodologies described will benefit both water quality modelers and decision makers in meeting challenging water resource and policy goals, particularly at the urban–rural interface.

Methods

Study Area and Site Selection

The study area focused on four HUC 10 catchments and eighteen HUC 12 catchments in the state of Vermont and their surrounding MS4 areas (Figure 35.1). MS4 areas were included in the classifications because improved estimates of IC will be useful in meeting or setting regulatory requirements at

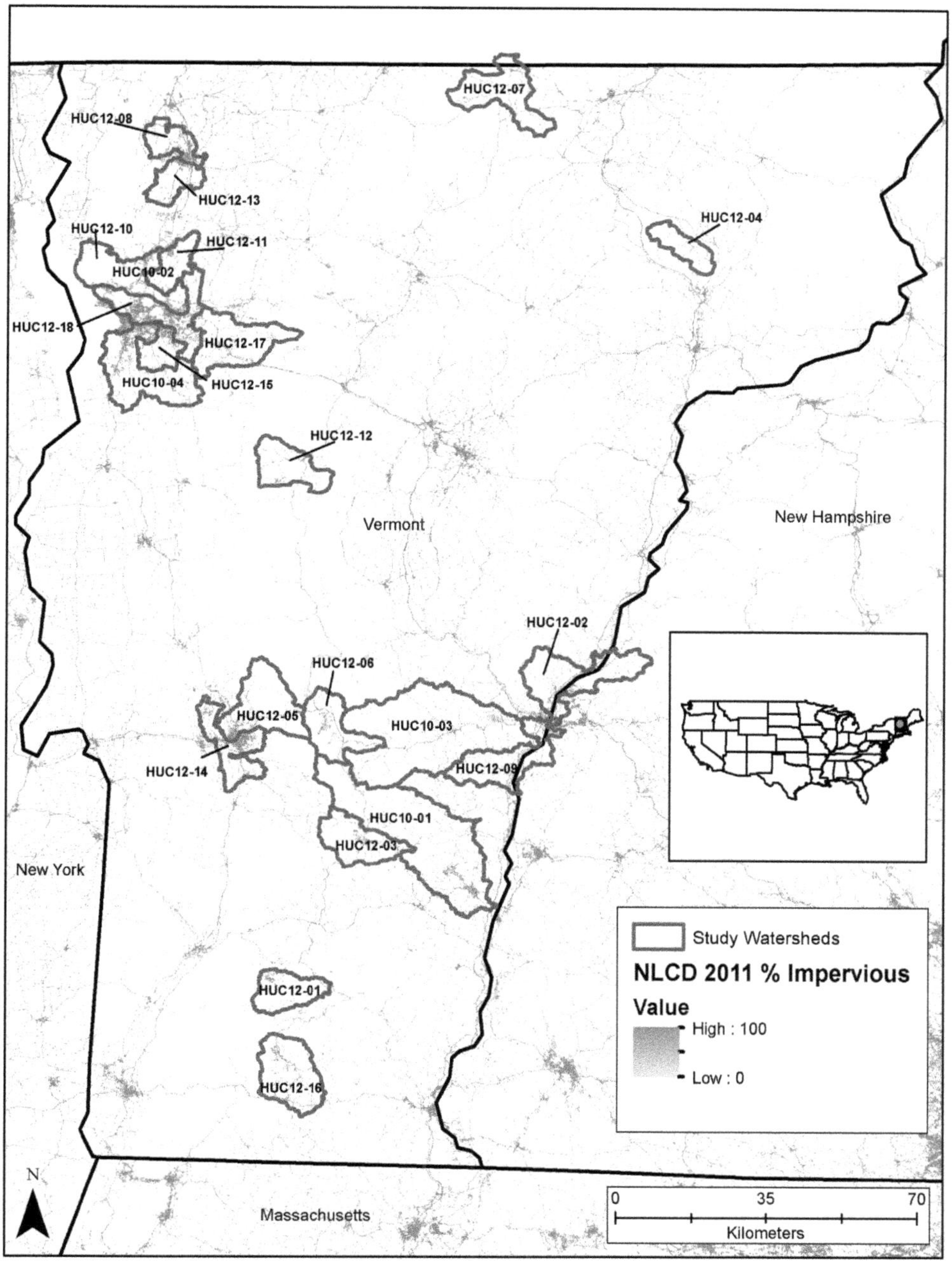

FIGURE 35.1 (See color insert.) Map of study area—watershed labels correspond to catchment and catchment number in Table 35.1.

TABLE 35.1 HUC 10 and HUC 12 Accuracy Assessments and % IC Comparison

Catchment Number	HUC 12 Catchment	NAIP (%IC)	NLCD (%IC)	UVM (%IC)	NAIP Overall Accuracy (%)	NAIP Producer's Accuracy (IC)	NAIP User's Accuracy (IC)
1	Ball Mountain Brook	2.33	1.20	[a]	98.5	100.0	97.0
2	Bloody Brook	2.88	2.71	[a]	96.5	100.0	93.0
3	Branch Brook	3.05	1.90	[a]	98.5	100.0	97.0
4	Calendar Brook	1.83	0.91	[a]	95.0	100.0	90.0
5	East Creek	3.35	3.08	3.17	98.0	100.0	96.0
6	HW Ottauquechee	2.40	1.72	[a]	96.5	100.0	93.0
7	Jay Branch	1.88	1.00	1.77	95.0	100.0	90.0
8	Jewett Brook	6.96	7.19	6.94	94.0	98.9	89.0
9	Lulls Brook	3.45	3.68	[a]	95.5	100.0	91.0
10	Mallets Bay	3.92	3.87	4.19	99.0	100.0	98.0
11	Mallets Creek	2.80	1.97	2.77	93.0	96.7	89.0
12	Mill Brook	2.30	1.05	2.01	97.0	100.0	94.0
13	Mill River	4.15	3.63	4.10	96.5	99.0	94.0
14	Moon Brook	6.81	7.56	6.61	95.5	98.9	92.0
15	Muddy Brook	6.89	7.34	7.22	96.5	100.0	93.0
16	N Branch Deerfield	3.28	1.78	[a]	94.0	100.0	88.0
17	Snipe Island	2.25	1.58	2.47	96.0	98.9	93.0
18	Winooski	10.95	13.15	11.69	97.0	99.0	95.0
Catchment Number	HUC 10 Catchment	NAIP (%IC)	NLCD (%IC)	UVM (%IC)	NAIP Overall Accuracy (%)	NAIP Producer's Accuracy (IC)	NAIP User's Accuracy (IC)
1	Black River	2.16	1.56	[a]	96.5	98.9	92.0
2	Mallets	3.58	3.27	3.75	93.8	98.8	85.0
3	Ottauquechee	1.66	1.03	[a]	94.9	98.9	88.0
4	Shelburne	5.06	5.45	6.12	96.1	98.9	91.0

NAIP, National Agriculture Imagery Program; NLCD, National Land Cover Database; UVM, University of Vermont.
[a] Data not available.

this scale but were not included in the analysis because they are administrative boundaries that cross catchment boundaries and lack ecological significance. HUCs are defined as drainage areas that are delineated into nested, hierarchical classifications of surface areas draining to a specific point, based on hydrologic principles (USGS and USDA 2013). HUCs range in size from HUC 2 to HUC 16, with HUC 2s representing the broadest national scale and HUC 16s representing the finest local scale. The HUC 10 units ranged in size from 190.3 to 578.4 km^2. HUC 12 units ranged in size from 55.1 to 165.4 km^2. The HUC 12 catchments were selected because they were determined to contain the highest number of stormwater best management practices in the state of Vermont, over a range of IC. All HUC 10 and HUC 12 units contained less than 13.15% IC, making comparisons at higher levels of IC impossible (Table 35.1).

In order to increase the number of samples over a wider range of IC and catchment size, IC was calculated for 902 NHDPlus catchments within the study area for both the high spatial resolution and NLCD data. There are 5,982 NHDPlus catchments fully or partially contained in Vermont; thus, approximately 15% of the catchments in Vermont were analyzed, primarily within urbanized areas. Selected NHDPlus catchments ranged in size from 0.001 to 55.2 km^2. A comparison of NAIP and NLCD classifications of NHDPlus catchments with low (<10%), intermediate (10%–25%), and high (>25%) levels of IC showed various levels of accuracy across different landscape types (Figure 35.2).

	NAIP	NLCD
NHD Catchment: 4578214 1.2% IC NAIP 0.24% IC NLCD <10% IC	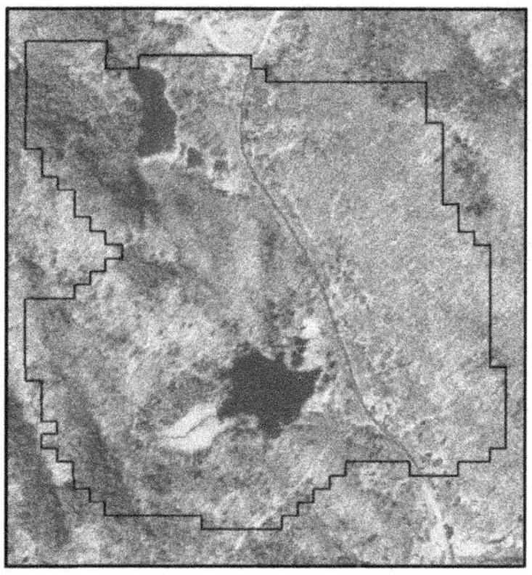	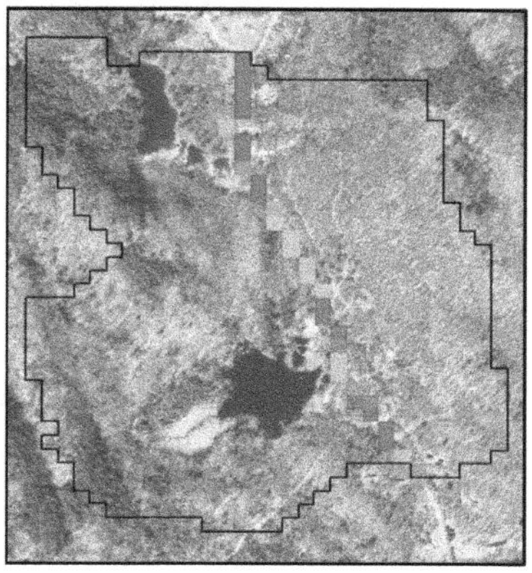
NHD Catchment: 22220891 22.5% IC NAIP 31.6% IC NLCD 10-25% IC	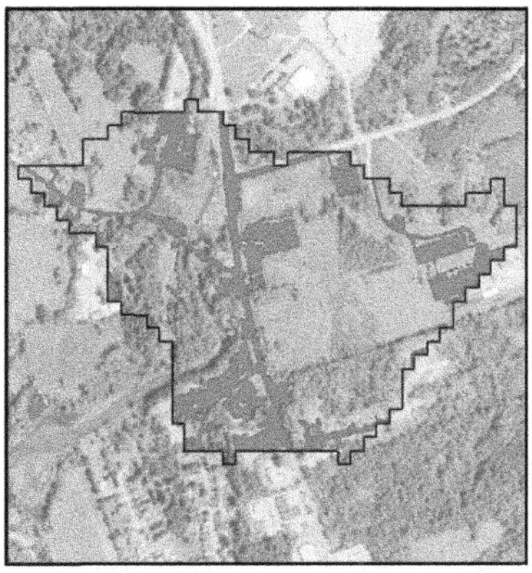	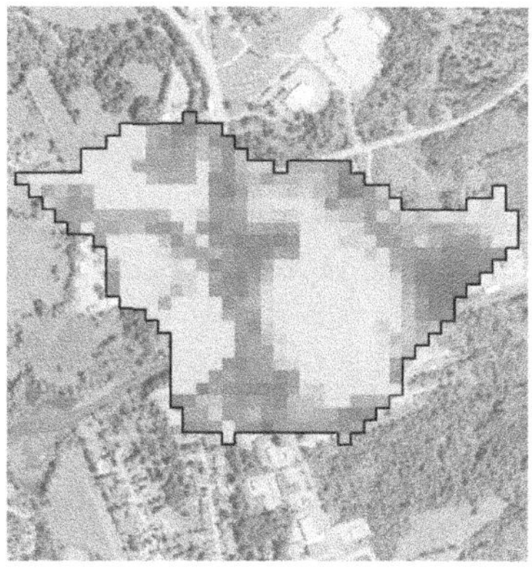
NHD Catchment: 6090067 28.3% IC NAIP 48.9% IC NLCD >25% IC	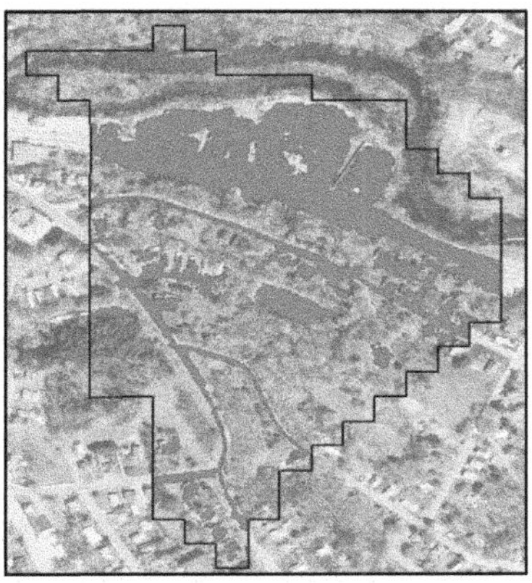	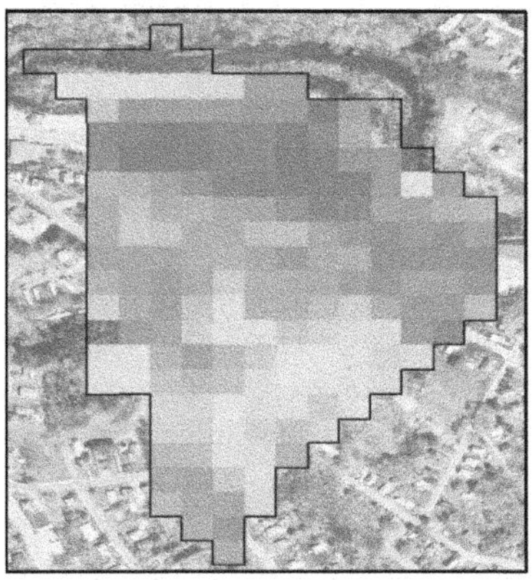

FIGURE 35.2 (See color insert.) Comparison of NAIP and NLCD IC classifications in select NHDPlus catchments.

Data Preparation

NAIP imagery was selected because it is readily available at a national scale at frequent temporal intervals; has a fine spatial resolution; supports red, green, blue (RGB), and near-infrared (NIR) spectral bands, is low or no cost, and has been acquired using consistent standardized United States Department of Agriculture (USDA) Farm Service Association protocols (USDA 2013). One challenge of working with NAIP data is that it is recorded during the growing season and tree canopy cover can complicate characterization of IC. We added road networks from the Vermont E911 dataset (VGCI 2015) to compensate for forest and tree overhang inherent in data collected during the growing season. We created Normalized Difference Vegetation Index (NDVI) bands and added to the five-band NAIP data in ArcGIS 10.1 (ESRI, Redlands, CA). Recent studies have suggested that adding an NDVI band to NAIP imagery greatly improves mapping of water/wetland areas as well as identifying impervious surfaces (O'Neil-Dunne 2013). NDVI is traditionally used for estimating vegetation properties from remote sensing data and is calculated using the equation:

$$\text{NDVI} = \frac{(\text{NIR} - \text{VIR})}{(\text{NIR} + \text{VIR})}$$

where NIR = the value of the near-infrared band and VIR = the value of the visible infrared band.

Impervious Cover Classifications

IC data were extracted utilizing GeniePro2.4 (Observera, Inc. Los Alamos, NM). GeniePro uses genetic algorithms and supervised classification to analyze multispectral remote sensing imagery. In addition to spectral input vectors, GeniePro also incorporates spatial relationships (texture, shape, proximity) and evolutionary algorithms to classify imagery (Harvey et al. 2002). IC classifications were validated against a high-resolution dataset from the University of Vermont (UVM) Spatial Analysis Laboratory, which was based on a vector classification of 2011 NAIP imagery using eCognition software (O'Neil-Dunne 2013; VGCI 2013). For this study, the UVM data were converted to raster format and then clipped to individual catchments for comparison to our data, where the two datasets overlapped. IC classifications were then compared to previously established estimates based on 30 m NLCD 2011 IC data (MRLC 2014).

The NHDPlus hydrologic framework includes attributes of the NHD, the National Elevation Dataset, and the Watershed Boundary Dataset (USGS and US EPA 2010). NHDPlus flowlines are based on medium-resolution 1:100,000 scale stream networks and have associated local catchments as well as full upstream drainage basins (USGS 2007). To compare NAIP and NLCD IC classifications within NHDPlus V1 catchments, landscape attributes (% 2011 NLCD IC) were allocated to the catchments using the NHDPlus Ca3T program (Horizon Systems, Herndon, VA). Mean percent IC was calculated for both the NAIP and UVM datasets using the zonal statistics tool in ArcGIS 10.1. The USGS cautions against using NLCD data in catchments less than tens of square kilometers (USGS 2012). However, a multiple extents accuracy assessment of NLCD land use and land cover suggests that NLCD may be accurate for spatial extents as small as 10 km^2 or larger, particularly for predominant land use classes or those with unique spectral signatures (Hollister et al. 2004). Fourteen NHDPlus catchments were eliminated from the analysis due to very small catchments for which characterization was limited by the resolution of the 30 m NLCD data (i.e., one or two pixels caused a large variation in the percent IC) for a total of 888 catchments (Figure 35.3). These 14 outlier catchments were less than 1 km^2, well below the recommended threshold for analysis of 30 m NLCD data (USGS 2012).

Post-Classification Accuracy Assessment

IC characterization was checked at various stages. Initial IC characterization of preliminary datasets in GeniePro was checked visually to confirm that mapping methods were accurately delineating

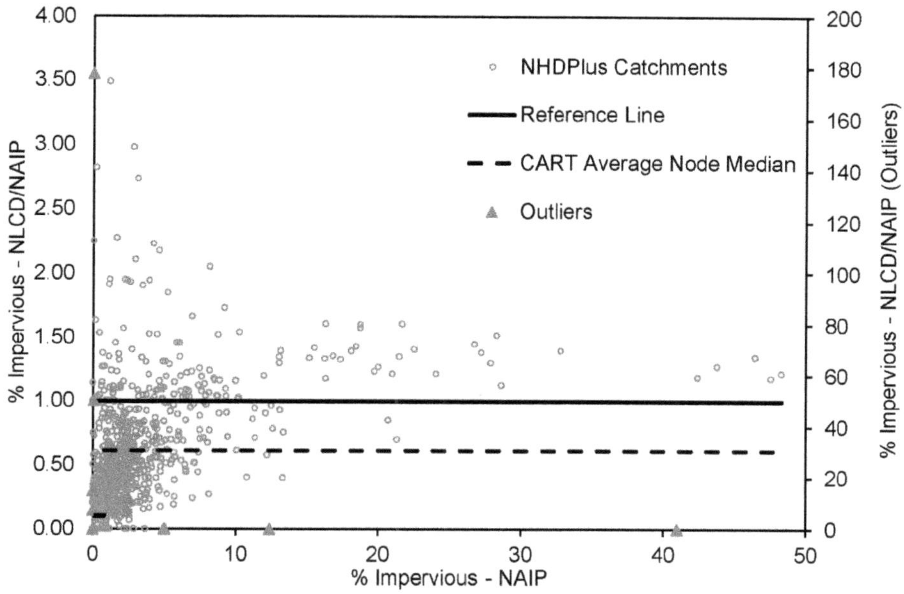

FIGURE 35.3 Ratio of medium-to-high spatial resolution IC classifications at the NHDPlus catchment scale over a range of % IC classified from NAIP data. Outliers removed from analysis are shown on the secondary y-axis. Solid line represents a perfect ratio of NLCD to NAIP data with slope=0. Dashed lines represent CART average node medians of 0.10 and 0.61 for the ratio of NLCD to NAIP IC. Median cut value=0.77% IC.

separate land use features. A post-classification accuracy assessment was performed for the 1 m NAIP data utilizing the methods of Stehman and Czaplewski (1998). Post-classification accuracy assessments were conducted at the HUC 10 and HUC 12 scales. Reference data included the original NAIP imagery and Google Earth historical imagery. Stratified random sampling of the reference data was performed with 100 points per class and compared to the mapped land cover classifications. Overall, user's and producer's accuracies were calculated to describe the accuracies of the IC classifications. Overall, accuracy describes the number of correctly identified pixels divided by the total number of classified pixels. Producer's accuracy measures the ratio between the number of correctly classified pixels and the total number of reference pixels for that category (probability that IC on the ground is correctly classified). User's accuracy measures the ratio of correctly classified pixels to the total number of pixels in that category (probability that a pixel labeled as IC is actually IC).

Statistical Analysis

A regression analysis was run to identify the relationship between NLCD and NAIP data. The ratio of NLCD to NAIP IC was then compared to the NAIP data within the NHDPlus catchments to identify where the NLCD data are adequate and where they may be overestimating or underestimating IC. Classification and Regression Tree (CART) analyses were run using Systat 13.1 (Systat Software, Inc. San Jose, CA) to quantitatively assess the level at which the NLCD underestimates or overestimates IC. CART analysis compares independent and dependent variables through a series of binary splits (Breiman et al. 1984). CART was run for all NHD catchments in the study area using the least absolute deviation option for choosing splits with bootstrapping (sample of 800 repeated 1,000 times, maximum = 2 splits, p=0.05 stopping rule). CART was rerun for the 41 catchments larger than 10 km^2 in the study area using the least absolute deviation option for choosing splits with bootstrapping (sample of 38 repeated 1,000 times, maximum = 2 splits, p = 0.05 stopping rule).

Results

Post-Classification Accuracy Assessment

Overall accuracies of HUC 10 and HUC 12 IC classifications for the NAIP data ranged from 93.0% to 99.0% with an average accuracy of 95.3% for HUC 10s and 96.2% for HUC 12s (Table 35.1). Producer's accuracies were close to 100% for all catchments and user's accuracies ranged from 85.0% to 98.0%. Accuracy assessments were not conducted for individual NHDPlus catchments because all catchments were contained within the HUC 12 and HUC 10 study areas. Accuracy assessment results did not vary by catchment size or by level of urban development.

Impervious Cover Classification Comparison

IC characterized from NAIP imagery ranged from 1.83% to 10.95% in the HUC 12 units and from 1.66% to 5.06% in HUC 10 units (Table 35.1). The NAIP data at the HUC 10 and HUC 12 level predicted NLCD classifications well but with a bias shown by the regression equation (Figure 35.4):

$$\text{NLCD} = 1.2 + 0.74(\text{NAIP}) \quad r^2 = 0.98 \ (n = 22,\ p < 0.001).$$

The ratio of NLCD IC to NAIP IC was plotted against NAIP IC (Figure 35.5). An idealized relationship would yield a slope of 0 across all levels of IC (ratio = 1), indicating no difference between the two classifications. The regression equation for the NLCD-to-NAIP ratio resulted in a nonlinear relationship, with NLCD underestimating at low levels of IC and overestimating at higher levels of IC (Figure 35.5):

$$\text{NLCD/NAIP} = 0.33 + 0.39\ln(\text{NAIP}) \quad r^2 = 0.71 \ (n = 22,\ p < 0.001).$$

A comparison of the ratio of the UVM to NAIP data shows a ratio closer to 1 across all levels of IC, resulting in a slope closer to 0, with 95% confidence intervals between −0.01 and 0.03 (Figure 35.5):

$$\text{UVM/NAIP} = 0.97 + 0.01(\text{NAIP}) \quad r^2 = 0.08 \ (n = 13,\ p = 0.34).$$

A pattern emerged of NLCD underestimating at low levels of IC (<5%) by up to 50% and overestimating by up to 20% at higher levels of IC (>6%), although there are fewer data points at the upper end of the scale (Figure 35.5).

Statistical Analysis

To further elucidate the relationship between the high-resolution estimates and the NLCD data across a range of IC at a finer scale, IC classifications were compared across NHDPlus catchments. There are IC in NHDPlus catchments which ranged from 0% to 48.31% (Figure 35.3). With more sample points, it is clear that the NLCD is underestimating IC at low levels of development, although there is a wider range of NLCD-to-NAIP ratios at lower levels of IC, suggesting greater uncertainty in the data. Across all NHDPlus catchments, the subset of classification trees with a single split had a median cut value of 0.77 for high-resolution IC (95% confidence interval 0.18–3.59), with average node medians of 0.10 and 0.61 for the ratio of NLCD to NAIP IC and an average 16.2% reduction in error (Figure 35.3). The median cut value shows that the data shifted at 0.77% IC, with an average NLCD-to-NAIP ratio of 0.10 below that level of IC and an average ratio of 0.61 above that level.

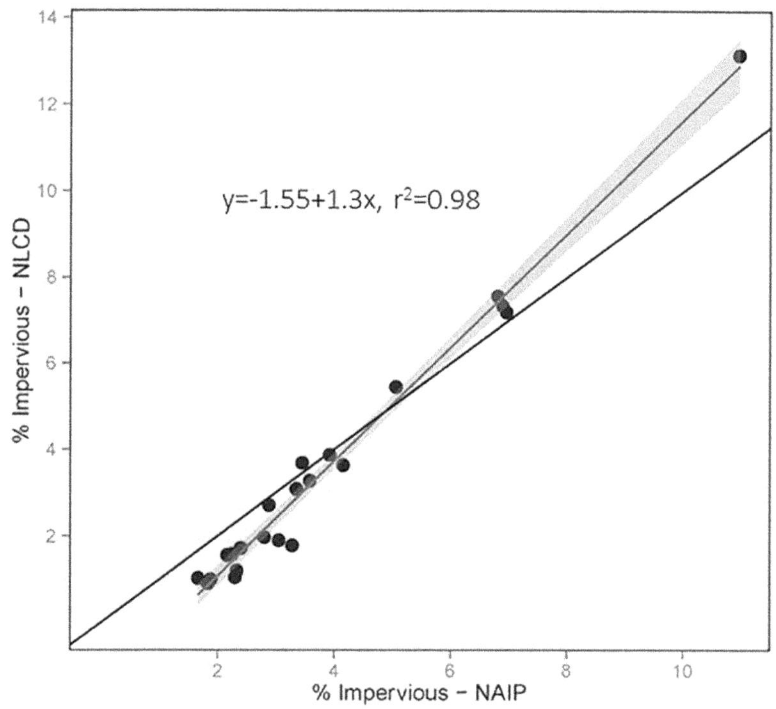

FIGURE 35.4 Relationship of NLCD to NAIP data at the HUC 10 and HUC 12 scale. Shaded area shows 95% confidence interval. Solid line through origin shows 1:1 ratio.

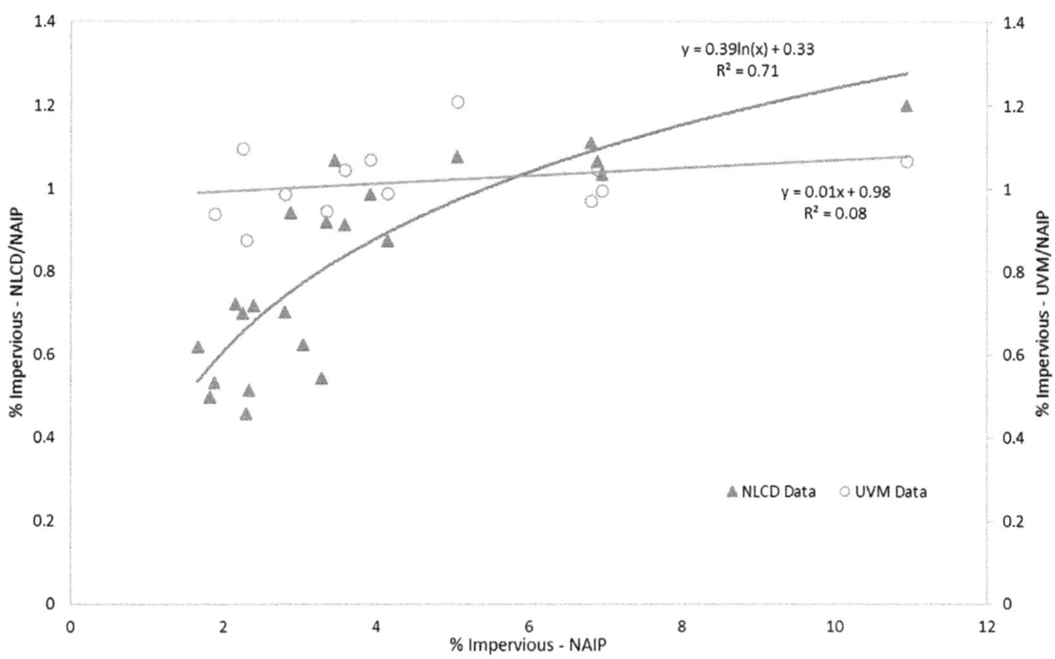

FIGURE 35.5 Ratio of medium to high spatial resolution IC classifications at the HUC 10 and HUC 12 scale over a range of % IC classified from NAIP data.

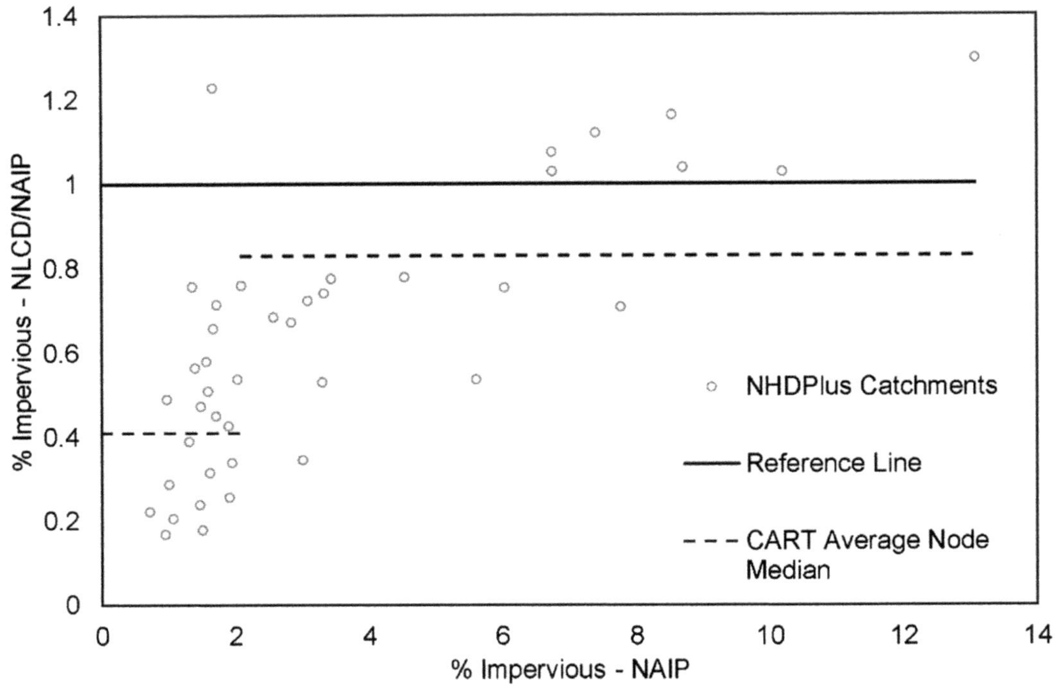

FIGURE 35.6 Ratio of medium to high spatial resolution IC classifications in NHDPlus catchments greater than 10km2. Solid line represents a perfect ratio of NLCD to NAIP data with slope=0. Dashed lines represent CART average node medians of 0.41 and 0.83 for the ratio of NLCD to NAIP IC.

For NHDPlus catchments larger than 10 km^2, the subset of classification trees with a single split had a median cut value of 2.09 for high spatial resolution IC (95% confidence interval 1.36–7.40), with average node medians of 0.41 and 0.83 for the ratio of NLCD to NAIP IC and an average 25.3% reduction in error (Figure 35.6). The median cut value shows that the data shifted at 2.09% IC, with an average NLCD-to-NAIP ratio of 0.41 below 2.09% IC and 0.83 above. Even with the smaller sample size, the data follow a similar trend of underestimating IC by 20%–80% at low levels of development, with less noise in the data than in the sample containing all NHD catchments (Figure 35.6). There were too few samples with greater than 10% IC to identify trends in the data at higher levels of development.

Discussion

Classification of IC from high-resolution NAIP data produced a highly accurate dataset for eighteen HUC 12 units and four HUC 10 units in Vermont. Accuracy levels were quite high in all catchments and supported by a comparison to the UVM high spatial resolution dataset from the same 2011 NAIP data. Producer's accuracy was expected to be very high due to the fact that most pixels within the catchments are not impervious; thus, the chance of correctly identifying a pixel as not impervious is somewhat biased. The user's accuracies were expected to be somewhat lower than the producer's accuracies but were still high and met or exceeded the USGS standard for accuracy assessment. Key areas with higher levels of IC included the MS4 areas in Burlington, Rutland, and St. Albans as well as isolated resort areas. Calculating IC by NHDPlus catchment provided a broader gradient of urbanization to compare differences between highly developed catchments and catchments with low levels of IC. The data clearly indicate that NLCD is underestimating IC at low levels of development and overestimating at higher levels of development.

High-resolution classifications of IC are needed for effective modeling and management of water resources at the catchment scale. Medium-resolution data, such as the NLCD, have traditionally been utilized for regional analyses. However, recent regional analyses have shown impacts to stream macroinvertebrate and periphyton populations at very low levels of IC. Baker and King (2010) found taxon-specific change points of declining macroinvertebrates at 0.81%–3.3% developed land. Declines in macroinvertebrate community metrics have been found at 1%–2% IC and as low as 0.6% IC for diatom communities (Detenbeck et al. 2013; Smucker et al. 2013). Data from the current study suggest that the NLCD data are underestimating IC in this range of low levels of IC. Corrected values using the regression equation for Figure 35.4 would suggest that macroinvertebrate community metrics would be affected at 1.94%–2.68% IC and diatom communities would be affected at as low as 1.65% IC. This bias would be magnified in areas where community metrics are affected at even slightly higher levels of IC (i.e., 3%–10% IC). Further research is needed to identify the impact of the spatial resolution of IC data on regional analyses of stream response to urbanization.

Municipalities across United States are increasingly exploring the concept of stormwater utilities to fund stormwater infrastructure implementation, repair, and maintenance. The IC TMDL developed by the Connecticut Department of Energy and Environmental Protection identifies a target level of 11% IC for both regulated and unregulated sources of stormwater (12% IC target—1% margin of safety) for the Eagleville Brook catchment in Mansfield, CT (Arnold et al. 2010). Both stormwater utilities and IC TMDLs require high spatial resolution estimates of IC for accurate rate setting and target levels of IC. A difference of even 1% accuracy, as shown in this study, could potentially result in significant legal and policy ramifications.

While the data suggest that NLCD is underestimating IC at the catchment scale for low levels of IC and overestimating at higher levels of IC, there are many catchments where there is minimal difference between the two datasets. A predictive model of where the NLCD data are adequate for analyses and where higher resolution estimates are required may be of particular interest to watershed organizations and communities attempting to find effective stormwater management solutions with limited budgets and resources. Residential and commercial properties developed from agricultural and non-forested landscapes may yield more accurate catchment IC estimates than those developed from forested landscapes, due to the lack of overhanging tree cover that can impede classification accuracy. Evaluating the age of housing developments may identify areas of urban fringe and expansion, which may differ in classification accuracies than older, more established neighborhoods with more plentiful canopy cover. The data from this study will be utilized to create a predictive model of high-resolution IC values, utilizing readily available land cover and US Census data.

There are several possible sources of error in the data. First, differences in spatial resolution and methodologies may impact the classification of IC based on the format of the source data and not misclassification (Hollister et al. 2004; Loveland et al. 2005). NLCD IC was characterized from 30 m spatial resolution raster data while the NAIP data were characterized as rasters from 1 m digital orthophotographs. The UVM dataset was classified as a vector and then rasterized for the purposes of this study. All raster data exhibit some degree of "stair stepping" effects, and the comparison and conversion of data from different formats may introduce some error. Second, the reference data used in the accuracy assessments included the source data as well as historical imagery from Google Earth. Congalton and Green (2008) do not recommend using the source data as reference data as it may introduce additional error. However, the imagery was not analyzed until several years after it was taken, thus it was impossible to set up an accurate system for ground referencing in the field. Finally, the analysis of the NHDPlus catchment data includes catchments that are smaller than the spatial extent recommended by the USGS (2012). However, removal of catchments less than 10 km^2 results in a similar trend of underestimating IC at low levels of development by 20%–80%.

Acknowledgments

This document has been reviewed by the US Environmental Protection Agency, Office of Research and Development, and approved for publication. Mention of trade names or commercial products does not constitute endorsement or recommendation for use. This manuscript is contribution number ORD-016058.

References

Allan JD (2004) Landscapes and riverscapes: The influence of land use on stream ecosystems. *Annu Rev Ecol Evol S* 35:257–284.

Arnold CL, Bellucci CJ, Collins K, Claytor R (2010) Responding to the first impervious cover-based TMDL in the nation. *Watershed Sci Bull, J Assoc Watershed Stormwater Prof* 1:11–18.

Arnold CL, Gibbons CJ (1996) Impervious surface coverage: The emergence of a key environmental indicator. *J Am Plann Assoc* 62:243–258.

Baker ME, King RS (2010) A new method for detecting and interpreting biodiversity and ecological community thresholds. *Methods Ecol Evol* 1:25–37.

Brabec E (2002) Impervious surfaces and water quality: A review of current literature and its implications for watershed planning. *J Plan Lit* 16:499–514.

Breiman L, Friedman J, Stone CJ, Olshen RA (1984) *Classification and Regression Trees*. CRC Press, Boca Raton, FL.

Claggett PR, Irani FM, Thompson RL (2013) Estimating the extent of impervious surfaces and turf grass across large regions. *J Am Water Resour As* 49:1057–1077.

Congalton R., Green K (2008) *Assessing the Accuracy of Remotely Sensed Data: Principles and Practices*, 2nd edn. Lewis Publishers, Boca Raton, FL.

Detenbeck NE, Rosiu C, Hayes L, Legros J (2013) Differentiating impacts of watershed development from Superfund sites on stream macroinvertebrates (EPA Report Number EPA/600/R-13/352). U.S. Environmental Protection Agency, Office of Research and Development, National Health and Environmental Effects Research Laboratory, Atlantic Ecology Division, Narragansett, RI.

Harding J, Benfield E, Bolstad P, Helfman G, Jones E (1998) Stream biodiversity: The ghost of land use past. *P Natl Acad Sci USA* 95:14843–14847.

Harvey NR, Theiler J, Brumby SP, Perkins S, Szymanski JJ, Bloch JJ, Porter RB, Galassi M, Young AC (2002) Comparison of GENIE and conventional supervised classifiers for multispectral image feature extraction. *IEEE T Geosci Remote* 40:393–404.

Hollister JW, Gonzalez MJ, Paul JF, August PV, Copeland JL (2004) Assessing the accuracy of National Land Cover Dataset area estimates at multiple spatial extents. *Photogramm Eng Rem S* 70:405–414.

Homer CG, Dewitz JA, Yang L, Jin S, Danielson P, Xian G, Coulston J, Herold ND, Wickham JD, Megown K (2015) Completion of the 2011 National Land Cover Database for the conterminous United States-Representing a decade of land cover change information. *Photogramm Eng Rem S* 81(5):345–354.

Johnson LB, Host, GE (2010) Recent developments in landscape approaches for the study of aquatic ecosystems. *J N Am Benthol Soc* 29:41–66.

Jones J, Jarnagin T (2009) Evaluation of a moderate resolution, satellite-based impervious surface map using an independent, high-resolution validation data set. *J Hydrol Eng* 14(4):369–376.

Loveland TR, Gallant AL, Vogelmann, JE (2005) Perspectives on the use of land cover data for ecological investigations. In: Wiens J, Moss M (eds) *Issues and Perspectives in Landscape Ecology*, Cambridge University Press, Cambridge, UK, pp. 120–128.

Morse CC, Huryn AD, Cronan C (2003) Impervious surface area as a predictor of the effects of urbanization on stream insect communities in Maine, USA. *Environ Monit Assess* 89:95–127.

Multi-Resolution Land Characteristics Consortium (MRLC) (2014) National land cover database 2011 (NLCD2011). Available from www.mrlc.gov/nlcd11_data.php (accessed February 2016).

Nelson KC, Palmer MA (2007) Stream temperature surges under urbanization and climate change: Data, models, and responses. *J Am Water Resour As* 43:440–452.

New Hampshire Department of Environmental Services (NH DES) (2014) Stormwater utilities: Funding municipal stormwater programs. Available from http://des.nh.gov/organization/divisions/water/stormwater/utilities.htm (accessed February 2016).

Nowak DJ, Greenfield EJ (2010) Evaluating the National Land Cover Database tree canopy and impervious cover estimates across the conterminous United States: A comparison with photo-interpreted estimates. *Environ Manage* 46(3):378–390.

O'Neil-Dunne J (2013) Mapping impervious surfaces in the Lake Champlain Basin. Lake Champlain Basin Program Final Report, Technical Report No. 76. University of Vermont Spatial Analysis Laboratory, Burlington, VT. 32 pp.

Parikh P, Taylor MA, Hoagland T, Thurston H, Shuster W (2005) Application of market mechanisms and incentives to reduce stormwater runoff: An integrated hydrologic, economic and legal approach. *Environ Sci Policy* 8:133–144.

Roy AH, Rosemond A, Paul M, Leigh D, Wallace J (2003) Stream macroinvertebrate response to catchment urbanisation (Georgia, USA). *Freshwater Biol* 48:329–346.

Roy AH, Shuster WD (2009) Assessing impervious surface connectivity and applications for watershed management. *J Am Water Resour Assoc* 45:198–209.

Schueler TR (1994) The importance of imperviousness. *Watershed Prot Tech* 1(3):100–111.

Schueler TR (2003) *Impacts of Impervious Cover on Aquatic Systems. Watershed Protection Research Monograph No. 1*, Center for Watershed Protection, Ellicott City, MD.

Schueler TR, Claytor R (1997) Impervious cover as an urban stream indicator and a watershed management tool. In: Roesner LA (ed.) *Proceedings of an Engineering Foundation Conference*, American Society for Civil Engineers, Snowbird, UT, 1997.

Schueler TR, Fraley-McNeal L, Cappiella K (2009) Is impervious cover still important? Review of recent research. *J Hydrol Eng* 14:309–315.

Smucker NJ, Detenbeck NE, Morrison AE (2013) Diatom responses to watershed development and potential moderating effects of near-stream forest and wetland cover. *Freshwater Sci* 32:230–249.

Smucker NJ, Kuhn A, Charpentier MA, Cruz-Quinones CJ, Elonen CM, Whorley SB, Jicha TM, Serbst JR, Hill BH, Wehr JD (2016) Quantifying urban watershed stressor gradients and evaluating how different land cover datasets affect stream management. *Environ Manage* 57:683–695.

Stehman SV, Czaplewski RL (1998) Design and analysis for thematic map accuracy assessment: Fundamental principles. *Remote Sens Environ* 64:331–344.

Stueve KM, Hollenhorst TP, Kelly JR, Johnson LB, Host GE (2015) High-resolution maps of forest-urban watersheds present an opportunity for ecologists and managers. *Landscape Ecol* 30:313–323.

United States Department of Agriculture (USDA) (2013) NAIP imagery information sheet. Available from www.fsa.usda.gov/Internet/FSA_File/naip_info_sheet_2013.pdf (accessed February 2016).

United States Environmental Protection Agency (US EPA) (2002) *National Water Quality Inventory: 2000 Report (EPA Report Number EPA 841-R-02-001)*. U.S. EPA Office of Water, Washington, DC.

United States Environmental Protection Agency (US EPA) (2016) NPDES Phase II Small MS4 General Permit Annual Reporting Requirements. Available from www3.epa.gov/region1/npdes/stormwater/ms4-annual-rpt.html (accessed February 2016).

United States Geological Survey (USGS) (2007) National Hydrography Dataset. Available from http://nhd.usgs.gov (accessed February 2016).

United States Geological Survey (USGS) (2012) Accuracy Assessment of 1992 National Land Cover Data. Available from http://landcover.usgs.gov/accuracy/ (accessed February 2016).

United States Geological Survey (USGS) and United States Department of Agriculture (USDA) (2013) *Federal Standards and Procedures for the National Watershed Boundary Dataset (WBD)*, 4th edn. United States Geological Survey and United States Department of Agriculture, Reston, VA.

United States Geological Survey (USGS) and United States Environmental Protection Agency (US EPA) (2010) NHDPlus Version 1 User Guide. Available from www.horizon-systems.com/NHDPlus/NHDPlusV1_documentation.php (accessed February 2016).

van der Tak L, Matichich M, Fraley R, Virts M (2014) User-Fee Funded Stormwater Programs, 2013 Update to the 1994 WEF Manual. *Proc Water Environ Federation* 2014:1–25.

Vermont Center for Geographic Information (VGCI) (2013) LandLandcov_IMPERVLCB2011 Dataset. Available from http://vcgi.vermont.gov/opendata (accessed February 2016).

Vermont Center for Geographic Information (VGCI) (2015) EmergencyE911_RDS Dataset. Available from http://vcgi.vermont.gov/opendata (accessed February 2016).

Walsh CJ, Roy AH, Feminella JW, Cottingham PD, Groffman PM, Morgan RP (2005) The urban stream syndrome: Current knowledge and the search for a cure. *J N Am Benthol Soc* 24:706–723.

Wang L, Kanehl, P (2003) Influences of watershed urbanization and instream habitat on macroinvertebrates in cold water streams. *J Am Water Resour Assoc* 39:1181–1196.

Wickham J, Neale A, Mehaffey M, Jarnagin T, Norton D (2016) Temporal trends in the spatial distribution of impervious cover relative to stream location. *J Am Water Resour Assoc* 52(2):409–419.

Zhang G, Guhathakurta S, Dai G, Wu L, Yan L (2013) The control of land-use patterns for stormwater management at multiple spatial scales. *Environ Manage* 51:555–570.

Zhou Y, Wang, Y (2008) Extraction of impervious surface areas from high spatial resolution imagery by multiple agent segmentation and classification. *Photogramm Eng Rem S* 74:857–868.

Zhou Y, Wang Y, Gold AJ, August, PV (2010) Modeling watershed rainfall–runoff relations using impervious surface-area data with high spatial resolution. *Hydrogeol J* 18:1413–1423.

Zhou Y, Wang Y, Gold AJ, August PV, Boving TB (2014) Assessing impact of urban impervious surface on watershed hydrology using distributed object-oriented simulation and spatial regression. *GeoJournal* 79:155–166.

36

Analysis of Impervious Cover in Riparian Zones

Jessica Morgan
United States Coast Guard Academy

Yeqiao Wang
University of Rhode Island

Naomi Detenbeck
U.S. Environmental Protection Agency

Introduction 291
Methods 292
Defining the Riparian Zone • Delineating Riparian Zones • Impervious Cover Mapping • Statistical Analysis
Results 295
Riparian Buffer Zones • Statistical Analysis
Discussion 300
Acknowledgments 300
References 300

Introduction

Riparian buffer areas provide a number of functions for maintaining the integrity of stream ecosystems threatened by encroaching urbanization, including water temperature regulation; bank stability; sources of organic and inorganic material; energy dissipation; and nutrient, sediment, and contaminant retention (Allan 2004, Chen et al. 1998, Gregory et al. 1991, Naiman and Decamps 1997, Naiman et al. 2005). Further, forested and wetland riparian buffer areas have been shown to help to mitigate the effects of urbanization, even when the natural functions of these zones are altered by stormwater drainage infrastructure (Smucker et al. 2013). Recognizing the key role that buffer areas play in preserving water quality, municipalities across the Northeast have enacted regulations to protect and restore riparian areas using a variety of methods for determining optimal buffer widths.

Riparian buffer zone regulations vary by state and municipality, and many analyses for informing stormwater regulations utilize fixed-width buffers. Most commonly, these analyses create 30–120 m buffer zones, based on the 30 m pixel width of the National Land Cover Database (NLCD) (Goetz 2006). Previous analyses have defined 120 m fixed-width buffer distances along flow paths and summarized land use types (urban, agriculture, etc.) at the riparian and catchment scale (Detenbeck et al. 2013, Smucker et al. 2013). Several studies have found that buffer distances smaller than 100 m were not accurate predictors of fish or macroinvertebrate populations (Lammert and Allan 1999, Roth et al. 1996, Van Sickle et al. 2004, Vølstad et al. 2003). In another study, a 30 m buffer was found to be a weak secondary predictor of stream health, with regional land use undermining the ability of vegetation to maintain high quality habitat (Roth et al. 1996). Fixed-width buffers may underestimate the actual riparian boundary by up to 2.5 times the distance from the stream (Skally and Sagor 2001). Further, it has been suggested that medium spatial resolution data such as the NLCD may not be adequate for mapping riparian zone features (Baker et al. 2007, Fernández et al. 2014, Hollenhorst et al. 2006).

Accurate mapping of riparian zone condition from remote sensing data presents several challenges, including the determination of functional buffer width (Baker et al. 2006) and the spatial resolution of elevation (Abood et al. 2012), land use and land cover (Hollenhorst et al. 2006), and stream map data (Baker et al. 2007). Fixed-width buffers may not correspond to relevant ecotones and can include areas that are irrelevant to the functions of buffers, causing the misinterpretation of landscape patterns (Baker et al. 2006). Functional riparian metrics have been developed based on the connectivity of source land cover area to the stream channel (Baker et al. 2006), and distance weighting has been used to examine land use/land cover influence on stream health (Van Sickle and Johnson 2008). More recent studies have used elevation and flood height data to delineate riparian buffer zones based on readily available digital elevation models (DEMs) and the 50-year flood plain (Abood et al. 2012, Fernández et al. 2012).

Urbanization expansion increasingly affects riparian zones, causing hydrologic changes, which then cause changes to soil, vegetation, and the ability to filter pollutants (Groffman et al. 2003). While urbanization and increasing amounts of IC are known stressors on stream ecosystem health (Schueler et al. 2009, Walsh et al. 2005), proximal IC, i.e., IC that is adjacent to streams, may disproportionally affect water quality compared to IC distributed throughout the watershed (Wickham et al. 2016). While 27% of streams in the United States have proximally distributed IC, in watersheds that had spatial changes over time, most had increases in IC across the entire watershed compared to increases near surface waters (Wickham et al. 2016). This may have important implications for IC Total Maximum Daily Load (TMDL) allocations that are increasingly being developed to restore water quality (Wickham et al. 2016). Further, NLCD data have been shown to underestimate IC in less developed catchments and overestimate IC in more developed catchments (Claggett et al. 2013, Roy and Shuster 2009, Smucker et al. 2016) while others have found that NLCD underestimates IC regardless of development intensity (Jones and Jarnagin 2009). Higher resolution estimates of IC in the riparian zone are needed for more accurate estimates of in-stream response to urbanization.

This chapter presents a study designed to (i) determine the effects of spatial resolution of IC estimates in the riparian zone and (ii) compare IC derived from fixed-width and elevation-based buffers. Given the smaller spatial area of delineated riparian buffers, it was expected that the NLCD would underestimate proximally distributed IC. Based on the literature, it was expected that fixed-width buffers would underestimate IC when compared to elevation-based buffers. Both spatial resolution of IC data and delineation method are critical considerations for accurate modeling of water resources.

Methods

Defining the Riparian Zone

The study area included riparian buffer areas of four Hydrologic Unit Code (HUC) 10 catchments and eighteen HUC 12 catchments in the state of Vermont (Figure 36.1). Definitions of the riparian zone are many, varied, and dependent on the resource being evaluated. Functional definitions of riparian areas focus on the ecosystem services provided by riparian zones rather than their individual components (Ilhardt et al. 2000, Verry et al. 2004). For the purposes of this study, a functional riparian zone is defined as "a three-dimensional space of interaction that includes terrestrial and aquatic ecosystems that extend down into the groundwater, up above the canopy, outward across the floodplain, up the near slopes that drain to the water, laterally into the terrestrial ecosystem, and along the water course at a variable width." (Verry et al. 2004). This study focuses on the riparian zone up to the 50-year flood height.

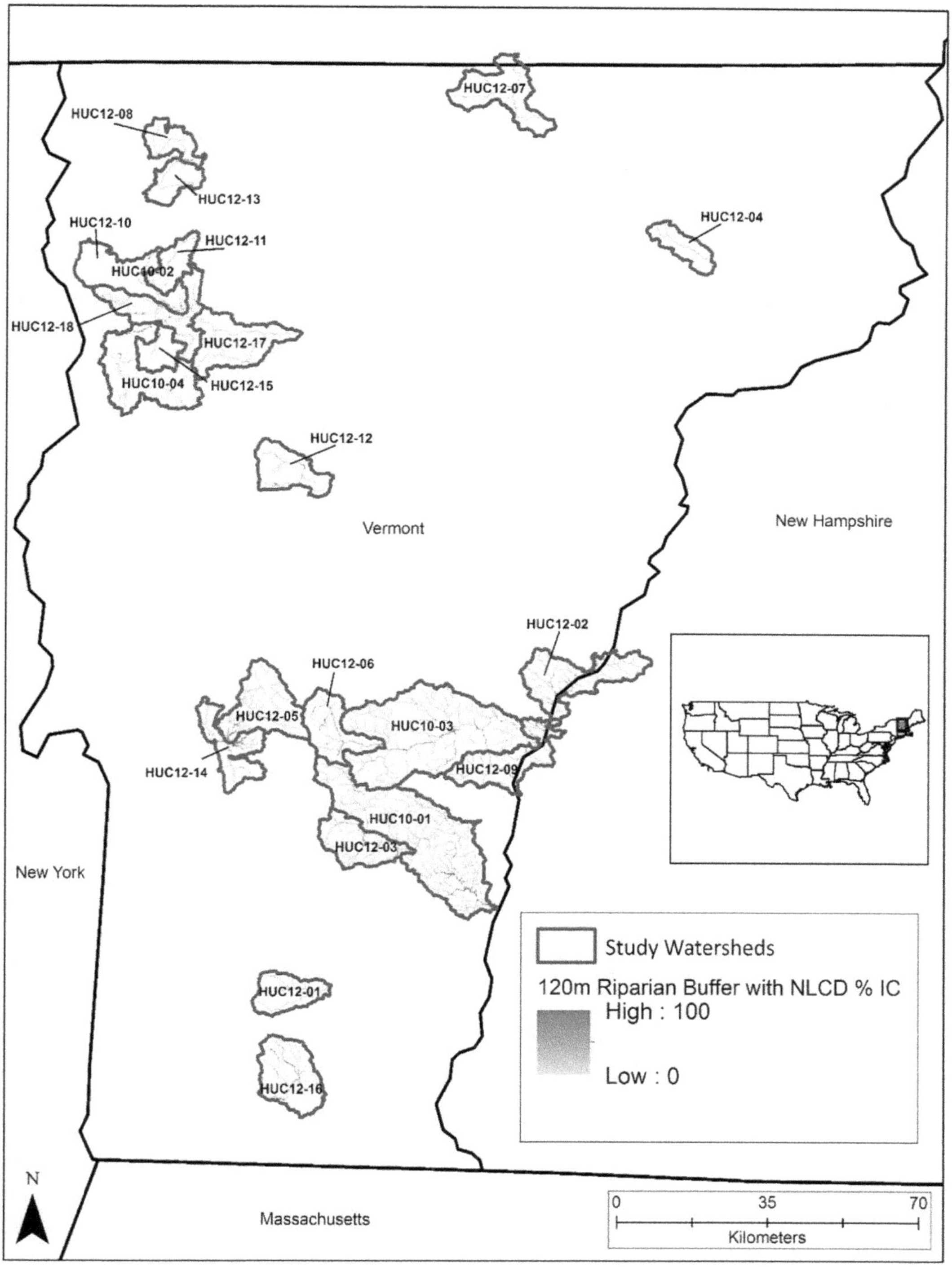

FIGURE 36.1 Map of study area—catchment labels correspond to catchment and catchment number in Tables 36.1 and 36.2.

Delineating Riparian Zones

Fixed-Width Buffers

Fixed-width buffers were generated along flow paths from the National Hydrography Dataset Plus (NHDPlus) Version 1 stream network (USGS 2007). The NHDPlus hydrologic framework includes attributes of the NHD, the National Elevation Dataset, and the Watershed Boundary Dataset (USGS and US EPA 2010). NHDPlus flowlines are based on medium-resolution 1:100,000 scale stream networks and have associated local catchments as well as full upstream drainage basins (USGS 2007). Fixed-width buffers of 120 m were generated using the ArcGIS FLOWPATH tool based on the distance along the flow path rather than perpendicular to the flow path (Detenbeck et al. 2013, Detenbeck et al. 2016). This generated a distance from the NHDPlus flowline, which was then classified into two categories: buffer and non-buffer areas. This process resulted in a buffer along the flow path of 120 m on each side. A null conditional was set in ArcGIS Spatial Analyst to make the buffer area = 1 and the non-buffered areas = NoData.

Elevation-Based Buffers

Elevation buffers were created with the Riparian Buffer Delineation Model (RBDM) ArcGIS Toolbox (Abood et al. 2012) using ArcGIS 10.4.1(ESRI, Redlands, CA). The toolbox allows delineation of riparian buffers using DEMs and flood height data. Required inputs include a vector catchment boundary, the NHDPlus stream lines, NHDPlus water bodies, either a 10 m or 30 m DEM, and optionally, classified land use/land cover data. The stream map requires the Strahler stream order (Strahler 1957) from the NHDPlus Flowline Value Added Attributes table and a specified flood depth for the 50-year floodplain. The 50-year floodplain was calculated as twice the bankfull depth, as that elevation is most commonly associated with the first stream terrace and likely to extend to areas of steep slopes (Bent and Waite 2013). Bankfull depth was estimated by the equation:

$$\text{Bankfull mean depth}(\text{ft}) = 0.9502\left[\text{Drainage area}\left(\text{mi}^2\right)\right]^{0.2960}.$$

The bankfull mean depths and associated 50-year floodplain elevations for each HUC 10 and HUC 12 unit are shown in Table 36.1. Bankfull depth was held constant over each HUC 10 and HUC 12.

Ten-meter DEMs were obtained from the 3D Elevation Program (3DEP) (formerly the National Elevation Dataset) 1/3 arc second (10 m) continuous dataset (Gesch et al. 2014). Data were projected into NAD83 Albers using nearest neighbor interpolation. It was necessary to reproject the data prior to processing because the RBDM automatically extracts percent IC by area based on a classified land use and land cover dataset. Two HUC 12 units overlapped into New Hampshire. In these cases, the NED data from each state were downloaded and mosaicked together with the mosaic operator set to take the values of the Vermont raster, where data overlapped. One HUC 12 unit overlapped into Canada so the catchment boundary was clipped at the Canadian border, as the NLCD does not extend into Canada.

The RBDM calculates 360° radial transects from sample points along the stream to calculate elevation data, based on a user-specified maximum transect length. Abood et al. (2012) suggested setting a maximum transect length of 250–500 m, while balancing floodplain characteristics and processing time. Preliminary analysis showed that setting a 500 m transect length had minimal improvement in riparian buffer delineation, resulting in a 0.1% change in IC calculations but greatly increased processing time, so the maximum transect length was set at 250 m.

Impervious Cover Mapping

Percent IC for both the fixed-width and elevation-based buffers were extracted from the 30 m NLCD 2011 IC data (MRLC 2014) and 1 m 2011 high spatial resolution data derived from the National Agriculture Imagery Program (NAIP) for the selected HUCs in Vermont. The resulting raster was used to calculate

TABLE 36.1 Bankfull Depth by HUC 10 and HUC 12 Catchment

Catchment Number	HUC 12 Catchment	Area (km²)	Bankfull Depth (ft)	2 × Bankfull Depth (m)
1	Ball Mountain Brook	87.18	2.79	1.70
2	Bloody Brook	165.43	3.37	2.05
3	Branch Brook	114.13	3.02	1.84
4	Calendar Brook	60.01	2.50	1.52
5	East Creek	159.06	3.33	2.03
6	HW Ottauquechee	114.96	3.03	1.84
7	Jay Branch	136.69	3.19	1.94
8	Jewett Brook	59.96	2.50	1.52
9	Lulls Brook	139.51	3.20	1.95
10	Mallets Bay	136.55	3.19	1.94
11	Mallets Creek	61.45	2.51	1.53
12	Mill Brook	98.69	2.89	1.76
13	Mill River	60.01	2.50	1.52
14	Moon Brook	100.49	2.91	1.77
15	Muddy Brook	55.14	2.43	1.48
16	N Branch Deerfield	144.93	3.24	1.98
17	Snipe Island	152.20	3.29	2.00
18	Winooski	127.87	3.12	1.90
Catchment Number	HUC 10 Catchment	Area (km²)	Bankfull Depth (ft)	2 × Bankfull Depth (m)
1	Black River	528.94	4.76	2.90
2	Mallets	197.99	3.56	2.17
3	Ottauquechee	578.36	4.88	2.98
4	Shelburne	190.34	3.51	2.14

the percent IC within the buffer. Zonal statistics in the Spatial Analyst toolbox were used to calculate NLCD 2011 IC within the buffer area for each HUC 10 and HUC 12. Zonal statistics were run to calculate the mean percent IC for each buffer zone in the HUC 12 and HUC 10 units for the fixed-width buffers. The RBDM tool automatically calculated % IC from the high spatial resolution data as an optional input. To maintain consistency with the RBDM tool, which removes water bodies (e.g., lakes) from the final output, water bodies were removed from the fixed-width buffers as well.

Statistical Analysis

A regression analysis was run to identify the relationship between NLCD and NAIP data among HUC 10 and HUC 12 units for both fixed-width and elevation-based buffers. The ratio of NLCD to NAIP IC was then compared to the NAIP data within the buffer types to identify where the NLCD data are adequate and where they may be overestimating or underestimating IC. Finally, high spatial resolution IC data were compared among buffer types to identify differences in the total percent IC.

Results

Riparian Buffer Zones

Elevation buffers generally had greater spatial extents than those generated by the fixed-width method (Figure 36.2). IC characterized from NAIP imagery by both methods ranged from 1.58% to 8.67% in the buffer zone of HUC 12 units and from 3.79% to 5.42% in HUC 10 units (Table 36.2). IC characterized from NLCD imagery by both methods ranged from 1.40% to 8.81% in the HUC 12 units and from 3.07%

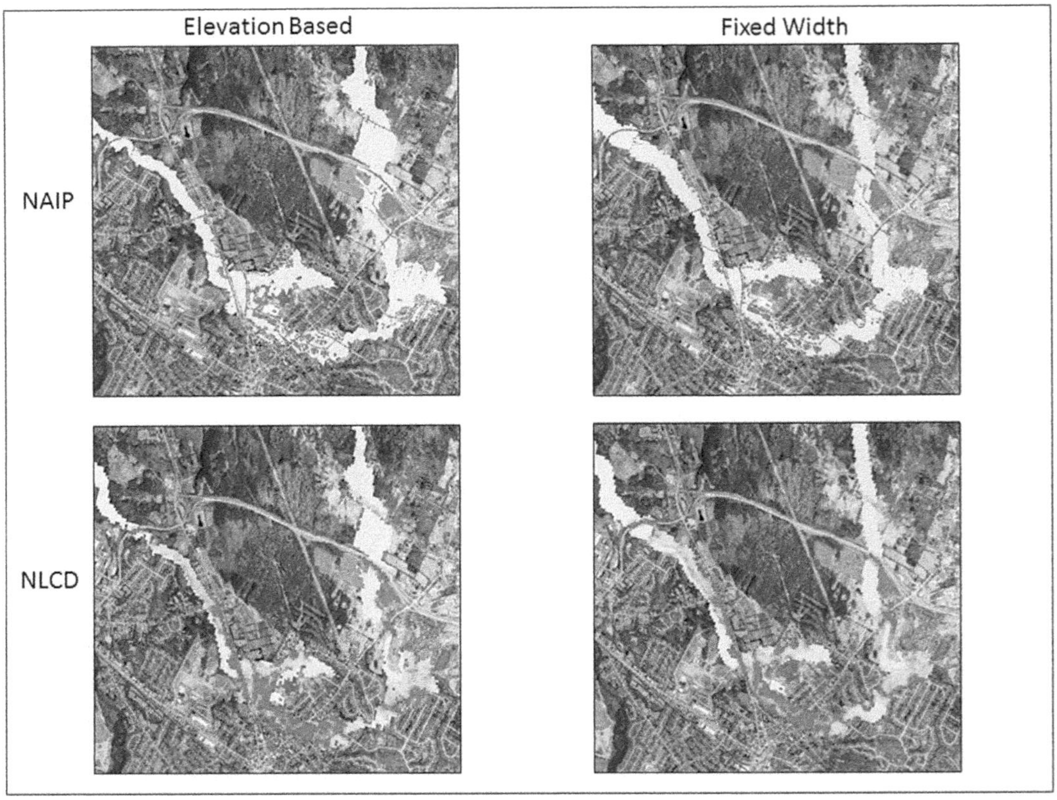

FIGURE 36.2 **(See color insert.)** Comparison of delineated riparian buffer zone by data (NAIP, NLCD) and buffer type (elevation based, fixed width).

to 5.48% in HUC 10 units (Table 36.2). As expected, greater percentages of IC in both types of buffer zones were found in the more urbanized catchments such as those located in Burlington, Rutland, and St. Albans.

Statistical Analysis

The NAIP data at the HUC 10 and HUC 12 level predicted NLCD classifications well for both types of buffers, with some noise in the data (Figures 36.3 and 36.4). The fixed-width buffer percent IC classifications trended toward NLCD underestimating the amount of IC in the buffer area at the lowest levels of IC but the difference was not significant (95% confidence interval or CI of slope 0.77–1.41, intercept $p = 0.295$). The elevation-based buffers showed an opposite trend of overestimating at higher lower levels of IC, but again, the difference was not significant (95% CI 0.57–1.03 of slope, intercept $p = 0.254$). The regression equations for each buffer type are shown below.

$$\textit{Fixed-width}(\text{Figure 36.3}): \text{NLCD} = -0.80 + 1.09(\text{NAIP})$$

$$r^2 = 0.72,\ n = 22,\ p < 0.001,\ 95\%\ \text{CI for slope} = 0.77 - 1.41$$

$$\textit{Elevation based}(\text{Figure 36.4}): \text{NLCD} = 0.68 + 0.80(\text{NAIP})$$

$$r^2 = 0.72,\ n = 22,\ p < 0.001,\ 95\%\ \text{CI for slope} = 0.57 - 1.03$$

TABLE 36.2 HUC 10 and HUC 12 Percent Impervious Cover (IC) Comparison by Buffer Type

Catchment Number	HUC 12 Catchment	Area (km²)	NAIP (%IC) Fixed-Width	NAIP (%IC) Elevation Based	NLCD (%IC) Fixed-Width	NLCD (%IC) Elevation Based
1	Ball Mountain Brook	87.18	3.00	3.83	1.40	1.80
2	Bloody Brook	165.43	2.84	3.21	2.13	2.71
3	Branch Brook	114.13	5.80	7.86	4.42	6.22
4	Calendar Brook	60.01	2.56	2.83	1.76	2.12
5	East Creek	159.06	3.94	4.72	3.77	4.78
6	HW Ottauquechee	114.96	6.04	8.67	5.69	7.89
7	Jay Branch	136.69	2.48	2.30	1.89	2.31
8	Jewett Brook	59.96	8.16	7.92	8.56	8.81
9	Lulls Brook	139.51	3.52	4.50	3.68	4.87
10	Mallets Bay	136.55	6.15	6.88	7.06	8.09
11	Mallets Creek	61.45	2.88	2.31	1.57	1.58
12	Mill Brook	98.69	6.55	8.15	3.95	5.37
13	Mill River	60.01	3.86	3.87	3.80	3.89
14	Moon Brook	100.49	7.13	5.14	8.72	6.09
15	Muddy Brook	55.14	5.32	1.58	7.60	3.21
16	N Branch Deerfield	144.93	7.24	7.74	5.20	5.74
17	Snipe Island	152.20	2.35	2.00	1.75	1.97
18	Winooski	127.87	3.55	2.95	4.65	4.08
Catchment Number	**HUC 10 Catchment**	**Area (km²)**	**NAIP (%IC) Fixed Width**	**NAIP (%IC) Elevation Based**	**NLCD (%IC) Fixed Width**	**NLCD (%IC) Elevation Based**
1	Black River	528.94	3.79	4.50	3.15	3.72
2	Mallets	197.99	4.58	4.20	3.07	4.63
3	Ottauquechee	578.36	4.21	5.42	3.28	4.10
4	Shelburne	190.34	4.83	4.50	5.48	5.21

NAIP, National Agriculture Imagery Program; NLCD, National Land Cover Database.

The ratio of NLCD IC to NAIP IC was plotted against NAIP IC for both buffer types (Figures 36.5 and 36.6). An idealized relationship would yield a slope of 0 across all levels of IC (ratio = 1), indicating no difference between the two image classification products. The relationship between IC classifications in both the fixed-width and elevation buffers were not significantly different from 0 (95% CI −0.02 to 0.11 for fixed width and −0.11 to 0.02 for elevation based), suggesting that there was little difference in the percent IC based on spatial resolution. However, there is a trend of NLCD underestimating the percent IC in the fixed-width buffers and less of a trend of NLCD overestimating IC at low levels of IC (Figures 36.5 and 36.6). The regression equations for both buffer types are shown below:

Fixed-width (Figure 36.5):

$$NLCD = 0.66 + 0.05(NAIP)$$
$$r^2 = 0.11,\ n = 22,\ p = 0.13,\ 95\%\ CI = -0.02 - 0.11$$

Elevation based (Figure 36.6):

$$NLCD = 1.19 - 0.04(NAIP)$$
$$r^2 = 0.09,\ n = 22,\ p = 0.19,\ 95\%\ CI = -0.11 - 0.02$$

The elevation-based buffers showed an outlier (Muddy Brook) that severely overestimated the amount of IC in the buffer area, with a ratio 2. This outlier is not indicated in the fixed-width buffer analysis. With the exception of the Muddy Brook outlier, the NAIP percent IC was comparable between buffer types at lower levels of IC (<5.5%) but showed more scatter at higher levels of IC (Figure 36.7).

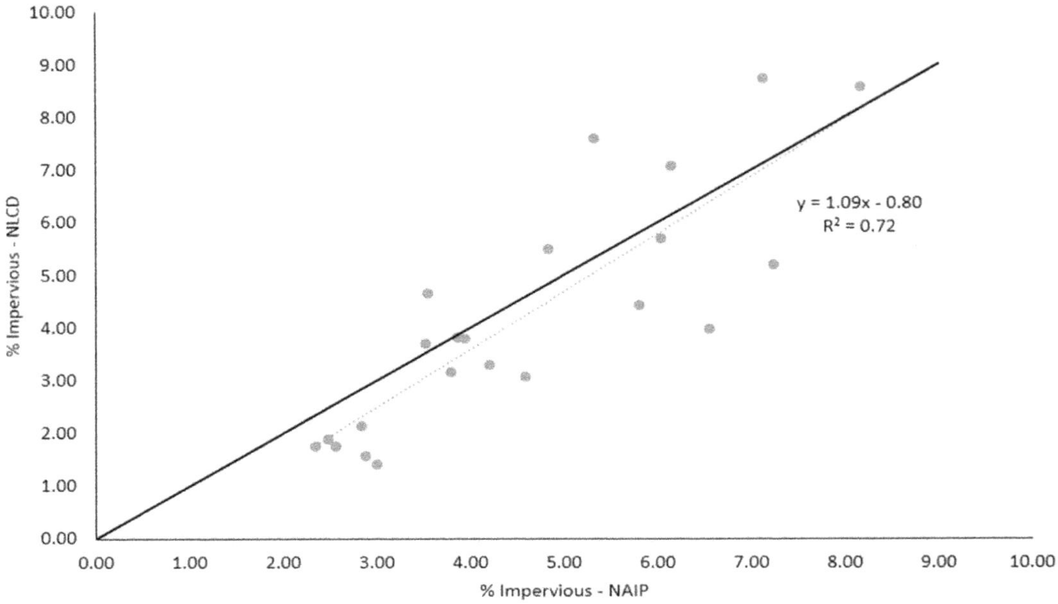

FIGURE 36.3 Relationship of NLCD to NAIP data at the HUC 10 and HUC 12 scale for fixed-width riparian buffers. Solid line through origin shows 1:1 ratio.

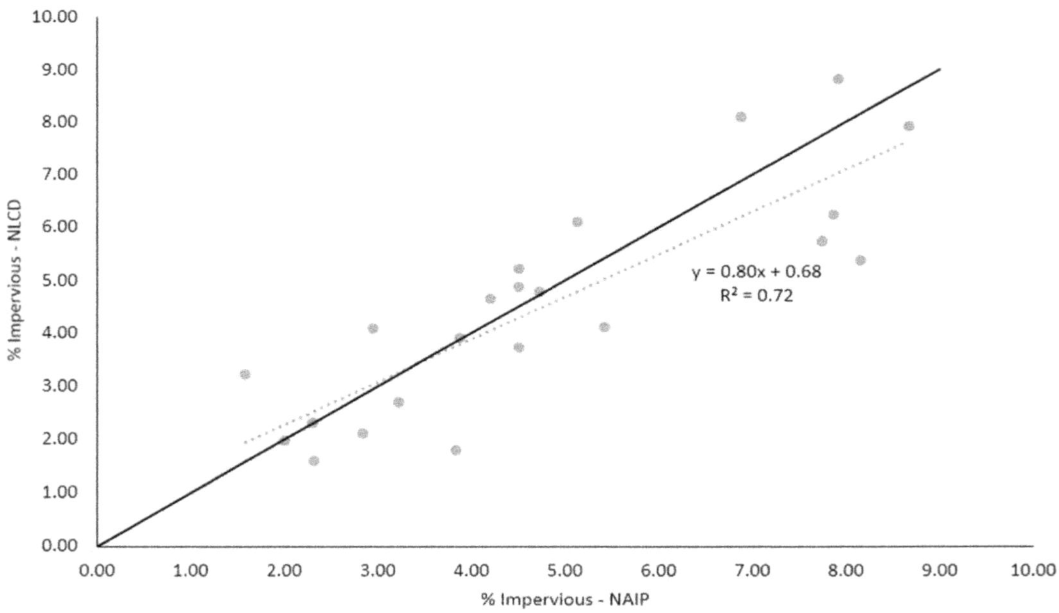

FIGURE 36.4 Relationship of NLCD to NAIP data at the HUC 10 and HUC 12 scale for elevation-based riparian buffers. Solid line through origin shows 1:1 ratio.

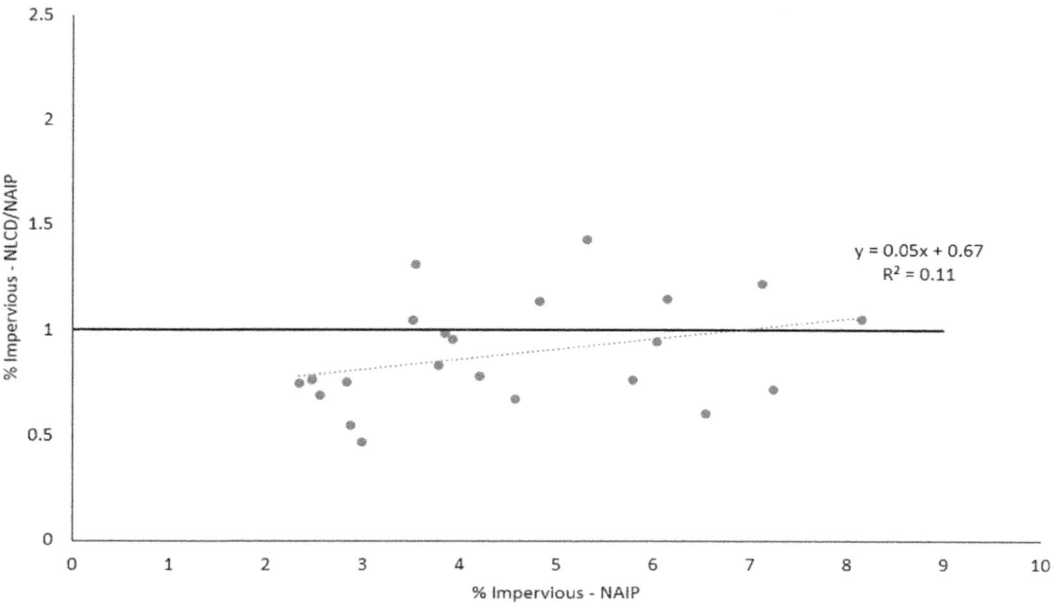

FIGURE 36.5 Ratio of medium to high spatial resolution IC classifications for fixed-width riparian buffers. Solid line represents a perfect ratio of NLCD to NAIP data with slope = 0.

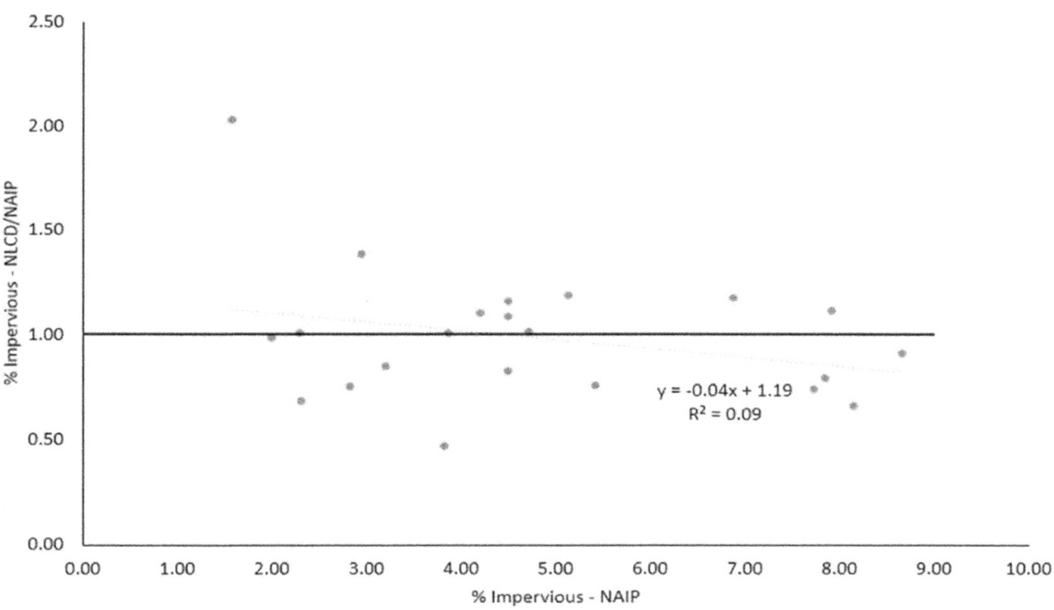

FIGURE 36.6 Ratio of medium to high spatial resolution IC classifications for elevation-based riparian buffers. Solid line represents a perfect ratio of NLCD to NAIP data with slope = 0.

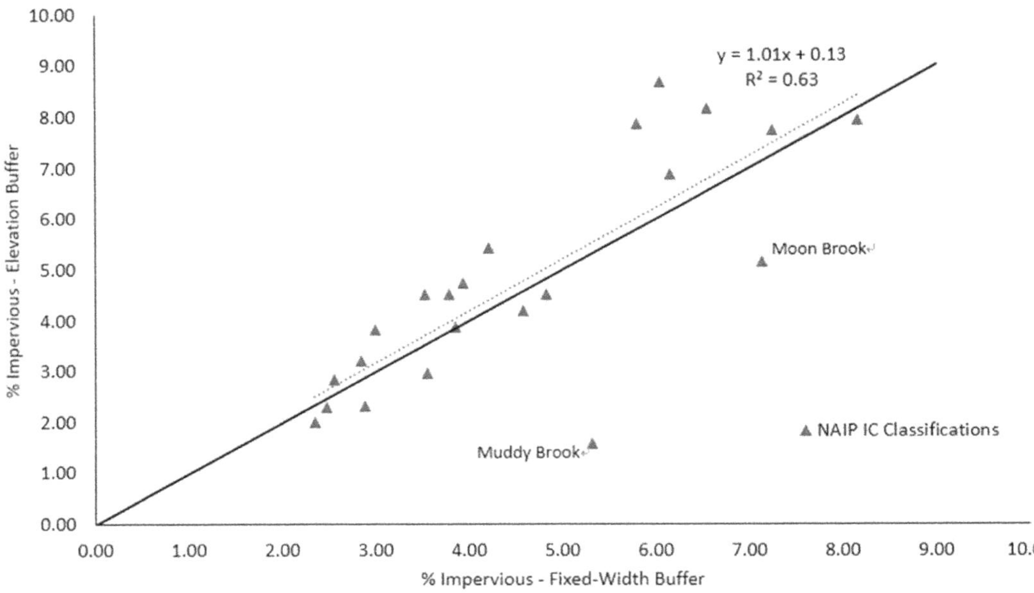

FIGURE 36.7 Comparison of fixed-width and elevation-based buffer IC classifications for high spatial resolution NAIP data. Solid line through origin shows 1:1 ratio. Outliers are labeled by name.

Discussion

Choosing the appropriate spatial scale and methodology for delineating riparian buffer zones are critical considerations for defining riparian buffer regulations. Many municipalities create regulations based on fixed-width buffers because they are easy to define and enforce. Recent studies suggest that functional riparian zones may be more cost-effective than fixed-width buffers (Tiwari et al. 2016). The spatial resolution of IC data had less of an effect in the riparian zone than it did on the full catchments within the same HUC 10 and HUC 12 units. Buffer type had minimal impact on percent IC, except in areas of unconfined valleys where there were notable differences between fixed-width and elevation-based buffers.

Acknowledgments

This document has been reviewed by the US Environmental Protection Agency, Office of Research and Development, and approved for publication. Mention of trade names or commercial products does not constitute endorsement or recommendation for use.

References

Abood, S.A., A.L. Maclean, and L.A. Mason. 2012. Modeling riparian zones utilizing DEMS and flood height data. *Photogrammetric Engineering & Remote Sensing* 78: 259–269.

Allan, J.D. 2004. Landscapes and riverscapes: The influence of land use on stream ecosystems. *Annual Review of Ecology, Evolution, and Systematics* 35: 257–284 doi:10.2307/30034117.

Baker, M., D. Weller, and T. Jordan. 2006. Improved methods for quantifying potential nutrient interception by riparian buffers. *Landscape Ecology* 21: 1327–1345. doi:10.1007/s10980-006-0020-0.

Baker, M.E., D.E. Weller, and T.E. Jordan. 2007. Effects of stream map resolution on measures of riparian buffer distribution and nutrient retention potential. *Landscape Ecology* 22: 973–992.

Bent, G.C. and A.M. Waite. 2013. Equations for estimating bankfull channel geometry and discharge for streams in Massachusetts. U.S. Geological Survey, Scientific Investigations Report 2013-5155, p. 62. U.S. Geological Survey, Reston, VA. doi:10.3133/sir20135155 (accessed September 2016).

Chen, Y., R. Carsel, S. McCutcheon, and W. Nutter. 1998. Stream temperature simulation of forested riparian areas: Watershed-scale model development. *Journal of Environmental Engineering* 124: 304–315.

Claggett, P.R., F.M. Irani, and R.L. Thompson. 2013. Estimating the extent of impervious surfaces and turf grass across large regions. *Journal of the American Water Resources Association* 49: 1057–1077.

Detenbeck, N.E., C. Rosiu, L. Hayes, and J. Legros. 2013. Differentiating impacts of watershed development from Superfund sites on stream macroinvertebrates. EPA Report Number EPA/600/R-13/352. U.S. Environmental Protection Agency, Office of Research and Development, National Health and Environmental Effects Research Laboratory, Atlantic Ecology Division, Narragansett, RI.

Detenbeck, N.E., A.C. Morrison, R.W. Abele, and D. Kopp. 2016. Spatial statistical network models for stream and river temperature in New England, USA. *Water Resources Research* 52(8): 6018–6040.

Fernández, D., J. Barquín, M. Álvarez-Cabria, and F. Peñas. 2012. Quantifying the performance of automated GIS-based geomorphological approaches for riparian zone delineation using digital elevation models. *Hydrology and Earth System Sciences* 16: 3851–3862.

Fernández, D., J. Barquín, M. Álvarez-Cabria, and F. Peñas. 2014. Land-use coverage as an indicator of riparian quality. *Ecological Indicators* 41: 165–174.

Gesch, D.B., M.J. Oimoen, and G.A. Evans. 2014. Accuracy assessment of the U.S. geological survey national elevation dataset, and comparison with other large-area elevation datasets-SRTM and ASTER. U.S. Geological Survey Open-File Report 2014-1008, p. 10. U.S. Geological Survey, Reston, VA.

Goetz, S.J. 2006. Remote sensing of riparian buffers: Past progress and future prospects. *Journal of the American Water Resources Association* 42: 133–143.

Gregory, S.V., F.J. Swanson, W.A. McKee, and K.W. Cummins. 1991. An ecosystem perspective of riparian zones. *BioScience* 41: 540–551.

Groffman, P.M., D.J. Bain, L.E. Band, K.T. Belt, G.S. Brush, J.M. Grove, R.V. Pouyat, I.C. Yesilonis, and W.C. Zipperer. 2003. Down by the riverside: Urban riparian ecology. *Frontiers in Ecology and the Environment* 1: 315–321.

Hollenhorst, T.P., G.E. Host, and L.B. Johnson. 2006. Scaling issues in mapping riparian zones with remote sensing data: Quantifying errors and sources of uncertainty. In: Wu, J. et al. (Ed.) *Scaling and Uncertainty Analysis in Ecology*. Springer, New York, pp. 275–295.

Ilhardt, B.L., E.S. Verry, and B.J. Palik. 2000. Defining riparian areas. *Forestry and the Riparian Zone Conference Proceedings*. Orono, ME.

Jones, J.W., and T. Jarnagin. 2009. Evaluation of a moderate resolution, satellite-based impervious surface map using an independent, high-resolution validation data set. *Journal of Hydrologic Engineering* 14: 369–376.

Lammert, M., and J.D. Allan. 1999. Assessing biotic integrity of streams: Effects of scale in measuring the influence of land use/cover and habitat structure on fish and macroinvertebrates. *Environmental Management* 23: 257–270

Multi-Resolution Land Characteristics Consortium (MRLC). 2014. National Land Cover Database 2011 (NLCD2011). Available from www.mrlc.gov/nlcd11_data.php (accessed February 2016).

Naiman, R.J. and H. Decamps. 1997. The ecology of interfaces: Riparian zones. *Annual Review of Ecology and Systematics* 28: 621–658.

Naiman, R.J., H. Decamps, and M.E. McClain. 2005. *Riparia: Ecology, Conservation, and Management of Streamside Communities. Aquatic Ecology*. Academic Press, London.

Roth, N., J.D. Allan, and D. Erickson. 1996. Landscape influences on stream biotic integrity assessed at multiple spatial scales. *Landscape Ecology* 11: 141–156.

Roy, A.H. and W.D. Shuster. 2009. Assessing impervious surface connectivity and applications for watershed management. *Journal of the American Water Resources Association* 45: 198–209.

Schueler, T.R., L. Fraley-McNeal, and K. Cappiella. 2009. Is impervious cover still important? Review of recent research. *Journal of Hydrologic Engineering* 14: 309–315.

Skally, C. and E. Sagor. 2001. Comparing riparian management zones to riparian areas in Minnesota: A pilot study. Minnesota Forest Resources Council, Research Report # RR-1001. St. Paul, MN.

Smucker, N.J., N.E. Detenbeck, and A.C. Morrison. 2013. Diatom responses to watershed development and potential moderating effects of near-stream forest and wetland cover. *Freshwater Science* 32: 230–249.

Smucker, N.J., A. Kuhn, M.A. Charpentier, C.J. Cruz-Quinones, C.M. Elonen, S.B. Whorley, T.M. Jicha, J.R. Serbst, B.H. Hill, and J.D. Wehr. 2016. Quantifying urban watershed stressor gradients and evaluating how different land cover datasets affect stream management. *Environmental Management* 57: 683–695.

Strahler, A.N. 1957. Quantitative analysis of watershed geomorphology. *Transactions American Geophysical Union* 38(6): 913–920.

Tiwari, T., J. Lundström, L. Kuglerová, H. Laudon, K. Öhman, and A. Ågren. 2016. Cost of riparian buffer zones: A comparison of hydrologically adapted site-specific riparian buffers with traditional fixed widths. *Water Resources Research* 52: 1056–1069.

United States Geological Survey (USGS). 2007. National hydrography dataset. Available from http://nhd.usgs.gov (accessed February 2016).

United States Geological Survey (USGS) and United States Environmental Protection Agency (US EPA). 2010. NHDPlus version 1 user guide. Available from www.horizon-systems.com/NHDPlus/NHDPlusV1_documentation.php (accessed February 2016).

Verry, E.S., C.A. Dolloff, and M.E. Manning. 2004. Riparian ecotone: A functional definition and delineation for resource assessment. *Water, Air, and Soil Pollution: Focus* 4: 67–94.

Van Sickle, J., J. Baker, A. Herlihy, P. Bayley, S. Gregory, P. Haggerty, L. Ashkenas, and J. Li. 2004. Projecting the biological condition of streams under alternative scenarios of human land use. *Ecological Applications* 14: 368–380.

Van Sickle, J. and C.B. Johnson. 2008. Parametric distance weighting of landscape influence on streams. *Landscape Ecology* 23: 427–438.

Vølstad, J.H., N.E. Roth, G. Mercurio, M.T. Southerland, and D.E. Strebel. 2003. Using environmental stressor information to predict the ecological status of Maryland non-tidal streams as measured by biological indicators. *Environmental Monitoring and Assessment* 84: 219–242. doi:10.1023/A:1023374524254.

Walsh, C.J., A.H. Roy, J.W. Feminella, P.D. Cottingham, P.M. Groffman, and R.P. Morgan II. 2005. The urban stream syndrome: Current knowledge and the search for a cure. *Journal of the North American Benthological Society* 24: 706–723.

Wickham, J., A. Neale, M. Mehaffey, T. Jarnagin, and D. Norton. 2016. Temporal trends in the spatial distribution of impervious cover relative to stream location. *Journal of the American Water Resources Association* 52(2): 409–419.

37

Cyanobacteria in Inland Waters: Remote Sensing

Introduction 303
Spectral Characteristics of Inland Waters 304
Basic Radiative Transfer Equation 304
Empirical Algorithms for PC Estimation 305
Semi-Empirical Algorithms for PC Estimation 308
TBM • DTBB • FBM • FBBM
Semi-Analytical Algorithms for PC Estimation 311
NBR • QAA_{pc} • EIIMIW
Performance Comparison among Semi-Empirical and Semi-Analytical Models 313
Datasets for Three Reservoirs of Central Indiana, USA • Datasets for Reservoirs in the USA, the Netherlands, and China • Datasets for Baekje Reservoir in South Korea • Datasets for Lake Balaton, Hungary • Algorithms Recommended for Further Examination
Practical Considerations and Current Challenges 315
Satellite- and Airborne-Based Cyanobacteria Monitoring • Variability of PC Specific Absorption Coefficient • Variability of Specific Absorption Coefficients of Non-PC Constituents • Variability of Suspended Matter Backscattering Coefficients
Conclusions 317
References 317

Lin Li
Indiana University—Purdue University Indianapolis

Introduction

Cyanobacterial (CYB) blooms are one of the most important issues concerning environmental agencies, water authorities, and public health organizations. CYB in surface water systems pose a health concern for humans, livestock, and native wildlife across the globe. Ecological effects of CYB blooms include changes in phytoplankton community structure, fish community structures, and lake anoxia. Nuisance and harmful CYB blooms in water bodies can result in a series of social-economic issues such as aesthetic degradation of lakes and reservoirs due to the presence of surface scums and earthy smells; recreational degradation due to hypolimnetic anoxia and kills of desirable sport fish; and human health impairment due to the production of toxins such as anatoxins, microcystins, and cylindrospermopsins. The occurrence of toxic CYB blooms is of great concern due to embedded implications for alternated biodiversity, public health, and for overall ecosystem health of inland waters. It is critical for water resource managers and policy makers to monitor toxic CYB blooms effectively.

Monitoring CYB blooms via *in situ* water sampling is time and labor intensive and is otherwise limited to infrequent water sample collection. To overcome this limitation, remote sensing has been used in estimating and mapping CYB blooms, and three remote sensing algorithms, namely empirical, analytical, semi-analytical, have been proposed to quantify phycocyanin (PC), an accessory pigment unique to CYB blooms in inland waters, as well as chlorophyll-a (Chl-*a*), a primary CYB pigment. The empirical approach aims at establishing a statistical relationship between spectral variables (e.g., reflectance, reflectance ratio, or derivative) and the CYB pigments [1–12]. However, the performance of varying empirical approaches is dataset dependent because of the variation of imaging and water conditions [e.g., 13,14]. The semi-analytical approach to quantifying CYB blooms utilizes the correlation between remote sensing reflectance and water inherent optical properties (IOPs), which are referred to as the total absorption (a) and backscattering (b_b) of optically active constituents (OACs): phytoplankton, color dissolved organic matter (CDOM), and non-algal particle (NAP). Quantification of CYB concentration via the semi-analytical approach is primarily achieved by first isolating the CYB absorption signal from the total OACs' absorption derived from remote sending reflectance, then removing the interference of overlapping Chl-*a* absorption from CYB absorption, and finally deriving PC concentrations from the remaining CYB absorption spectra. The semi-empirical approach focuses on analyzing the relationship between the CYB pigments and one or some spectral indices with the latter being parameterized by the IOPs of the relevant OACs. Insightful reviews of these PC estimation algorithms were given by Matthews et al. [15], Li and Song [16], and Yan et al. [17].

Over the past several years, a quite few literatures have been published to compare or assess various PC remote sensing algorithms when applied to different inland waters around the world, but the performance of a specific PC algorithm is inconsistent. Therefore, it is necessary to revisit several typical PC algorithms to examine their assumptions and suitability for different inland waters. Beside this main goal, these PC algorithms are assessed for their performance accuracy, ability for removal of the non-PC constituent interference, as well as their adaptability for different sensors. At last, scientifically important investigations to develop PC remote sensing algorithms in future are recommended.

Spectral Characteristics of Inland Waters

Algae contain colored pigments and show characteristic spectral features. Figure 37.1 shows *in situ* reflectance spectra of Eagle Creek Reservoir measured in August of 2004. The spectral signature signifying the presence of algal pigments in the water includes (a and b) low reflectance at 440 and 500 nm resulting from algal chlorophyll and carotenoid, respectively [18,19]; (c) maximum reflectance between 560 and 570 nm due to the lack of absorption by algal chlorophyll, thus giving algae a green color to our eyes [20]; (d) a strong PC absorption at 620 nm unique to CYB due to PC absorbing primarily green and red light [20,21]; (e) a weak reflection at 640 nm ascribed to backscattering from dissolved organic matter or fluorescence of accessory pigment [21]; (f) strong Chl-*a* absorption at about 675 nm [18,19]; (g) strong reflectance peak around 690–700 nm caused by an interaction of algal-cell scattering and a combined effect of pigment and water absorption [18–21]; and (h) a weak reflectance peak at about 810 nm likely due to backscattering from algal cells combined with the general absorption of near infrared in clear water [18,19]. These spectral characteristics of pigments and other water constituents provide a physical basis for quantifying the concentrations of CYB using remote sensing.

Basic Radiative Transfer Equation

Remote sensing reflectance above water surface ($R_{rs}(\lambda)$) is a function of two IOP variables $a(\lambda)$ and $b_b(\lambda)$, and the function can be written as Equation 37.1 [22]:

$$R_{rs}(\lambda)=\frac{f(\lambda)}{Q(\lambda)}\times\frac{b_b(\lambda)}{a(\lambda)+b_b(\lambda)} \tag{37.1}$$

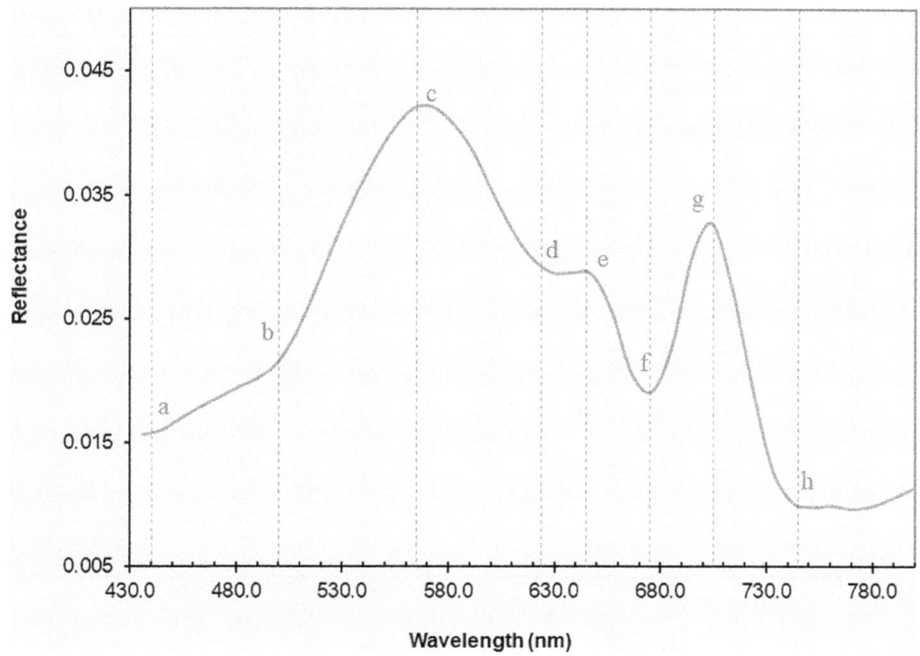

FIGURE 37.1 Reflectance spectrum measured with ASD spectrometer in Geist Reservoir, USA. Vertical dashed lines represent wavelength locations at which optically active compoenents are spectrally diagnostic, and their spectral features are described in text.

where $f(\lambda)$ describes the sensitivity of the reflectance to variations in the solar zenith angle and $Q(\lambda)$ is the bidirectional properties of the reflectance. For inland turbid waters, $a(\lambda)=a_w(\lambda)+a^*_{\text{Chl}}(\lambda)[\text{CHL}]+a^*_{\text{pc}}(\lambda)[\text{PC}]+a^*_{\text{NAP}}(\lambda)[\text{NAP}]+a^*_{\text{CDOM}}(\lambda)[\text{CDOM}]$ denotes total absorption coefficient of four water substances in the water column at a given wavelength, each term on the right side of this equation is the bulk absorption of the corresponding constituent which is defined as the product of the specific absorption coefficient (e.g., $a^*_{\text{Chl}}(\lambda)$) of a constituent and concentration of that constituent (e.g., [CHL]). $b_b(\lambda)$ represents the total backscattering coefficient (m^{-1}) of OACs and water and can be described as a linear combination of OACs' bulk backscattering coefficient, each of which is the product of the specific scattering coefficient (e.g., $b^*_{b,\text{phy}}(\lambda)$) and concentration of that constituent (e.g., [Chl-*a*]). Furthermore, $f(\lambda)$ and $Q(\lambda)$ are often assumed to be weakly dependent on wavelength [23,24]. This assumption is significant because a reflectance band ratio (BR) can then be written as:

$$\frac{R_{rs}(\lambda_1)}{R_{rs}(\lambda_2)}=\frac{b_b(\lambda_1)\left[a(\lambda_2)+b_b(\lambda_2)\right]}{b_b(\lambda_2)\left[a(\lambda_1)+b_b(\lambda_1)\right]} \tag{37.2}$$

A BR as shown by Equation 37.2 is important for developing most PC algorithms which are listed in Table 37.1 and described below. Table 37.2 lists the symbols and acronyms which are used to describe the models.

Empirical Algorithms for PC Estimation

Empirical algorithms for estimation of PC are commonly established based on its absorption feature around 620 nm. Table 37.3 shows a summary of typical empirical algorithms. The simplest empirical algorithm in this group is the reflectance ratio of a near-infrared wavelength to 620 nm. Schalles and

TABLE 37.1 Site Conditions for Developing Representative PC Models and Their Performance Description

Model	Sampling Site	PC (μg L^{-1}) Range	R^2	RMSE (μg L^{-1}) or rRMSE (%)	References
$BR_{709/620}$	Lab culture	-	0.95	6.35	[25]
$BR_{700/600}$	Lab culture	506–273,883*	0.94†	19,957†	[26]
$BR_{650/625}$	Carter Lake, USA	0–530	0.612	-	[8]
MRBI	Lakes	7–130	0.99	2.34	[3]
SSA	Bear Lake	-	-	-	[27]
PCI	Lake Taihu	1–300	0.64	85.4%	[28]
TBM	Lake Loch Leven and Esthwaite Water, UK	0–93.7	0.92	2.65	[29]
DTBB	Eagle Creek and Geist Reservoirs, USA	6.6–140	0.857	31.4%	[30]
FBM	Lake Taihu, China	2.67–107.67	0.86	4.83	[31]
FBBM	Lakes and reservoirs in China, the Netherlands, and the USA	0–710.28	0.73	27.69	[32]
NBR	Lakes Loosdrecht and IJsselmeer in the Netherlands	0.8–79.8	0.94	6.5	[11]
QAA_{pc}	Aquaculture ponds, USA	68.13–3032.47	0.99	30.7%	[33]
EIIMIW	Eagle Creek, Geist, and Morse reservoirs, USA	0.73–370.95	0.81†	33.6%†	[34]

* Cyanobacteria cell number per mL.
† Results for validation.
- Unavailable data.

TABLE 37.2 List of Symbols and Acronyms

Symbol/Acronym	Description	Units
$b_b(\lambda)$	Total backscattering coefficients at wavelength λ	m^{-1}
$b_{bp}(\lambda)$	Backscattering coefficients of particles at wavelength λ	m^{-1}
$a^*_{pc}(\lambda)$	PC-specific absorption coefficients at wavelength λ	m^2 (mg PC)$^{-1}$
$a_i(\lambda)$	Absorption coefficients of compound i at wavelength λ. Subscripts used: t-w = non-water; phy = phytoplankton; cdm = colored detritus matter; PC = phycocyanin; ph-pc = phytoplankton pigments excluding phycocyanin; cdm + pc = colored detritus matter plus phycocyanin	m^{-1}
S_{cdm}	Exponential slope of $a_{cdm}(\lambda)$; $a_{cdm}(\lambda) = a_{cdm}(\lambda_0)\exp[-S_{cdm} \times (\lambda - \lambda_0)]$	nm^{-1}
$R_{rs}(\lambda)$	Remote sensing reflectance above water surface at wavelength λ	sr^{-1}
$r_{rs}(\lambda)$	Remote sensing reflectance below water surface at wavelength λ	sr^{-1}
η	Linear interpolation weight	-
nLw	Normalized leaving water radiance	mW cm^{-2}/μm sr^{-1}
[PC]	Phycocyanin (concentration)	mg m^{-3}
[Chl-*a*]	Chlorophyll-a concentration	mg m^{-3}
[TSM]	Total suspended matter concentration	g m^{-3}
[NAP]	Non-algal particles	g m^{-3}
CDOM	Colored dissolved organic matter	-
CDM	Colored detritus matter; CDOM and NAP combined	-
rRMSE	Relative root mean square error	-
AOP	Apparent optical properties	-
IOP	Inherent optical properties	-
OAC	Optically active constituents	-
BR	Band ratio	-
MRBI	Midpoint reflectance baseline index	-
SSA	Spectral shape algorithm	-

(*Continued*)

TABLE 37.2 (*Continued*) List of Symbols and Acronyms

Symbol/Acronym	Description	Units
PCI	Phycocyanin index	-
TBM	Three band model	-
DTBB	Double three band baseline	-
FBM	Four band model	-
FBBM	Four band baseline model	-
NBR	Nested band ratio	-
QAA_{pc}	Quasi-analytical algorithms for PC	-
EIIMIW	Extension of IOP inversion model for inland waters	-
MERIS		-
OLCI	Ocean and land color imager	-
AISA	Airborne Imaging Spectrometer for Applications	-
EnMap	Environmental Mapping And Analysis Program	-
HyspIRI	Hyperspectral Infrared Imager	-
PRISMA	PRecursore IperSpettrale della Missione Applicativa	-

Yacobi [8] proposed to use the ratio of radiance reflectance at 650 nm to that at 620 nm ($BR_{650/620}$) and obtained an R^2 of 0.612 when applied to *in situ* reflectance data measured by a portable Ocean Optics ST1000 spectraradiometer from September 1994 to July 1998 for Carter Lake, NE, USA. The strong performance of this BR can be attributed to the dominance of CYB indicated by the range of PC concentrations (10–530 μg L^{-1}) and strong correlation of PC to Chl-*a* [8]. Hunter et al. [25] proposed a $\log_{10}$-transformed ratio of radiance at 710 nm to that at 620 nm ($BR_{710/620}$) and obtained an R^2 of 0.95 when applied to time series images acquired by the Compact Airborne Spectrographic Imager (CASI-2) over the shallow eutrophic waters of Barton Broad, UK, to monitor diurnal changes in the spatial distribution of the potentially toxic cyanobacterium *Microcystis aeruginosa* with a PC range of 36–139 μg L^{-1} (N = 13). Li et al. [26] examined the performance of $BR_{650/620}$ and $BR_{710/620}$ in the estimation of PC concentrations with the Airborne Imaging Spectrometer for Application (AISA) image spectra of Geist Reservoir, IN, USA (PC: 25–185 μg L^{-1}) and showed weak relationships of the two ratios to PC concentration ranging from 25 to 185 μg L^{-1} as a result of the effects of Chl-*a* and suspended sediment. Mishra et al. [35] observed the performance of $BR_{700/600}$ ($R^2 = 0.97$) was superior to reflectance ratios $BR_{650/620}$ ($R^2 = 0.71$) and $BR_{710/620}$ ($R^2 = 0.88$) in the estimation of PC concentrations (506–273,883 cells L^{-1}) with lab measured USB 4000 radiometer (Ocean Optics, Inc., Dunedin, FL, USA) spectra of cultured CYB species *Synechocystis* and *Anabaena* and concluded that the strongest performance of $BR_{700/600}$ was due to its insusceptibility to varying Chl-*a* concentrations while the poor performance of $BR_{650/620}$ and $BR_{710/620}$ resulted from the effect of Chl-*a* on the reflectance at 620 nm. Ogashawara et al. [36] evaluated the accuracy and sensitivity of $BR_{650/620}$, $BR_{710/620}$, $BR_{700/600}$, $BR_{709/600}$, and $BR_{724/600}$ for retrieval of PC concentrations from the reflectance spectra measured by the RAMSES hyperspectral radiometers (TriOS GmbH, Oldenburg, Germany) in Funil Reservoir, Brazil, and the spectra measured with Ocean Optics spectroradiometers (Ocean Optics Inc., Dunedin, FL, USA) in catfish ponds, USA. In terms of R^2 from high to low, these ratios were ranked as $BR_{709/600}$, $BR_{700/600}$, $BR_{710/620}$, $BR_{650/620}$, and $BR_{724/600}$ for the Funil dataset (PC: 9–35 μg L^{-1}); $BR_{700/600}$, $BR_{724/600}$, $BR_{650/620}$, $BR_{709/600}$, and $BR_{710/620}$ for the catfish ponds (PC: 68–857 μg L^{-1}); and $BR_{700/600}$, $BR_{724/600}$, $BR_{650/620}$, $BR_{710/620}$, and $BR_{709/600}$. These BR results indicate the dependence of their performance on datasets for estimating PC concentration [37].

A midpoint reflectance baseline index (MRBI) using bands 600, 624, and 648 nm was proposed to describe the difference between reflectance R_{rs} at 624 nm and the reflectance midpoint for bands 600 and 648 nm [3]. This index gave rise to a strong relationship ($R^2 = 0.99$) to PC measurements for the dataset collected from nine shallow eutrophic lakes in the central western part of the Netherlands, but the application of MRBI to other datasets has a mixed success. For example, Li et al. [26] reported

TABLE 37.3 A Summary of Empirical Algorithms for PC Estimation

	Spectral Predictor	References
$BR_{709/620}$	$R_x(709)/R_x(620)$	[25]
$BR_{700/600}$	$R_x(700)/R_x(600)$	[26]
$BR_{650/625}$	$R_{rs}(650)/R_{rs}(625)$	[8]
MRBI	$0.5\left[R_{rs}(665)+R_{rs}(560)\right]-R_{rs}(624)$	[3]
SSA(681)	$\text{nLw}(681)-\left(\text{nLw}(709)-\text{nLw}(665)\right)\left(\frac{681-665}{709-665}\right)-\text{nLw}(665)$	[27]
PCI(620)	$R_{rs}(560)+\left(R_{rs}(665)-R_{rs}(560)\right)\left(\frac{620-560}{665-560}\right)-R_{rs}(620)$	[28]

a stronger performance of the MRBI than $BR_{650/620}$ and $BR_{710/620}$ with the AISA image spectra, but an inferior PC estimation relative to $BR_{650/620}$ was observed with the radiance reflectance spectra collected from Carter Lake [8] and relative to all the ratios examined by Ogashawara et al. [36] with the Funil dataset. Explanation for these inconsistent results is not straightforward, but the relationship $a_{\text{pc}}(620)=\text{MRBI}\Big/\left(\frac{R_{rs}(\lambda)-\text{MRBI}}{a(\lambda)+b_b(\lambda)}\right)$ can be derived from Equations 5.13a–c in [3], which illustrates the effects of both the reflectance and the total light extinction $(a(\lambda)+b_b(\lambda))$ at 620 nm on the correlation between MRBI and PC concentration.

Wynne et al. [27] developed a spectral shape algorithm (SSA) and applied to the MEdium Resolution Imaging Spectrometer (MERIS) data to identify CYB blooms in Bear Lake, MI, USA. It is interesting to note that the MRBI is a special case of SSA for a set of three evenly spaced wavelengths. Unlike other empirical methods, this SSA does not rely on PC absorption around 620 nm, instead uses the Chl-*a* fluorescence at 681 nm to construct second derivatives with reflectance at wavelengths 665 and 709 nm based on the difference of the fluorescence peak between CYB (weak fluorescence) and non-CYB species (strong fluorescence). The SSA was later adjusted for Moderate Imaging Spectrometer (MODIS) bands 678, 667, and 748 nm to monitor CYB blooms in Lake Erie [38,39], but Wynne et al. [38] pointed out that a weak Chl-*a* fluorescence signal could be due to other processes instead of the CYB presence.

Qi et al. [28] proposed a PC index (PCI) using the spectral shape for the derivatives at 550, 620, and 665 nm to estimate PC concentrations in Lake Taihu in China. Despite its mathematical equivalent to SSA and MRBI, PCI is expected to be more reliable than SSA for estimating PC concentration and more suitable than MRBI for satellite multispectral sensors. However, like MRBI and other empirical models, PCI's performance can be affected by sediment, other phytoplankton assemblage, and a strong absorption of water between 550 and 620 nm [40]. These drawbacks are more or less addressed by the semi-empirical algorithms to be discussed below.

Semi-Empirical Algorithms for PC Estimation

Table 37.4 shows PC algorithms in this group including the models established using three bands [25,29,41], double three bands and the absorption baseline [30], four bands [31], and four bands and reflectance baseline [32]. For the convenience of the description below, these algorithms are called the three band model (TBM), double three band baseline (DTBB), four band model (FBM), and four band baseline model (FBBM), respectively.

TABLE 37.4 A Summary of Semi-Empirical Algorithms for PC Estimation

Model	Spectral Predictor	References
TBM	$\left(R_{rs}^{-1}(\lambda_1) - R_{rs}^{-1}(\lambda_2)\right) \times R_{rs}(\lambda_3)$	[29]
DTBB	$0.5 \times \left\{[a_w(725) + b_b(725)](R31 + R32) - 2 \times a_w(624) + a_w(600) + a_w(648)\right\}$	[30]
FBM	$\left(R_{rs}(\lambda_1)^{-1} - R_{rs}(\lambda_2)^{-1}\right) / \left(R_{rs}(\lambda_4)^{-1} - R_{rs}(\lambda_3)^{-1}\right)$	[31]
FBBM	$\left(R_{rs}(\lambda_1)^{-1} - \eta_{\lambda_3}^{\lambda_2}(\lambda_1) R_{rs}(\lambda_2)^{-1} - \left(1 - \eta_{\lambda_3}^{\lambda_2}(\lambda_1)\right) R_{rs}(\lambda_3)^{-1}\right) R_{rs}(\lambda_4)$	[32]

TBM

These three algorithms are proposed to estimate PC using three spectral bands denoted as λ_1, λ_2, λ_3, respectively. Selection of these three wavelengths or spectral bands is based on the following assumptions: (i) λ_1 is sensitive to PC absorption; (ii) λ_2 is less sensitive to PC absorption, but the absorption coefficients of NAP and CDOM at λ_2 should be similar to those at λ_1; (iii) the backscattering coefficients of OACs are spectrally neutral across λ_1, λ_2, and λ_3. The first TBM was constructed using $\lambda_1 = 630$ nm, $\lambda_2 = 660$ nm, and $\lambda_3 = 725$ nm to estimate PC and obtained an R^2 of 0.95 and a root mean square error (RMSE) of 6.35 μg L^{-1} when applied to cultured CYB spectra measured with an ASD FieldSpec® HandHeld Spectroradiometer (Analytical Spectral Devices Inc., Boulder, CO, USA) [25]. Later the TBM for airborne hyperspectral sensors was established with $\lambda_1 = 615$ nm, $\lambda_2 = 600$ nm, and $\lambda_3 = 725$ nm for hyperspectral sensors CASI-2 (ITRES Research Ltd., Calgary, AB, Canada) and AISA Eagle (SPECIM, Oulu, NO, Finland) and performed well with $R^2 = 0.92$ and RMSE = 2.65 μg L^{-1} [29]. Duan et al. [41] established a TMB with three MERIS bands at $\lambda_1 = 620$ nm, $\lambda_2 = 709$ nm, and $\lambda_3 = 754$ nm for estimating PC concentrations in three inland lakes in China (Lake Taihu, Lake Dongjiu, and Lake Gehu). Nonetheless, the TBM is found to perform poorly for highly turbid waters or when CYB blooms do not dominate [26,42,43], and other semi-analytical models are proposed using the assumptions for the TBM and additional spectral characteristics.

DTBB

The DTBB algorithm is composed of double TBMs and a midpoint baseline index. The first TBM (Equation 37.3) uses bands at $\lambda_1 = 624$ nm, $\lambda_2 = 600$ nm, and $\lambda_3 = 725$ nm, while the second TBM (Equation 37.4) uses $\lambda_1 = 624$ nm, $\lambda_2 = 648$ nm, and $\lambda_3 = 725$ nm.

$$R31 = \frac{b_b(725)}{a_w(725) + a_{\text{cdm}}(725) + a_{\text{pigs}}(725) + b_b(725)} \left[\frac{a_w(624) + a_{\text{cdm}}(624) + a_{\text{pigs}}(624) + b_b(624)}{b_b(624)} - \frac{a_w(600) + a_{\text{cdm}}(600) + a_{\text{pigs}}(600) + b_b(600)}{b_b(600)} \right] \quad (37.3)$$

$$R32 = \frac{b_b(725)}{a_w(725) + a_{\text{cdm}}(725) + a_{\text{pigs}}(725) + b_b(725)} \left[\frac{a_w(624) + a_{\text{cdm}}(624) + a_{\text{pigs}}(624) + b_b(624)}{b_b(624)} - \frac{a_w(648) + a_{\text{cdm}}(648) + a_{\text{pigs}}(648) + b_b(648)}{b_b(648)} \right] \quad (37.4)$$

The selection of these spectral bands for the construction of the two three band indices was inspired by the TBM in [29] with $\lambda_1 = 624$ nm being sensitive to PC, but $\lambda_2 = 600$ or 648 nm being less sensitive to PC and able to compensate for the NAP and CDOM absorption (i.e. a_{cdm}, hereafter cdm represents both NAP and CDOM) at λ_1. A neutral backscattering is assumed for wavelengths λ_1 and λ_2. Additionally, the assumption of dominant water absorption $\lambda_3 = 725$ nm implies $a_{cdm}(725) = a_{pigs}(725) \approx 0$ [11,44], and rearrangement of Equations 37.3 and 37.4 results in Equations 37.5 and 37.6 [30]:

$$a_{pigs}(624) - a_{pigs}(600) = \left[a_w(725) + b_b(725)\right]R31 - a_w(624) + a_w(600) - a_{cdm}(624) + a_{cdm}(600) \quad (37.5)$$

$$a_{pigs}(624) - a_{pigs}(648) = \left[a_w(725) + b_b(725)\right]R32 - a_w(624) + a_w(648) - a_{cdm}(624) + a_{cdm}(648) \quad (37.6)$$

Based on the MRBI by Dekker [3], a midpoint absorption baseline was proposed to form the DTBB model by use of the phytoplankton absorption at 600, 624, and 648 nm [30]:

$$a_{pc}(624) = a_{pig}(624) - 0.5\left(a_{pig}(600) + a_{pig}(648)\right) \quad (37.7)$$

The advantage of DTBB is capable of removing the colored detritus matter (cdm) interference with PC absorption at 624 nm using $0.5\left[a_{cdm}(600) + a_{cdm}(648)\right] - a_{cdm}(624) \approx 0$ and the interference of Chl-*a* by subtraction of $0.5[a_{pigs}(600) + a_{pigs}(648)]$ from $a_{pigs}(624)$. Application of Equations 37.5–37.7 results in an expression for $a_{pc}(624)$.

Compared with the TBM, the advantage of DTBB is able to compensate for the interference of Chl-*a* with PC absorption at 624 nm. However, the backscattering at λ_3 is accounted for by use of an equation by Simis et al. [11] and Gons [45]. Application of DTBB for estimating PC concentrations in Eagle Creek and Geist Reservoirs (USA) resulted in an R^2 of 0.85 and a relative RMSE (rRMSE) of 31.4%. As expected, DTBB is less sensitive to the absorption interference of NAP, CDOM, and Chl-*a* and performed better than classical three band algorithms at low PC concentrations ($PC \leq 50 \mu g L^{-1}$). However, one caveat is that the $a_{pc}(624)$ expressed by Equation 37.7 is relative to the baseline between 600 and 648 nm where PC has nonzero absorption values, implying that the derived $a_{pc}(624)$ should have a relatively smaller value than the actual $a_{pc}(624)$. This may explain why the $a^*_{pc}(620)$ values used in Dekker [3] and Li et al. [30] are lower than the values used in Simis et al. [11,44] as described later.

FBM

The first FBM for PC estimation was proposed by Le et al. [31]. Selection of the first two bands is based on the same assumption as those for the three band algorithm, but bands three and four are selected in the near-infrared region so that non-water absorption is eliminated. Assuming the neutral backscattering across four bands, the FBM is expressed as:

$$\begin{aligned} \text{FBM} &= \left[R_{rs}(\lambda_1)^{-1} - R_{rs}(\lambda_2)^{-1}\right] \Big/ \left[R_{rs}(\lambda_4)^{-1} - R_{rs}(\lambda_3)^{-1}\right] \\ &= a_{pc} \Big/ \left[a_w(\lambda_4) - a_w(\lambda_3)\right] \end{aligned} \quad (37.8)$$

A set of four bands $\lambda_1 = 630$ nm, $\lambda_2 = 645$ nm, $\lambda_3 = 695$ nm, and $\lambda_4 = 730$ nm was used by the FBM resulting in an RMSE of 4.83 mg m^{-3} and 6.8 mg m^{-3} when applied to spectral datasets collected for Lake Taihu in 2007 and 2008, respectively.

FBBM

The FBBM is the second four band algorithm proposed for estimating PC [32]. This four band baseline algorithm has the assumptions similar to those for the DTBB, but there are two differences. One is that the FBBM uses the interpolation of non-PC component absorption at bands λ_2 ($<\lambda_1$) and λ_3 ($>\lambda_1$) with a weight value $\eta_{\lambda_3}^{\lambda_2}(\lambda_1)$ ranging from −0.1 to 1.1 to approximate non-PC component absorption at wavelength λ_1, and this linear interpolation absorption baseline differs from the DTBB which uses a weight value of 0.5; the other difference is that the FBBM assumes a neutral backscattering across bands λ_1, λ_2, λ_3, and λ_4, whereas *the* DTBB uses $b_b(\lambda_1) \approx b_b(\lambda_2)$ and an explicit expression for $b_b(\lambda_3)$. The FBBM is expressed as:

$$\text{FBBM} = \left[R_{rs}(\lambda_1)^{-1} - \eta_{\lambda_3}^{\lambda_2}(\lambda_1)R_{rs}(\lambda_2)^{-1} - \left(1 - \eta_{\lambda_3}^{\lambda_2}(\lambda_1)\right)R_{rs}(\lambda_3)^{-1}\right]R_{rs}(\lambda_4)$$
$$\propto \left[a_{\text{pc}}(\lambda_1) + a_w(\lambda_1) - a_w(\lambda_2) - a_w(\lambda_3)\right]/a_w(\lambda_4) \quad (37.9)$$

This expression indicates an inherent correlation of this four band index to PC concentration and has been evaluated with the spectral bands of MERIS. The FBBM has been tested with datasets collected from aquatic systems located in the United States, the Netherlands, and in China [32].

Semi-Analytical Algorithms for PC Estimation

Table 37.5 shows three semi-analytical PC algorithms to date. Simis et al. [11] proposed the earliest semi-analytical PC algorithm using nested band ratios (NBRs) and three empirical relationships. The second algorithm was developed by Mishra et al. [33,46] using the classical quasi-analytical algorithm (QAA) and an empirical procedure for removal of overlapping Chl-*a* absorption at 620 nm. The third algorithm, an extension of IOP Inverse Model for Inland Waters (IIMIW) was developed by Li et al. [34] to partition non-water absorption coefficient ($a_{t\text{-}w}(\lambda)$) into the contribution of NAP, CDOM, non-PC pigments, and PC. These three algorithms, hereafter called NBR, QAA_{pc} and EIIMIW, respectively, are used to derive the PC absorption at 620 nm and then divide with the specific absorption coefficient of PC to calculate its concentration. The use of some empirical relations makes it appropriate to consider these algorithms being semi-analytical.

NBR

NBR uses the first two assumptions on PC absorption at wavelengths λ_1 and λ_2 that are used to develop other semi-empirical algorithms and assumes a neutral backscattering between these wavelengths. The total absorption coefficients at 665 and 620 nm are derived by applying the relationship between the IOP and remote sensing reflectance to BRs $R_{rs}(709)/R_{rs}(665)$ ($\lambda_1 = 665$ nm and $\lambda_2 = 709$ nm) and $R_{rs}(709)/R_{rs}(620)$ ($\lambda_1 = 620$ nm and $\lambda_2 = 709$ nm), respectively, in which water absorption and backscattering across these wavelengths are considered. Two empirical constants γ and δ are applied to the total absorption coefficients derived for 665 and 620 nm, respectively, to correct for the effect of CDOM and NAP, resulting in the absorption coefficient of Chl-*a* at 665 nm and phytoplankton at 620 nm. Finally, the absorption of Chl-*a* at 665 nm derived from $R_{rs}(709)/R_{rs}(665)$ is converted to its absorption at 620 nm with a correlation coefficient between the two wavelengths of *in vivo* Chl-*a* absorption (ε) and then nested into the absorption of phytoplankton at 620 nm derived from $R_{rs}(709)/R_{rs}(620)$, resulting in the absorption coefficient of PC at 620 nm as shown in Table 37.5. Where γ is a factor for correcting for CDOM and NAP absorption, $\lambda_1 = 665$ nm and pigment absorption at $\lambda_2 = 709$ nm.

The most obvious advantage of NBR is to compensate for the interference of Chl-*a* with PC absorption at 620 nm, but the correction for the absorption of CDOM and NAP is empirical and required for different water bodies [47,48]. Le et al. [31] showed that the NBR algorithm did not perform as well as the four band algorithm, but Duan et al. [41] recommended to use the NBR algorithm instead of the three band algorithm.

TABLE 37.5 A Summary of Semi-Analytical Algorithms for PC Estimation

Model	Spectral Predictor	References
NBR	$a_{chl\text{-}a}(665)=\{[R_{rs}(709)/R_{rs}(665)\times(a_w(709)+b_b)-b_b-a_w(665)]\times\gamma^{-1}\}$	[11]
	$a_{phy}(620)=(\{R_{rs}(709)/R_{rs}(620)\times[a_w(709)+b_b]\}-b_b-a_w(620))\times\delta^{-1}$	
	$a_{pc}(620)=(\{R_{rs}(709)/R_{rs}(620)\times[a_w(709)+b_b]\}-b_b-a_w(620))$ $\times\delta^{-1}-\varepsilon\times a_{Chl\text{-}a}(665)$	
QAA_{pc}	QAA to derive $a_{phy}(\lambda)$	[33]
	$a_{phy}(665)=a_{Chl\text{-}a}(665)+a_{pc}(665)$	
	$a_{phy}(620)=a_{Chl\text{-}a}(620)+a_{pc}(620)$	
	$a_{pc}(620)=(a_{phy}(620)\psi_1-a_{phy}(665))/(\psi_1-\psi_2)$	
EIIMIW	$r_{rs}(\lambda)=R_{rs}(\lambda)/(0.52+1.7R_{rs}(\lambda))$	[34]
	$b_b(778)=r_{rs}(778)a_w(778)/(0.082-r_{rs}(778))$	
Deviation of non-water OACs' absorption	$Y=2.0(1-1.2\exp(-0.9r_{rs}(443)/r_{rs}(560)))$	
	$b_{bp}(560)=(b_p(778)-b_w(778))/0.1798^Y$	
	$b_b(\lambda)=b_{bp}(560)(560/\lambda)^Y+b_{bw}(778)$	
	$a_{t\text{-}w}(\lambda)=r_{rs}(709)b_b(\lambda)(a_w(709)+b_b(709))/(r_{rs}(\lambda)b_b(709))-b_b(\lambda)-a_w(\lambda)$	
Derivation of non-algal particle and CDOM absorption with and without PC	$a_{phy\text{-}pc}(\lambda)=1.1872C_1(\lambda)a_{t\text{-}w}(665)+C_2(\lambda)$	
	$a_{cdm+pc}(\lambda)=a_{t\text{-}w}(\lambda)-a_{phy\text{-}pc}(\lambda)$	
	$a_{cdm}(\lambda)=a_{cdm}(412)\exp\left(-\ln\left(\frac{a_{cdm}(412)}{a_{cdm}(510)}\right)(\lambda-412)/98\right)$	
	where $a_{cdm}(412)=a_{cdm+pc}(412), a_{cdm}(510)=a_{cdm+pc}(510)$, and $a_{cdm}(709)$ should be forced to be zero	
	$a_{pc}(620)=a_{cdm+pc}(620)-a_{cdm}(620)$	

QAA_{pc}

QAA_{pc} first utilizes the QAA to derive the $a_{phy}(\lambda)$ at 665 and 620 nm, respectively, each of which is considered to the summation attributed to the absorption of PC and Chl-*a* as shown by the two relationships in Table 37.5. Solving these two relations for the PC absorption at 620 nm (a_{pc} (620)) becomes straightforward when ψ_1 and ψ_2 are defined to be $a_{Chl\text{-}a}(665)/a_{Chl\text{-}a}(620)$ and $a_{pc}(665)/a_{pc}$ (620), respectively. QAA_{pc} performance depends on optimizing ψ_1 and ψ_2 factors. When applied to MERIS spectra of hypereutrophic catfish ponds, QAA_{pc} gave rise to a strong PC estimation with an R^2 of 0.99 and a relative error (RE) of 30.7%.

EIIMIW

The EIIMIW aims at partitioning non-water absorption coefficient ($a_{t\text{-}w}(\lambda)$) into the contribution of NAP, CDOM, non-PC pigments, and PC [34,49]. The first assumption is the dominance of water absorption at 709 nm over cdm and phytoplankton, i.e., $a(709) \approx a_w(709)$, which is used to derive the total non-water constituent absorption at a wavelength from the ratio of $R_{rs}(709)$ to $R_{rs}(\lambda)$ and using $b_b(709)$

which is derived from particle backscattering at 560 nm ($b_{bp}(560)$) and $b_b(778)$. The second assumption is $a_{cdm}(\lambda) = a_{pc}(\lambda) = 0$ for $\lambda = 665$ nm or larger, implying $a_{ph\text{-}pc}(665) \approx a_{t\text{-}w}(665)$ and $a_{ph\text{-}pc}(675) \approx a_{t\text{-}w}(675)$. Li et al. [30] observed a strong correlation between $a_{t\text{-}w}(665)$ and $a_{t\text{-}w}(675)$, implying a relationship $a_{ph\text{-}pc}(\lambda) = C_1(\lambda) a_{ph\text{-}pc}(675) + C_2(\lambda)$ where $C_1(\lambda)$ and $C_2(\lambda)$ are spectral constants derived from $a_{phy\text{-}pc}(\lambda)$ measured in pigment extraction using acetone in laboratory and shown in the Appendix B of Li et al. [46]. Therefore, the absorption of NAP, CDOM, and PC at a wavelength $\lambda(a_{cdm+pc}(\lambda))$ can be calculated as the difference between $a_{t\text{-}w}(\lambda)$ and $a_{ph\text{-}pc}(\lambda)$. The third assumption is $a_{pc}(\lambda) = 0$ for $\lambda = 510$ nm or 412, i.e., $a_{cdm}(412) = a_{t\text{-}w}(412) - a_{phy\text{-}pc}(412)$ and $a_{cdm}(510) = a_{t\text{-}w}(510) - a_{phy\text{-}pc}(510)$, then the cdm absorption at a wavelength $\lambda(a_{cdm}(\lambda))$ can be derived from $a_{cdm}(412)$ and $a_{cdm}(510)$. At last, $a_{pc}(620)$ is calculated to be $a_{pc}(620) = a_{t\text{-}w}(620) - a_{phy\text{-}pc}(620) - a_{cdm}(620)$.

The EIIMIW was applied to water samples collected from reservoirs of Northeast China, Lake Tai of southern China, rivers and lakes of South Australia in addition to the three Central Indiana reservoirs for deriving the IOPs of phytoplankton, PC and CDM. When calibrated with data collected in 2010 from three Indiana reservoirs, the EIIMIW estimated the absorption spectra of both CDM ($a_{cdm}(\lambda)$) and phytoplankton ($a_{ph}(\lambda)$) with $R^2 \geq 0.80$ and a rRMSE ≤31.79% for $a_{cdm}(412)$, $a_{ph}(443)$, $a_{ph}(620)$, and $a_{ph}(665)$. The EIIMIW achieved more accurate PC estimation with $R^2 = 0.81$, rRMSE = 33.60%, and mean RE = 49.11% than the widely used semi-empirical algorithm with $R^2 = 0.73$, rRMSE = 45.09%, and mean RE = 182.29% for the same dataset. The EIIMIW validation on data collected from 2005 to 2008 from three Indiana reservoirs, USA, and a strong performance of EIIMIW is evident compared to the NBR model, particularly for low PC range samples ([PC] ≤ 50 μg L^{-1}) [46].

Performance Comparison among Semi-Empirical and Semi-Analytical Models

Recently, several studies have been devoted to comparison of various semi-empirical and semi-analytical models for the estimation of PC concentration [42,50,51]. The discussion below will be focused on describing the research by Li and Song [16], Liu et al. [32], Pyo et al. [52], and Riddick et al. [53], in which the comparison of PC algorithms was carried out either based on a large dataset [32] or for a relatively complete list of currently available semi-empirical and semi-analytical algorithms [52,53]. Table 37.4 shows the comparison results from these studies.

Datasets for Three Reservoirs of Central Indiana, USA

Li and Song [16] systematically compared the performance of various empirical, semi-empirical, and semi-analytical PC algorithms with *in situ* measured reflectance spectra for three central Indiana reservoirs: Eagle Creek, Geist, and Morse. For these three reservoirs, an ASD FieldSpec ultraviolet/visible and near-infrared (UV/VNIR) spectroradiometer (Analytical Spectral Devices, Inc., Boulder, CO, USA) was used to measure the remote sensing reflectance above the water surface $R_{rs}(\lambda)$ in 2005 and 2006, and an Ocean Optics USB4000 unit (Ocean Optics, Inc., Dunedin, FL, USA) with dual radiometers to measure remote sensing reflectance below the water surface $r_{rs}(\lambda)$ in 2007, 2008, and 2010, and resulted in a total of 649 water samples for which both *in situ* spectra and PC concentrations were available. $BR_{650/620}$, MRBI, NBR, and EIIMIW were compared for their performance, and $R^2 = 0.54$, 0.21, 0.73, and 0.74 were obtained, respectively.

Datasets for Reservoirs in the USA, the Netherlands, and China

Liu et al. [32] compared the performance of various empirical, semi-empirical, and semi-analytical PC algorithms with *in situ* reflectance spectra measured with an Ocean Optics USA400 unit (Ocean Optics, Inc., Dunedin, FL, USA) radiometer in 2010 for Eagle Creek, Geist, and Morse Reservoirs located in central Indiana, USA, with a PR-650 (Photo Research) in the summers of 2004 and 2005 for Lake IJsselmeer (LIJ) in the Netherlands, and with a FieldSpec spectroradiometer (Analytical Spectral Devices, Inc., Boulder, CO, USA) in 2016 for Lake Taihu (LTH), Lake Chaohu (LCH) and Lake Hengshui (LHS) in China. Algorithms that were compared in this study were $BR_{650/620}$, PCI, TBM, and FBBM and $R^2 = 0.63$, 0.178, 0.701, and 0.73 were obtained, respectively.

Datasets for Baekje Reservoir in South Korea

Pyo et al. [52] compared the performance of semi-analytical PC algorithms with *in situ* reflectance spectra measured from June to October, 2016, with a FieldSpec HandHeld 2 spectroradiometer (ASD Inc., Boulder, CO, USA) having a wavelength range of 325–1075 nm for Baekje reservoir located in the main stream of Geum River, South Korea. Algorithms that were optimized and then assessed against a total of 160 samples with PC concentration ranging from 0 to 1014 μg L^{-1} included NBR and EIIMIW, for which $R^2 = 0.53$ and 0.83 were obtained, respectively. Therefore, the EIIMIW was favorable over NBR for estimating PC concentrations in this reservoir. Nonetheless, it is important to note that the original EIIMIW forced a zero a_{cdm} at 709 nm before removing the cdm absorption from the total non-water absorption [34]. Table S2 that was provided by Pyo et al. [52] does not indicate whether this step was taken into consideration, but this could explain why the EIIMIW resulted in so dramatically different R^2 values with ($R^2 = 0.83$) and without being recalibrated ($R^2 = 0$).

Datasets for Lake Balaton, Hungary

Riddick et al. [53] performed a comprehensive comparison among various PC algorithms with a MERIS and *in situ* reflectance dataset of Lake Balaton, Hungary. The PC concentrations were measured by Balaton Limnological Institute (BLI) at five stations of this reservoir from 2010 to 2011 at a biweekly or monthly basis as well as at 30 stations from August 18 to 26, 2010, giving a PC range of 2.34–113.0 μg L^{-1}. These PC datasets called the BLI and the August 2010 datasets, respectively, were used to validate PC algorithms with the image spectra of a MERIS overpass within a time window of ±1 day.

Algorithms that were compared on 22 data pairs of PC concentration and MERIS spectrum at ±1 day windows were $BR_{650/620}$, MRBI, PCI, TBM, NBR, QAA_{pc}, EIIMIW, and FBBM, which resulted in an R^2 of 0.595, 0.0992, 0.091, 0.662, 0.71, 0.00836, 0.716, and 0.634, respectively, and an RMSE of 14.4, 21.6, 715, 17.8, 11.8, 22.9, 46.3, and 16.5 μg L^{-1}, respectively. These 22 samples were also separated based on the PC threshold 50 μg L^{-1}. Schalles00, MRBI, PCI, TBM, NBR, QAA_{pc}, EIIMIW, and FBBM were only validated on 19 samples having PC concentration <50 μg L^{-1} and resulted in an R^2 of 0.554, 0.0719, 0.0588, 0.77, 0.793, 0.44, 0.697, and 0.814, respectively and an RMSE of 13.57, 17.84, 766, 10.4, 10.7, 9.76, 20.2, and 15.3 μg L^{-1}, respectively. Based on the R^2 and RMSE values for the PC algorithms subject to the comparison performed by this study, NBR was determined to be the optimal PC algorithm instead of EIIMIW and FBBM because NBR resulted in lower RMSE values than EIIMIW and FBBM. However, additional investigations are still needed to compare NBR, EIIMIW, and FBBM with each other because of the limitation to the work performed by Riddick et al. [53]. First, a small set of water samples from one lake was used in the comparison by Riddick et al. [53], whereas the FBBM was optimized using a large number of water samples from inland waters in the Netherlands, central Indiana, and typical lakes in China and should be more transferable geographically. Second, the EIIMIW implemented by Riddick et al. [53] used a constant $S_{CDOM} = 0.02\,nm^{-1}$ to characterize cdm absorption, whereas the original EIIMIW used the natural log transformed ratio of NAP absorption at 412 nm to that at 510 nm for describing cdm absorption. Giving the high concentration of inorganic particle in Lake Balaton, it is not unexpected that the EIIMIW implemented by Riddick et al. [53] resulted in a relatively large RMSE for Lake Balaton.

Algorithms Recommended for Further Examination

Based on the results described above for the performance of various PC algorithms, it is evident that the semi-analytical algorithms NBR and EIIMIW and the semi-empirical FBBM performed stronger than empirical algorithms MRBI and PCI showing the weakest performance along with another semi-analytical algorithm QAA_{pc}. Empirical algorithms $BR_{650/620}$ and TBM showed stronger performance than MRBI, PCI, and QAA_{pc}, which can be explained by a high dataset-dependent correlation between the former ratios and PC concentrations. Therefore, future investigation should focus on assessing, comparing, and further improving the performance of NBR, EIIMIW, and FBBM and developing completely new algorithms in order to determine optimal algorithms for monitoring CYB blooms in regional and global scales.

Practical Considerations and Current Challenges

Satellite- and Airborne-Based Cyanobacteria Monitoring

Detection or monitoring of inland water CYB blooms relies on the PC absorption feature around 620 nm, but very few satellite sensors own this spectral channel. Satellite sensors such as those on Hyperion, HICO, and MERIS used to have bands around 620 nm, and these sensors ended their data acquisition service in 2017, 2014, and 2012. Launched in April 28, 2018, the Ocean and Land Color Instrument (OLCI) on board Sentinel-3 has similar spectral configuration to the MERIS and is able to maintain data continuity of ENVISAT [54]. Italian PRecursore IperSpettrale della Missione Applicativa (PRISMA) was developed by the Italian Space Agency (ASI in Italian) and launched in March 22, 2019. This sensor is able to acquire hyperspectral images at 30 m spatial resolution and 250 bands from 400 to 2,500 nm [55]. Both the OLCI and PRISMA capable of acquiring images at 620 nm provide opportunities to assess the algorithms NBR, EIIMIW, and FBBM and determine the best for monitoring CYB blooms of inland waters around the world. In addition, another two satellite hyperspectral sensors such as German Environmental Mapping and Analysis Program (EnMAP) and NASA's Hyperspectral Infrared Imager (HyspIRI) have been scheduled to launch in 2020 and 2022, respectively [56,57]. Hyperspectral sensors PRISMA EnMAP and HysPIRI will provide opportunities to determine the best PC algorithm among NBR, EIIMIW, and FBBM. Furthermore, PRISMA EnMAP and HysPIRI data would allow for assessing the DTBB, which cannot be done with the OLCI data in spite of its better performance than NBR and EIIMIW.

Variability of PC Specific Absorption Coefficient

The convention of practicing semi-analytical algorithms for estimating PC is to divide the bulk PC absorption coefficient at 620 nm by the corresponding specific absorption coefficient, and different $a^*_{pc}(624)$ values result in different PC concentrations. For example, Dekker [3] reported that $a^*_{pc}(624) = 0.0032 \pm 0.0012\,m^2\,(mg\,PC)^{-1}$ for the Netherlands inland waters, Li et al. [30] determined an average $a^*_{pc}(624)$ to be $0.0024\,m^2\,(mg\,PC)^{-1}$ for three central Indiana reservoirs, and these low values for $a^*_{pc}(620)$ have been attributed to the use of the baseline between 600 and 648 nm for determining $a^*_{pc}(620)$. However, high $a^*_{pc}(620)$ values of $0.0095\,m^2\,(mg\,PC)^{-1}$ and $0.007\,m^2\,(mg\,PC)^{-1}$ were used by Simis et al. [11,44], respectively, which have been attributed to the absorption of intracellular water-soluble compounds akin to sheath pigments found in CYB [30]. When assessing and comparing different PC algorithms, it is important to ensure that the performance assessment and comparison for different PC algorithms are performed with the same PC-specific absorption coefficient.

Variability of Specific Absorption Coefficients of Non-PC Constituents

Semi-empirical and semi-analytical models are commonly built upon the assumption that the water constituent IOPs are spatially and seasonally stable or that the spectral shapes of individual constituent absorption and backscattering are known. These assumptions are often violated in real-world applications. For example, the specific phytoplankton absorption coefficient $\left(a^*_{pc}\right)$ is defined to be the ratio of the phytoplankton absorption coefficient (a_{ph}) to Chl-*a* concentration and is known to vary across individual species grown in culture and among natural phytoplankton assemblages because of varying pigment composition and the package effect [58–60]. Some studies suggested 0.02 m^2 mg^{-1} for the average value of a^*_{ph} (using 675 nm as reference) [60,61], but this value could have a range of 0.01–0.033 m^2mg^{-1} for New Zealand lakes [62], 0.005–0.05 m^2mg^{-1} for Nebraska and Iowa lakes [63], and 0.01–0.025 m^2mg^{-1} for three water reservoirs in north-eastern Australia [64]. Investigators often use different power laws to describe a^*_{ph} as a function of Chl-*a* concentration. Although this varying power law relationship may not have a significant influence on the performance of the NBR and EIIMIW algorithms because the removal of Chl-*a* interference on PC estimation is performed using the correlation between Chl-*a* absorption coefficients at bands 665 and 620 nm, this varying power law relationship should affect the performance of FBBM in which the linear interpolation weight needs to be optimized. It is worth pointing out that QAA_{pc} did not perform as strong as NBR, EIIMIW, and FBBM, but the two summations of Chl-*a* and PC absorption coefficients at 665 and 620 nm in QAA_{pc} can be useful for improving the FBBM.

NAP and CDOM are another two substances interfering remote estimation for PC concentrations. The exponential function is often used to describe the spectral dependence of dissolved matter and detritus absorption [65,66], but the exponent varies as a result of seasonal and/or spatial changes in the origin and type of dissolved materials (humic vs. fulvic) [67] and non-alga particles (mineral sediments, non-algal organic detritus such as fecal matter, degrading phytoplankton cells, and living non-algal particulates) [68,69]. The exponent of CDOM S_{CDOM} can range from 0.01 to 0.02 nm^{-1} [70,71] or even a wider range [72]. However, a true synthesis of the variability in S_{CDOM} for inland waters is difficult to obtain from existing *in situ* data because the investigations were conducted over limited seasons and regional regions. The exponential slope for NAPs S_{NAP} has been reported to be 0.011 nm^{-1} for productive and CDOM-rich marine waters [73], 0.0123 nm^{-1} (a range of 0.0089–0.0178 nm^{-1}) for European coastal waters [70], and 0.011 nm^{-1} (a range of 0.0077–0.017 nm^{-1}) for Lake Erie [69]. This variability in S_{NAP} may result from a wide range of particle types and a relative extreme proportion of mineral or organic matter observed by these studies [69]. The variability described for S_{CDOM} and S_{NAP} should affect the performance of NBR in which two empirical constants are used to correct for the effect of CDOM and NAP on estimating the PC and Chl-*a* absorption at 620 and 665 nm, respectively, implying the necessity for recalibrating NBR.

Variability of Suspended Matter Backscattering Coefficients

The main scattering contributors in inland waters are organic suspended matter (OSM) and inorganic suspended matter (ISM). Morel [74] showed that the variation of non-absorbing particle scattering coefficient follows a power law decay, which is often a valid relationship for coastal and inland waters [75–79]. This power law relationship has two implications for assessing the semi-analytical PC algorithms NBR and FBBM and the semi-empirical algorithm DTBB because they all assume a neutral backscattering. First, a neutral backscattering could be invalid when the power relationship shows a rapid decay and the interband space for these three algorithms is very wide such as 620 nm versus 709 nm in NBR, 530 nm versus 750 nm in FBBM, and 600 nm versus 725 nm in DTBB. Second, spectral scattering coefficients have been shown as a significant departure from a power law decay at strong spectral absorption bands [80–82], and the power law model has been modified to account for the effect of particulate absorption [78]. These observations imply that a further correction for the effect of $b_b(\lambda)$ on the performance of NBR, FBBM, and DTBB is necessary for achieving a reliable PC retrieval for very turbid water bodies.

Conclusions

Toxic CYB blooms are of great concern due to embedded implications for alternated biodiversity, public health, and for overall ecosystem health of inland waters. The development of remote sensing approaches to efficiently monitoring CYB blooms has important implications for effectively managing CYB blooms, and this critically relies on highly performing PC estimation algorithms.

Remote sensing algorithms for PC estimation are grouped into empirical, semi-empirical, and semi-analytical methods. In this chapter, 13 of these methods have been described with 12 of them being reviewed and compared with each other based on their performances on *in situ* measured field reflectance spectra and airborne or satellite sensor collected image spectra. Five empirical PC algorithms show data-dependent performances with the MRBI and PCI being consistently weaker than the three BR models, which can be attributed to a strong interference of non-PC constituents with a relatively weaker PC absorption at 620 nm. Nonetheless, when these BRs are improved with the baseline idea embedded in the MRBI and PCI, semi-empirical models especially DTBB and FBBM can be established and show stronger performance than the TBM with DTBB even performing stronger than the NBR, a semi-analytical model. Assessing three semi-analytical models indicates that the NBR and EIIMIW consistently performed well compared to the QAA_{pc}. However, given a large variability of non-PC constituents in their IOPs, one caveat for applying the EIIMIW is to carefully follow the original model step for calculating the parameter a_{cdm}, while for applying the NBR, it is necessary to recalibrate two empirical parameters for removing the interference of NAP and CDOM.

Despite being shown to perform stronger than the NBR for estimating PC, so far neither DTBB nor FBM has been examined extensively with satellite MERIS and OLCI images because neither sensor has a spectral channel at 645 nm required for implementing semi-empirical DTBB and FBM. With the availability of hyperspectral satellite images from PRISMA, EnMAP, and HyspIRI, both DTBB and FBM should be evaluated together with EIIMIW and NBR for a possibly improved CYB detection and management at large or regional scales.

References

1. Millie, D.F.; Baker, M.C.; Tucker, C.S.; Vinyard, B.T.; Dionigi, C.P. High-resolution airborne remote sensing of bloom-forming phytoplankton. *Journal of Phycology* 1992, 28, 281–290.
2. Mittenzwey, K.H.; Gitelson, A.A.; Kondratiev, K.Y. Determination of chlorophyll a of inland waters on the basis of spectral reflectance. *Limnology and Oceanography* 1992, 37, 147–149.
3. Dekker, A.G. Detection of Optical Water Quality Parameters for Eutrophic Waters by High Resolution Remote Sensing. Ph.D. Dissertation, Vrije University Amsterdam, 1993.
4. Gitelson, A.A.; Laorawat, S.; Keydan, G.P.; Vonshank, A. Optical properties of dense algal culture outdoors and their application to remote estimation of biomass and pigment concentration in Spirulina Platensis (Cyanobacteria). *Journal of Phycology* 1995, 31, 828–834.
5. Yacobi, Y.Z.; Gitelson, A.; Mayo, M. Remote sensing of chlorophyll in Lake Kinneret using high-spectral-resolution radiometer and Landsat TM, spectral features of reflectance and algorithm development. *Journal of Plankton Research* 1995, 17, 2155–2173.
6. Kuster, T.; Hannonen, T.; Kallio, K.; Koponen, S.; Pulliainen, J.; Pyhalahti, T. Monitoring of turbid coastal and inland water by airborne imaging spectrometer AISA. *Geoscience and Remote Sensing Symposium Proceedings* 1998, 5, 2597–2599.
7. Fraser, R.N. Hyperspectral remote sensing of turbidity and chlorophyll a among Nebraska Sand Hills lakes. *International Journal of Remote Sensing* 1998, 19, 1579–1589.
8. Schalles, J.F.; Yacobi, Y.Z. Remote detection and seasonal patterns of phycocyanin, carotenoid and chlorophyll pigments in eutrophic waters. *Arch. Hydrobiol. Spec. Issues Advanc. Limnol.* 2000, 55, 153–168.

9. Kallio, K.; Kuster, T.; Hannonen, T.; Koponen, S.; Pullianinen, J.; Vepsalainen, J. Retrieval of water quality from airborne imaging spectrometry of various lake types in different season. *Science of the Total Environment* 2001, 268, 59–77.
10. Vincent, R.K.; Qin, X.; McKay, R.M.L.; Miner, J.; Czajkowski, K.; Savino, J.; Bridgeman, T. Phycocyanin detection from Landsat TM data for mapping cyanobacterial blooms in Lake Erie. *Remote Sensing of Environment* 2004, 89, 381–392.
11. Simis, S.G.; Peters, S.W.; Gons, H. Remote sensing of the cyanobacterial pigment phycocyanin in turbid inland water. *Limnology and Oceanography* 2005, 50, 237–245.
12. Zimba, P.V.; Gitelson, A. Remote estimation of chlorophyll concentration in hyper-eutrophic aquatic systems: Model tuning and accuracy optimization. *Aquaculture* 2006, 256, 272–286.
13. Li, L.; Pascual, D.L.; Tedesco, L.P.; Randolph, K.L.; Sengpiel, R.E.; Hall, B.E. Developing a Survey Tool for the Rapid Assessment of Blue-Green Algae in Central Indiana's Reservoirs. Central Indiana Water Resources Partnership/Center for Earth and Environmental Science, IUPUI, Indianapolis, IN, 2006.
14. Robertson, A.L. Using Band Ratio, Semi-empirical, Curve Fitting, and Partial Least Squares (PLS) Models to Estimate Cyanobacterial Pigment Concentration from Hyperspectral Reflectance. M.S. Thesis, Indiana University-Purdue University Indianapolis, 2009, p. 108.
15. Matthews, M.W. A current review of empirical procedures of remote sensing in inland and near-coastal transitional waters. *International Journal of Remote Sensing* 2011, 32, 6855–6899.
16. Li, L.; Song, K. Bio-optical modeling of phycocyanin. In: Mishra, D.R.; Ogashawara, I.; Gitelson, A.A. (eds.) *Bio-Optical Modeling and Remote Sensing of Inland Waters*. Elsevier: Amsterdam, 2017, pp. 233–262. ISBN: 978-0-12-804644-9.
17. Yan, Y.; Bao, Z.; Shao, J. Phycocyanin concentration retrieval in inland waters: A comparative review of the remote sensing techniques and algorithms. *Journal of Great Lakes Research* 2018, 44, 748–755.
18. Han, L.; Rundquist, D.C. Comparison of NIR/red ratio and first derivative of reflectance in estimating algal-chlorophyll concentration: A case study in a turbid reservoir. *Remote Sensing of Environment* 1997, 62, 253–261.
19. Rundquist, D.C.; Han, L.; Schalles, J.F.; Peake, J.S. Remote measurement of algal chlorophyll in surface waters: The case for the first derivative of reflectance near 690 nm. *Photogrammetric Engineering and Remote Sensing* 1996, 62, 195–200.
20. Dekker, A.G.; Malthus, T.J.; Seyhan, E. Quantitative modeling of inland water quality for high resolution MSS system. *IEEE Transactions on Geoscience and Remote Sensing* 1991, 29, 89–95.
21. Gitelson, A. The peak near 700 nm on radiance spectra of algae and water: Relationships of its magnitude and position with chlorophyll concentration. *International Journal of Remote Sensing* 1992, 13, 3367–3373.
22. Gordon, H.R.; Brown, O.B.; Jacobs, M.M. Computed relationships between the inherent and apparent optical properties of a flat homogeneous ocean. *Applied Optics* 1975, 14, 417–427.
23. Morel, A.; Gentili, B. Diffuse-reflectance of oceanic waters. II. Bidirectional aspects. *Applied Optics* 1993, 32, 6864–6879.
24. Morel, A.; Gentili, B. Diffuse reflectance of oceanic waters. III. Implication of bidirectionality for the remote-sensing problem. *Applied Optics* 1996, 35, 4850–4862.
25. Hunter, P.D.; Tyler, A.N.; Willby, N.J.; Gilvear, D.J. The spatial dynamics of vertical migration by Microcystis aeruginosa in a eutrophic shallow lake: A case study using high spatial resolution time-series airborne remote sensing. *Limnology and Oceanography* 2008, 53, 2391–2406.
26. Li, L.; Sengpiel, R.E.; Pascual, D.L.; Tedesco, L.P.; Wilson, J.S.; Soyeux, E. Using hyperspectral remote sensing to estimate chlorophyll-a and phycocyanin in a mesotrophic reservoir. *International Journal of Remote Sensing* 2010, 31, 4147–4162.
27. Wynne, T.; Stumpf, R.; Tomlinson, M.; Warner, R.; Tester, P.; Dyble, J.; Fahnenstiel, G. Relating spectral shape to cyanobacterial blooms in the Laurentian Great Lakes. *International Journal of Remote Sensing* 2008, 29, 3665–3672.

28. Qi, L.; Hu, C.; Duan, H.; Cannizzaro, J.; Ma, R. A novel MERIS algorithm to derive cyanobacterial phycocyanin pigment concentrations in a eutrophic lake: Theoretical basis and practical considerations. *Remote Sensing of Environment* 2014, 154, 298–317.
29. Hunter, P.D.; Tyler, A.N.; Carvalho, L.; Codd, G.A.; Maberly, S.C. Hyperspectral remote sensing of cyanobacterial pigments as indicators for cell populations and toxins in eutrophic lakes. *Remote Sensing of Environment* 2010, 114, 2705–2718.
30. Li, L.; Li, L.; Shi, K.; Li, Z.; Song, K. A semi-analytical algorithm for remote estimation of phycocyanin in inland waters. *Science of the Total Environment* 2012, 435–436, 141–150.
31. Le, C.; Li, Y.; Zha, Y.;Wang, Q.; Zhang, H.; Yin, B. Remote sensing of phycocyanin pigment in highly turbid inland waters in Lake Taihu, China. *International Journal of Remote Sensing* 2011, 32, 8253–8269.
32. Liu, G.; Simis, S.G.H.; Li, L.; Wang, Q.; Li, Y.; Song, K.; Lyu, H.; Zheng, Z.; Shi, K. A four-band semi-analytical model for estimating phycocyanin in inland waters from simulated MERIS and OLCI Data. *IEEE Transactions on Geoscience and Remote Sensing* 2018, 56, 1374–1385.
33. Mishra, S.; Mishra, D.R.; Lee, Z.P.; Tucker, C.S. Quantifying cyanobacterial phycocyanin concentration in turbid productive waters: A quasi-analytical approach. *Remote Sensing of Environment* 2013, 133, 141–151.
34. Li, L.H.; Li, L.; Song, K. Remote sensing of freshwater cyanobacteria: An extended IOP Inversion Model of Inland Waters (IIMIW) for partitioning absorption coefficient and estimating phycocyanin. *Remote Sensing of Environment* 2015, 157, 9–23.
35. Mishra, S.; Mishra, D.R.; Schluchter, W.M. A novel algorithm for predicting phycocyanin concentrations in cyanobacteria: A proximal hyperspectral remote sensing approach. *Remote Sensing* 2009, 1, 758–775.
36. Ogashawara, I.; Mishra, D.; Mishra, S.; Curtarelli, M.; Stech, J. A performance review of reflectance based algorithms for predicting phycocyanin concentrations in inland waters. *Remote Sensing* 2013, 5, 4774–4798.
37. Song, K.; Li, L.; Li, Z.; Tedesco, L.; Hall, B.; Shi, K. Remote detection of cyanobacteria through phycocyanin for water supply source using three-band model. *Ecological Informatics* 2013, 15, 22–33.
38. Wynne, T.T.; Stumpf, R.P.; Briggs, T.O. Comparing MODIS and MERIS spectral shapes for cyanobacterial bloom detection. *International Journal of Remote Sensing* 2013, 34, 6668–6678.
39. Wynne, T.T.; Stumpf, R.P. Spatial and temporal patterns in the seasonal distribution of toxic cyanobacteria in Western Lake Erie from 2002–2014. *Toxins* 2015, 7, 1649–1663. doi: 10.3390/toxins7051649.
40. Stumpf, R.P.; Davis, T.W.; Wynne, T.T.; Graham, J.L.; Loftin, K.A.; Johengen, T.H.; Gossiaux, D.; Palladino, D.; Burtner, A. Challenges for mapping cyanotoxin patterns from remote sensing of cyanobacteria. *Harmful Algae* 2016, 54, 160–173.
41. Duan, H.; Ma, R.; Hu, C. Evaluation of remote sensing algorithms for cyanobacterial pigment retrievals during spring bloom formation in several lakes of East China. *Remote Sensing of Environment* 2012, 126, 126–135.
42. Wozniak, M.; Bradtke, K.M.; Darecki, M.; Krezel, A. Empirical model for phycocyanin concentration estimation as an indicator of cyanobacterial bloom in the optically complex coastal waters of the Baltic Sea. *Remote Sensing* 2016, 8, 212.
43. Ogashawara, I. The use of Sentinel-3 imagery to monitor cyanobacterial blooms. *Environments* 2019, 6, 60. doi: 10.3390/environments6060060.
44. Simis, S.G.H.; Ruiz-Verdu, A.; Dominguez-Gomez, J.A.; Pena-Martinez, R.; Peters, S.W.M.; Gons, H.J. Influence of phytoplankton pigment composition on remote sensing of cyanobacterial biomass. *Remote Sensing of Environment* 2007, 106, 414–427.
45. Gons, H. Optical teledetection of chlorophyll a in turbid inland waters. *Environmental Science and Technology* 1999, 33, 1127–1132.

46. Mishra, S.; Mishra, D.R.; Lee, Z.P. Bio-optical inversion in highly turbid and cyanobacteria-dominated waters. *IEEE Transactions on Geoscience and Remote Sensing* 2014, 52 (1), 375–388.
47. Randolph, K.L.; Wilson, J.S.; Tedesco, L.P.; Li, L.; Pascual, D.L.; Soyeux, E. Hyperspectral remote sensing of cyanobacteria in turbid productive water using optically active pigments, chlorophyll a and phycocyanin. *Remote Sensing of Environment* 2008, 112, 4009–4019.
48. Lyu, H.; Wang, Q.; Wu, C.; Zhu, L.; Yin, B.; Li, Y.M.; Huang, J.Z. Retrieval of phycocyanin concentration from remote-sensing reflectance using a semi-analytic model in eutrophic lakes. *Ecological Informatics* 2013, 18, 178–187.
49. Li, L.H.; Li, L.; Song, K.; Li, Y.; Tedesco, L.P.; Shi, K.; Li, Z.C. An inversion model for deriving inherent optical properties of inland waters: Establishment, validation and application. *Remote Sensing of Environment* 2013, 135, 150–166.
50. Ruiz-Verdu, A.; Simis, S.G.H.; de Hoyos, C.; Gons, H.J.; Pena-Martinez, R. An evaluation of algorithms for the remote sensing of cyanobacterial biomass. *Remote Sensing of Environment* 2008, 112, 3996–4008.
51. Ogashawara, I.; Mishra, D.R.; Mishra, S.; Curtarelli, M.P.; Stech, J.L. A performance review of reflectance based algorithms for predicting phycocyanin concentrations in inland waters. *Remote Sensing* 2013, 5 (10), 4774–4798.
52. Pyo, J.C.; Yachepsky, Y.; Baek, S.S.; Kwon, Y.S.; Kim, M.J.; Lee, H; Park, S.; Cha, Y.K.; Ha, R.; Nam, G.; Park, Y.G.; Cho, K.H. Optimizing semi-analytical algorithms for estimating chlorophyll-a and phycocyanin concentrations in inland waters in Korea. *Remote Sensing* 2017, 9, 542. doi: 10.3390/rs9060542.
53. Riddick, C.A.L.; Hunter, P.D.; Dominguez-Gómez, J.A.; Martinez-Vicente, V.; Présing, M.; Horváth, H.; Kovács, A.W.; Vörös, L.; Zsigmond, E.; Tyler, A.N. Optimal cyanobacterial pigment retrieval from ocean colour sensors in a highly turbid, optically complex lake. *Remote Sensing* 2019, 11, 1613. doi: 10.3390/rs11131613.
54. Donlon, C.; Berruti, B.; Buongiorno, A.; Ferreira, M.H.; Femenias, P.; Frerick, J.; Goryl, P.; Klein, U.; Laur, H.; Mavrocordatos, C.; Nieke, J.; Redhan, H.; Seitz, B.; Stroede, J.; Sciarra, R. The global monitoring for environment and security (GMES) Sentinel-3 mission. *Remote Sensing of Environment* 2012, 120, 37–57. doi: 10.1016/j.rse.2011.07.024.
55. Labate, D.; Ceccherini, M.; Cisbani, A.; De Cosmo, V.; Galeazzi, C.; Giunti, L.; Melozzim, M.; Pieraccini, S.; Stagi, M. The PRISMA payload optomechanical design, a high performance instrument for a new hyperspectral mission. *Acta Astronautica* 2009, 65, 1429–1436. doi: 10.1016/j.actaastro.2009.03.077.
56. Stuffler, T.; Förster, K.; Hofer, S.; Leipold, M.; Sang, B.; Kaufmann, H.; Penné, B.; Mueller, A.; Chlebek, C. Hyperspectral imaging-An advanced instrument concept for the EnMAP mission (Environmental Mapping and Analysis Programme). *Acta Astronautica* 2009, 65 (7–8), 1107–1112. doi: 10.1016/j.actaastro.2009.03.042.
57. National Research Council (NCR). *Earth Science and Applications from Space: National Imperatives for the Next Decade*. National Academy Press, Division on Engineering and Physical Sciences: Washington, DC, 2017.
58. Bricaud, A.; Babin, M.; Morel, A.; Claustre, H. Variability in the chlorophyll-specific absorption coefficients of natural phytoplankton: Analysis and parameterization. *Journal of Geophysical Research* 1995, 100, 13321–13332.
59. Bricaud, A.; Stramski, D. Spectral absorption coefficients of living phytoplankton and noalgal biogenous matter: A comparison between the Peru upwelling area and the Sargasso Sea. *Limnology and Oceanography* 1990, 35, 562–582.
60. Le, C.F.; Li, Y.M.; Zha, Y.; Sun, D.Y.; Yin, B. Validation of a quasi-analytical algorithm for highly turbid eutrophic water of Meiliang Bay in Taihu Lake, China. *IEEE Transactions on Geoscience and Remote Sensing* 2009, 47 (8), 2492–2500.

61. Bricaud, A.; Morel, A.; Prieur, L. Optical efficiency factors of some phytoplankters. *Limnology and Oceanography* 1983, 28, 816–832.
62. Daviescolley, R.J.; Vant, W.N. Absorption of light by yellow substance in fresh-water lakes. *Limnology and Oceanography* 1987, 32, 416–425.
63. Dall'olmo, G.; Gitelson, A.A. Effect of bio-optical parameter variability on the remote estimation of chlorophyll-a concentration in turbid productive waters: Experimental results. *Applied Optics* 2005, 44, 3342–3342.
64. Campbell, G.; Phinn, S.R.; Daniel, P. The specific inherent optical properties of three sub-tropical and tropical water reservoirs in Queensland, Australia. *Hydrobiologia* 2011, 658, 233–252.
65. Bricaud, A.; Babin, M.; Claustre, H.; Ras, J.; Tieche, F. Light absorption properties and absorption budget of Southeast Pacific waters. *Journal of Geophysical Research-Oceans* 2010, 115. doi: 10.1029/2009JC005517.
66. Bricaud, A.; Morel, A.; Babin, M.; Allali, K.; Claustre, H. Variations of light absorption by suspended particles with chlorophyll a concentration in oceanic (case 1) waters: Analysis and implications for bio-optical models. *Journal of Geophysical Research* 1998, 103, 31033–31044.
67. Song, K.S.; Zang, S.Y.; Zhao, Y.; Li, L.; Du, J.; Zhang, N.N.; Wang, X.D.; Shao, T.T.; Guan, Y.; Liu, L. Spatiotemporal characterization of dissolved carbon for inland waters in semi-humid/semiarid region, China. *Hydrology and Earth System Sciences* 2013, 10, 6559–6597. doi: 10.5194/hessd-10-6559-2013.
68. Stramski, D.; Babin, M.; Wozniak, S. Variation in the optical properties of terrigenous mineral-rich particulate matter suspended in sea water. *Limnology and Oceanography* 2007, 52 (2), 418–2433.
69. Binding, C.E.; Jerome, J.H.; Bukata, R.P.; Booty, W.G. 2008. Spectral absorption properties of dissolved and particulate matter in Lake Erie. *Remote Sensing of Environment* 2008, 112, 1702–1711.
70. Babin, M.; Morel, A.; Fournier-Sicre, V.; Fell, F.; Stramski, D. Light scattering properties of marine particles in coastal and open ocean waters as related to the particle mass concentration. *Limnology and Oceanography* 2003, 48, 843–859.
71. Kirk, J.T.O. *Light and Photosynthesis in Aquatic Ecosystem*. Cambridge University Press: Cambridge, 1994.
72. Kowalczuk, P. Seasonal variability of yellow substance absorption in the surface layer of the Baltic Sea. *Journal of Geophysical Research-Oceans* 1999, 104, 30047–30058.
73. Roesler, C.S.; Perry, M.J.; Carder, K.L. Modeling in situ phytoplankton absorption from total absorption spectra in productive inland marine waters. *Limnology and Oceanography* 1989, 34, 1510–1523.
74. Morel, A. Optical modeling of the upper ocean in relation to its biogenous matter content (case-I waters). *Journal of Geophysical Research-Oceans* 1988, 93, 10749–10768.
75. Morel, A.; Gentili, B.; Chami, M.; Ras, J. Bio-optical properties of high chlorophyll case 1 waters and of yellow-substance-dominated case 2 waters. *Deep-Sea Research Part I* 2006, 53, 1439–1459.
76. Roesler, C.S.; Boss, E. Spectral beam attenuation coefficient retrieved from ocean color inversion. *Geophysical Research Letters* 2003, 30, 1468.
77. Song, Q.J.; Tang, J.W. The study on the scattering properties in the Huanghai Sea and East China Sea. *Acta Oceanologica Sinica* 2006, 28, 56–63.
78. Doxaran, D.; Ruddick, K.; McKee, D.; Gentili, B.; Tailliez, D.; Chami, M. Spectral variations of light scattering by marine particles in coastal waters, from visible to near infrared. *Limnology and Oceanography* 2009, 54, 1257–1271.
79. Sun, D.; Li, Y.M.; Wang, Q.; Le, C.F.; Lyv, H.; Huang, C. Specific inherent optical quantities of complex turbid inland waters from the perspective of water classification. *Photochemical and Photobiological Sciences* 2012, 11 (8), 1299–1312.
80. Barnard, A.H.; Pegau, W.S.; Zaneveld, J.R.V. Global relationships of the inherent optical properties of the oceans. *Journal of Geophysical Research-Oceans* 1998, 103, 24955–24968.

81. Stramski, D.; Bricaud, A.; Morel, A. Modeling the inherent optical properties of the ocean based on the detailed composition of the planktonic community. *Applied Optics* 2001, 40, 2929–2945.
82. Doxaran, D.; Babin, M.; Leymarie, E. Near-infrared light scattering by particles in coastal waters. *Optics Express* 2007, 15, 12834–12849.

38

Bibliometric and Visualized Analysis of Water Footprint Research

Peili Duan
Ludong University

Lijie Qin
Northeast Normal University

Peng Yin
Ocean University of China

Introduction 323
Methodology and Data Sources 324
General Statistical Analysis 324
Temporal Distribution of Publications Analysis • Comparison of Research Outputs between Different Countries • Analysis by Most Productive and Influential Organizations • Co-Authorship Network Analysis among Productive Authors • Quantitative Analysis of Main Source Journals
Research Hotspots Analysis 329
The Subject Categories and Research Areas of Water Footprint Papers • Analysis of High-Frequency Keywords
Terms Co-Occurrence and Cluster Analysis 330
Conclusions 331
References 332

Introduction

Fresh water is a basic natural resource for human, socioeconomic, and environmental development (Tian et al., 2018). Under the increasing pressure of population growth, economic development, and climate change, the available global freshwater resources are gradually decreasing (Dalin et al., 2012). Reported by World Bank, as of 2014, renewable internal freshwater resources per capita data in the world was reported at 5,920.508 m^3. However, data was 13,401.912 m^3 in 1962 (The World Bank, 2016). This reflects a decrease of almost 1,500 m^3 every 10 years (Zhang et al., 2018). To meet future freshwater supply challenges, it is important to evaluate water resource utilization in the production and consumption process for human beings.

The water footprint concept provides a new approach for assessing water resource utilization and addressing the water crisis (Hoekstra and Chapagain, 2008). The concept of virtual water proposed by Tony Allan in 1993 was regarded as the amount of water required to produce goods and services (Allan, 1993). At the foundation of virtual water, Hoekstra first introduced the concept of water footprint in 2002 (Hoekstra, 2003). Subsequently, elaborated by Hoekstra and Chapagain in 2008, the water footprint is an indicator that accounts for both the direct and indirect water used, consumed, and/or polluted (Vanham and Bidoglio, 2013; Castellanos et al., 2016).

The concept of water footprint has been put forward for more than 10 years. Scholars all over the world have paid extensive attention to and applied it. At present, the research on water footprint mainly focuses on the analysis of water footprint content of single product (Mekonnen and Hoekstra,

2014; Hosseinian et al., 2017; Ma et al., 2018), the study of influencing factors of water footprint (Zhao et al.,2014; Jin et al.,2016; Shrestha et al., 2017), the assessment of water footprint at regional or national level (Yu et al., 2010; Rodriguez et al.,2015; Cao et al., 2018), and the analysis of water footprint structure (Chapagain and Hoekstra, 2011; Zhuo et al., 2016; Bosire et al., 2017). Among them, there are also many reviews on the research results of water footprint (Paterson et al., 2015; Lovarelli et al., 2016; Sabli et al., 2017). Previous review papers on water footprint research were either quantitative or qualitative in nature, while no attempt had been made to visualize knowledge maps. Bibliometric analysis methods are frequently used to provide quantitative analysis of academic literature (Chen et al., 2016). This study was aimed at making an overview and giving the development of water footprint research through bibliometric analysis in the last few years.

Methodology and Data Sources

In order to analyze the development of water footprint, bibliometric analysis was carried out to evaluate the characteristics and trend in water footprint area. Bibliometric analysis is a mathematical and statistical approach to analyze pertinent literature and understand the global research trends in a specific area (Khudzari et al., 2018; Zou et al., 2018). It has been applied to environmental engineering and science, soil science, ecology, food safety, new energy utilization, and other aspects.

Bibliometric indicators were analyzed in this research and directly obtained from the Web of Science. These quantity indicators mainly included the number of publications, countries, organizations, authors, journals, keywords, subject categories, and research areas. VOSviewer software was used to develop co-authorship collaboration networks of organizations and authors, terms co-occurrence analysis. Invented by Van Eck and Waltman (Leiden University) in the Netherlands, VOSviewer is a freely available texting software for generating bibliometric maps and analyzing trends in the scientific literature (van Eck and Waltman, 2010). It is a robust tool that calculates the present clustering algorithms and functionalities of the network based on the strength of the connections between items. VOSviewer software uses a circle and label to represent an element. The circle size represents the importance, and circles with the same color belong to the same cluster.

The relevant documents were retrieved from the Science Citation Index Expanded (SCI-Expanded) and Social Science Citation Index (SSCI) of the Web of Science database. The document types include article, review, proceedings paper, meeting abstract, editorial material, letter, book chapter, note, news item, book review. Only articles published in international journals were collected in this study, because they represented the majority of documents with complete research results and outcomes. The research of water footprint is developed on the basis of virtual water. Therefore, the time span of the study was from 1993 to 2018, since the concept of virtual water was put forward in 1993. And the search items were being set as "water footprint" or "virtual water." After searching, a total of 8,118 articles downloaded on March 16, 2019, were retrieved as the data for this study.

General Statistical Analysis

Temporal Distribution of Publications Analysis

The distribution of annual publications can reflect the development speed and progress of research and is a key indicator to evaluate the concentration and exciting periods of research in relevant fields. Figure 38.1 showed the distribution of all the publications on water footprint from 1993 to 2018. On the whole, the number of publications on water footprint had shown an upward trend, with small fluctuations in individual years.

According to the dates, the evolution of the published article outputs could be divided into three stages. The first stage was from 1993 to 2001 and annual publications fluctuated within 100 articles. The articles in this stage were mainly related to virtual water. The reason was that the concept of water

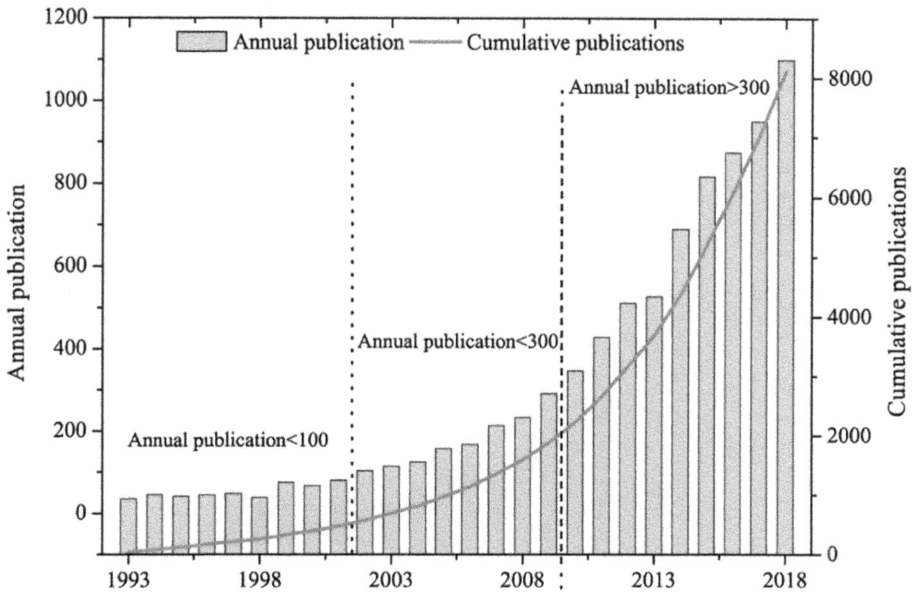

FIGURE 38.1 **(See color insert.)** Annual publication and cumulative number in water footprint studies, 1993–2018.

footprint had not yet been formally proposed. The second stage was from 2002 to 2009. The number of documents started to rise significantly during this time and increased steadily, which resulted in the rapid increase in cumulative publications. In this stage, the conference on water footprint had attracted much attention from researchers, since the concept of water footprint was put forward in 2002. Then the third stage was from 2010 to 2018. As can be seen, research on water footprint had entered the rapid development stage with an average annual growth of 83 articles.

Comparison of Research Outputs between Different Countries

According to the retrieved results, a total of 134 countries (or territories; hereafter referred to as "countries" for simplification) contributed to water footprint research. Twenty-four countries published more than 100 articles. The top 15 productive countries for the overall study period are shown in Table 38.1. As expected, the United States ranked first with a dominant output of 2,621 papers or a share of 32.29%. China with 1,075 (13.24%) and Germany with 712 (8.77%) ranked second and third, respectively. It can be concluded that scientific researchers in the United States, China, and Germany had already paid great attention to water footprint research and published a great number of related papers in recent years. Other productive countries were England (7.59%), Canada (6.23%), Italy (6.1%), and Australia (5.94%). In terms of total citations, the United States was the highest country with more than 74,000 citations and occupied the leading position. From the perspective of the average citation, Sweden ranked first in the top 15 countries and was the most powerful nation. It can be seen that the articles of this country had great influence.

Analysis by Most Productive and Influential Organizations

According to the analysis, research papers on water footprint studies came from 6,417 organizations. The top 15 research institutions with the largest number of documents were listed in Table 38.2. The top productive institution was Chinese Academy of Sciences in China, which was far ahead with 256 publications. The University of Twente in the Netherlands was in second place with 130 publications and was

TABLE 38.1 Top 15 Productive Countries in Water Footprint Studies from 1993 to 2018

Rank	Name	Number	Percentage %	Total Citations	Average Citation
1	USA	2,621	32.29	74,450	28.41
2	China	1,075	13.24	14,528	13.51
3	Germany	712	8.77	26,547	37.29
4	England	616	7.59	15,460	25.10
5	Canada	506	6.23	12,627	24.95
6	Italy	495	6.10	10,941	22.10
7	Australia	482	5.94	11,618	24.10
8	Netherlands	478	5.89	17,107	35.79
9	France	464	5.72	9,749	21.01
10	Spain	426	5.25	7,742	18.17
11	Japan	294	3.62	6,090	20.71
12	Switzerland	270	3.33	9,076	33.61
13	India	226	2.78	3,563	15.77
14	Sweden	202	2.49	15,192	75.21
15	Brazil	179	2.20	21,43	11.97

TABLE 38.2 Top 15 Productive Organizations in Water Footprint Research

Rank	Organization	Country	Number	Citations
1	Chinese Academy of Sciences	China	256	3,762
2	University of Twente	Netherlands	130	7,935
3	Beijing Normal University	China	96	1,933
4	University of Chinese Academy of Sciences	China	86	564
5	NASA	USA	83	3,561
6	University of Calif Berkeley	USA	81	2,677
7	Texas A&M University	Germany	81	1,459
8	Caltech	USA	71	3,008
9	University of Maryland	USA	67	2,429
10	University of Illinois	USA	66	1,461
11	University of Wisconsin	USA	63	4,215
12	Delft University of Technology	Netherlands	61	1,636
13	University of Colorado	USA	60	1,569
14	University of Michigan	USA	60	1,027
15	NOAA	USA	59	2,364

the research institution of Hoekstra AY, the father of water footprint. However, considering the number of citations, University of Twente ranked first, followed by University of Wisconsin, Chinese Academy of Sciences, NASA, and Caltech. It is noteworthy that the United States accounted for three fifths of the top 15 institutions. The universities and institutions in China were also the main contributors on water footprint studies.

The bibliometric mapping of Figure 38.2 presented the three clusters of collaboration among prolific organizations with 53 or more publications. The largest cluster 1 contained ten institutions, nine from the United States and one from Germany. The 2nd cluster included two Netherlandish institutions and one Australian institutions. The 3rd cluster, unlike the others, composed of only Chinese institutions. The organization with the greatest total link strength was Chinese Academy of Sciences, which presented the largest international collaboration network.

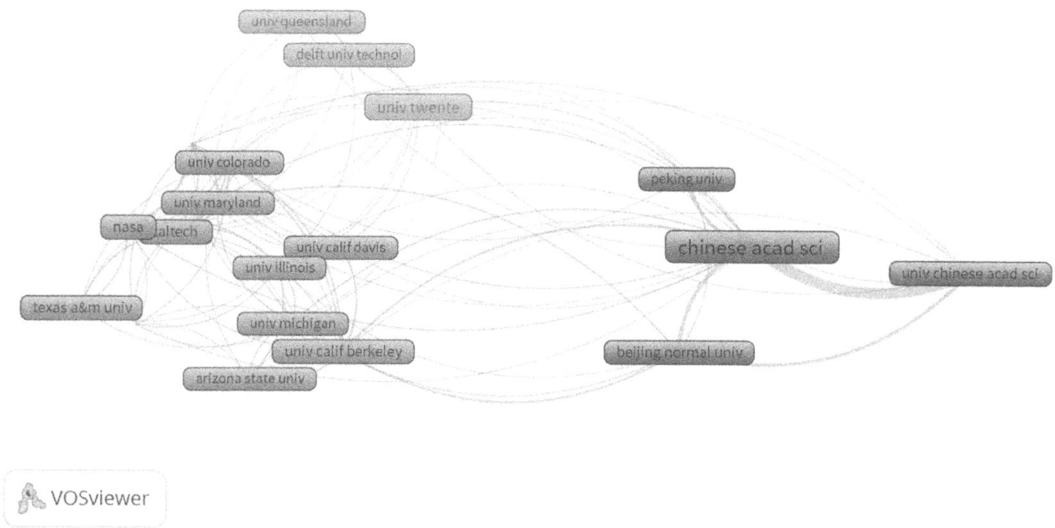

FIGURE 38.2 **(See color insert.)** Co-authorship network among productive organizations.

Co-Authorship Network Analysis among Productive Authors

As far as authorship was concerned, the number of researchers with a minimum productivity of 15 documents was 35 authors. The network of visualization map of 26 authors was shown in Figure 38.3. Despite the large number of articles written by the remaining nine authors, there has been no previous collaboration with these authors. In the network, each circle represented one author and the size of circles represented the number of articles published.

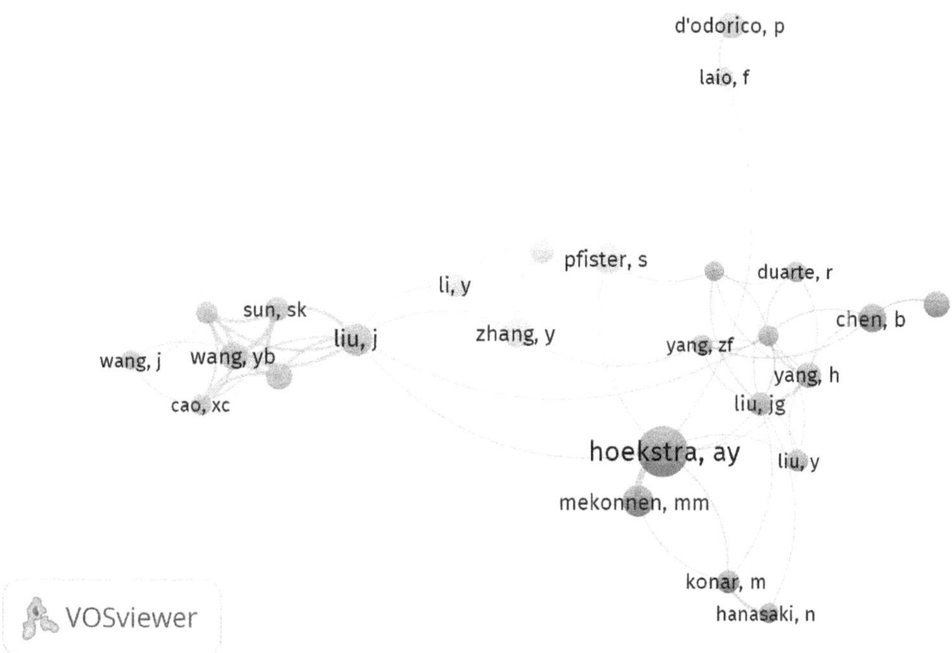

FIGURE 38.3 Co-authorship network among productive authors.

TABLE 38.3 Top 26 Author Collaboration in Water Footprint Research

Rank	Author	Number	Citations	Institution	Country	Cluster
1	Hoekstra AY	94	7,018	University of Twente	Netherlands	1
2	Liu J	38	510	Hohai University	China	2
3	Mekonnen MM	35	2,862	University of Nebraska	USA	1
4	Pfister S	33	1,400	Swiss Federal Institutes of Technology	Switzerland	3
5	Zhang Y	33	423	Beijing Forestry University	China	3
6	Chen B	30	1,126	Beijing Normal University	China	4
7	Wang YB	29	382	Northwest A&F University	China	2
8	Wu PT	26	384	University of Chinese Academy of Sciences	China	2
9	D'Odorico P	26	793	University of Virginia	USA	5
10	Chen GQ	26	777	Peking University	China	4
11	Yang H	25	1,444	Swiss Federal Institute of Aquatic Science and Technology	Switzerland	6
12	Sun SK	22	277	Northwest A&F University	China	2
13	Ridoutt BG	22	802	Commonwealth Scientific and Industrial Research Organization	Australia	3
14	Liu JG	21	1,020	Beijing Forestry University	China	6
15	Konar M	20	623	University of Illinois	USA	1
16	Li Y	20	316	Zhejiang Sci-Tech University	China	3
17	Zhao XN	19	338	Northwest A&F University	China	2
18	Liu Y	19	194	University of Chinese Academy of Sciences	China	6
19	Duarte R	17	284	University of Zaragoza	Spain	6
20	Cao XC	16	153	Hohai University	China	2
21	Zhao X	16	475	Beijing Normal University	China	6
22	Yang ZF	16	422	Beijing Normal University	China	6
23	Hanasaki N	15	807	National Institute for Environmental Studies	Japan	1
24	Hubacek K	15	1,053	University of Maryland	USA	6
25	Laio F	15	365	Politecnico di Torino	Italy	5
26	Wang J	15	594	Northwest A&F University	China	2

Table 38.3 displayed the list of top 26 active authors with strong collaborative relationships. In general, these authors were grouped in six clusters of research productions. Two large clusters (2 and 6) had seven authors, which presented the most characteristic collaboration networks. Clusters 1 and 3 had four researchers. Two other small clusters (4 and 5) had two authors. Hoekstra from the Netherlands was the most influential author in water footprint research with the highest papers (94) and citations (7,018). With regard to the number of publications, Liu J occupied the second position. Considering only the ranking of citations, Mekonnen appeared in the second place. In terms of origin, the authors were affiliated to eight countries as follows: China (15 authors), the United States (4 authors), Switzerland (2 authors), the Netherlands (1 author), Australia (1 author), Spain (1 author), Japan (1 author), and Italy (1 author). It should be noted that clusters 2 and 4 were composed only by members from China.

Quantitative Analysis of Main Source Journals

All related articles on water footprint were published in 2,067 different journals. Nevertheless, approximately 93% of these journals published less than ten articles. Table 38.4 shows the top ten journals in terms of publication. According to the rankings, *Journal of Cleaner Production* ranked first and

TABLE 38.4 Top Ten Main Source Journals in Water Footprint Research Field, 1993–2018

Rank	Name	Country	Number	Percentage%	h-Index
1	*Journal of Cleaner Production*	USA	360	4.44	34
2	*Science of the Total Environment*	Netherlands	132	1.63	22
3	*Environmental Science & Technology*	USA	107	1.32	40
4	*Water*	Switzerland	101	1.24	11
5	*Sustainability*	Switzerland	99	1.22	11
6	*Journal of Hydrology*	Netherlands	97	1.20	33
7	*Water Resources Research*	USA	90	1.11	27
8	*Ecological Indicators*	Netherlands	87	1.07	25
9	*Plos One*	USA	81	1.00	20
10	*Hydrology and Earth System Sciences*	Germany	80	0.99	24

was far ahead with 360 articles covering 4.44% of the total publications. *Science of the Total Environment* ranked second with 132 articles accounting for 1.63%; while *Environmental Science & Technology* ranked third with 107 articles accounting for 1.31%. Ranked at 4th and 5th places, *Water and Sustainability* accounted for 1.24% and 1.22%, respectively. Four journals in the top ten list were issued from the United States, three from the Netherlands, two from Switzerland, and one from Germany. *Environmental Science & Technology* had the highest h-index, which indicated that 40 papers in this journal had effectively 40 citations or more. Regarding the h-index, *Environmental Science & Technology* had the highest h-index (40).

Research Hotspots Analysis

The Subject Categories and Research Areas of Water Footprint Papers

Based on the classification of Web of Science database, water footprint papers covered a total of 206 subject categories or 125 research areas. The top ten subject categories and research areas were listed in Table 38.5. As can be noted, Environmental Sciences was the most common subject category, accounting for 25.72% of all publications, followed by Water Resources (14.24%), Engineering Environmental (11.46%), Green Sustainable Science Technology (7.80%), Meteorology Atmospheric Sciences (7.22%), Geosciences Multidisciplinary (6.94%), Engineering Civil (5.33%), Energy Fuels (4.92%), Engineering Chemical (4.85%), and Ecology (3.79%). As for research areas, Environmental Sciences Ecology, Engineering, and Water Resources were the top three research areas of water footprint studies, with over

TABLE 38.5 Top Ten Subject Categories and Research Areas in Water Footprint Research

Rank	Subject Categories	Number (Percentage %)	Research Areas	Number (Percentage %)
1	Environmental Sciences	2,088 (25.72)	Environmental Sciences Ecology	2,324 (28.63)
2	Water Resources	1,156 (14.24)	Engineering	2,230 (27.47)
3	Engineering Environmental	930 (11.46)	Water Resources	1,156 (14.24)
4	Green Sustainable Science Technology	633 (7.80)	Science Technology other Topics	914 (11.26)
5	Meteorology Atmospheric Sciences	586 (7.22)	Chemistry	605 (7.45)
6	Geosciences Multidisciplinary	563 (6.94)	Geology	605 (7.45)
7	Engineering Civil	433 (5.33)	Meteorology Atmospheric Sciences	586 (7.22)
8	Energy Fuels	399 (4.92)	Agriculture	525 (6.47)
9	Engineering Chemical	394 (4.85)	Energy Fuels	399 (4.92)
10	Ecology	308 (3.79)	Physics	337 (4.15)

TABLE 38.6 Top 20 High-Frequency Keywords in Research Field of Water Footprint in 1993–2018

Rank	Keywords	Occurrence	Rank	Keywords	Occurrence
1	Water footprint	450	11	Food security	64
2	Virtual water	213	12	Agriculture	63
3	Life cycle assessment	195	13	Irrigation	63
4	Sustainability	181	14	Virtual water trade	60
5	Climate change	138	15	Water consumption	56
6	Water scarcity	103	16	Water resources	55
7	Water	97	17	Water use	53
8	China	95	18	Water management	48
9	Footprint	78	19	Environmental impact	48
10	Input-out analysis	66	20	Water-energy nexus	47

1,000 articles and 70% of the total articles. Moreover, Science Technology Other Topics (914), Chemistry (605), Geology (605), Meteorology Atmospheric Sciences (586), Agriculture (525), Energy Fuels (399), and Physics (337) were also major research areas.

Analysis of High-Frequency Keywords

High-frequency keywords are representative terms in the field of water footprint research, which can represent hot topics and development directions in this field. The keywords in the literature include author keywords, keywords plus, title keywords, and abstract keywords. Among them, author keywords are the most accurate judgment basis for analyzing research hotspots. Table 38.6 listed the top 20 high-frequency keywords in water footprint research. Obviously, it can be noticed that the keyword "water footprint" and "virtual water" appeared with the highest occurrence frequency. As the accounting method of water footprint, the keyword "life cycle assessment" ranked 3rd in author keywords. Sustainability had become an issue of increasing environmental concern. Water footprint can be used as a means to evaluate the sustainable utilization of water resources. Therefore, the keyword "sustainability" ranked 4th. In recent years, scholars had paid more and more attention to the impact of climate change on water footprint. The keyword "climate change" ranked 5th. In addition, China focused on agriculture and food security in water footprint research. International research on virtual water trade and water resources consumption was also concentrated.

Terms Co-Occurrence and Cluster Analysis

There were 184,530 terms extracted from the titles and abstracts of the output data of publications. A density visualization map including 39 terms was created, with the minimum number of co-occurrence of 200 times (Figure 38.4). In the term map, the size of circles reflected the co-occurrence times of terms. Moreover, the terms were clustered according to interrelated tightness, and the closely related terms were divided into the same cluster in respect of a research topic. It can be seen that two clusters were formed in water footprint research field. Cluster 1 (red) was mainly about the study of water footprint accounting methods and models. Through setting specific parameters of accounting, the results of water footprint in different regions are compared. Based on this, explore the influence and response mechanism of water footprint. The co-occurrence terms included model, method, data, simulation, condition, comparison, performance, parameter, distribution, influence, response, etc. Cluster 2 (green) was mainly about the assessment of water footprint evaluation for specific products or regions, human occupancy of water resources from the perspective of water footprint, and the rationality of regional virtual water trade in alleviating water resources pressure. The co-occurrence terms included

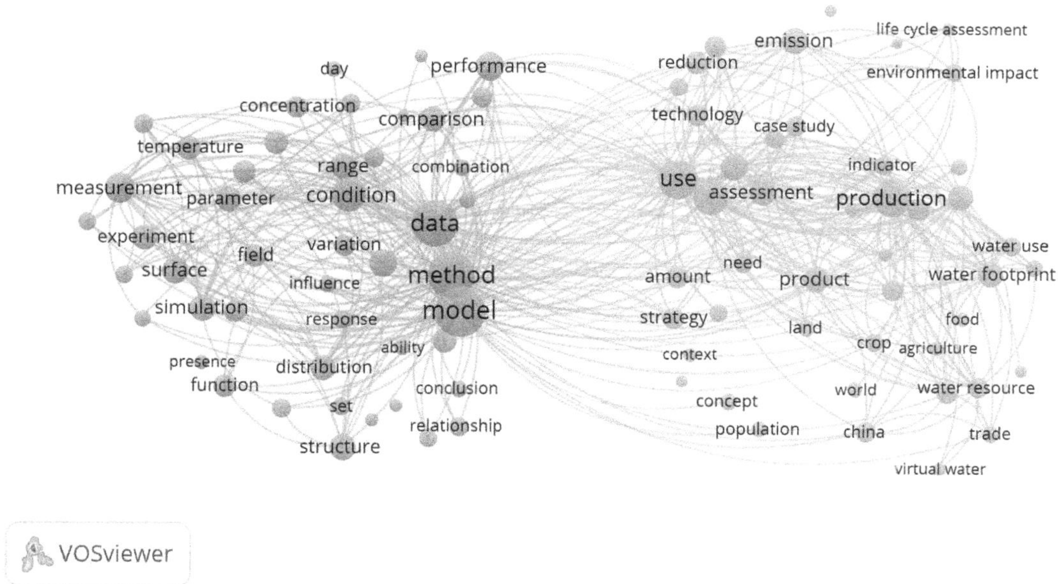

FIGURE 38.4 **(See color insert.)** The terms co-occurrence network map of water footprint.

production, assessment, use, product, amount, water footprint, virtual water, trade, reduction, environmental impact, water use, agriculture, food, crop, etc.

Conclusions

In terms of research approach, a general study of water footprint research has been evaluated utilizing bibliometric analysis methods to understand the developing status. For this, 8,118 articles were obtained from the SCI-Expanded and SSCI of the Web of Science database during 1993–2018. The following indicators were analyzed, including temporal distribution of publications, country, organization, author, journal keyword, subject category, and research area. VOSviewer was used to conduct analysis of co-authorships among organizations and authors and co-occurrence of terms.

The growth of the number of articles continued to rise in the period of 1993–2018 and could be divided in three stages. Specially, publications grew fast since the last 9 years. With regard to the contributing countries, the major driving force of water footprint was from the United States. Comprehensive influences of Chinese Academy of Sciences, Twente University, and Beijing Normal University were remarkable, which had become the core of global water footprint research. Hoekstra was the highest contributor, which stood out in terms of number of publications and citations. Authors from China showed the most characteristic collaboration network. International cooperation of water footprint research shows certain regional cooperation characteristics, and the research team cooperation within the institution is relatively close. In the future, exchanges and cooperation between countries and academic institutions should be strengthened. Articles from *Journal of Cleaner Production* had the highest number of publications, and *Environmental Science & Technology* had the highest h-index.

When examining research hotspots of water footprint, the Web of Science categories were mainly related to *Environmental Sciences and Water Resources. Environmental Sciences Ecology, Engineering and Water Resources* were the three major research areas. In addition to water footprint and virtual water, high-frequency keywords were life cycle assessment, sustainability, climate change, water scarcity, water, China, footprint, and input-out analysis. By clustering network map for terms co-occurrence, it can be found that all the topics of water footprint could be clustered into two groups, "methods and models" and "assessment and evaluation." The concept of water footprint has introduced a new way about thinking water resources management and the solution of water resources problems. Under the background of the increasing pressure of global water resources, water footprint research will play a more important role in future water cycle research and evaluation and continue to strengthen the relevant theoretical research and application research.

References

Allan, J.A. 1993. Fortunately there are substitutes for water: otherwise our hydropolitical futures would be impossible. In: ODA (Ed.), *Priorities for Water Resources Allocation and Management.* ODA, London, pp. 13–26.

Bosire, C.K., Lannerstad, M., Leeuw, J.D., et al. 2017. Urban consumption of meat and milk and its green and blue water footprints—Patterns in the 1980s and 2000s for Nairobi, Kenya. *Sci Total Environ* 579:786–796.

Cao, X.C., Wu, M.Y., Shu, R., et al. 2018. Water footprint assessment for crop production based on field measurements: A case study of irrigated paddy rice in East China. *Sci Total Environ* 610:84–93.

Castellanos, M.T., Cartagena, M.C., Requejo, M.I. 2016. Agronomic concepts in water footprint assessment: A case of study in a fertirrigated melon crop under semiarid conditions. *Agr Water Manage* 170:81–90.

Chapagain, A.K. and Hoekstra, A.Y. 2011. The blue, green and grey water footprint of rice from production and consumption perspectives. *Ecological Economics* 70:749–758.

Chen, D., Liu, Z.H., Luo, Z.H., et al. 2016. Bibliometric and visualized analysis of energy research. *Ecol Eng* 90:285–293.

Dalin, C., Konar, M., Hanasaki, N., et al. 2012. Evolution of the global virtual water trade network. *P Natl ACAD Sci USA* 109:5989–5994.

Hoekstra, A.Y., 2003. Virtual water trade. In: *Proceedings of the International Expert Meeting on Virtual Water Trade*, 12–13 December 2002. Value of Water Research Report Series No. 12, UNESCO-IHE, Delft.

Hoekstra, A.Y., Chapagain, A.K. 2008. *Globalization of Water: Sharing the Planet's Freshwater Resources.* Blackwell Publishing, Oxford

Hosseinian, S.M. and Nezamoleslami, R. 2017. Water footprint and virtual water assessment in cement industry: A case study in Iran. *J Clean Prod* 172:2454–2463.

Jin, C., Huang, K., Yu, Y.J., et al. 2016. Analysis of influencing factors of water footprint based on the STIRPAT model: Evidence from the Beijing agricultural sector. *Water* 8:513.

Khudzari, J.M., Kurian, J., Tartakovsky, B., et al. 2018. Bibliometric analysis of global research trends on microbial fuel cells using Scopus database. *Biochem Eng J* 136:51–60.

Lovarelli, D., Bacenetti, J., Fiala, M. 2016. Water Footprint of crop productions: A review. *Sci Total Environ* 548:236–251.

Ma, X., Ye, L.P., Qi, C.C., et al. 2018. Life cycle assessment and water footprint evaluation of crude steel production: A case study in China. *J Environ Manage* 224:10–18.

Mekonnen, M.M. and Hoekstra, A.Y. 2014. Water footprint benchmarks for crop production: A first global assessment. *Ecol Indic* 46:214–223.

Paterson, W., Rushforth, R., Ruddell, B.L. 2015. Water footprint of cities: A review and suggestions for future research. *Sustainability* 7:8461–8490.

Rodriguez, C.I., de Galarreta, V.R., Kruse E.E. 2015. Analysis of water footprint of potato production in the pampean region of Argentina. *J Clean Prod* 90:91–96.

Sabli, N.S.M., Noor, Z.Z., Kanniah, K. 2017. Developing a methodology for water footprint of palm oil based on a methodological review. *J Clean Prod* 146:173–180.

Shrestha, S., Chapagain, R., Babel M.S. 2017. Quantifying the impact of climate change on crop yield and water footprint of rice in the Nam Oon Irrigation Project, Thailand. *Sci Total Environ* 599:689–699.

Tian, X., Sarkis, J., Geng, Y., et al. 2018. Evolution of China's water footprint and virtual water trade: A global trade assessment. *Environ Int* 121:178–188.

The World Bank. 2016. Renewable internal freshwater resources per capita (cubic meters). http://data.worldbank.org/indicator/ER.H2O.INTR.PC, Accessed date: 4 November 2016.

van Eck, N.J. and Waltman, L. 2010. Software survey: VOSviewer, a computer program for bibliometric mapping. *Scientometrics* 84(2):523–538.

Vanham, D. and Bidoglio, G. 2013. A review on the indicator water footprint for the EU28. *Ecol Indic* 26:61–75.

Yu, Y., Hubacek, K., Feng, K.S., et al. 2010. Assessing regional and global water footprints for the UK. *Ecol Econ* 69:1140–1147.

Zhang, Y., Zhang, J. H., Tian, Q., et al. 2018. Virtual water trade of agricultural products: A new perspective to explore the Belt and Road. *Sci Total Environ* 622:988–996.

Zhao, C.F., Chen, B., Hayat, T., et al. 2014. Driving force analysis of water footprint change based on extended STIRPAT model: Evidence from the Chinese agricultural sector. *Ecol Indic* 47:43–49.

Zhuo, L., Mekonnen, M.M., Hoekstra, A.Y. 2016. The effect of inter-annual variability of consumption, production, trade and climate on crop-related green and blue water footprints and inter-regional virtual water trade: A study for China (1978–2008). *Water Res* 94:73–85.

Zou, X., Yue, W. L., Vu H. L. 2018. Visualization and analysis of mapping knowledge domain of road safety studies. *Accident Anal Prev* 118:131–145.

39

Calculation Method of Crop Water Footprint under Rain-Fed Condition

Lijie Qin and
Jianqin Wang
Northeast Normal University

Introduction 335
Calculation of Crop Water Footprint under No Irrigation 336
Blue Water Footprint • Green Water Footprint • Gray Water Footprint
Calculation of Crop Water Footprint under Inadequate Irrigation 338
Blue Water Footprint • Green Water Footprint • Gray Water Footprint
Conclusions 339
Acknowledgment 339
References 339

Introduction

Crop production is the guarantee of food security, in which rain-fed agriculture supports majority of the population and plays crucial roles in national economies [1]. At present, rain-fed agriculture, covering 84% of the arable land in use, accounts for 56% of all crop production and some 42% of cereal production in the world [2]. And it provides food for 70% of the world's population [3]. Therefore, it is essential to evaluate the water consumption during crop production under rain-fed condition, which is beneficial to understand the characteristics of crop water consumption and provide more effective methods for agricultural water management.

The water footprint concept and its methodological framework have provided a fresh perspective on assessing water consumption for crop growth [4,5]. The water footprint for crop growth is defined as the volume of fresh water required for growth and for diluting pollutants during crop growth, including blue, green, and gray water footprints [6–11], which can reflect both water consumption and water source [12–16]. Blue water footprint refers to the surface and ground water from rivers, lakes, and water contained in aquifers, which is typically viewed as irrigation water during crop growth. Green water footprint is the rainfall that infiltrates into the unsaturated soil layer and is available for crop growth [17–20]. Green water is the source of water in rain-fed farmland, which is an important water resource that supports agricultural production, especially in the arid areas [21]. Gray water footprint refers to the fresh water used for diluting pollutants during crop growth.

CROPWAT model is a major calculation method to evaluate green and blue water footprints. There are two ways that can be chosen under different agricultural conditions, which are the "crop water requirement option" (CWR option) and the "irrigation schedule option" [22]. When the crop can gain full irrigation during the period of its growth, CWR option should be chosen. However, full irrigation is realized for only a few crops in some areas, the "irrigation schedule option" should be widely applied from full irrigation to inadequate or no irrigation.

To understand the water sources utilized for crop growth, a distinction is made between the blue and green water. The blue water is typically viewed as irrigation water during crop growth. Green water is defined as the effective precipitation (EP) during the growing season. When a crop receives full irrigation, green water is calculated first and is equal to the EP during the growing season, and blue water equals the crop water use (CWU) for growth minus green water [7,11,22,23], which is the CWR option in the CROPWAT model. Because the rainfall conditions can meet the needs of crop growth in most cases in the rain-fed agricultural area, there is no irrigation, that is to say, crop production in the rain-fed agricultural area relies on rainwater. However, a severe drought occurs during the growing period, and a little irrigation will be needed there. For instance, spring drought may happen in some semiarid or arid region and causes soil moisture deficiency, resulting in delay in sowing date. In order to address the challenge of spring drought and ensure seedling emergence in northeast China, the production technology of crop involves sowing with water, in which the seed is irrigated with a small quantity of water at the time of sowing, creating a microenvironment with sufficient soil water to ensure germination and seedling establishment and having no irrigation water at any other stage. Thus, the "irrigation schedule option" in the CROPWAT model should be applied, which allows the specification of the actual irrigation water application and reflects the situations of water consumption of crop production in rain-fed agricultural areas. In this option, blue water (irrigation water) is first defined and then green water [22].

Calculation of Crop Water Footprint under No Irrigation

CWU during the growing season was equal to evapotranspiration, which was calculated using the Penman–Monteith model.

First, the reference evapotranspiration (ET_0, mm) under the influence of climatic factors was calculated [24,25] as follows:

$$ET_0 = \frac{0.408\Delta(R_n - G) + \gamma \frac{900}{T+273} U_2(e_s - e_a)}{\Delta + \gamma(1 + 0.34U_2)} \tag{39.1}$$

where R_n is the net radiation (MJ m^{-2}day^{-1}), G is the soil heat flux (MJ m^{-2}day^{-1}), T is the mean temperature (°C), U_2 is the wind speed at 2 m height (m s^{-1}), e_s is the saturation vapor pressure (kPa), e_a is the actual vapor pressure (kPa), Δ is the slope of saturation vapor pressure versus the air temperature curve (kPa/°C), and γ is the hygrometer constant (kPa/°C).

The reference evapotranspiration (ET_0) was then multiplied by the water stress coefficient (K_s) and crop parameter (K_c) to calculate crop evapotranspiration (ET_a, mm):

$$ET_a = K_s \times K_c \times ET_0 \tag{39.2}$$

The ET_a was then converted into the volume of CWU (m^3 hm^{-2}):

$$CWU = 10 \times ET_a \tag{39.3}$$

such that the factor of 10 converts water depth (mm) into water volume per area (m^3 hm^{-2}).

Blue Water Footprint

Under no irrigation condition, the blue water footprint of crop was zero.

Green Water Footprint

Under no irrigation condition, the volume of CWU is equal to the green water use (GWU), CWU=GWU. And there are two scenarios for the sources of green water. One is the EP sufficient for water use, that is the EP during the growing season can meet the GWU, i.e., EP>CWU, the green water equals the small value between ET_c and EP.

$$ET_{green} = \min(ET_c, EP) \tag{39.4}$$

EP is the sum of the daily effective precipitation (P_{eff}) during the crop growing period (mm), and if the daily EP is greater than the daily water requirement, any remaining EP will be used for the next day. P_{eff} is calculated according to the method developed by the US Department of Agriculture (USDA) as follows:

$$P_{eff} = \begin{cases} P_{total}(4.17 - 0.2P_{total})/4.17 & P_{total} < 8.3\text{mm} \\ 4.17 + 0.1P_{total} & P_{total} \geq 8.3\text{mm} \end{cases} . \tag{39.5}$$

where P_{total} is the total precipitation (mm) at a daily time step.

Then green water footprint of crop is calculated by the following formula:

$$WF_{green} = GWU/Y = CWU/Y \tag{39.6}$$

where WF_{green} is the green water footprint of crop (m^3kg^{-1}), and GWU is the green water use (m^3 hm^{-2}), and Y is the crop yield (kg hm^{-2}).

The other scenario is that the EP is not sufficient for water use during the crop growing season. Thus, the difference, called the moisture deficit, is needed to supply the remainder of the water use in crop production, which may come from the EP during the nongrowing season or the condensation of atmospheric water.

If drought conditions are common in a study area, the soil water balance is not maintained during crop growth. When the EP during the growing season cannot meet the GWU, i.e., EP<CWU, the soil water is not balanced, and there is a moisture deficit. The formula used for the moisture deficit is as follows:

$$MD = CWU - EP \tag{39.7}$$

where MD is the moisture deficit during the growing season (m^3 hm^{-2}).

In this case, the green water comes from two sources: the EP and moisture stored in soil during the nongrowing season or the condensation of atmospheric water. Thus, the green water footprint can be further defined as follows:

$$WF_{green} = WF_{EP} + WF_{MD} = EP/Y + MD/Y \tag{39.8}$$

where WF_{EP} is the water footprint of the EP (m^3kg^{-1}), and WF_{MD} is the water footprint of the moisture deficit (m^3kg^{-1}).

Gray Water Footprint

Applying fertilizer and pesticides can lead to water pollution. To prevent water pollution and ensure that water quality meets safety standards, fresh water is required to dilute various nutrient elements (such as N, P, and K) and pesticides. The fresh water used for the dilution of leached pollutants is called gray water [22,26].

Usually, nitrogen was used as the most critical pollutant to calculate gray water footprint, which was the maximum value of water dilution demand among the pollutants [22]. In general, 10% of the nitrogen fertilizer is assumed to represent the amount of leached nitrogen. If the rainfall produces runoff, the surface water pollution will be potentially caused. If the leached nitrogen infiltrates the soil, the groundwater pollution will be potentially caused. Therefore, it is necessary to calculate the amount of dilution water required to meet the nitrogen standard for surface water and groundwater. The environmental quality standards for local surface water and groundwater are chosen according to the local environmental function zoning.

The gray water footprint of crop is calculated as:

$$WF_{grey} = \frac{(\alpha \times AR)/(C_{max} - C_{nat})}{Y} \tag{39.9}$$

where WF_{gray} is the gray water footprint of crop ($m^3 kg^{-1}$), AR is the quantity of nitrogen fertilizer applied ($kg\ hm^{-2}$), α is the leaching-runoff fraction of nitrogen fertilizer, C_{max} is the nitrogen standard in surface water or groundwater ($kg\ m^{-3}$), and C_{nat} is the natural concentration of nitrogen ($kg\ m^{-3}$).

Calculation of Crop Water Footprint under Inadequate Irrigation

Rain-fed agriculture means that crops rely only on rainfall and are not replenished by soil moisture through irrigation during the period of crop growth [27]. In fact, when droughts occur, a little irrigation will be carried out to guaranteed crop production in the rain-fed agricultural area. Therefore, the calculation of crop water footprint under a little irrigation condition will be different from that under no irrigation condition.

Blue Water Footprint

Under inadequate irrigation conditions, the blue water footprint was defined first. The volume of blue water is the irrigation water, and the formula of blue water footprint of crop is as follows:

$$WF_{blue} = IU / Y \tag{39.10}$$

where WF_{blue} is the blue water footprint of crop ($m^3 kg^{-1}$), IU is the water use for irrigation ($m^3\ hm^{-2}$), and Y is the crop yield ($kg\ hm^{-2}$).

Green Water Footprint

When the EP during the growing season can meet the GWU, the green water equals the crop water use for growth minus the irrigation water consumption. The formula used for the green water footprint of crop is as follows:

$$WF_{green} = GWU / Y = (CWU - IU) / Y \tag{39.11}$$

where WF_{green} is the green water footprint of crop ($m^3 kg^{-1}$) and GWU is the green water use ($m^3\ hm^{-2}$).

When the EP during the growing season cannot meet the GWU, i.e., $EP < CWU - IU$, the moisture deficit is as follows:

$$MD = CWU - IU - EP \tag{39.12}$$

where MD is the moisture deficit during the growing season ($m^3\ hm^{-2}$) and EP is the effective precipitation during the growing season ($m^3\ hm^{-2}$).

The green water footprint can be further defined as follows:

$$WF_{green} = WF_{EP} + WF_{MD} = EP / Y + MD / Y \tag{39.13}$$

where WF_{EP} is the water footprint of the EP (m^3kg^{-1}) and WF_{MD} is the water footprint of the moisture deficit (m^3kg^{-1}).

Gray Water Footprint

The calculation of gray water footprint under inadequate irrigation conditions is the same as that under no irrigation condition.

Conclusions

The calculation of crop water footprint based on the actual water used for crop growth will highlight the characteristics of water consumption of crops in different rain-fed agricultural areas. If there is irrigation water, the blue water footprint will be considered in rain-fed agricultural areas. And the green water footprint consists of the water footprints of the EP and moisture deficit when the EP during the growing season cannot meet the GWU. Both the surface water pollution and groundwater pollution are measured to compute gray water footprint. The study on crop water footprint in rain-fed agricultural area will provide a basis for further understanding of water consumption and optimal management of water resources.

Acknowledgment

This work was funded by the National Natural Science Foundation of China (41571526).

References

1. Ogutu, G.E.O.; Franssena, W.H.P.; Supita, I.; Omondic, P.; Hutjes, R.W.A. Probabilistic maize yield prediction over East Africa using dynamic ensemble seasonal climate forecasts. *Agr. Forest Meteorol.* 2018, 250–251, 243–261.
2. Alexandratos, N.; Bruinsma, J. World agriculture towards 2030/2050: The 2012 revision. ESA Working paper No. 12-03. Rome, FAO, 2012.
3. Wu, H.T.; Wu, C.Y.; Hao, F.H.; Jin, Y.X. Assessment of green water from multi-angle view at catchment scale. *Resour. Sci.* 2009, 31, 420–428.
4. Chapagain, A.K.; Hoekstra, A.Y. The water footprint of coffee and tea consumption in the Netherlands. *Ecol. Econ.* 2007, 64, 109–118.
5. Hoekstra, A.Y.; Chapagain, A.K. The water footprints of Morocco and the Netherlands: Global water use as a result of domestic consumption of agricultural commodities. *Ecol. Econ.* 2007, 64, 143–151.
6. Bocchiola, D.; Nana, E.; Soncini, A. Impact of climate change scenarios on crop yield and water footprint of maize in the Po valley of Italy. *Agric. Water Manage.* 2013, 116, 50–61.
7. Chapagain, A.K.; Hoekstra, A.Y. The blue, green and grey water footprint of rice from production and consumption perspectives. *Ecol. Econ.* 2011, 70, 749–758.
8. Mekonnen, M.M.; Hoekstra, A.Y. The green, blue and grey water footprint of crops and derived crop products. *Hydrol. Earth Syst. Sci.* 2011, 15, 1577–1600.
9. Mekonnen, M.M.; Hoekstra, A.Y. A global assessment of the water footprint of farm animal products. *Ecosystems* 2012, 15, 401–415.
10. Shrestha, S.; Pandey, V.; Chanamai, C.; Ghosh, D. Green, blue and grey water footprints of primary crops production in Nepal. *Water Resour. Manage.* 2013, 27, 5223–5243.

11. Sun, S.K.; Wu, P.T.; Wang, Y.B.; Zhao, X.N. Temporal variability of water footprint for maize production: The case of Beijing from 1978 to 2008. *Water Resour. Manage.* 2013, 27, 2447–2463.
12. Chapagain, A.K.; Orr, S. An improved water footprint methodology linking global consumption to local water resources: A case of Spanish tomatoes. *J. Environ. Manage.* 2009, 90, 1219–1228.
13. Ercin, A.E.; Aldaya, M.M.; Hoekstra, A.Y. Corporate water footprint accounting and impact assessment: The case of the water footprint of a sugar-containing carbonated beverage. *Water Resour. Manage.* 2011, 25, 721–741.
14. Ge, L.Q.; Xie, G.D.; Zhang, C.X.; Li, S.M.; Qi, Y.; Cao, S.Y. An evaluation of China's water footprint. *Water Resour. Manage.* 2011, 25, 2633–647.
15. Ma, J.; Wang, D.X.; Lai, H.L.; Wang, Y. Water footprint: An application in water resources research. *Resour. Sci.* 2005, 27, 96–100.
16. Sun, S.K.; Wu, P.T.; Wang, Y.B.; Zhao, X.N.; Liu, J.; Zhang, X.H. The impacts of interannual climate variability and agricultural inputs on water footprint of crop production in an irrigation district of China. *Sci. Total Environ.* 2013, 444, 498–507.
17. Aldaya, M.M.; Allan, J.A.; Hoekstra, A.Y. Strategic importance of green water in international crop trade. *Ecol. Econ.* 2010, 69, 887–894.
18. Cheng, G.D.; Zhao, W.Z. Green water and its research progresses. *Adv. Earth. Sci.* 2006, 21, 221–227.
19. Falkenmark, M.; Rockstrom, J. The new blue and green water paradigm: Breaking new ground for water resources planning and management. *J. Water Resour. Plann. Manage.* 2006, 132, 129–132.
20. Liu, C.M.; Li, Y.C. Green water and water saving: Discussion on connotation of Chinese water resources. *Impact Sci. Soc.* 2006, 1, 16–20.
21. Cao, X.C.; Wu, M.Y.; Guo, X.P.; Zheng, Y.L.; Gong, Y.; Wu, N.; Wang, W.G. Assessing water scarcity in agricultural production system based on the generalized water resources and water footprint framework. *Sci. Total Environ.* 2017, 609, 587–597.
22. Hoekstra, A.Y.; Chapagai, A.K.; Aldaya, M.M.; Mekonnen, M.M. *Water Footprint Assessment Manual: Setting the Global Standard.* Earthscan: London, 2011.
23. Tian, T.H.; Zhu, D.J.; Wang, H.M.; Zang, M.D. Water footprint calculation of China's main food crops: 1987–2010. *China Population Resour. Environ.* 2013, 23, 122–128.
24. Allen, R.G.; Pereira, L.S.; Raes, D.; Smith, M. Crop Evapotranspiration Guidelines for Computing Crop Water Requirements, FAO Irrigation and drainage paper 56. FAO: Rome, 1998.
25. Allen, R.G.; Pruitt, W.O.; Wright, J.L.; Howell, T.A.; Ventura, F.; Snyder, R. A recommendation on standardized surface resistance for hourly calculation of reference ETO by the FAO56 Penman-Monteith method. *Agric. Water Manage.* 2006, 81, 1–22.
26. Bulsink, F.; Hoekstra, A.Y.; Booij, M.J. The water footprint of Indonesian provinces related to the consumption of crop products. *Hydrol. Earth Syst. Sci.* 2010, 14, 119–128.
27. Jia, Y.Y. Rain-fed agriculture in Asia-Pacific regions. *Agric. Res. Arid Area.* 1990, 2, 72–81.

40

Optical Properties of Colored Dissolved Organic Matter (CDOM): Poyang Lake

Jian Xu and
Dan Gao
Jiangxi Normal University

Introduction 341
Water Sampling and Analysis 342
Absorption Characteristics of CDOM 343
Conclusions 345
References 345

Introduction

Chromophoric dissolved organic matter (CDOM), an important component in natural waters, plays a number of key roles in biogeochemical cycle [1–3]. CDOM derived from water environment represents an essential link between aquatic and terrestrial ecosystems. Phytoplankton and macrophytes released by biological degradations are the main autochthonous sources of CDOM in inland aquatic ecosystems [4]. CDOM also plays an important role in the carbon cycle because of its relationship with dissolved organic carbon (DOC), which is the largest organic carbon component in water systems [5]. CDOM strongly absorbs ultraviolet (UV) spectrum [6] and, therefore, protects aquatic ecosystems by blocking harmful UV radiation. On the other hand, CDOM absorbs visible spectrum, especially blue light, which influences the primary production in deep water owing to light limitation. As a consequence, CDOM has a direct impact on water color remote sensing by controlling water optical properties [7]. Absorption spectral analysis has been used for tracing CDOM sources and chemical composition [8]. In recent years, optical properties of CDOM in inland waters have gained attention [9–12]. However, the characteristics of CDOM vary in diverse aquatic ecosystems because of various sources, especially for inland rivers and lakes.

Poyang Lake is the largest fresh water lake in China. It is a complex floodpath lake with significant annual water level and extent variations caused by hydrological conditions and the monsoon climate [13]. The lake also exerts an irreplaceable role for drinking water supply, flood control, waterway shipping, and conservation of biological diversity [14]. However, the water environment of Poyang Lake has been affected by anthropogenic impacts such as sand mining and major hydrologic engineering in recent years [15,16].

Water condition of Poyang Lake affects the downstream of the middle and lower Yangtze River, as well as the East China seashore. Poyang Lake is a large lake with complexity and differences in benthic topography and variation of hydrological characteristics [17]. It is necessary to investigate the spatial and temporal patterns through representative sections that contain different constituents and turbidity from contributing terrestrial sources to improve our understanding of optical properties of CDOM of Poyang Lake.

Water Sampling and Analysis

Field measurements were carried out on four cruises, including April 14, 2015 in Section A and B, April 17, 2015 in Section D, and August 5, 2015 in Section C (Table 40.1, Figure 40.1). At each sampling site, surface water between 0 and 50 cm in depth was collected in a 2,500 ml cleaned plastic bottle. The samples were held on ice and stored in dark while in field, and then analyzed in laboratory for measuring absorption coefficients and concentrations of water constituents. Water depths of sampling sites were measured by a depth sounder (SM-5, USA).

Water samples for DOC measuring were filtered under low vacuum through pre-combusted (4 h at 450°C) Whatman GF/F filters (0.7 μm) and stored in 60 ml bottles at 4°C in the dark until analysis. DOC concentrations were analyzed by high-temperature catalytic oxidation method using a total organic carbon analyzer (Shimadzu TOC L-CPN, Japan). Total suspended solids (TSS) collected on Whatman GF/C filters (1.2 μm) were determined gravimetrically before and after drying at 105°C for 4 h. Water samples after GF/F filtering were refiltered through Millex GP filters (0.22 μm) for CDOM measurement. The absorption coefficients of CDOM were measured between 240 and 750 nm (1 nm in interval) using a Shimadzu UV-2600PC UV-vis with Milli-Q water as a blank. The absorption coefficient of CDOM was obtained using the following Equations (40.1) and (40.2):

$$a_g(\lambda) = 2.303 * D(\lambda) / r \tag{40.1}$$

$$a_g(\lambda) = a_g'(\lambda) - a_g'(700) * \lambda / 700 \tag{40.2}$$

where $a'_g(\lambda)$ and $a_g(\lambda)$ stand for the uncorrected CDOM absorption coefficient and the modified absorption coefficient that has been corrected for scattering both at wavelength λ, respectively. $D(\lambda)$ refers to the optical density at wavelength λ; and r is the cuvette light path length in m. In this chapter, the concentration of CDOM was represented by the absorption coefficient at wavelength 440 nm ($a_g(440)$).

The spectral slope of CDOM was calculated using a single exponential function according to the nonlinear regression of Equation (40.3):

$$a_g(\lambda) = a_g(\lambda_0) * \exp\left[-S_g(\lambda - \lambda_0)\right] \tag{40.3}$$

where λ_0 is a reference wavelength and S_g refers to the fit parameter for the exponential decay of $a_g(\lambda)$ with increasing wavelength λ. In this study, λ_0 was at 440 nm, and the spectral slope ($S_{275-295}$) between the wavelengths of 275–295 nm was analyzed using the Origin Lab 8.0 software. The coefficient of determination (r^2) for the fitting of S_g was uniformly higher than 0.99.

TABLE 40.1 Cruise Descriptions of Four Sections

Cruise Area	Date (DD-MM-YYYY)	Number of Sampling Sites	Water Level at Hukou Gauging Station (m)	Average of Water Depth of Sampling Sites (m)
Section A	14-04-2015	15	13.76	8.33
Section B	14-04-2015	9	13.76	8.13
Section C	05-08-2015	9	15.53	3.17
Section D	17-04-2015	9	13.26	0.76

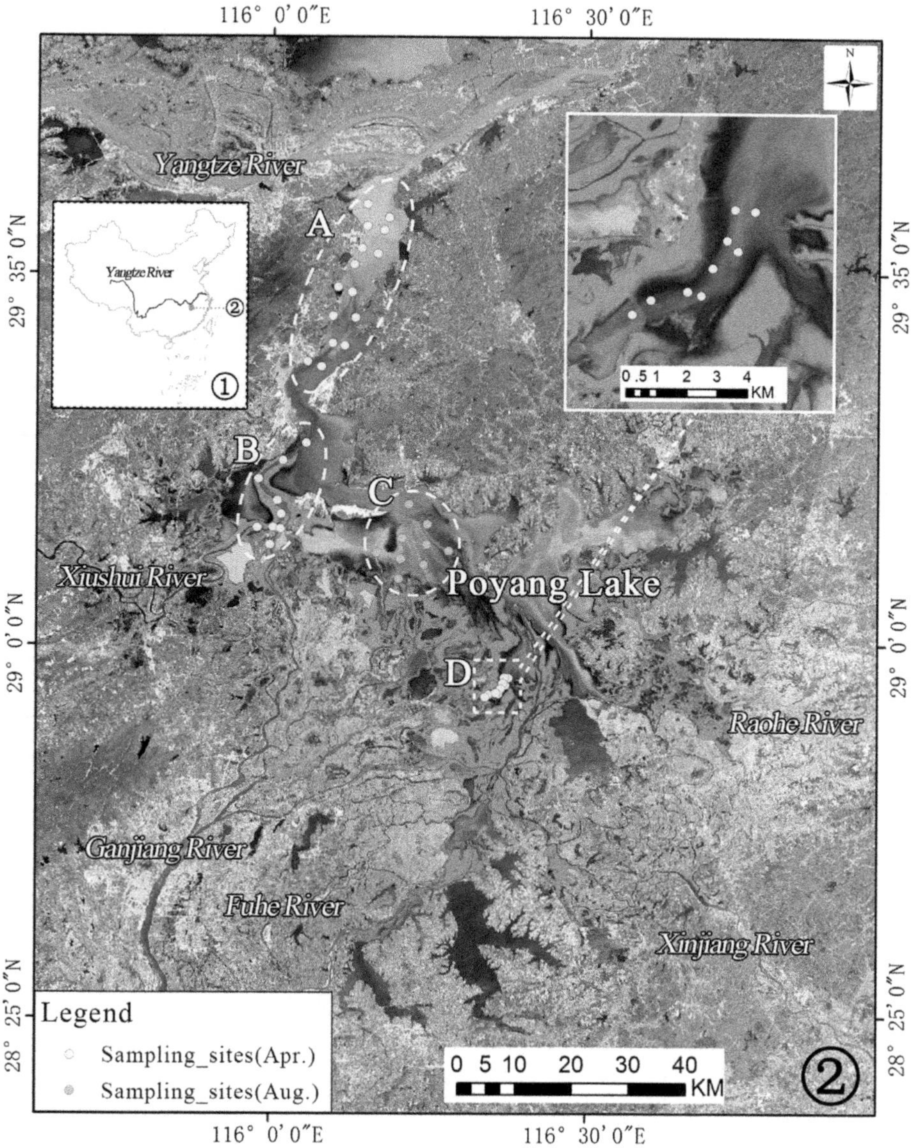

FIGURE 40.1 **(See color insert.)** Locations of representative sections and sampling sites. The background display is the false color image (red: 770–890 nm, green: 630–690 nm, blue: 520–590 nm) of Chinese high spatial resolution satellite data acquired on April 14, 2015.

Absorption Characteristics of CDOM

The absorption spectra demonstrated that CDOM spectral curves were exponentially decreased as wavelength increased. The pattern was especially significant through the UV radiation b (UV-B) (280–320 nm) and with near-zero absorption at the end of the visible spectrum (700 nm) (Figure 40.2a). The absorption coefficients of CDOM varied significantly and were apparently higher in Section D, which is identified as wetland waters from the Nanji wetland national natural reserve, than in waters in other sections (Figure 40.2b). The absorption coefficient of CDOM at 355 nm (a_g(355)) or 440 nm (a_g(440)) also showed a similar trend (Table 40.2).

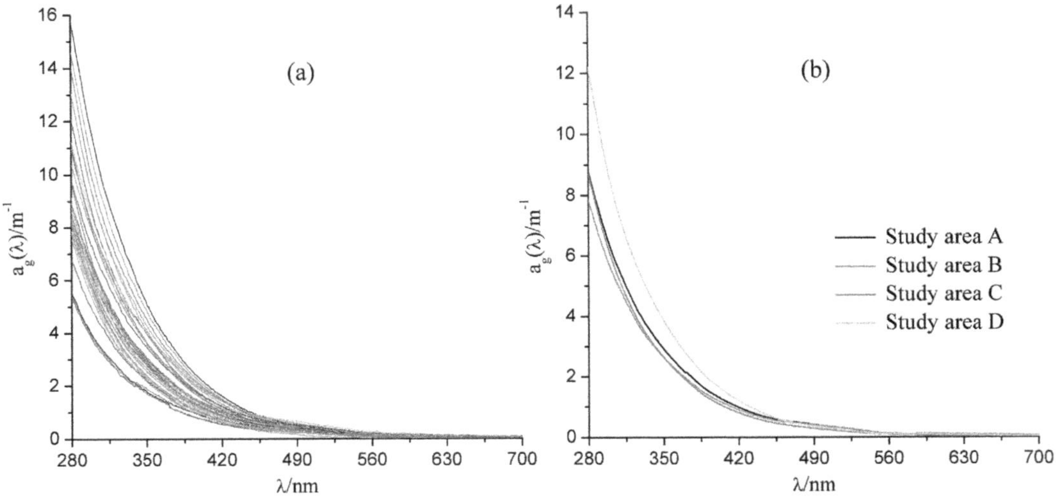

FIGURE 40.2 **(See color insert.)** Absorption spectra of CDOM: (a) CDOM absorption spectra of all sampled sites; (b) averaged CDOM absorption spectra of each section.

TABLE 40.2 Measured Parameters of Water Samples Summarized as the Descriptive Statistics in Sampled Waters: (All) All Sites; (A) Sites in Section A; (B) Sites in Section B; (C) Sites in Section C; (D) Sites in Section D

Study Area	Statistics	$a_g(355)$ (m^{-1})	$a_g(440)$ (m^{-1})	$S_{275-295}$	TSS	DOC
All (n = 42)	Min – max	1.59 – 4.71	0.39 – 1.14	0.0143 – 0.0184	5.6 – 126.8	1.49 – 4.10
	Mean ± SD	2.77 ± 0.76	0.70 ± 0.19	0.0162 ± 0.0009	32.5 ± 31.0	2.48 ± 0.59
A (n = 15)	Min – max	2.10 – 3.72	0.55 – 1.03	0.0144 – 0.0165	15.5 – 126.8	2.11 – 2.66
	Mean ± SD	2.70 ± 0.38	0.71 ± 0.12	0.0157 ± 0.0005	51.0 ± 37.9	2.38 ± 0.16
B (n = 9)	Min – max	1.59 – 3.67	0.39 – 1.05	0.0143 – 0.0164	14.8 – 23.8	1.49 – 2.58
	Mean ± SD	2.43 ± 0.75	0.67 ± 0.23	0.0154 ± 0.0007	20.9 ± 2.75	2.04 ± 0.42
C (n = 9)	Min – max	1.97 – 3.33	0.45 – 0.79	0.0165 – 0.0178	5.8 – 98.4	1.86 – 2.93
	Mean ± SD	2.42 ± 0.52	0.58 ± 0.13	0.0170 ± 0.0005	36.6 ± 31.2	2.22 ± 0.37
D (n = 9)	Min – max	2.07 – 4.71	0.43 – 1.14	0.0160 – 0.0184	5.6 – 15.7	2.55 – 4.10
	Mean ± SD	3.56 ± 0.93	0.83 ± 0.25	0.0169 ± 0.0007	9.2 ± 3.12	3.36 ± 0.49

Absorption data from each curve can be fit as a single exponential function, which determined the spectral slope parameter (S_g). Reported studies used a variety of selections of wavelength range for determining S_g, which presented difficulty in comparison of the S_g values for different wavelength ranges in literatures. The spectral slope between 275 and 295 nm ($S_{275-295}$) was commonly used recently, which indicated source and molecular weight of CDOM, with lower value indicating higher molecular weight and increasing aromaticity. The $S_{275-295}$ values in this study ranged from 0.0143 to 0.0184 nm^{-1} with an average value of 0.0162 ± 0.0009 nm^{-1} (Table 40.2). In Poyang Lake, the mean values of $S_{275-295}$ were relatively lower compared with other studies. For example, for the prairie lakes in the US Great Plains, the $S_{275-295}$ values were in the range between 0.0221 and 0.0408 nm^{-1} with mean value of 0.0302 nm^{-1}, which reflected CDOM from autochthonous-dominated sources [18]. The analysis indicated that CDOM in Poyang Lake contained relatively high molecular weight and aromatic moieties.

Conclusions

This study reveals that optical properties of CDOM have significant difference in waters of a large lake along the Yangtze River. CDOM concentrations in Poyang Lake presented a pattern of spatial variation. The absorption coefficients of CDOM were higher in wetland contributed waters than in open water areas. Spectral analysis indicated that CDOM had relatively high aromaticity and molecule weight dominated by terrestrial allochthonous inputs. This study established a field-based knowledge for remote sensing measurements of spatial and temporal distribution of CDOM in Poyang Lake which experienced annual variations in water level, coverage area, and turbidity. The findings from this study have some potential application for remote sensing modeling of CDOM, especially for algorithm parameterization.

References

1. Coble, P.G., Marine optical biogeochemistry: The chemistry of ocean color. *Chemical Reviews* 2007, 107, 402–418.
2. Green, S.A.; Blough, N.V., Optical absorption and fluorescence properties of chromophoric dissolved organic matter in natural waters. *Limnology and Oceanography* 1994, 39, 1903–1916.
3. Nelson, N.B.; Siegel, D.A., The global distribution and dynamics of chromophoric dissolved organic matter. *Annual Review of Marine Science* 2013, 5, 447–476.
4. Zhang, Y.; Liu, X.; Wang, M.; Qin, B., Compositional differences of chromophoric dissolved organic matter derived from phytoplankton and macrophytes. *Organic Geochemistry* 2013, 55, 26–37.
5. Carlson, C.A.; Ducklow, H.W.; Michaels, A.F., Annual flux of dissolved organic carbon from the euphotic zone in the northwestern Sargasso Sea. *Nature* 1994, 371, 405–408.
6. Bricaud, A.; Morel, A.; Prieur, L., Absorption by dissolved organic matter of the sea (yellow substance) in the UV and visible domains. *Limnology and Oceanography* 1981, 26, 43–53.
7. Effler, S.W.; Perkins, M.G.; Peng, F.; Strait, C.; Weidemann, A.D.; Auer, M.T., Light-absorbing components in lake superior. *Journal of Great Lakes Research* 2010, 36, 656–665.
8. Helms, J.R.; Stubbins, A.; Ritchie, J.D.; Minor, E.C.; Kieber, D.J.; Mopper, K., Absorption spectral slopes and slope ratios as indicators of molecular weight, source, and photobleaching of chromophoric dissolved organic matter. *Limnology and Oceanography* 2008, 53, 955–969.
9. Spencer, R.G.; Butler, K.D.; Aiken, G.R., Dissolved organic carbon and chromophoric dissolved organic matter properties of rivers in the USA. *Journal of Geophysical Research: Biogeosciences* 2012, 117.
10. Zhang, Y.; Qin, B.; Zhu, G.; Zhang, L.; Yang, L., Chromophoric dissolved organic matter (CDOM) absorption characteristics in relation to fluorescence in Lake Taihu, China, a large shallow subtropical lake. *Hydrobiologia* 2007, 581, 43–52.
11. Wen, Z.; Song, K.; Zhao, Y.; Du, J.; Ma, J., Influence of environmental factors on spectral characteristics of chromophoric dissolved organic matter (CDOM) in inner Mongolia Plateau, China. *Hydrology and Earth System Sciences* 2016, 20, 787–801.
12. Binding, C.E.; Jerome, J.H.; Bukata, R.P.; Booty, W.G., Spectral absorption properties of dissolved and particulate matter in Lake Erie. *Remote Sensing of Environment* 2008, 112, 1702–1711.
13. Wang, Y.; Qi, S.; Xu, J., Multitemporal remote sensing for inland water bodies and wetland monitoring. In: Ban, Y. (Ed.), *Multitemporal remote sensing*, Cham, Switzerland: Springer, 2016; pp. 357–371.
14. Gao, J.H.; Jia, J.; Kettner, A.J.; Xing, F.; Wang, Y.P.; Xu, X.N.; Yang, Y.; Zou, X.Q.; Gao, S.; Qi, S., Changes in water and sediment exchange between the Changjiang river and Poyang Lake under natural and anthropogenic conditions, China. *Science of the Total Environment* 2014, 481, 542–553.

15. Mei, X.; Dai, Z.; Du, J.; Chen, J., Linkage between three gorges dam impacts and the dramatic recessions in China's largest freshwater lake, Poyang Lake. *Scientific Reports* 2015, 5, 18197.
16. de Leeuw, J.; Shankman, D.; Wu, G.; De Boer, W.F.; Burnham, J.; He, Q.; Yesou, H.; Xiao, J., Strategic assessment of the magnitude and impacts of sand mining in Poyang Lake, China. *Regional Environmental Change* 2010, 10, 95–102.
17. Wu, G.; Liu, Y., Capturing variations in inundation with satellite remote sensing in a morphologically complex, large lake. *Journal of Hydrology* 2015, 523, 14–23.
18. Osburn, C.L.; Wigdahl, C.R.; Fritz, S.C.; Saros, J.E., Dissolved organic matter composition and photoreactivity in prairie lakes of the us great plains. *Limnology and Oceanography* 2011, 56, 2371–2390.

Index

A

Absorption characteristics, 343–345
Africa, 122
Airborne-based monitoring, 315
Algal bloom, 265
Algal bloom monitoring, 265
  bio-optical models, 269
  bloom initiation and duration, 269
  bloom start dates (BSDs), 269
  drinking water reservoirs, 269
  estimate bloom timing, 269
  forecasting chlorophyll concentrations, 268
  freshwater water bodies, 270
  harmful algal blooms, 270
  inland lakes, 269
  phytoplankton abundance, 266
  ocean color algorithms, 268
  reflectance band ratio algorithms, 268
  reflectance classification algorithms, 268
  satellite monitoring, 270
  spectral band difference algorithms, 268
  spectral signature of chlorophyll, 266
Aquifers recharge, 11
  Darcian method, 13
  definition, 13
  ground water data methods, 13
  other methods, 14
  tracer method, 13–14
  water budget methods, 12
Aquifer storage, 3
  compressibility, 3–5
  confined, 5
  confining units, 8
  description, 3
  three-dimensional storage, 8–9
  unconfined, 6–7
Aquifer types, 120
  basaltic and volcanic-rock, 121
  carbonate rock, 120
  consolidated/fractured sedimentary, 120
  unconsolidated sand and gravel, 120
Artificial reservoirs, 17
  Convective Available Potential Energy (CAPE), 20
  global dams, 18
  impact detection, 20
  land cover and land use (LCLU), 19
  land cover change, 17–24
  National Performance of Dams Program (NPDP), 19
  Three Gorges Dam (TGD), 19
Asia, 124
Associated behavior, 142

B

Basaltic and volcanic-rock, 121
Basin irrigation efficiency, 131
Bibliometric analysis, 323
  analysis, 325–329
  bibliometric indicators, 324
  cluster analysis, 330
  comparison, 325
  high frequency keywords, 330
  hotspots, 329
  science citation index expanded (SCI-Expanded), 324
  social science citation index (SSCI), 324
  temporal distribution of publications analysis, 324
  water footprint, 323–324
Bibliometric indicators, 324
Bio-optical models, 269
Bioretention, 157–158
Bloom initiation and duration, 269
Bloom start dates (BSDs), 269
Blue water footprint, 336
Buffer zone, 294
  elevation-based, 294
  fixed-width, 294

C

Calibration and validation, 72–75
Canopy transpiration rate, 208–209
Carbonate rock, 120
Caribbean, 122

Catchments, 277–279
Central America, 122
Channel flow, 176
Chromophoric dissolved organic matter (CDOM), 341
Climate change, 166
Cluster analysis, 330
Color dissolved organic matter (CDOM), 304
Comparison, 325
Compressibility, 3–5
Confined, 5
Confining units, 8
Consolidated/fractured sedimentary, 120
Contaminants, 115
  fate of, 117
  sources of, 115
  transport of, 117
  type of, 115
Convective Available Potential Energy (CAPE), 20
Crop water footprint, 335
  blue water footprint, 336
  calculation, 336, 338
    inadequate irrigation, 338
    no irrigation, 336
  crop production, 335
  gray water footprint, 337–338
  green, water footprint, 337
Cyanobacteria, 303
Cyanobacterial (CYB) blooms, 303–304

D

Darcian method, 13
Denitrification, 102
Detection frequency, 147
Distributed integrated management practices, 110
Double three band baseline (DTBB), 308
Drainage, 27
  environmental conditions, 27
    climate/precipitation, 27–28
    soils/hydrology, 27
  pollutant properties, 28
    absorption/filtration, 28–29
    persistence, 28
  management practices, 29
    controlled drainage, 30
    cropping, 30
    tillage system, 29
  water quality, 27
Drainage of water, 62
Drinking water reservoirs, 269

E

Efficiency and comparison, 212
Efficiency units, 212
Endorheic lakes, 33
  description, 33–34
  global distribution, 34–35
  remote sensing, 35
    lake area change, 37
    lake level change, 37–39
    monitoring, 35–37
    paleolake change, 41
    water salinity, 39–41
Energy balance, 53, 61
  description, 53–54
  frozen surface layers, 55
  snow covered, 55
  surface energy balance, 56
  vegetation control, 55
  water availability, 54
  water scarcity, 55
Environmental conditions, 27
  climate/precipitation, 27–28
  soils/hydrology, 27
Environmental influence, 213
Environmental Protection Agency (EPA), 178
  low impact development (LID) modeling, 178
  stormwater management model (SWMM), 178
EPA SWMM Applications, 184
  LID implementations, 184
  overland and channel flow, 179
Estimate bloom timing, 269
Europe, 121–122
Eutrophication, 45
  cultural, 45–46
  definition, 45
  effects of, 46–47
  rate of, 46
  reduction and management, 47
Evaporation, 49
  definition, 49
  energy balance, 53
    description, 53–54
    frozen surface layers, 55
    snow covered, 55
    surface energy balance, 56
    vegetation control, 55
    water availability, 54
    water scarcity, 55
Evaporation and transpiration, 61
Evapotranspiration basis, 211
Extreme event simulation, 76–78

F

Field water, 59
  applications, 62
    irrigation scheduling, 62
    natural resource modeling, 62
    plant growth, 62
    rainfed cropping, 62
  balance, 59
  drainage of water, 62

energy balance, 61
evaporation and transpiration, 61
horizontal movement, 60
lower boundary, 62
soil water balance, 59
precipitation and irrigation, 60
radiationbalance, 61
soil heat flux, 61
soil water content, 60
supply, 59
Flow attenuation, 110
Forecasting chlorophyll concentrations, 268
Four band model (FBM), 308
Four band baseline model (FBBM), 311–312
Freshwater water bodies, 270
Frozen surface layers, 55
Fumigants, 143
Fungicides, 143

G

Global dams, 18
Grassed swales, 158
Gray water footprint, 337–338
Green roofs, 158
Green water footprint, 337
Groundwater, 119
aquifer types, 120
basaltic and volcanic-rock, 121
carbonate rock, 120
consolidated/fractured sedimentary, 120
unconsolidated sand and gravel, 120
availability and use, 121
world resources, 121
Africa, 122
Asia, 124
Caribbean, 122
Central America, 122
Europe, 121–122
Middle East, 124
North America, 123
South America, 122
Groundwater contamination, 115, 144
contaminants, 115
fate of, 117
sources of, 115
transport of, 117
type of, 115
effect of climate, 116–117
in soils and water, 142
fumigants, 143
fungicides, 143
herbicides, 143
insecticides, 142
management non-point sources, 144
management point sources, 144
Ground water data methods, 13

H

Herbicides, 143
High frequency keywords, 330
Horizontal movement, 60
Hotspots, 329
Hydrograph effects, 89–90
Hydrograph modification, 128
irrigation efficiencies, 130
irrigation methods, 128–130
irrigation return flows, 128
reservoir storage, 128
Hydrologic modeling, 69
calibration and validation, 72–75
extreme event simulation, 76–78
model organization and parameterization, 71
site description, 71
TREX model, 69
channel flow, 71
overland flow, 70
upland processes, 70
Hydrologic processes, 87–88
Hydrologic unit code (HUC), 278
Hydrological model, 187
nonstationary time series analysis, 167–175
overland flow, 175
rainfall analysis, 166–167
Hydrology, 81
environmental, 82
definition, 81–82
history, 81
water quality, 82–83
water quantity, 83–84
urban, 87
alteration, 87
hydrograph effects, 89–90
hydrologic processes, 87–88
mitigation, 90–91
Hydrology cycle, 65
concept of, 65
energy sources, 66
implications for humans, 66–67
processes and pathways, 65
reservoirs, 66

I

Impact detection, 20
Impervious cover (IC), 93–97, 277–281
concept, 93
description, 93
environmental impacts, 94–95
extending utility of, 97
land cover change, 93–94
measuring and mapping, 95
National Land Cover Dataset (NLCD), 96
on-screen digitizing, 96

Impervious cover mapping, 277
accuracy assessment, 284
aerial imagery, 280
catchments, 277–279
comparison, 284–286
high spatial resolution, 277–279
hydrologic unit code (HUC), 278
National Agriculture Imagery Program (NAIP), 278
National Land Cover Database (NLCD), 278
Normalized Difference Vegetation Index (NDVI), 282
riparian zone, 291
buffer zone, 294
elevation-based, 294
fixed-width, 294
definition, 292–293
delineating, 292
statistical analysis, 284–286, 296–300
Impervious surface area (ISA), 93
In-stream flow requirements, 132
Infiltration basins, 159
Infiltration systems, nitrate removal, 101
microbial removal process, 102
denitrification, 102
nitrification, 102
uptake, 102
permeable reactive barriers, 104
wastewater treatment, 103
advanced systems, 103
conventional systems, 103–104
on site, 103
Infiltration trenches, 159
Inherent optical properties (IOP), 306
Inland waters, 269, 303
airborne-based monitoring, 315
challenges, 315
color dissolved organic matter (CDOM), 304
cyanobacteria, 303
cyanobacterial (CYB) blooms, 303–304
empirical algorithms, 305–308
midpoint reflectance baseline index (MRBI), 307
optically active constituents (OACs), 304
performance comparison, 313
phycocyanin (PC), 304
practical considerations, 315
radiative transfer equation, 304–305
satellite monitoring, 315
semi-analytical algorithms, 311
inherent optical properties (IOP), 306
IOP Inverse Model for Inland Waters (IIMIW), 311
quasi-analytical algorithm (QAA), 311
quasi-analytical algorithms for PC (QAA_{pc}), 307
semi-empirical algorithms, 308–311
double three band baseline (DTBB), 308
four band baseline model (FBBM), 311–312
four band model (FBM), 308
three band model (TBM), 309–310
spectral characteristics, 304
variability, 315
backscattering coefficients, 316
coefficient of non-PC constituents, 316
PC absorption coefficient, 315
specific absorption, 315–316
suspended matter, 316
Insecticides, 142
Integrated Water Resources Management (IWRM), 235
Irrigation, 127
basin irrigation efficiency, 131
depletion, 127
environmental concerns, 131
hydrograph modification, 128
irrigation efficiencies, 130
irrigation methods, 128–130
irrigation return flows, 128
reservoir storage, 128
in-stream flow requirements, 132
salt loading pick-up, 131
water quality implications, 131
Irrigation efficiencies, 130
Irrigation methods, 128–130
Irrigation return flows, 128
Irrigation scheduling, 62
natural resource modeling, 62
plant growth, 62
rainfed cropping, 62

L

Lake area change, 37
Lake level change, 37–39
Land change, water resource vulnerability, 133
agriculture, 134
Land use and land cover change (LULCC), 133–134, 221
urban, 136
United Nation Environment Program (UNEP), 133
World Water Development Report 4 (WWDR4), 134
Land cover and land use (LCLU), 19
Land cover change, 17–24
Land use and monitoring, 262
Leaf transpiration rate, 207
LID implementations, 184
Low-impact development (LID), 107, 154–155, 178
characteristics, 107–108
definition, 107
limitations, 108–109
site design and management objectives, 109–110
conservation measures, 110
distributed integrated management practices, 110
flow attenuation, 110

minimization techniques, 110
pollution prevention measures, 110
Low impact development (LID) modeling, 178
Lower boundary, 62
LULCC impact, 221

M

Management non-point sources, 144
Management point sources, 144
Microbial removal process, 102
denitrification, 102
nitrification, 102
uptake, 102
Middle East, 124
Midpoint reflectance baseline index (MRBI), 307
Minimization techniques, 110
Mitigation, 90–91
Monitoring network design, 258

N

National Agriculture Imagery Program (NAIP), 278
National Land Cover Database (NLCD), 96, 278
National Performance of Dams Program (NPDP), 19
Natural resource modeling, 62
Nitrate removal, 101
infiltration systems, 101
microbial removal process, 102
denitrification, 102
nitrification, 102
uptake, 102
permeable reactive barriers, 104
wastewater treatment, 103
advanced systems, 103
conventional systems, 103–104
on site, 103
Nitrification, 102
Nitrogen enrichment, 191
nitrogen levels, 191–195
reduce enrichment, 195
Nitrogen fertilizer pollution, 197
biochemical processes, 198
fertilizers, 198
fertilizer use efficiency, 201
hydrologic processes, 198–200
hydrologic process models, 201
problems caused by, 197–198
reducing nitrogen transport, 202
regional nitrogen input–output analysis, 200
sources of pollution, 198
spatial variability, 200
watershed scale analysis, 200
Nonstationary time series analysis, 167–175
Normalized Difference Vegetation Index (NDVI), 282
North America, 123

O

Ocean color algorithms, 268
Optical properties, 341
absorption characteristics, 343–345
chromophoric dissolved organic matter (CDOM), 341
Poyang Lake, 341–345
water sampling, 342–343
Optically active constituents (OACs), 304
Overland and channel flow, 175, 179
Overland flow, 175

P

Paleolake change, 41
Permeable reactive barriers, 104
Persistence, 28
Pervious pavement, 159–162
Pesticide contamination, 141, 147
associated behavior, 142
groundwater contamination, 144
in soils and water, 142
fumigants, 143
fungicides, 143
herbicides, 143
insecticides, 142
management point sources, 144
management of non-point sources, 144
irrigation management, 144–145
pesticide use, 141–142
surface water, 147
affecting factors, 149
concentrations, 147
detection frequency, 147
pesticides toxicity, 148
seasonal cycles, 147
Pesticides toxicity, 148
Pesticide use, 141–142
Phycocyanin concentration (PC), 304
Phytoplankton abundance, 266
Plant–water relations, 218
Pollutant properties, 28
absorption/filtration, 28–29
persistence, 28
Pollution prevention measures, 110
Poyang Lake, 341–345
Precipitation and irrigation, 60
Precipitation-use efficiency (PUE), 225
Processes and pathways, 65

Q

Quasi-analytical algorithm (QAA), 307
Quasi-analytical algorithms for PC (QAA_{pc}), 307

R

Radiation balance, 61
Radiative transfer equation, 304–305
Rain use efficiency (RUE), 225
  calculation, 226–228
  evapotranspiration (ET), 225
  precipitation-use efficiency (PUE), 225
  satellite observation, 226–228
  water-use efficiency (WUE), 225
Rainfall analysis, 166–167
Rainfed cropping, 62
Rainwater collection/reuse, 162
Reduce peak flow rate, 155–156
Reduce runoff, 156–157
Reflectance band ratio algorithms, 268
Reflectance classification algorithms, 268
Remote sensing, 35
  lake area change, 37
  lake level change, 37–39
  monitoring, 35–37
  paleolake change, 41
  water salinity, 39–41
Reservoirs, 66
Reservoir storage, 128
Riparian zone, 291
  buffer zone, 294
    elevation-based, 294
    fixed-width, 294
  definition, 292–293
  delineating, 292
  statistical analysis, 284–286, 296–300
Rising CO_2, 213

S

Salt loading pick-up, 131
Sand filters, 162
Satellite monitoring, 270
Satellite observation, 226–228
Science citation index expanded (SCI-Expanded), 324
Seasonal cycles, 147
Semi-analytical algorithms, 311
  inherent optical properties (IOPs), 306
  IOP Inverse Model for Inland Waters (IIMIW), 311
  non-algal particle (NAP), 304
  phycocyanin concentration (PC), 304
  quasi-analytical algorithm (QAA), 311
  quasi-analytical algorithms for PC (QAA_{pc}), 307
Semi-empirical algorithms, 308–311
  double three band baseline (DTBB), 308
  four band model (FBM), 308
  four band baseline model (FBBM), 311–312
  three band model (TBM), 309–310
Social science citation index (SSCI), 324
Soil heat flux, 61
Soil–plant–atmosphere continuum, 215–216
Soil water balance, 59
Soil water content, 60
Sources of pollution, 198
South America, 122
Specific absorption, 315–316
Spectral band difference algorithms, 268
Spectral signature of chlorophyll, 266
Spectral variability, 315
  backscattering coefficients, 316
  coefficient of non-PC constituents, 316
  PC absorption coefficient, 315
  specific absorption, 315–316
  suspended matter, 316
Stomata, 205
Stormwater causes, 153
Stormwater control, 154
Stormwaters management, 153
  best practices, 155
  bioretention, 157–158
  grassed swales, 158
  green roofs, 158
  infiltration basins, 159
  infiltration trenches, 159
  low-impact development, 154–155
  operation and maintenance, 162–163
  pervious pavement, 159–162
  rainwater collection/reuse, 162
  reduce peak flow rate, 155–156
  reduce runoff, 156–157
  sand filters, 162
  stormwater causes, 153
  stormwater control, 154
  water quality characteristics, 154
Stormwater urban modeling, 165
  channel flow, 176
  low impact development, 177
  overland flow, 175
Storm water management model (SWMM), 178
  best management practices (BMPS), 177
  channel flow, 176
  climate change, 166
  Environmental Protection Agency (EPA), 178
    low impact development (LID) modeling, 178
    stormwater management model (SWMM), 178
  EPA SWMM Applications, 184
    LID implementations, 184
    overland and channel flow, 179
  Hydrological model, 187
    nonstationary time series analysis, 167–175
    overland flow, 175
    rainfall analysis, 166–167
Surface energy balance, 56
Surface water, 147, 191
  land use controls, 195
  nitrogen enrichment, 191
    nitrogen levels, 191–195
    reduce enrichment, 195

nitrogen fertilizer pollution, 197
biochemical processes, 198
fertilizers, 198
fertilizer use efficiency, 201
hydrologic processes, 198–200
hydrologic process models, 201
problems caused by, 197–198
reducing nitrogen transport, 202
regional nitrogen input–output analysis, 200
sources of pollution, 198
spatial variability, 200
watershed scale analysis, 200
Suspended matter, 316

T

Temporal distribution of publications analysis, 324
Three band model (TBM), 309–310
Three-dimensional storage, 8–9
Three Gorges Dam (TGD), 19
Tracer method, 13–14
Transpiration, 205, 211
canopy transpiration rate, 208–209
leaf transpiration rate, 207
stomata, 205
water use efficiency, 211
development of concept, 211
drought research, 213
efficiency and comparison, 212
efficiency units, 212
environmental influence, 213
evapotranspiration basis, 211
individual leaf level, 212
instantaneous measured, 212
physiology of, 212
rising CO_2, 213
transpiration basis, 212
transpiration units, 212
water deficit effect, 213
TREX model, 69
channel flow, 71
overland flow, 70
upland processes, 70

U

Unconfined, 6–7
Unconsolidated sand and gravel, 120
United Nation Environment Program (UNEP), 133

V

Visualized analysis, 323
analysis, 325–329
bibliometric indicators, 324
cluster analysis, 330
comparison, 325
high frequency keywords, 330
hotspots, 329
science citation index expanded (SCI-Expanded), 324
social science citation index (SSCI), 324
temporal distribution of publications analysis, 324
water footprint, 323–324

W

Wastewater treatment, 103
advanced systems, 103
conventional systems, 103–104
Water availability, 54
Water budget methods, 12
Water deficits, 215
effects of, 218–291
mechanisms, transport control, 217–218
plant–water relations, 218
soil–plant–atmosphere continuum, 215–216
water status, crop plant, 217
Water footprint, 323–324, 335
blue water footprint, 336
calculation, 336, 338
inadequate irrigation, 338
no irrigation, 336
crop production, 335
gray water footprint, 337–338
green, water footprint, 337
Water management, 233
implementation-Bangladesh, 236–240
implementation-Nepal, 240–250
Integrated Water Resources Management (IWRM), 235
South Asia, 233–235
Water quantity, 27, 257
land use and monitoring, 262
monitoring network design, 258
objective, 258
problem, 258
sampling, 260
collection and analysis, 261
frequency and duration, 261
location, 260
preservation and transport, 261
quality assurance, 261
type, 260
scale of study, 260
statistical designs, 259–260
variable selection, 260
Water quality characteristics, 154
Water quality implications, 131
Water status of crop plant, 217
Water use efficiency, 211, 225
development of concept, 211
drought research, 213
efficiency and comparison, 212

Water use efficiency (*cont.*)
efficiency units, 212
environmental influence, 213
evapotranspiration basis, 211
individual leaf level, 212
instantaneous measured, 212
physiology of, 212
rising CO_2, 213
transpiration basis, 212
transpiration units, 212
water deficit effect, 213
Water salinity, 39–41
Water scarcity, 55
Watershed hydrology, 221
hydrologic impact, 222
hydrologic modeling, 223
land-use and land-cover change (LULCC), 221
LULCC impact, 221
World Water Development Report 4 (WWDR4), 134

For Product Safety Concerns and Information please contact our EU
representative GPSR@taylorandfrancis.com
Taylor & Francis Verlag GmbH, Kaufingerstraße 24, 80331 München, Germany

www.ingramcontent.com/pod-product-compliance
Ingram Content Group UK Ltd.
Pitfield, Milton Keynes, MK11 3LW, UK
UKHW050057200926
475702UK00019B/2886

* 9 7 8 1 0 3 2 4 7 4 3 9 7 *